Lecture Notes in Computer Science

Commenced Publication in 1973
Founding and Former Series Editors:
Gerhard Goos, Juris Hartmanis, and Jan van Leeuwen

T0250774

Beniamino Di Martino   Dieter Kranzlmüller
Jack Dongarra (Eds.)

Recent Advances in Parallel Virtual Machine and Message Passing Interface

12th European PVM/MPI Users' Group Meeting
Sorrento, Italy, September 18-21, 2005
Proceedings

 Springer

Volume Editors

Beniamino Di Martino
Second University of Naples
Dipartimento di Ingegneria dell'Informazione
Real Casa dell'Annunziata, via Roma, 29, 81031 Aversa (CE), Italy
E-mail: beniamino.dimartino@unina.it

Dieter Kranzlmüller
Johannes Kepler University Linz
GUP – Institute of Graphics and Parallel Processing
Altenbergerstr. 69, 4040 Linz, Austria
E-mail: kranzlmueller@gup.jku.at

Jack Dongarra
University of Tennessee
Computer Science Department
1122 Volunteer Blvd., Knoxville, TN 37996-3450, USA
E-mail: dongarra@cs.utk.edu

Library of Congress Control Number: 2005932205

CR Subject Classification (1998): D.1.3, D.3.2, F.1.2, G.1.0, B.2.1, C.1.2

ISSN   0302-9743
ISBN-10   3-540-29009-5 Springer Berlin Heidelberg New York
ISBN-13   978-3-540-29009-4 Springer Berlin Heidelberg New York

Springer is a part of Springer Science+Business Media

springeronline.com

© Springer-Verlag Berlin Heidelberg 2005
Printed in Germany

Typesetting: Camera-ready by author, data conversion by Scientific Publishing Services, Chennai, India
Printed on acid-free paper   SPIN: 11557265   06/3142   5 4 3 2 1 0

Preface

The message passing paradigm is the most frequently used approach to developing high performance computing applications on parallel and distributed computing architectures. The Parallel Virtual Machine (PVM) and Message Passing Interface (MPI) are the two main representatives in this domain.

This volume comprises 61 selected contributions presented at the 12th European PVM/MPI Users' Group Meeting, which was held in Sorrento, Italy, September 18–21, 2005. The conference was organized by the Dipartimento di Ingegneria dell'Informazione of the Second University of Naples, Italy in collaboration with CREATE and the Institute of Graphics and Parallel Processing (GUP) of the Johannes Kepler University Linz, Austria.

The conference was previously held in Budapest, Hungary (2004), Venice, Italy (2003), Linz, Austria (2002), Santorini, Greece (2001), Balatonfüred, Hungary (2000), Barcelona, Spain (1999), Liverpool, UK (1998), and Krakow, Poland (1997). The first three conferences were devoted to PVM and were held in Munich, Germany (1996), Lyon, France (1995), and Rome, Italy (1994).

In its twelfth year, this conference is well established as the forum for users and developers of PVM, MPI, and other message passing environments. Interactions between these groups have proved to be very useful for developing new ideas in parallel computing, and for applying some of those already existent to new practical fields. The main topics of the meeting were evaluation and performance of PVM and MPI, extensions, implementations and improvements of PVM and MPI, parallel algorithms using the message passing paradigm, parallel applications in science and engineering, and cluster and grid computing.

Besides the main track of contributed papers, the conference featured the fourth edition of the special session "ParSim 2005 — Current Trends in Numerical Simulation for Parallel Engineering Environments". The conference also included two tutorials, one on "Using MPI-2: A Problem-Based Approach" by William Gropp and Ewing Lusk, and one on "Advanced Message Passing and Threading Issues" by Graham Fagg and George Bosilea; and invited talks on "New Directions in PVM/Harness Research" by Al Geist, "Towards a Productive MPI Environment" by William Gropp, "Components of Systems Software for Parallel Systems" by Ewing Lusk, and "Virtualization in Parallel Distributed Computing" by Vaidy Sunderam. These proceedings contain papers on the 61 contributed presentations together with abstracts of the invited and tutorial speakers' presentations.

We would express our gratitude for the kind support of our sponsors (see below) and we thank the members of the Program Committee and the additional reviewers for their work in refereeing the submitted papers and ensuring the high quality of Euro PVM/MPI. Finally, we would like to express our gratitude to our colleagues at the Second University of Naples and GUP, JKU Linz for their help and support during the conference organization, in particular Bernhard Aichinger, Valentina Casola, Domenico Di Sivo, Francesco Moscato, Patrizia Petrillo, Günter Seiringer, Salvatore Venticinque and Mariella Vetrano.

September 2005

Beniamino Di Martino
Dieter Kranzlmüller
Jack Dongarra

Organization

General Chair

Jack Dongarra   University of Tennessee,
Knoxville, USA

Program Chairs

Beniamino Di Martino   DII, Second University of Naples, Italy
Dieter Kranzlmüller   GUP, Joh. Kepler University Linz,
Austria

Program Committee

David Abramson	Monash University, Australia
Vassil Alexandrov	University of Reading, UK
Ranieri Baraglia	Italian National Research Council, Italy
Arndt Bode	Technical University of Munich, Germany
Marian Bubak	AGH, Cracow, Poland
Barbara Chapman	University of Houston, USA
Jacques Chassin de Kergommeaux	LSR-IMAG, France
Yiannis Cotronis	University of Athens, Greece
Jose C. Cunha	New University of Lisbon, Portugal
Marco Danelutto	University of Pisa, Italy
Frederic Desprez	INRIA, France
Erik D'Hollander	Ghent University, Belgium
Beniamino Di Martino	Second University of Naples, Italy
Jack Dongarra	University of Tennessee, Knoxville, USA
Graham Fagg	University of Tennessee, Knoxville, USA
Thomas Fahringer	University of Innsbruck, Austria
Al Geist	Oak Ridge National Laboratory, USA
Michael Gerndt	Technical University of Munich, Germany
Andrzej Goscinski	Deakin University, Australia
William Gropp	Argonne National Laboratory, USA
Rolf Hempel	DLR, Simulation Aerospace Center, Germany
Ladislav Hluchy	Slovak Academy of Sciences, Slovakia
Peter Kacsuk	MTA SZTAKI, Hungary
Dieter Kranzlmüller	Joh. Kepler University Linz, Austria
Jan Kwiatkowski	Wroclaw University of Technology, Poland
Domenico Laforenza	Italian National Research Council, Italy
Erwin Laure	CERN, Switzerland

Organizing Committee

Bernhard Aichinger	GUP, Joh. Kepler University Linz, Austria
Valentina Casola	University "Frederico II" of Naples, Italy
Beniamino Di Martino	DII, Second University of Naples, Italy
Domenico Di Sivo	DII, Second University of Naples, Italy
Dieter Kranzlmüller	GUP, Joh. Kepler University Linz, Austria
Francesco Moscato	DII, Second University of Naples, Italy
Patrizia Petrillo	DII, Second University of Naples, Italy
Günter Seiringer	GUP, Joh. Kepler University Linz, Austria
Salvatore Venticinque	DII, Second University of Naples, Italy
Mariella Vetrano	CREATE, Italy

Sponsoring Institutions

HP
IBM
Intel
Microsoft
Myricom
Quadrics
NEC
Centro di Competenza sull' ICT — Regione Campania

Table of Contents

Extensions and Improvements

Cluster and Grid

Tools and Environments

Performance

Applications

Special Session: ParSim 2005

New Directions in PVM/Harness Research

Al Geist

Oak Ridge National Laboratory,
PO Box 2008,
Oak Ridge, TN 37831-6016
gst@ornl.gov
http://www.csm.ornl.gov/~geist

Abstract. The PVM development team continues to do distributed virtual machine research. Today that research revolves around the PVM project follow-on called Harness. Every three years the team chooses a new direction to explore. This year marks the beginning of a new cycle and this talk will describe the new directions and software that the PVM/Harness research team will developing over the next few years.

The first direction involves the use of Harness technology in a DOE project to develop a scalable Linux OS suited to petascale computers. Harness distributed control will be leveraged to increase the fault tolerance of such an OS. The second direction involves the use of the Harness runtime environment in the new Open MPI software project. Open MPI is an integration of several previous MPI implementations, including LAM-MPI, the LA-MPI package, and the Fault Tolerant MPI from the Harness project. The third research direction is called the "Harness Workbench" and will investigate the creation of a unified and adaptive application development environment across diverse computing platforms. This research effort will leverage the dynamic plug-in technology developed in our Harness research. Each of these three research efforts will be described in detail.

Finally the talk will describe the latest news on DOE's National Leadership Computing Facility, which will house a 100 TF Cray system at ORNL, and an IBM Blue Gene system at Argonne National Lab. We will describe the scientific missions of this facility and the new concept of "computational end stations" being pioneered by the Facility.

1 Background

The PVM/Harness team has gone through several cycles of research directions each following logically from the results of the previous cycles.

PVM was first developed as a framework for our team to do research in heterogeneous distributed machines. We first looked at heterogeneous OS and CPUs then expanded into heterogeneous parallel architectures: Cray, CM5, clusters, and eventually mixed shared and distributed memories. In the meantime the parallel computing community found PVM to be very robust and useful for teaching and solving science problems. PVM's popularity grew exponentially

B. Di Martino et al. (Eds.): EuroPVM/MPI 2005, LNCS 3666, pp. 1–3, 2005.

and the team got many requests to add new features and capabilities into the software. Often these were conflicting and they often couldn't be supported in the broad heterogeneous environment that PVM worked in.

These requests lead the team to develop Harness[1] as a framework to do research in parallel software plug-ins. This would allow users to customize their distributed virtual machine with new features or tune existing features by replacing standard components with their own hardware-tuned functions.

The Harness project included research on a scalable, self-adapting core called H2O, and research on fault tolerant MPI. The Harness software framework provides parallel software "plug-ins" that adapt the run-time system to changing application needs in real time. We have demonstrated Harness' ability to self-assemble into a virtual machine specifically tailored for particular applications[2].

Harness is a more adaptable distributed virtual machine package than PVM. To illustrate this, there is a PVM plug-in for Harness that allows Harness to behave like PVM. PVM is just one of many instantiations of Harness. There is also an MPI plug-in and an RPC plug-in each instantiating a different parallel programming environment in Harness.

Many other environments and future uses can be envisioned for the Harness framework, which can dynamically customize, adapt, and extend a virtual machine's features to more closely match the needs of an application or optimize the virtual machine for the underlying computer resources.

2   New Research Directions

This past year the FastOS project was begun in the USA and several research teams were funded to create a modular and configurable Linux system that allows customized changes based on the requirements of the applications, runtime systems, and cluster management software. The teams are also building runtime systems that leverage the OS modularity and configurability to improve efficiency, reliability, scalability, ease-of-use, and provide support to legacy and promising programming models. The talk will describe how the distributed control and dynamic adaptability features in Harness are being used by one of the research teams in their OS development[3].

The Open MPI project[4] is developing a completely new open source MPI-2 compliant implementation. The project combines the technologies and teams from FT-MPI, LA-MPI, LAM-MPI, and PACX-MPI in order to build the best MPI library possible. Open MPI offers advantages for system and software vendors, application developers and computer science researchers. The Harness team is involved in the Open MPI effort through our Fault Tolerant MPI (FT-MPI) implementation. The runtime system plays a big role in the monitoring and recovery of applications that experience faults. The fault tolerance features in the Harness runtime are being incorporated into the Open MPI runtime environment. The talk will describe how the new cycle of Harness research is going to produce a next generation runtime environment that could be used as the Open MPI runtime.

In the USA, the Department of Energy (DOE) has a significant need for high-end computing, and is making substantial investments to support this endeavor. The diversity of existing and emerging architectures however, poses a challenge for application scientists who must expend considerable time and effort dealing with the requisite development, setup, staging and runtime interfacing activities that are significantly different on each platform. The Harness workbench research will provide a common view across diverse HPC systems for application building and execution. The second, complementary, component of the Harness workbench is a next generation runtime environment that provides a flexible, adaptive framework for plugging in modules dealing with execution-time and postprocessing activities. The talk will briefly describe the plans and goals of the Harness workbench project, which started in September 2005.

3   DOE National Leadership Computing Facility

For the past year Oak Ridge National Laboratory (ORNL) had been working to establish the National Leadership Computing Facility (NLCF)[5]. The NLCF will use a new approach to increase the level of scientific productivity. The NLCF computing system will be a unique world-class research resource (100 TF Cray supercomputer by 2007) dedicated to a small number of research teams. This appraoch is similar to other large-scale experimental facilities constructed and operated around the world. At these facilities, scientists and engineers make use of "end stations"- best-in-class instruments supported by instrument specialists-that enable the most effective use of the unique capabilities of the facilities. In similar fashion the NLCF will have "Computational End Stations" that offer access to best-in-class scientific application codes and world-class computational specialists. The Computational End Stations will engage multi-national, multi-disciplinary teams undertaking scientific and engineering problems that can only be solved on the NLCF computers and who are willing to enhance the capabilities of the NLCF and contribute to its effective operation. Teams will be selected through a competitive peer-review process that was announced in July 2005. It is envisioned that there will be computational end stations in climate, fusion, astrophysics, nanoscience, chemistry, and biology as these offer great potential for breakthrough science in the near term.

References

1. G. Geist, et al, "Harness", (www.csm.ornl.gov/harness)(2003)
2. D. Kurzyniec, et al, "Towards Self-Organizing Distributed Computing Frameworks: The H2O Approach", International Journal of High Performance Computing (2003).
3. S. Scott, et al, "MOLAR: Modular Linux and Adaptive Runtime Support for High-end Computing Operating and Runtime Systems", (http://forge-fre.ornl.gov/molar) (2005)
4. E. Gabriel, et al, "Open MPI: Goals, Concept, and Design of a Next Generation MPI Implementation." , In Proceedings, 11th European PVM/MPI Users' Group Meeting, Budapest, Hungary, September 2004 (http://www.open-mpi.org)
5. J. Nichols, et al, "National Leadership Computing Facility" (www.ccs.ornl.gov) (2005)

Towards a Productive MPI Environment

William D. Gropp*

Mathematics and Computer Science Division,
Argonne National Laboratory,
Argonne, IL
gropp@mcs.anl.gov
http://www.mcs.anl.gov/~gropp

Abstract. MPI has revolutionized parallel computing in science and engineering. But the MPI specification provides only an application programming interface. This is merely the first step toward an environment that is seamless and transparent to the end user as well as the developer. This talk discusses current progress toward a productive MPI environment.

To expert users of MPI, one of the major impediments to a seamless and transparent environment is the lack of a application binary interface (ABI) that would allow applications using shared libraries to work with any MPI implementation. Such an ABI would ease the development and deployment of tools and applications. However, defining a common ABI requires careful attention to many issues. For example, defining the contents of the MPI header file is insufficient to provide a workable ABI; the interaction of an MPI program with any process managers needs to be defined independent of the MPI implementation. In addition, some solutions that are appropriate for modest-sized clusters may not be appropriate for massively parallel systems with very low latency requirements or even for large conventional clusters.

To novice users of MPI, the relatively low level of the parallel abstractions provided by MPI is the greatest barrier to achieving high productivity. This problem is best addressed by developing a combination of compile-time and run-time tools that aid in the development and debugging of MPI programs. One well-established approach is the use of libraries and frameworks written by using MPI. However, libraries limit the user to the data structures and operations implemented as part of the library. An alternative is to provide source-to-source transformation tools that bridge the gap between a fully compiled parallel language and the library-based parallelism provided by MPI.

This talk will discuss both the issues in a common ABI for MPI and some efforts to provide better support for user-defined distributed data structures through simple source-transformation techniques.

---

* This work was supported by the Mathematical, Information, and Computational Sciences Division subprogram of the Office of Advanced Scientific Computing Research, Office of Science, U.S. Department of Energy, under Contract W-31-109-ENG-38.

B. Di Martino et al. (Eds.): EuroPVM/MPI 2005, LNCS 3666, p. 4, 2005.

Components of Systems Software
for Parallel Systems*

Ewing Lusk

Mathematics and Computer Science Division,
Argonne National Laboratory, Argonne, Illinois 60439

Abstract. Systems software for clusters and other parallel systems affects multiple types of users. End users interact with it to submit and interact with application jobs and to avail themselves of scalable system tools. Systems administrators interact with it to configure and build software installations on individual nodes, schedule, manage, and account for application jobs and to continuously monitor the status of the system, repairing it as needed. Libraries interact with system software as they deal with the host environment. In this talk we discuss an ongoing research project devoted to an architecture for systems software that promotes robustness, flexibility, and efficiency. We present a component architecture that allows great simplicity and flexibility in the implementation of systems software. We describe a mechanism by which systems administrators can easily customize or replace individual components independently of others. We then describe the introduction of parallelism into a variety of both familiar and new system tools for both users and administrators. Finally, we present COBALT (COmponent-BAsed Lightweight Toolkit), an open-source, freely available preliminary implementation of the systems software components and scalable user tools, currently in production use in a number of environments.

* This work was supported by the Mathematical, Information, and Computational Sciences Division subprogram of the Office of Advanced Scientific Computing Research, Office of Science, U.S. Department of Energy, SciDAC Program, under Contract W-31-109-ENG-38.

B. Di Martino et al. (Eds.): EuroPVM/MPI 2005, LNCS 3666, p. 5, 2005.

Virtualization in Parallel Distributed Computing

Vaidy Sunderam

Department of Math & Computer Science,
Emory University, Atlanta, GA 30322, USA
vss@emory.edu
http://www.mathcs.emory.edu/dcl/

Abstract. A decade and a half ago, the Parallel Virtual Machine abstraction strived to virtualize a collection of heterogeneous components into a generalized concurrent computing resource. PVM presented a unified programming, model backed by runtime subsystems and scripts that homogenized machine dependencies. This paradigm has had mixed success when translated to computing environments that span multiple administrative and ownership domains. We argue that multidomain resource sharing is critically dependent upon decoupling provider concerns from client requirements, and suggest alternative methodologies for virtualizing heterogeneous resource aggregates. We present the H2O framework, a core subsystem in the Harness project, and discuss its alternative approach to metacomputing. H2O is based on a "pluggable" software architecture to enable flexible and reconfigurable distributed computing. A key feature is the provisioning of customization capabilities that permit clients to tailor provider resources as appropriate to the given application, without compromising control or security. Through the use of uploadable "pluglets", users can exploit specialized features of the underlying resource, application libraries, or optimized message passing subsystems on demand. The current status of the H2O system, recent experiences, and planned enhancements are described, in the context of evolutionary directions in virtualizing distributed resources.

B. Di Martino et al. (Eds.): EuroPVM/MPI 2005, LNCS 3666, p. 6, 2005.
© Springer-Verlag Berlin Heidelberg 2005

Advanced Message Passing and Threading Issues

Graham E. Fagg and George Bosilca

Dept. of Computer Science, 1122 Volunteer Blvd., Suite 413,
The University of Tennessee, Knoxville, TN 37996-3450, USA

Tutorial

Today the challenge of programming current machines on the Top500 is a well understood problem. The next generation of super-computers will include hundreds of thousands of processors, each one composed of by several semi-specialized cores. From a pure performance view point such architectures appear very promising, even if the most suitable programming methods from the users view point have yet to be discovered. Currently both message passing and threading techniques each have their own domain and occasionally they are used together. Future architectures will force users to always combine these programming paradigms until new languages or better compilers become available.

The current trend in HPC area is toward clusters of multi-core processors exhibiting a deep hierarchical memory model. Simultaneously, some classes of application require a variable number of processors during their execution time. The latest version of the MPI standard introduces several advanced topics that fit well in these environments, this includes threading and process management which when combined allow for much better potential performance in future systems.

This tutorial will focus from a users perspective on how to extract the best performance out of different parallel architectures by mixing these MPI-2 capabilities. It will cover new derived data-types, dynamic process management and multi-threading issues as well as hybrid programming with Open MP combined with MPI. In particular we will cover:

1. Dynamic process management such as; connect/accept, spawning and disconnection issues.
2. Threading issues starting with MPIs threading support, moving onto threading and dynamic processes for load balancing and server client applications and then ending with the mixing of MPI and Open MP for real problems with solutions.

The examples shown will utilize the Open MPI library, and will explicitly demonstrate the ability of this particular MPI implementation to satisfy the performance needs of a large variety of parallel applications. However, the tutorial will cover only MPI features included in the MPI-2 standard. These features are available (at various degrees) on several other commonly available MPI implementations.

B. Di Martino et al. (Eds.): EuroPVM/MPI 2005, LNCS 3666, p. 7, 2005.
© Springer-Verlag Berlin Heidelberg 2005

Using MPI-2: A Problem-Based Approach[*]

William Gropp and Ewing Lusk
{gropp, lusk}@mcs.anl.gov

Mathematics and Computer Science Division,
Argonne National Laboratory, Argonne, Illinois 60439

Abstract. This tutorial will cover topics from MPI-2, the collection of advanced features added to the MPI standard specification by the second instantiation of the MPI Forum. Such topics include dynamic process management, one-sided communication, and parallel I/O. These features are now available in multiple vendor and freely available MPI implementations. Rather than present this material in a standard "reference manual" sequence, we will provide details of designing, coding and tuning solutions to specific problems. The problems will be chosen for their practical use in applications as well as for their ability to illustrate specific MPI-2 topics. Familiarity with basic MPI usage will be assumed.

William Gropp and Ewing Lusk are senior computer scientists in the Mathematics and Computer Science Division at Argonne National Laboratory. They have been involved in both the specification and implementation of MPI since its beginning, and have written and lectured extensively on MPI.

[*] This work was supported by the Mathematical, Information, and Computational Sciences Division subprogram of the Office of Advanced Scientific Computing Research, Office of Science, U.S. Department of Energy, SciDAC Program, under Contract W-31-109-ENG-38.

Some Improvements to a Parallel Decomposition Technique for Training Support Vector Machines

Thomas Serafini[1], Luca Zanni[1], and Gaetano Zanghirati[2]

[1] Department of Mathematics, University of Modena and Reggio Emilia
[2] Department of Mathematics, University of Ferrara

Abstract. We consider a parallel decomposition technique for solving the large quadratic programs arising in training the learning methodology Support Vector Machine. At each iteration of the technique a subset of the variables is optimized through the solution of a quadratic programming subproblem. This inner subproblem is solved in parallel by a special gradient projection method. In this paper we consider some improvements to the inner solver: a new algorithm for the projection onto the feasible region of the optimization subproblem and new linesearch and steplength selection strategies for the gradient projection scheme. The effectiveness of the proposed improvements is evaluated, both in terms of execution time and relative speedup, by solving large-scale benchmark problems on a parallel architecture.

Keywords: Support vector machines, quadratic programs, decomposition techniques, gradient projection methods, parallel computation.

1 Introduction

Support Vector Machines (SVMs) are an effective learning technique [13] which received increasing attention in the last years. Given a training set of labelled examples

$$D = \{(\boldsymbol{z}_i, y_i), \ i = 1, \dots, n, \quad \boldsymbol{z}_i \in \mathbb{R}^m, \ y_i \in \{-1, 1\}\},$$

the SVM learning methodology performs classification of new examples $\boldsymbol{z} \in \mathbb{R}^m$ by using a decision function $F : \mathbb{R}^m \to \{-1, 1\}$, of the form

$$F(\boldsymbol{z}) = \text{sign}\left(\sum_{i=1}^n x_i^* y_i K(\boldsymbol{z}, \boldsymbol{z}_i) + b^*\right), \tag{1}$$

where $K : \mathbb{R}^m \times \mathbb{R}^m \to \mathbb{R}$ denotes a special kernel function (linear, polynomial, Gaussian, ...) and $\boldsymbol{x}^* = (x_1^*, \dots, x_n^*)^T$ is the solution of the convex quadratic programming (QP) problem

$$
\begin{aligned}
\min \ & f(\boldsymbol{x}) = \frac{1}{2} \boldsymbol{x}^T G \boldsymbol{x} - \boldsymbol{x}^T \mathbf{1} \\
\text{sub. to } & \boldsymbol{y}^T \boldsymbol{x} = 0, \quad 0 \le x_j \le C, \quad j = 1, \dots, n,
\end{aligned}
\tag{2}
$$

B. Di Martino et al. (Eds.): EuroPVM/MPI 2005, LNCS 3666, pp. 9–17, 2005.

ALGORITHM PGPDT   (Parallel SVM Decomposition Technique)

---

1. Let $\boldsymbol{x}^{(0)}$ be a feasible point for (2), let n_{sp} and n_c be two integer values such that $n \geq n_{sp} \geq n_c \geq 2$, n_c even, and set $i = 0$. Arbitrarily split the indices $\{1, \ldots, n\}$ into the set B of *basic* variables, with $\#B = n_{sp}$, and the set $N = \{1, \ldots, n\} \setminus B$ of *nonbasic* variables. Arrange the arrays $\boldsymbol{x}^{(i)}$, $\boldsymbol{y}$ and G with respect to B and N:

$$\boldsymbol{x}^{(i)} = \begin{bmatrix} \boldsymbol{x}_B^{(i)} \\ \boldsymbol{x}_N^{(i)} \end{bmatrix}, \qquad \boldsymbol{y} = \begin{bmatrix} \boldsymbol{y}_B \\ \boldsymbol{y}_N \end{bmatrix}, \qquad G = \begin{bmatrix} G_{BB} & G_{BN} \\ G_{NB} & G_{NN} \end{bmatrix}.$$

2. **Compute in parallel** the Hessian matrix G_{BB} of the subproblem

$$\min_{\boldsymbol{x}_B \in \Omega_B} f_B(\boldsymbol{x}_B) = \frac{1}{2}\boldsymbol{x}_B^T G_{BB} \boldsymbol{x}_B - \boldsymbol{x}_B^T (1 - G_{BN}\boldsymbol{x}_N^{(i)}) \qquad (3)$$

where $\Omega_B = \{\boldsymbol{x}_B \in \mathbb{R}^{n_{sp}} \mid \boldsymbol{y}_B^T \boldsymbol{x}_B = -\boldsymbol{y}_N^T \boldsymbol{x}_N^{(i)},\ 0 \leq \boldsymbol{x}_B \leq C\mathbf{1}\}$ and **compute in parallel** the solution $\boldsymbol{x}_B^{(i+1)}$ of the above problem. Set $\boldsymbol{x}^{(i+1)} = \begin{bmatrix} \boldsymbol{x}_B^{(i+1)T} & \boldsymbol{x}_N^{(i)T} \end{bmatrix}^T$.

3. **Update in parallel** the gradient $\nabla f(\boldsymbol{x}^{(i+1)}) = \nabla f(\boldsymbol{x}^{(i)}) + [G_{BB}\ G_{BN}]^T (\boldsymbol{x}_B^{(i+1)} - \boldsymbol{x}_B^{(i)})$ and terminate if $\boldsymbol{x}^{(i+1)}$ satisfies the KKT conditions for problem (2).

4. Update B by changing at most n_c elements through the strategy described in [12]. Set $i \leftarrow i + 1$ and go to step 2.

---

where G has entries $G_{ij} = y_i y_j K(\boldsymbol{z}_i, \boldsymbol{z}_j)$, $i, j = 1, 2, \ldots, n$, $\mathbf{1} = (1, \ldots, 1)^T$ and C is a parameter of the SVM algorithm. Once the vector $\boldsymbol{x}^*$ is computed, $b^* \in \mathbb{R}$ in (1) is easily derived. The matrix G is generally dense and in many real-world applications its size is very large ($n \gg 10^4$). Thus, strategies suited to exploit the special features of the problem become a need, since standard QP solvers based on explicit storage of G cannot be used. Among these strategies, decomposition techniques are certainly the most investigated (see for instance [1,2,4,5] and the references therein). They consist in splitting the original problem into a sequence of QP subproblems sized $n_{sp} \ll n$ that can fit into the available memory. An effective parallel decomposition technique is proposed in [14] (see also [11]). It is based on the Joachims' decomposition idea [4] and on a special variable projection method [7,8] as QP solver for the inner subproblems. In contrast with other decomposition algorithms, that are tailored for very small-size subproblems (typically less than 10^2), the proposed technique is appropriately designed to be effective with subproblems large enough (typically more than 10^3) to produce few decomposition steps. Due to the effectiveness of the inner QP solver, this method is well comparable with the most effective decomposition approaches on scalar architectures. However, its straightforward parallelization is an innovative feature. In fact, the expensive tasks (kernel evaluations and QP subproblems solution) of the few decomposition steps can be efficiently performed in parallel and promising computational results are reported in [14]. The new version of this approach, equipped with the gradient projection method GVPM introduced in [10] as inner QP solver and with the acceleration strategies suggested in [12], is called Parallel Gradient Projection-based Decomposition Technique (PGPDT) and its main steps are summarized in Algorithm PGPDT .

In this paper we examine two improvements to the PGPDT. The improvements are suggested by the recent work [3] where a new algorithm for computing the projection of a vector onto the feasible region of the SVM QP problem and a special nonmonotone gradient projection method for (3) are proposed. We will show that these strategies give rise to an efficient inner QP solver for the decomposition techniques and imply better performance for PGPDT, both in terms of execution time and relative speedup.

2   About Gradient Projection-Type Inner Solvers

In order to explain the PGPDT improvements we are going to introduce, we need to briefly discuss about the numerical behaviour of the parallel decomposition technique. To this end, we show the PGPDT performance on two well-known benchmark problems: the QP problem sized $n = 60000$ arising when a classifier for digit "8" is trained through a Gaussian SVM ($K(z_i, z_j) = e^{-\|z_i - z_j\|^2/(2\sigma^2)}$, $\sigma = 1800$, $C = 10$) on the MNIST database of handwritten digits[1] and the QP problem sized $n = 49749$ arising in training Gaussian SVM ($\sigma = \sqrt{10}$, $C = 5$) on the Web data set[2]. We solve these problems by PGPDT on a IBM CLX/768, a Linux Cluster equipped with Intel Xeon 3GHz processors and 1GB of memory per processor. Table 1 shows the results obtained for different numbers of processing elements (PEs): the *time* column shows the overall training time, sp_r is the relative speedup and it is the number of decomposition iterations. t_{prep}, t_{solv} and t_{grad} are respectively the time for computing the Hessian G_{BB} of the subproblem (step 2.), to solve the subproblem (step 2.) and to update the gradient (step 3.). Finally, it_{in} shows the total number of iterations of the inner QP solver and t_{fix} is the execution time of the non-parallelized code, which is obtained by $t_{fix} = time - t_{prep} - t_{solv} - t_{grad}$, i.e. by subtracting the execution time of the parallelized parts from the total execution time. It can be observed that t_{fix} is very small compared to the total execution time, and this shows that steps 2 and 3 are the core computational tasks. This table shows satisfactory speedups on the MNIST data set; for 2 and 4 processing elements we even have

Table 1. PGPDT performance scaling on MNIST ($n = 60000$) and Web ($n = 49749$) data sets

	MNIST set, $n_{sp} = 2000, n_c = 600$							Web set, $n_{sp} = 1500, n_c = 750$						
PEs	time	sp_r	it	t_{prep}	t_{solv}	t_{grad}	t_{fix}	it_{in}	time	sp_r	it	t_{prep}	t_{solv}	t_{grad} t_{fix}	it_{in}
1	598.2		15	14.6	81.9	501.4	0.3	5961	242.7		23	3.2	182.6	56.6 0.3	17955
2	242.2	2.5	15	13.6	49.0	179.3	0.3	6091	125.3	1.9	26	3.9	90.1	30.8 0.5	18601
4	129.6	4.6	15	10.0	28.8	90.2	0.6	6203	73.2	3.3	25	3.0	53.6	16.1 0.5	18759
8	75.5	7.9	15	7.9	21.3	45.7	0.6	5731	46.4	5.2	22	2.2	35.7	7.9 0.6	17408
16	43.8	13.7	17	5.5	13.7	24.2	0.4	5955	32.8	7.4	25	1.5	25.9	5.0 0.4	17003

---

[1] Available at http://yann.lecun.com/exdb/mnist.

[2] Available at http://research.microsoft.com/~jplatt/smo.html.

a superlinear behaviour due to the increased amount of memory available for caching the elements of G. For the Web data set, the speedup sp_r is not as good as for the MNIST case. This can be explained as follows: the gradient updating (t_{grad}) has always a good speedup in both the data sets, while the inner solver is not able to achieve the same good speedup. In the MNIST-like data sets, where $t_{\mathrm{grad}} \gg t_{\mathrm{solv}}$, the suboptimal speedup of the inner solver is compensated by the good speedup of the gradient updating, and the overall behaviour is good. On the other hand, when $t_{\mathrm{grad}} < t_{\mathrm{solv}}$, as for the Web data set, the speedup of the inner solver becomes the main bottleneck.

In the following we will introduce an improved inner solver able to reduce the above drawbacks. The inner QP subproblems (3) have the following general form:

$$\min_{w \in \Omega} \quad \bar{f}(w) = \frac{1}{2} w^T A w + b^T w \tag{4}$$

where the matrix $A \in \mathbb{R}^{n_{sp} \times n_{sp}}$ is symmetric and positive semidefinite, $w, b \in \mathbb{R}^{n_{sp}}$ and the feasible region Ω is defined by

$$\Omega = \{w \in \mathbb{R}^{n_{sp}}, \quad 0 \le w \le C\mathbf{1}, \quad c^T w = d, \quad d \in \mathbb{R}\}. \tag{5}$$

We recall that now the size n_{sp} allows the matrix A to be stored in memory. The special gradient projection method (GVPM) used by PGPDT [10] combines a monotone linesearch strategy with an adaptive steplength selection based on the Barzilai–Borwein rules. Gradient projection methods are appealing approaches for problems (4) since they consist in a sequence of projections onto the feasible region, that are nonexpensive due to the special constraints (5), as it will be described in the next section. As a consequence, the main task of each iteration remains a matrix-vector product for computing $\nabla \bar{f}(w)$ that can be straightforwardly parallelized by a row block-wise distribution of the entries of A.

Recently, Dai and Fletcher [3] have proposed a new gradient projection method for singly linearly constrained QP problems subject to lower and upper bounds. In the computational experiments reported in [3], this method has shown better convergence rate in comparison to GVPM on some medium-scale SVM test problems.

Here we are interested in evaluating the PGPDT performance improvements due to this new inner solver. We recall our implementation of the Dai and Fletcher method in Algorithm DF.

For this method the same considerations given for GVPM about the computational cost per iteration and the parallelization still hold true. Nevertheless, the linesearch step and the steplength selection rule are very different.

The algorithm DF uses an adaptive nonmonotone linesearch in order to allow the objective function value $\bar{f}(w^{(k)})$ to increase on some iterations. Its main feature is the adaptive updating of the reference function value f_{ref}. The purpose of the updating rule is to cut down the number of times the linesearch is brought into play and, consequently, to frequently accept the iterate $w^{(k+1)} = w^{(k)} + d^{(k)}$ obtained through an appropriate steplength α_k.

ALGORITHM DF Gradient Projection Method

1. *Initialization.* Let $\quad w^{(0)} \in \Omega, \quad 0 < \alpha_{min} < \alpha_{max}, \quad \alpha_0 \in [\alpha_{min}, \alpha_{max}], \quad L = 2$;
   set $\quad f_{ref} = \infty, \quad f_{best} = f_c = \bar{f}(w^{(0)}), \quad \ell = 0, \quad k = 0, \quad s^{(k-1)} = y^{(k-1)} = 0$.

2. *Projection.* Terminate if $w^{(k)}$ satisfies a stopping criterion; otherwise compute the descent direction

$$d^{(k)} = P_\Omega(w^{(k)} - \alpha_k(Aw^{(k)} + b)) - w^{(k)}.$$

3. *Linesearch.*

   If $\left(k = 0 \text{ and } \bar{f}(w^{(k)} + d^{(k)}) \geq \bar{f}(w^{(k)})\right)$ or $\left(k > 0 \text{ and } \bar{f}(w^{(k)} + d^{(k)}) \geq f_{ref}\right)$
   then

$$w^{(k+1)} = w^{(k)} + \lambda_k d^{(k)}, \qquad \text{with} \qquad \lambda_k = \arg \min_{\lambda \in [0,1]} \bar{f}(w^{(k)} + \lambda d^{(k)});$$

   else

$$w^{(k+1)} = w^{(k)} + d^{(k)};$$

   end.

4. *Update.* Compute $\quad s^{(k)} = w^{(k+1)} - w^{(k)}; \qquad y^{(k)} = A(w^{(k+1)} - w^{(k)})$.
   If $\quad s^{(k)T} y^{(k)} \leq 0 \quad$ then
       set $\quad \alpha_{k+1} = \alpha_{max}$;
   else
       If $\quad s^{(k-1)T} y^{(k-1)} \leq 0 \quad$ then
           set $\quad \alpha_{k+1} = \min\left\{\alpha_{max}, \max\left\{\alpha_{min}, \frac{s^{(k)T} s^{(k)}}{s^{(k)T} y^{(k)}}\right\}\right\}$;
       else
           set $\quad \alpha_{k+1} = \min\left\{\alpha_{max}, \max\left\{\alpha_{min}, \frac{s^{(k)T} s^{(k)} + s^{(k-1)T} s^{(k-1)}}{s^{(k)T} y^{(k)} + s^{(k-1)T} y^{(k-1)}}\right\}\right\}$;
       end.
   end.
   If $\quad \bar{f}(w^{(k+1)}) < f_{best} \quad$ then
       set $\quad f_{best} = \bar{f}(w^{(k+1)}), \quad f_c = \bar{f}(w^{(k+1)}), \quad \ell = 0$;
   else
       set $\quad f_c = \max\left\{f_c, \bar{f}(w^{(k+1)})\right\}, \quad \ell = \ell + 1$;
       If $\quad \ell = L \quad$ then
           set $\quad f_{ref} = f_c, \quad f_c = \bar{f}(w^{(k+1)}), \quad \ell = 0$;
       end.
   end.
   Set $k \leftarrow k + 1$, and go to step 2.

For the steplength updating in DF, the rule

$$\alpha_{k+1} = \frac{\sum_{i=0}^{m-1} s^{(k-i)T} s^{(k-i)}}{\sum_{i=0}^{m-1} s^{(k-i)T} y^{(k-i)}}, \qquad m \geq 1,$$

is used with the choice $m = 2$ because it is observed to be the best one for the SVM QP problems. We conclude the introduction to DF by recalling that its global convergence can be proved by proceeding as in [9].

Table 2. PGPDT performance scaling with DF as inner solver

| | MNIST set, $n_{sp} = 2000, n_c = 600$ | | | | | | Web set, $n_{sp} = 1500, n_c = 750$ | | | | | |
PEs	time	sp_r	it	t_{prep}	t_{solv}	t_{grad}	it_{in}	time	sp_r	it	t_{prep}	t_{solv}	t_{grad}	it_{in}
1	591.1		15	14.8	73.0	503.0	5254	206.5		24	3.2	145.9	57.0	12247
2	232.4	2.5	15	13.7	38.9	178.2	4937	91.2	2.3	23	3.7	59.0	28.1	12068
4	124.9	4.7	15	10.3	23.8	90.5	5061	59.9	3.4	29	3.3	39.6	16.6	13516
8	70.1	8.4	15	7.6	16.2	45.7	5044	40.6	5.1	26	2.4	29.1	8.8	13554
16	41.4	14.3	15	5.1	11.9	24.0	5062	27.3	7.6	23	1.5	20.7	4.7	12933

In order to analyze the behaviour of this new solver within PGPDT, we can refer to Table 2 which shows the same test of Table 1 but using DF as the inner solver, in place of GVPM. Looking at the it_{in} column, one can observe a great reduction of the overall inner iterations; considering that the cost per iteration of the DF method is almost the same as for GVPM, this implies a more efficient solution of the inner subproblem. This fact is confirmed by comparing the t_{solv} columns, which report the total execution time of the inner solvers. As a result, the overall execution time of the PGPDT and its speedup sp_r are improved by the introduction of this new inner solver.

3   About the Projection onto the Feasible Region

The gradient projection algorithms used as inner solvers in the PGPDT require at each iteration to project a vector onto a feasible region Ω of the form (5). This feasible region has a special structure defined by a single linear equality constraint and box constraints. This section is about methods for computing efficiently a projection onto such a feasible region.

The orthogonal projection of a vector z onto Ω is the solution $P_\Omega(z)$ of the following separable strictly convex quadratic program:

$$\min \ \tfrac{1}{2}w^T w - z^T w$$
$$\text{sub. to } \ 0 \le w \le C1, \quad c^T w = d. \tag{6}$$

By exploiting the KKT conditions of (6) it is possible to prove [3,6] that $P_\Omega(z)$ can be derived from the solution of the piecewise linear monotone nondecreasing equation in one variable

$$r(\lambda) = c^T w(\lambda) - d = 0, \tag{7}$$

where

$$w(\lambda) = \text{mid}(\ 0, \ (z + \lambda c), \ C1 \ )$$

in which $\text{mid}(v_1, v_2, v_3)$ is the componentwise operation that gives the median of its three arguments. Once the solution λ^* of (7) is computed, we obtain $P_\Omega(z)$ by setting $P_\Omega(z) = w(\lambda^*)$.

Thus, the main computational task for computing $P_\Omega(z)$ consists in solving the equation (7). The gradient projection methods tested in the previous section solve this root finding problem by the $O(n)$ bisection-like algorithm proposed

Table 3. PGPDT performance scaling with DF and secant-based projector

PEs	MNIST set, $n_{sp}=2000, n_c=600$							Web set, $n_{sp}=1500, n_c=750$						
	time	sp_r	it	t_{prep}	t_{solv}	t_{grad}	it_{in}	time	sp_r	it	t_{prep}	t_{solv}	t_{grad}	it_{in}
1	590.0		15	14.9	67.4	507.5	5107	200.8		23	3.0	143.4	54.1	12824
2	234.1	2.5	16	14.1	39.4	179.2	5432	99.1	2.0	26	4.0	63.7	30.5	14090
4	121.0	4.9	15	10.1	20.1	89.7	4913	46.6	4.3	23	3.0	28.0	15.2	11477
8	67.7	8.7	16	8.1	13.2	46.1	5004	35.5	5.7	25	2.0	24.6	8.3	14500
16	39.2	15.1	15	5.2	9.6	23.9	5476	22.1	9.1	24	1.5	15.1	5.1	13200

in [6]. Here, we consider an alternative approach introduced in [3] based on a $O(n)$ secant-type method. In particular, we are interested in evaluating how the use of this new projection strategy within the DF gradient projection method can improve the PGPDT performance. Our implementation of the secant-type algorithm for (7) follows the one proposed in [3] and, since the projection is required at each iteration of the DF scheme, we use to hot-start the algorithm by providing the optimal λ of the previous projection as the initial approximation. Finally, we stop the secant-type algorithm if one of the following conditions is satisfied: $|r(\lambda_i)| < 10^{-10}\sqrt{Cn}$ or $|\Delta\lambda_i| < 10^{-11}(1+|\lambda_i|)$, where λ_i is the current approximation and $\Delta\lambda_i = \lambda_i - \lambda_{i-1}$.

Table 3 shows the performance results of the PGPDT equipped with the DF gradient projection method as inner QP solver and the secant-type algorithm [3] for computing the projections. By comparing Tables 3 and 2, we can evaluate the different behaviour due to the new secant-based projector with respect to the bisection-based projector previously used. First of all, by considering the total number of the inner solver iterations (it_{in}) we may observe a very similar behaviour in terms of convergence rate of the DF gradient projection method. This confirms that the two projectors works with a well comparable accuracy. Furthermore, by comparing the total scalar time, slightly better results are obtained with the new secant-based projector. In spite of these similarities, since the projector is not parallelized and, consequently, it gives rise to a fixed time independent on the number of PEs, the time saving due to the new more efficient projector implies a significantly better speedup of the inner solver and of the overall PGPDT, especially when many PEs are used.

By comparing Tables 3 and 1 it is possible to evaluate the overall improvements of the new strategies presented in this paper which, especially in the case of Web-like data sets, consist in a promising reduction of solution times and an increase of the speedup when many processors are used.

4   Test on a Large Data Set

In this section we briefly present the performance of the improved PGPDT on a large-size data set: the KDDCUP-99.

The KDDCUP-99 is the Intrusion Detection data set[3], used for the Third International Knowledge Discovery and Data Mining Tools Competition, which

---

[3] Available at http://kdd.ics.uci.edu/databases/kddcup99/kddcup99.html

was held in conjunction with KDD-99. The training set consists in binary TCP dump data from seven weeks of network traffic. Each original pattern has 34 continuous features and 7 symbolic features. We normalize each continuous feature to the range $[0, 1]$, and transform each symbolic feature to integer data and then normalize it to the range $[0, 1]$. The original training set consists in 4898431 examples containing repeated examples. We removed the duplicate data and extracted the first 200000 and the last 200000 examples, thus obtaining a training set of size 400000. We used a Gaussian kernel with parameters $\sigma^2 = 0.625, C = 0.5$ and 512MB of caching area.

Table 4. Performance of the improved PGPDT on a large-scale data set

	KDDCUP-99, $n_{sp} = 600, n_c = 200$						
PEs	$time$	sp_r	it	t_{prep}	t_{solv}	t_{grad}	$MKer$
1	46301		1042	62.0	526.0	45711	170469
2	19352	2.4	1031	56.0	300.0	18994	97674
4	8394	5.5	1003	40.1	177.7	8175	74478
8	4821	9.6	1043	30.4	150.1	4639	77465
16	2675	17.3	1016	27.8	113.6	2532	75423

Table 4 shows the training results using PGPDT algorithm on the IBM Linux cluster. The column $MKer$ counts the millions of total Kernel evaluations, which is the total sum of the kernel evaluations over all the processing elements.

A superlinear speedup can be observed also for the largest number of processors. As we mentioned, this is mainly due to the PGPDT ability to efficiently exploit the larger amount of memory for caching the G's entries, which yields a significant kernel evaluations reduction. The good scalability shown in these experiments confirms that the proposed improved PGPDT is very suitable to face in parallel large and even huge data sets.

5 Conclusions

In this paper we presented some improvements to the PGPDT decomposition algorithm for training large-scale SVMs. The changes concern the solution of the inner QP subproblems, which is a crucial point for the performance of the PGPDT. We experimentally show that a new gradient projection scheme, based on the adaptive nonmonotone linesearch in combination with an averaged Barzilai-Borwein steplength rule, improves the performance of both the sequential and the parallel version of the code over the monotone gradient projection solver previously used by PGPDT. Besides, a secant-based projector gives a further improvement with respect to the bisection-based projector currently available in the PGPDT software. The combination of these new strategies gives rise to an improved PGPDT version for large-scale SVM problems, which is available for download at *http://dm.unife.it/gpdt*.

References

1. C.C. Chang, C.J. Lin (2002), LIBSVM: a Library for Support Vector Machines, available at http://www.csie.ntu.edu.tw/~cjlin/libsvm.
2. R. Collobert, S. Benjo (2001), SVMTorch: Support Vector Machines for Large-Scale Regression Problems, *Journal of Machine Learning Research* 1, 143–160.
3. Y.H. Dai, R. Fletcher (2003), New Algorithms for Singly Linearly Constrained Quadratic Programs Subject to Lower and Upper Bounds, Research Report NA/216, Department of Mathematics, University of Dundee.
4. T. Joachims (1998), Making Large-Scale SVM Learning Practical, *Advances in Kernel Methods*, B. Schölkopf *et al.*, eds., MIT Press, Cambridge, MA.
5. C.J. Lin (2001), On the Convergence of the Decomposition Method for Support Vector Machines, *IEEE Transactions on Neural Networks* 12(6), 1288–1298.
6. P.M. Pardalos, N. Kovoor (1990), An Algorithm for a Singly Constrained Class of Quadratic Programs Subject to Upper and Lower Bounds, *Math. Programming* 46, 321–328.
7. V. Ruggiero, L. Zanni (2000), A Modified Projection Algorithm for Large Strictly Convex Quadratic Programs, *J. Optim. Theory Appl.* 104(2), 281–299.
8. V. Ruggiero, L. Zanni (2000), Variable Projection Methods for Large Convex Quadratic Programs, *Recent Trends in Numerical Analysis*, D. Trigiante, ed., Advances in the Theory of Computational Mathematics 3, Nova Science Publ., 299–313.
9. T. Serafini (2005), Gradient Projection Methods for Quadratic Programs and Applications in Training Support Vector Machines, Ph.D. Thesis, Dept. of Mathematics, University of Modena and Reggio Emilia.
10. T. Serafini, G. Zanghirati, L. Zanni (2005), Gradient Projection Methods for Large Quadratic Programs and Applications in Training Support Vector Machines, *Optim. Meth. and Soft.* 20, 353–378.
11. T. Serafini, G. Zanghirati, L. Zanni (2004), Parallel Decomposition Approaches for Training Support Vector Machines, *Parallel Computing: Software Technology, Algorithms, Architectures and Applications*, G.R. Joubert, W.E. Nagel, F.J. Peters and W.V. Walter, ed., Advances in Parallel Computing 13, Elsevier, Amsterdam, The Netherlands, 259–266.
12. T. Serafini, L. Zanni (2005), On the working set selection in Gradient Projection-based Decomposition Techniques for Support Vector Machines, Optim. Meth. and Soft., to appear. (http://cdm.unimo.it/home/matematica/zanni.luca/).
13. V.N. Vapnik (1998), Statistical Learning Theory, John Wiley and Sons, New York.
14. G. Zanghirati, L. Zanni (2003), A Parallel Solver for Large Quadratic Programs in Training Support Vector Machines, *Parallel Computing* 29, 535–551.

Nesting OpenMP in MPI to Implement a Hybrid Communication Method of Parallel Simulated Annealing on a Cluster of SMP Nodes

Agnieszka Debudaj-Grabysz[1] and Rolf Rabenseifner[2]

[1] Silesian University of Technology, Department of Computer Science,
Akademicka 16, 44-100 Gliwice, Poland
`agrabysz@star.iinf.polsl.gliwice.pl`
[2] High-Performance Computing-Center (HLRS), University of Stuttgart,
Nobelstr 19, D-70550 Stuttgart, Germany
`rabenseifner@hlrs.de`
`www.hlrs.de/people/rabenseifner`

Abstract. Concurrent computing can be applied to heuristic methods for combinatorial optimization to shorten computation time, or equivalently, to improve the solution when time is fixed. This paper presents several communication schemes for parallel simulated annealing, focusing on a combination of OpenMP nested in MPI. Strikingly, even though many publications devoted to either intensive or sparse communication methods in parallel simulated annealing exist, only a few comparisons of methods from these two distinctive families have been published; the present paper aspires to partially fill this gap. Implementation for VRPTW—a generally accepted benchmark problem—is used to illustrate the advantages of the hybrid method over others tested.

Keywords: Parallel processing, MPI, OpenMP, communication, simulated annealing.

1 Introduction

The paper presents a new algorithm for parallel simulated annealing—a heuristic method of optimization—that uses both MPI [9] and OpenMP [12] to achieve significantly better performance than a pure MPI implementation. This new hybrid method is compared to other versions of parallel simulated annealing, distinguished by varying level of inter-process communication intensity. Defining the problem as searching for the optimal solution given a pool of processors available for a specified period of time, the hybrid method yields distinctively better optima as compared to other parallel methods. The general reader (i.e., not familiar with simulated annealing) will find the paper interesting as it refers to a practical parallel application run on a cluster of SMPs with the number of processors ranging into hundreds.

Simulated annealing (SA) is a heuristic optimization method used when the solution space is too large to explore all possibilities within a reasonable amount of time. The vehicle routing problem with time windows (VRPTW) is an example

B. Di Martino et al. (Eds.): EuroPVM/MPI 2005, LNCS 3666, pp. 18–27, 2005.

of such a problem. Other examples of VRPTW are school bus routing, newspaper and mail distribution or delivery of goods to department stores. Optimization of routing lowers distribution costs and parallelization allows to find a better route within the given time constraints.

The SA bibliography focuses on the sequential version of the algorithm (e.g., [2,15]), however parallel versions are investigated too, as the sequential method is considered to be slow when compared with other heuristics [16]. In [1,3,8,10,17] and many others, directional recommendations for parallelization of SA can be found. The only known detailed performance analyses of intensive versus sparse communication algorithms are in [4,11,13].

VRPTW—formally formulated by Solomon [14], who also proposed a suite of tests for benchmarking, has a rich bibliography as well (e.g., [16]). Nevertheless, parallel SA to solve the VRPTW is discussed only in [4,6,7].

The parallel implementation of SA presented in this paper had to overcome many practical issues in order to achieve good parallel speedups and efficiency. Tuning of the algorithms for distributed as well as for shared memory environment was conducted.

The plan of the paper is as follows: Section 2 presents the theoretical basis of the sequential and parallel SA algorithm. Section 3 describes how the MPI and OpenMP parallelization was done, while Section 4 presents the results of the experiments. Conclusions follows.

2   Parallel Simulated Annealing

In simulated annealing, one searches for the optimal state, i.e., the state that gives either the minimum or maximum value of the *cost function*. It is achieved by comparing the current solution with a random solution from a specific *neighborhood*. With some probability, worse solutions could be accepted as well, which can prevent convergence to local optima. However, the probability of accepting a worse solution decreases during the process of annealing, in sync with the parameter called *temperature*. An outline of the SA algorithm is presented in

```
01   S ← GetInitialSolution();
02   T ← InitialTemperature;
03   for i ← 1 to NumberOfTemperatureReduction do
04       for j ← 1 to EpochLength do
05           S′ ← GetSolutionFromNeighborhood();
06           ΔC ← CostFunction(S′) - CostFunction(S);
07           if (ΔC < 0 or AcceptWithProbabilityP(ΔC, T))
08               S ← S′;    {i.e., the trial is accepted}
09           end if;
10       end for;
11       T ← λT;    {with λ < 1}
12   end for;
```

Fig. 1. SA algorithm

Figure 1, where a single execution of the innermost loop step is called a *trial*. The sequence of all trials within a temperature level forms a *chain*. The returned final solution is the best one ever found.

2.1   Decomposition and Communication

Although SA is often considered to be an inherently sequential process since each new state contains modifications to the previous state, one can isolate *serialisable sets* [8]—a collection of rejected trials which can be executed in any order, and the result will be the same (starting state). Independence of searches within a serialisable set makes the algorithm suitable for parallelization, where the creation of random solutions is decomposed among processors. From the communication point of view SA may require broadcasting when an acceptable solution is found. This communication requirement suggests message passing as the suitable paradigm of communication, particularly if intended to run on a cluster.

2.2   Possible Intensity of Communication in Parallel Simulated Annealing

Selection of both decomposition and communication paradigms seems to be naturally driven by the nature of the problem, but setting the right intensity of communication is not a trivial task. The universe of possible solutions is spanned by two extremes: communicating each event, where *event* means an *accepted trial*, and, independent runs method, where no event is communicated. The former method results in the *single chain algorithm*—only a single path in the search space is carried out, while the latter results in the *multiple chains algorithm*—several different paths are evaluated simultaneously (see Figure 2). The location of starting points depends on implementation.

Intensive Communication Algorithm—The Time Stamp Method. In current research the intensive communication algorithm is represented by its speed-up optimized version called the *time stamp method*. The communication model with synchronization at solution acceptance events proposed in [7] was the starting point. The main modification, made for efficiency reasons, is to let processes work in an asynchronous way, instead of frequent computation interruptions by synchronization requests that resulted in idle time. After finding

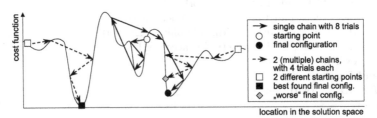

Fig. 2. One single chain versus multiple chains

an accepted trial, the process announces the event and continues its computation without any synchronization. In the absence of the mechanism which ensures that all processes are aware of the same, global state and choose the same, accepted solution, a single process can decide only locally, based on its own state and information included in received messages. Information about the real time when the accepted solution was found—the *time stamp*—is used as the criterion for choosing among a few acceptable solutions (known locally). The solution with the most recent time stamp is accepted, while older ones are rejected. From the global point of view the same solutions will be preferred.

Generally, the single chain approach is believed to have two main drawbacks: only limited parallelism is exploited due to the reduction to a single search path and noticeable communication overhead. The second drawback especially reduces the application of this method to a small number of engaged processes.

Non-communication Algorithm—Independent Runs. The main assumptions for independent runs were formulated in [2], where the *division algorithm* is proposed. The method uses all available processors to run basically sequential algorithms, where the original chain is split into subchains of *EpochLength* (see Figure 1) divided by the number of processes. At the end, the best solution found is picked up as the final one; thus the communication is limited to merely one reduction operation.

Although the search space is exploited in a better way than in the approach described previously, very likely only a few processes work in the "right" areas while the rest perform useless computation. Additionally, excessive shortening of the chain length negatively affects the quality of results, so application of this method is not suitable for a great number (e.g., hundreds) of engaged processes.

Lightweight Communication—Periodically Interacting Searches. Recognizing the extreme character of the independent runs method, especially when using a large number of processes, one is tempted to look for the golden mean in the form of periodic communication. The idea was fully developed in [11]. In that approach processes communicate after performing a subchain called a *segment*, and the best solution is selected and mandated for all of them. In this study a segment length is defined by a number of temperature decreases. As suggested in [11] to prevent search paths from being trapped in local minima areas as a result of communication, the period of the information exchange needs to be carefully selected. Additionally, the influence of the periodic exchange doesn't always result in a positive effect and varies according to the optimized problem.

Hybrid Communication Method—Nesting OpenMP in MPI. In this study a new approach is proposed, which tries to adopt the advantages of the methods mentioned above while minimizing their disadvantages. In contrast with these methods, this implementation is intended to run on modern clusters of SMP nodes. The parallelization is accomplished using two levels: the outer parallelization which uses MPI to communicate between SMP nodes, and the inner parallelization which uses OpenMP for shared memory parallelization within nodes.

Outer-Level Parallelization. It can be assumed that the choice of an appropriate algorithm should be made between independent runs or periodically interacting searches, as they are more suitable for more than few processes. The maximal number of engaged nodes is limited by reasonable shortening of the chain length, to preserve an acceptable quality of results.

Inner-Level Parallelization. Within a node a few threads can build one subchain of a length determined at the outer-level. Negligible deterioration of quality is a key requirement. If this requirement is met, the limit on the total number of processors to achieve both speed-up and preserve quality is determined by the product of the processes number limit at the outer level and the threads number limit at the inner level. An efficient implementation can also take advantage of the fact that CPUs on SMP nodes communicate by fast shared memory and communication overhead should be minimal relative to that between nodes. In this study a modified version of the *simple serialisable set* algorithm [8] was applied (see Section 3). For a small number of processors (i.e., 2 to 8), apart from preserving the quality of solutions, it should provide speed-up.

3   Implementation of Communication with MPI and OpenMP

3.1   Intensive Communication Algorithm

Every message contains a solution together with its time stamp. As the assumption was to let the processes work asynchronously polling is applied to detect moments when data is to be received. An outline of the algorithm is presented in Figure 3.

In practice, as described in [7], the implementation underwent a few stages of improvement to yield acceptable speed-up. Among others: a long message containing a solution was split into two, to test the differences in performance when sending different types of data, data structure was reorganized—an array of structures was substituted by a structure of arrays, MPICH2 was used since there was a bug in MPICH that prevented the program from running.

3.2   Non– and Lightweight Communication Algorithms

In case of both independent runs and periodically interacting searches methods, MPI reduction instructions (MPI_Bcast, MPI_Allreduce) are the best tools for exchanging the data.

3.3   Hybrid Communication Method

The duality of the method is extended to its communication environment: MPI is used for communication between the nodes and OpenMP for communication

```
01  MyData.TimeStamp ← 0;
02  do in parallel
03     do
04         MPI_Iprobe();    { check for incoming messages }
05         if (there is a message to receive)
06             MPI_Recv(ReceivedData, ...);
07             if (MyData.TimeStamp < ReceivedData.TimeStamp)
08                 update MyData and current TimeStamp;
09             end if;
10         end if;
11     while (there is any message to receive);
12     performTrial();
13     if (an acceptable solution was found, placed in MyData.Solution)
14         MyData.TimeStamp ← MPI_Wtime();
15         for all cooperating processors do
16             MPI_Send(MyData, ...);
17         end for;
18     end if;
19  while (not Finish);
```

Fig. 3. The outline of the intensive communication algorithm

```
01  for i ← 1 to NumberOfTemperatureReduction do
02     {entering OpenMP parallel region}
03     for j ← 1 to EpochLength do
04         {OpenMP parallel for loop worksharing}
05         for i ← 0 to set_of_trials_size do
06             performTrial();
07         end for;
08         {OpenMP entering master section}
09         select one solution, common for all threads, from all
               accepted ones, based on ΔC < 0 or AcceptWithProbabilityP(ΔC, T)
10         j ← j + set_of_trials_size;
11         {OpenMP end of master section}
12     end for;
13     {end of OpenMP parallel region}
14     T ← λT;
15  end for;
```

Fig. 4. Parallel SA algorithm within a single node

among processors within a single node. The former algorithm is implemented as described in the previous section (3.2), whereas an outline of the latter one is presented in Figure 4.

At the inner-level, the total number of trials (*EpochLength* from the outer level) in each temperature step is divided into short sets of trials All trials in such a set are done independently. This modification is the basis for the OpenMP parallelization with loop worksharing. To achieve an acceptable speed-up, the following optimizations are necessary:

- The parallel threads must not be forked and joined for each inner loop because the total execution time for a set of trials can be too short, compared to the OpenMP fork-join overhead;
- The size of such a set must be larger than the number of threads to minimize the load imbalance due to the potentially extremely varying execution time for each trial. Nevertheless, for keeping quality, the size of the set of trials should be as short as possible to minimize the number of accepted but unused trials;
- Each thread has to use its own independent random number generator to minimize OpenMP synchronization points.

4   Experimental Results

In the vehicle routing problem with time windows it is assumed that there is a warehouse, centrally located to n customers. The objective is to supply goods to all customers at the minimum cost. The solution with lesser number of route legs (the first goal of optimization) is better then a solution with smaller total distance traveled (the second goal of optimization). Each customer as well as the warehouse has a time window. Each customer has its own demand level and should be visited only once. Each route must start and terminate at the warehouse and should preserve maximum vehicle capacity. The sequential algorithm from [5] was the basis for parallelization.

Experiments were carried out on NEC Xeon EM64T Cluster installed at the High Performance Computing Center Stuttgart (HLRS). Additionally, for tests of the OpenMP algorithm, NEC TX-7 (ccNUMA) system was used. The numerical data were obtained by running the program 100 times for Solomon's [14] R108 set with 100 customers and the same set of parameters.

The quality of results, namely the number of final solutions with the minimal number of route legs generated by pure MPI-based algorithms in 100 experiments is shown in Table 1. Experiments stopped after 30 consecutive temperature decreases without improving the best solution. As can be seen in the table, the intensive communication method gives acceptable results only for a small number of cooperating processes. Secondly, excessively frequent periodical communication hampers the annealing process and deteriorates the convergence. The best algorithm for the investigated problem on a large number of CPUs, as far as the quality of results is concerned, is the algorithm of independent runs, so this one was chosen for the development of the hybrid method.

The results generated by the hybrid algorithm are shown in Table 2. It compares two methods using the same number of processors, e.g., 20 processor independent runs (MPI parallelization) versus computation on 10 nodes with 2 CPUs each or on 5 nodes with 4 CPUs each (MPI/OMP parallelization). For better comparison, a real time limit was fixed for each independent set of processors. The time limit is the average time needed by the sequential algorithm to find the minimal-leg solution divided by the number of processors. The hybrid version of the algorithm with 2 OMP threads per each node ran on NEC Xeon

Table 1. Comparison of quality results for MPI based methods

No. of processes	Percentage of solutions with minimal no. of route legs						
	Non-comm.	Periodic communication with the period of					Intensive comm.
		1	5	10	20	30	
seq	94	N/A	N/A	N/A	N/A	N/A	N/A
2	97	95.6	96.2	96.8	94.2	96	91
4	95	93	96	93	93	96	96
8	91	91	82	86	90	91	93
10	94	85	88	85	88	96	82
20	84	70	77	77	74	89	69
40	85	56	60	63	71	74	N/A
60	76	30	46	55	60	68	N/A
100	60	32	38	35	44	55	N/A
200	35	12	23	30	38	37	N/A

Table 2. Comparison of quality results for hybrid and independent runs methods

Total no. of used processors	Used time limit [s]	Speed -up	Hyb., 2 OMP threads		Hyb., 4 OMP threads		Non-comm.
			No. of MPI processes	No. of sol. with min. no. of route legs	No. of MPI processes	No. of sol. with min. no. of route legs	No. of sol. with min. no. of route legs
1	1830.0	N/A	N/A	N/A	N/A	N/A	97
4	457.5	4	2	92	1	96	95
16	114.4	16	8	93	4	97	93
20	91.5	20	10	93	5	93	94
32	57.2	32	16	90	8	90	85
40	45.8	40	20	92	10	93	85
60	30.5	60	30	86	15	91	76
80	22.9	80	40	78	20	88	69
100	18.3	100	50	87	25	87	64
120	15.3	120	60	67	30	85	62
200	9.2	200	100	55	50	78	34
400	4.6	400	200	27	100	57	9
600	3.1	600	300	9	150	31	0
800	2.3	800	400	N/A	200	13	0

Cluster, while the usage of 4 OMP threads per node was emulated, due to the lack of access to the desired machine. Because a separate set of experiments with 4 OMP threads demonstrated the speed-up of 2.7, then the emulation was carried out by multiplying the applied time limit by the this factor, as if undoing the speed-up to be observed on a real cluster of SMP nodes. The accumulated CPU time of a real experiment would be shortened by the factor $2.7/4 = 0.67$.

It should be noted that both variants of the hybrid method give a distinctively greater number of solutions with the minimal number of route legs if one uses 32 or more CPUs. Additionally, for smaller number of CPUs, the 4 OMP threads version could be competitive as well (up to 40 CPUs), despite the loss of CPU-time due to the limited efficiency of the parallelization inside of each SMP node. If one can accept a reduced quality, e.g. 85%, then only a speed-up of 40 can be achieved without SMP parallelization. With hybrid parallelization, the speed-up can be raised to 60 (with 2 threads) and 120 (with 4 threads), i.e., an interactive time-scale of about 15 sec can be reached.

5   Conclusions and Future Work

In this study a new implementation of the multiple chain parallel SA that uses OpenMP with MPI was developed. Additionally, within the framework of the single chain parallel SA, a time-stamp method was proposed. Based on experimental results the following conclusions may be drawn:

- Multiple chain methods outperform single chain algorithms, as the latter lead to a faster worsening of results quality and are not scalable. Single chain methods could be used only in environments with a few processors;
- The periodically interacting searches method prevails only in some specific situations; generally the independent runs method achieves better results;
- The hybrid method is very promising, as it gives distinctively better results than other tested algorithms and satisfactory speed-up;
- Emulated results shown need verification on a cluster of SMPs with 4 CPUs on a single node.

Specifying the time limit for the computation, by measurements of the elapsed time, gives a new opportunity to determine the exact moment to exchange data. Such a time-based scheduling could result in much better balancing than the investigated temperature-decreases-based one (used within the periodically interacting searches method). The former could minimize idle times, as well as enables setting the number of data exchanges. Therefore, future work will focus on forcing a data exchange (e.g., after 90% of specified limit time), when—very likely—the number of route legs was finally minimized (first goal of optimization). Then, after selecting the best solution found so far, all working processes—instead of only one—could minimize the total distance (the second goal of optimization), leading to significant improvement of the quality of results.

Acknowledgment

This work was supported by the EC-funded project HPC-Europa. Computing time was also provided within the framework of the HLRS-NEC cooperation.

References

1. Aarts, E., de Bont, F., Habers, J., van Laarhoven, P.: Parallel implementations of the statistical cooling algorithm. Integration, the VLSI journal (1986) 209–238
2. Aarts, E., Korst, J.: Simulated Annealing and Boltzman Machines, John Wiley & Sons (1989)
3. Azencott, R. (ed): Simulated Annealing Parallelization Techniques. John Wiley & Sons, New York (1992)
4. Arbelaitz, O., Rodriguez, C., Zamakola, I.: Low Cost Parallel Solutions for the VRPTW Optimization Problem, Proceedings of the International Conference on Parallel Processing Workshops, IEEE Computer Society, Valencia–Spain, (2001) 176–181
5. Czarnas, P.: Traveling Salesman Problem With Time Windows. Solution by Simulated Annealing. MSc thesis (in Polish), Uniwersytet Wrocławski, Wrocław (2001)
6. Czech, Z.J., Czarnas, P.: Parallel simulated annealing for the vehicle routing problem with time windows. 10th Euromicro Workshop on Parallel, Distributed and Network-based Processing, Canary Islands–Spain, (2002) 376–383
7. Debudaj-Grabysz, A., Czech, Z.J.: A concurrent implementation of simulated annealing and its application to the VRPTW optimization problem, in Juhasz Z., Kacsuk P., Kranzlmuller D. (ed), Distributed and Parallel Systems. Cluster and Grid Computing. Kluwer International Series in Engineering and Computer Science, Vol. 777 (2004) 201–209
8. Greening, D.R.: Parallel Simulated Annealing Techniques. Physica D, 42, (1990) 293–306
9. Gropp, W., Lusk, E., Doss, N., Skjellum, A.: A high-performance, portable implementation of the MPI message passing interface standard, Parallel Computing 22(6) (1996) 789–828
10. Lee, F.A.: Parallel Simulated Annealing on a Message-Passing Multi-Computer. PhD thesis, Utah State University (1995)
11. Lee, K.–G., Lee, S.–Y.: Synchronous and Asynchronous Parallel Simulated Annealing with Multiple Markov Chains, IEEE Transactions on Parallel and Distributed Systems, Vol. 7, No. 10 (1996) 993–1008
12. OpenMP C and C++ API 2.0 Specification, from www.openmp.org/specs/
13. Onbaoglu, E., Özdamar, L.: Parallel Simulated Annealing Algorithms in Global Optimization, Journal of Global Optimization, Vol. 19, Issue 1 (2001) 27–50
14. Solomon, M.: Algorithms for the vehicle routing and scheduling problem with time windows constraints, Operation Research 35 (1987) 254–265, see also http://w.cba.neu.edu/~msolomon/problems.htm
15. Salamon, P., Sibani, P., and Frost, R.: Facts, Conjectures and Improvements for Simulated Annealing, SIAM (2002)
16. Tan, K.C, Lee, L.H., Zhu, Q.L., Ou, K.: Heuristic methods for vehicle routing problem with time windows. Artificial Intelligent in Engineering, Elsevier (2001) 281–295
17. Zomaya, A.Y., Kazman, R.: Simulated Annealing Techniques, in Algorithms and Theory of Computation Handbook, CRC Press LLC, (1999)

Computing Frequent Itemsets in Parallel Using Partial Support Trees

Dora Souliou*, Aris Pagourtzis*, and Nikolaos Drosinos

School of Electrical and Computer Engineering,
National Technical University of Athens,
Heroon Politechniou 9, 15780 Zografou, Greece
{dsouliou, ndros}@cslab.ece.ntua.gr, pagour@cs.ntua.gr

Abstract. A key process in association rules mining, which has attracted a lot of interest during the last decade, is the discovery of frequent sets of items in a database of transactions. A number of sequential algorithms have been proposed that accomplish this task. In this paper we study the parallelization of the partial-support-tree approach (Goulbourne, Coenen, Leng, 2000). Results show that this method achieves a generally satisfactory speedup, while it is particularly adequate for certain types of datasets.

Keywords: Parallel data mining, association rules, frequent itemsets, partial support tree, set-enumeration tree.

1 Introduction

Mining of association rules between itemsets in transactional data is an important and resource demanding computational task, that calls for development of efficient parallelization techniques. Typically, the problem is described as follows: a database $\mathcal{D}$ of transactions is given, each of which consists of several distinct items. The goal is to determine association rules of the form $A \to B$, where A and B are sets of items (itemsets). A fundamental ingredient of this task is the generation of all itemsets the *support* (or frequency) of which exceeds a given threshold t.

Several sequential and parallel methods to tackle this problem have been proposed in the literature. The most widely known is the "A-priori" algorithm of Agarwal and Srikant [6] which generates frequent itemsets in order of increasing size, making use of the fact that supersets of infrequent itemsets cannot be frequent in order to avoid redundant frequency calculations. Goulbourne, Coenen and Leng [12] combine the A-priori method with the idea of storing the database in a structure called Partial Support Tree (P-tree for short) which allows to

---

* Dora Souliou and Aris Pagourtzis were partially supported for this research by "Pythagoras" grant of the Ministry of Education of Greece, co-funded by the European Social Fund (75%) and National Resources (25%), under Operational Programme for Education and Initial Vocational Training (EPEAEK II).

B. Di Martino et al. (Eds.): EuroPVM/MPI 2005, LNCS 3666, pp. 28–37, 2005.

compute the support of an itemset without re-scanning the database, usually searching a small part of the tree.

In this paper we describe a sequential algorithm for the problem of generating all frequent itemsets, which we call PS (Partial Support). Our algorithm is based on the algorithm of [12], augmented with several implementation-oriented details. We further present a parallel version of PS, which we call PPS (Parallel Partial Support). We implement PPS using MPI and present experimental results which show that PPS achieves a satisfactory speedup, especially in cases where the frequency calculation part is dominant.

2   The Sequential Algorithm

In this section we describe algorithm PS (Partial Support), which makes use of a special tree structure called the *Partial Support Tree* (or *P*-tree for short), introduced in [12]. The *P*-tree is a set-enumeration tree [14] that contains itemsets in its nodes. These itemsets represent either transactions of database $\mathcal{D}$ or com-

---

Algorithm PS (* Partial Support *)
Build *P*-tree from database $\mathcal{D}$;

(* *1st level construction* *)
for $i := 1$ **to** *nitems* **do**   (* *nitems* = number of items *)
      get total_support($\{i\}$) from *P*-tree;
      if total_support($\{i\}$) $\geq t$ **then** append $\{i\}$ to L_1;   (* t = support threshold *)
$k:=2$;
while L_{k-1} not empty and $k \leq$ *nitems* **do**
    (* *k-th level construction* *)

    set L_k to be the empty list;
    for each itemset $I \in L_{k-1}$ **do**
        (* *Generation of all k-element supersets of I that have I as prefix.*
        Such an itemset, if frequent, can be seen as the union of I and an
        itemset I′ that belongs to L_{k-1} and differs from I at the last position. *)
        $I' := next(I)$;
        while $I' \neq$ NULL **do**
            if I and I' differ only at the last item **then**
                $I_k := I \cup I'$;
                if all subsets of I_k of size $(k-1)$ are in L_{k-1} **then**
                    get total_support($\{I_k\}$) from *P*-tree;
                    if total_support($\{I_k\}$) $\geq t$ **then** insert I_k to L_k;
                $I' := next(I')$;
            else exit while;   (* *no other k-element superset of I, having*
                                    I as prefix needs to be considered *)
    $k:=k+1$;
(* *end of while-loop* *)

Fig. 1. The Sequential Algorithm

mon prefices of transactions of $\mathcal{D}$. An integer is stored in each tree node which represents the partial support of the corresponding itemset I, that is, the number of transactions that contain I as a prefix. Details of the construction of the P-tree were given in [12]; a more complete description can be found in [11]. Figure 2 shows a database with 16 transactions and 6 items, and the corresponding P-tree.

The partial support tree is constructed during one scan of database $\mathcal{D}$ as follows. For each transaction read, with itemset I, the tree is traversed in order to find a node with itemset I; if such node exists its partial support counter is increased by 1, otherwise a new node is created with itemset I and then it is inserted into the tree at an appropriate point. Supports of all ancestors of the new node are also increased by 1. For each inserted node, a second node may have to be created, containing the common prefix of I and some existing node (this happens if this prefix is not already present).

Once the P-tree is available, it is possible to count the total support of any itemset by traversing the nodes of P-tree and summing the partial supports of appropriate nodes. Such a tree traversal is described in detail in [12]. In algorithm PS we refer to this process for itemset I as 'get total_support(I) from P-tree'.

The algorithm first reads the database and creates the P-tree as described above. It then starts building lists of frequent itemsets in an *a-priori* manner [6], that is, a level-by-level strategy is followed: frequent itemsets with one item (singletons) are generated first, then frequent 2-itemsets (pairs) are generated, and so on. The key property is that if an itemset with k items is frequent then so are all its subsets with $k - 1$ items; therefore, there is no need to examine a k-itemset (itemset of size k) a subset of which was not found frequent at level $k - 1$. The lists of frequent itemsets for the database of Figure 2 with threshold $t = 4$ are shown below. Lists L_i, $1 \leq i \leq 3$, contain frequent itemsets of size i; there are no frequent itemsets with 4 items.

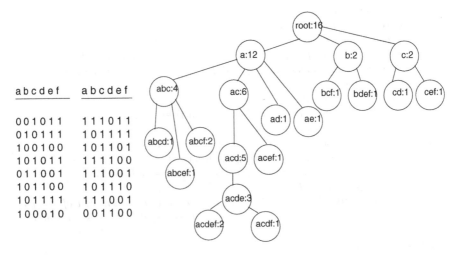

Fig. 2. A database with 16 transactions, which are subsets of $\{a, b, c, d, e, f\}$, and the corresponding tree of partial supports

L_1 a:12 b:6   c:13 d:9   e:8   f:10
L_2 ac:10 ad:7 ae:6 af:7 bc:5 bf:5 cd:7 ce:6 cf:9 ef:6
L_3 acd:6 ace:5 acf:7 cef:5

The difference between algorithm PS and the A-priori algorithm [6] lies in the generation of candidate itemsets and in the way total support is counted. For level-k computation, A-priori first generates a list C_k of all candidate itemsets, based on information of level $(k-1)$, and then reads the whole database and for each transaction I that it reads it updates all items in C_k that are subsets of I; once the total support of each itemset is computed it removes from C_k all infrequent itemsets, thus obtaining L_k. In contrast, PS generates candidate itemsets one by one and for each of them it traverses an appropriate part of the P-tree in order to count its total support; if the itemset is found frequent it is inserted in L_k, otherwise it is ignored. That is, PS creates L_k directly, without first creating the intermediate list C_k. This saves some time because no pass of the (potentially large) list C_k is needed.

A description of the algorithm, together with some implementation details, is given in Figure 1.

3   The Parallel Algorithm (PPS)

We now present a parallelization of the algorithm described in the previous section. Our approach follows the ideas of a parallel version of A-priori, called Count Distribution, which was described by Agrawal and Shafer [7]; the difference is, of course, that PPS makes use of partial support trees.

In the beginning, the root process distributes the database transactions to the processors in a round-robin fashion; then, each of the processors creates its own local P-tree based on the transactions that it has received. Next, each processor starts by computing the local support of all singletons (1-itemsets) by accessing its own P-tree. Then the total support of singletons is computed from local supports by an appropriate parallel procedure. The result is distributed to all processors so that each one ends up with the same list L_1 of frequent singletons by removing all infrequent singletons. During k-level computation (for each $k \geq 2$), all processors first generate the same list of candidate itemsets C_k from the common list L_{k-1}. Then the same procedure as that followed for the first level allows each processor to obtain the total support for all itemsets of C_k, and finally to derive list L_k by removing infrequent itemsets of C_k.

Note that here, in contrast to the sequential algorithm, we do not avoid creation of list C_k; it would be possible to do it but then all processors would have to synchronize after calculating the support of each itemset, thus resulting in a high synchronization cost.

A description of the algorithm is given in Figure 3. The lists of candidate itemsets and the lists of frequent itemsets for the database of Figure 2 are shown below, for threshold $t = 4$. Lists C_i, $1 \leq i \leq 3$, contain candidate itemsets of size i.

Algorithm PPS (Parallel Partial Support *)*

distribute the database $\mathcal{D}$ to the processors in a round-robin manner;
let $\mathcal{D}_j$ denote the part of D assigned to processor p_j;
in each processor p_j **in parallel do**
    build local P-tree from local database $\mathcal{D}_j$;

    (1st level construction *)*
    for $i := 1$ **to** *nitems* **do**
        get local_support$_j(\{i\})$ from local P-tree;

    (Global synchronized computation *)*
    for $i := 1$ **to** *nitems* **do**   *(* nitems = number of items *)*

        total_support$(\{i\})$:= parallel_sum$_{j=1}^{nprocs}$ local_support$_j(\{i\})$;   *(* nprocs =*
                                                  *number of processors *)*

    (Local computation continues *)*
    for $i := 1$ **to** *nitems* **do**
        if total_support$(\{i\}) \geq t$ **then** append $\{i\}$ to L_1;   *(* all processors obtain*
                                                     *the same list L_1 *)*

    $k:=2$;
    while L_{k-1} not empty and $k \leq$ *nitems* **do**
        (k-th level construction *)*

        set L_k to be the empty list;
        for each itemset $I \in L_{k-1}$ **do**
            $I' := next(I)$;
            while $I' \neq$ NULL **do**
                if I and I' differ only at the last item **then**
                    $I_k := I \cup I'$;
                    if all subsets of I_k are in L_{k-1} **then** insert I_k into C_k;
                    $I' := next(I')$;
                else exit while;

        for all itemsets $I_k \in C_k$ **do**
            get local_support$_j(\{I_k\})$ from local P-tree;
        (Global synchronized computation *)*
        for all itemsets $I_k \in C_k$ **do**
            total_support(I_k):= parallel_sum$_{j=1}^{nprocs}$(local_support$_j(I_k)$);
        (Local computation continues *)*
        for all itemsets $I_k \in C_k$ **do**
            if total_support$(\{I_k\}) \geq t$ **then** insert I_k to L_k;

    k:=k+1;
    (end of while-loop *)*

Fig. 3. The Parallel Algorithm

Lists L_i, $1 \leq i \leq 3$, contain frequent itemsets of size i. Note that there is no itemset of size 4 such that all its subsets of size 3 are frequent, therefore C_4 is not created at all.

C_1 a:12   b:6   c:13   d:9   e:8   f:10
L_1 a:12   b:6   c:13   d:9   e:8   f:10
C_2 ab:4 ac:10 ad:7 ae:6 af:7 bc:5 bd:2 be:2 bf:5 cd:7 ce:6 cf:9 de:4 df:4 ef:6
L_2 ac:10 ad:7 ae:6 af:7 bc:5 bf:5 cd:7 ce:6 cf:9 ef:6
C_3 acd:6 ace:5 acf:7 aef:4 bcf:4 cef:5
L_3 acd:6 ace:5 acf:7 cef:5

4 Experimental Results

Our experimental platform is an 8-node Pentium III dual-SMP cluster interconnected with 100 Mbps FastEthernet. Each node has two Pentium III CPUs at
800 MHz, 256 MB of RAM, 16 KB of L1 I Cache, 16 KB L1 D Cache, 256 KB
of L2 cache, and runs Linux with 2.4.26 kernel. We use MPI implementation
MPICH v.1.2.6, compiled with the Intel C++ compiler v.8.1.

We perform two series of experiments. In the first case, we consider randomly
generated datasets, where both the number of transactions and the number of
items are determined as user-defined parameters. Although transactions are generated in a pseudo-random manner, items are distributed over the various transactions in a reasonably uniform pattern. We consider a relatively low minimum
support threshold value of 5%.

Figure 4 reveals that the speedups attained at the parallel execution of the
algorithm on our SMP cluster are reasonably efficient, particularly in the computationally intensive case of 50K transactions and 500 items. We observe that
the speedup is improved as the number of items increases. This happens because the parallelized part of the computation (frequency calculation) occupies
a larger part of the whole computation as the number of items increases. We
also observe that efficiency of the computation for the 50K transactions / 50
items dataset drops when transisioning from 8 to 16 processors. This is due to
memory congestion overhead within an SMP node (see also the discussion for
similar slope reduction in Figure 5, below). However, this drop is not obvious
in the slopes corresponding to the two largest datasets (which have either more
transactions or more items), probably because the parallelization gains balance
this speedup loss.

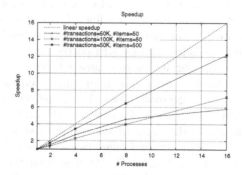

Fig. 4. Speedup obtained for randomly generated synthetic datasets

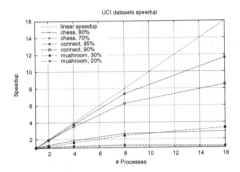

Fig. 5. Speedup obtained for three UCI datasets (chess, connect, mushroom)

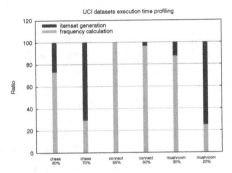

Fig. 6. Sequential execution time profiling for the three UCI datasets (chess, connect, mushroom)

The satisfactory results for the synthetic datasets have motivated us to extend the application of the parallel algorithm to widely used datasets, such as the ones contained in the UC Irvine Machine Learning repository (http://www.ics.uci.edu/~mlearn/ MLRepository.html).

We have used three UCI datasets, namely chess, connect and mushroom. The experimental results for a varying number of processes are depicted in Figure 5. While the obtained parallel efficiency ranges around 60-80% for connect, it drops heavily to about 10-20% for mushroom, and even 5-10% for chess. A slight reduction in the slope of the parallel efficiency line when transitioning from 8 to 16 processes can be ascribed to the memory congestion overhead inside each SMP node, which inevitably needs to host two MPI processes, given our experimental infrastructure. Clearly, the parallel performance achieved is directly associated with both the input datasets and the minimum support considered. For the latter, we considered typical suggested values, depending on the specific dataset (70% and 80% for chess, 90% and 95% for connect, 20% and 30% for mushroom).

In order to demonstrate the effect of the specific dataset on the efficiency of the proposed parallelization, we performed in each case extensive profiling of the sequential execution times (Figure 6). We measured the time required for all component parts of the parallel algorithm; nevertheless, in practice, the total execution time is exclusively determined by the following parts of the parallel algorithms:

- the frequency calculation of all itemsets against the local dataset of the specific process
- given the current list of frequent itemsets, the generation of the list of candidate itemsets of the next level

Other component times are essentially much smaller, e.g. the time associated with the MPI communication for the global frequency reduction, as well as the time required for the elimination of the infrequent itemsets. The execution time of connect is obviously dominated by the frequency calculation part, as indicated by Figure 6. Thus, the fact that the parallelization of the frequency calculation has proved to be quite beneficial for connect is not surprising. On the other hand, both chess and mushroom incur a significant overhead on the itemset generation

Dataset	Min Support (in %)	Procs	Total execution time (in sec)	Frequency calc. time (in sec)	Itemsets gen. time (in sec)
chess	80	1	7.21	5.28	1.93
		2	5.33	3.37	1.95
		4	3.87	1.89	1.95
		8	2.71	.74	1.94
		16	2.50	.25	2.18
	70	1	97.82	28.72	69.09
		2	88.57	18.46	69.61
		4	80.72	10.15	69.67
		8	74.11	3.41	70.19
		16	82.20	1.43	79.85
connect	95	1	72.82	72.69	.13
		2	37.29	37.16	.13
		4	19.22	19.05	.13
		8	9.87	9.69	.14
		16	6.25	5.96	.14
	90	1	880.24	850.29	29.94
		2	460.82	430.56	30.13
		4	249.70	218.96	30.08
		8	141.18	110.66	30.26
		16	103.78	67.75	34.52
mushroom	30	1	1.83	1.60	.22
		2	1.52	1.29	.23
		4	1.13	.89	.23
		8	.75	.50	.23
		16	.55	.24	.25
	20	1	122.69	30.65	92.03
		2	112.79	20.86	91.71
		4	105.58	13.02	92.26
		8	99.11	6.33	92.71
		16	109.56	2.58	105.25

process, which has not been parallelized, and is thus sustained as a constant offset in all parallel execution times. As our parallelization strategy involves the parallel implementation of certain parts of the algorithm, the maximum efficiency that can be obtained is limited according to Amdahl's law.

Conclusively, the experimental evaluation renders our parallelization technique meaningful in the case of

- relatively high support threshold values,
- relatively large datasets, that is, high number of transactions in respect to the number of items,
- dataset patterns that result to high frequency calculation needs, in respect to itemset generation overhead.

Finally, the following table displays the measured times for the three UCI datasets considered here. Note that the frequency calculation part scales well with the number of processes, as was anticipated. Note also the increase in the itemsets generation time from 8 to 16 processes, which reflects the memory congestion when two processes reside on the same SMP node.

5   Conclusions – Future Work

In this work we have investigated the parallelization of an algorithm for mining frequent itemsets from a database of transactions. The algorithm is based on the use of partial support trees that facilitates the process of support (frequency) calculation. In particular, each processor handles a part of the database and creates a small local tree that can be kept in memory, thus providing a practicable solution when dealing with extremely large datasets.

We have implemented the algorithm using message passing, with the help of MPI. Results show that the above described strategy results in quite satisfactory parallelization of the frequency calculation part of the algorithm; however, another part of the algorithm, namely that of itemset generation, remains sequential.

We are currently considering the parallelization of the itemset generation part, as well. To this end, we plan to explore the use of hybrid MPI-OpenMP parallelization; more specifically, MPI will maintain the inter-process communication, while at the same time OpenMP, which allows inter-thread parallelization and synchronization within a specific process, will take care of the incremental parallelization of computationally intensive loops.

References

1. R. Agrawal, C. Aggarwal and V. Prasad. Depth First Generation of Long Patterns. In KDD 2000, ACM, pp. 108-118.
2. S. Ahmed, F. Coenen, and P.H. Leng: A Tree Partitioning Method for Memory Management in Association Rule Mining. In Proc. DaWaK 2004, LNCS 3181, pp. 331-340, 2004.

3. F.Angiulli, G. Ianni, L. Palopoli. On the complexity of inducing categorical and quantitative association rules, arXiv:cs.CC/0111009 vol 1, Nov 2001.
4. R. Agrawal, T. Imielinski, and A. Swami. Mining Association Rules between Sets of Items in Large Databases. In Proc. of ACM SIGMOD Conference on Management of Data, Washington DC, May 1993.
5. R. Agrawal, T. Imielinski, and A. Swami. Database mining: a performance perspective. *IEEE Transactions on Knowledge and Data Engineering*, 5(6), pp. 914–925, Dec 1993. Special Issue on Learning and Discovery in Knowledge-Based Databases.
6. R. Agrawal and R. Srikant. Fast Algorithms for mining association rules. In Proc. VLDB'94, pp. 487–499.
7. R. Agrawal, J.C. Shafer. Parallel Mining of Association Rules. IEEE Trans. Knowl. Data Eng. 8(6), pp. 962-969, 1996.
8. R. J. Bayardo Jr. and R. Agrawal. Mining the Most Interesting Rules. In Proc. of the Fifth ACM SIGKDD Int'l Conf. on Knowledge Discovery and Data Mining, pp. 145–154, 1999.
9. E. Boros, V. Gurvich, L. Khachiyan, K. Makino. On the complexity of generating maximal frequent and minimal infrequent sets, in *STACS* 2002.
10. F. Coenen, G. Goulbourne, and P. Leng. Computing Association Rules using Partial Totals. In L. De Raedt and A. Siebes eds, *Principles of Data Mining and Knowledge Discovery* (Proc 5th European Conference, PKDD 2001, Freiburg, Sept 2001), Lecture Notes in AI 2168, Springer-Verlag, Berlin, Heidelberg: pp. 54–66.
11. F. Coenen, G. Goulbourne and P. Leng. Tree Structures for Mining Association Rules. *Data Mining and Knowledge Discovery*, 8 (2004), pp. 25-51
12. G. Goulbourne, F. Coenen and P. Leng. Algorithms for Computing Association Rules using a Partial-Support Tree. *Journal of Knowledge-Based Systems* 13 (2000), pp. 141–149.
13. J. Han, J. Pei, Y.Yin and R. Mao. Mining Frequent Patterns without Candidate Generation: A Frequent-Pattern Tree Approach. *Data Mining and Knowledge Discovery*, 8 (2004), pp. 53-87.
14. R. Raymon. Search Through Systematic Search Enumeration. In Proc. 3rd Intl Conf. on Principles of Knowledge Representation and Reasoning, pp. 539-550.
15. A.Savasere, E. Omiecinski and S. Navathe. An Efficient Algorithm for Mining Association Rules in Large Databases. In VLDB 1995, pp. 432-444.
16. H. Toivonen. Sampling Large Databses for Association Rules. In VLDB 1996, pp. 1-12.

A Grid-Aware Branch, Cut and Price Implementation

Emilio P. Mancini[1], Sonya Marcarelli[1], Pierluigi Ritrovato[2],
Igor Vasil'ev[3], and Umberto Villano[1]

[1] Università del Sannio, Dipartimento di Ingegneria, RCOST, Benevento, Italy
{epmancini, sonya.marcarelli, villano}@unisannio.it
[2] Centro di Ricerca in Matematica Pura ed Applicata, Università di Salerno,
Fisciano (SA), Italy
ritrovato@crmpa.unisa.it
[3] Institute of System Dynamics and Control Theory, SB RAS,
Irkutsk, Russia
vil@icc.ru

Abstract. This paper presents a grid-enabled system for solving large-scale optimization problems. The system has been developed using Globus and MPICH-G2 grid technologies, and consists of two BCP solvers and of an interface portal. After a brief introduction to Branch, Cut and Price optimization algorithms, the system architecture, the solvers and the portal user interface are described. Finally, some of the tests performed and the obtained results are illustrated.

1   Introduction

Most exact solution approaches to optimization problems are based on Branch and Bound, which solves optimization problems by partitioning the solution space. Unfortunately, most of the practical problems that can be solved by Branch and Bound are NP-hard, and, in the worst case, may require searching a tree of exponential size. At least in theory, Branch and Bound lends itself well to parallelization. Therefore, the use of a sufficiently high number of processors can make the solution of large-scale problems more practical.

Among the possible implementation methods for Branch and Bound, a powerful technique is Branch, Cut, and Price (BCP). BCP is an implementation of Branch and Bound in which linear programming is used to derive valid bounds during the construction of the search tree. Even if the parallelization of BCP is considerably more complex than basic Branch and Bound, currently there are many existing and widely known parallel BCP implementations, along with frameworks that allows quick development of customized code.

While parallel BCP solvers for "traditional" parallel systems are rather customary, the potential of computing Grids [1,2] seems to have been only partially exploited at the state of the art [3,4,5]. This paper represents a step in this direction, since it describes our experience in developing a grid-enabled platform for solving large-scale optimization problems. The developed system is composed of

B. Di Martino et al. (Eds.): EuroPVM/MPI 2005, LNCS 3666, pp. 38–47, 2005.

two solvers, BCP-G and Meta-PBC, and of a web portal, SWI-Portal. BCP-G is a customized version of COIN/BCP, an open source framework developed within the IBM COIN-OR project [6]. The original COIN/BCP framework, based on the use of PVM libraries, has been provided with a new MPI communication API able to exploit the MPICH-G2 system, a grid-enabled MPI implementation [7, 8]. MetaPBC is instead a brand new solver that we have developed, implementing a decentralized master-worker schema [9]. In order to make the system as user-friendly as possible, we have also developed a web portal (SWI-Portal) that manages users and jobs. All of them will be described here. The paper also presents two example solvers that have been developed for testing purposes, a solver of the p-median problem [10], and a solver of mixed integer programming problems.

In the next section, we introduce the Branch, Cut and Price algorithms and the COIN/BCP framework. Then, in section 3, we present the architecture of our grid-enabled system. Next, several case studies and the obtained results are presented. The paper closes with a discussion on our future work and the conclusions.

2 Branch, Cut and Price Algorithms and Frameworks

Branch and Bound algorithms are the most widely used methods for solving complex optimization problems [6]. An optimization problem is the task of minimizing (maximizing) an *objective function*, a function that associates a cost to each solution. Branch and Bound is a strategy of exploration of solution space based on implicit enumeration of solutions. Generally, it is an exact method, but it is also possible to stop the search when some prefixed condition is reached. As is well known, it is made up of two phases: a *branching* one, where disjoint subsets of solutions are examined, and a *bounding* one, where they are evaluated using an objective function and the subsets not including the optimal solution are deleted.

Branch and Cut algorithms use a hybrid approach, joining the Branch and Bound method, used to explore the solution space, and the method of cutting planes, used for the bounding phase. The cutting planes method finds the optimal solution introducing a finite number of cuts, that is, inequalities satisfied by all the feasible solutions, but not by the optimal current solution of the problem with some relaxed constraints (relaxed problem) [11,12].

Branch and Price algorithms are instead based on column generation. This method is used to solve problems with a very large number of variables. It uses initially only a small subset of the problem variables and of the respective columns in the constraints matrix, thus defining a reduced problem. In fact, in the original problem, there are too many columns and great part of them will have the respective variables equal to zero in an optimal solution.

Branch, Cut and Price joins the two methods used by Branch and Cut and Branch and Price, producing dynamically both cutting planes and variables [6].

2.1   COIN/BCP

COIN/BCP is an open-source framework that implements the Branch, Cut and Price algorithms for solving mixed integer programming problems, a class of problems where some of the variables must be integer [13]. It offers a parallel implementation of the algorithm using the message-passing library PVM (Parallel Virtual Machine). All its functions are grouped in four independent computational modules:

- Tree Manager (TM), which is the master process. It is responsible for the entire search process, starts new processes and checks their status, sends the problems to the slave processes and stores the best solutions. Finally, it recognizes the end of the search, stopping all processes.
- Linear Programming (LP), which is a slave process. It performs the most complex computational work, since it is responsible for the branching and bounding phases. It uses a sequential solver to solve the LP relaxation through the Open Solver Interface (OSI). This is a uniform API to various LP solvers, such as Ilog Cplex, used in the tests that will be presented next.
- Cut Generator (CG), a slave process that creates globally-valid inequalities not satisfied by the current solution of LP relaxation, sending them to the LP that requested them.
- Variable Generator (VG), which performs the Cut Generation. It creates variables with reduced costs, and sends them to the requester LP.

COIN/BCP implements a Branch, Cut and Price single-node pool algorithm, where there is a single central list of candidate sub-problems to be processed, owned by the tree manager. The modules communicate with each other by exchanging messages through a message-passing protocol defined in a separate communications API. In the standard version of the framework, this exploits the PVM run-time system. The first phase of our work was to implement a new parallel interface based on MPI, in order to make it possible the use of the framework in a grid environment using the Globus Toolkit and MPICH-G2 [7,8].

3   System Description

The architecture of the grid-enabled platform developed is shown in Fig. 1. In the figure, the upper layer is the portal interface, in the middle there are the two solvers BCP-G and MetaPBC, all of which rely on the lower layer (the Globus and MPICH-G2 frameworks).

3.1   BCP-G

BCP-G is the optimization solver that we have implemented extending COIN/ BCP. As mentioned before, this required the development of a new communication interface written in MPI. The new interface is implemented by the

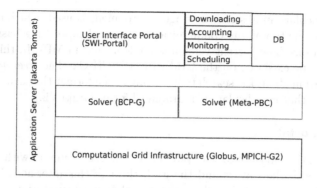

Fig. 1. System architecture

two classes BCP_mpi_environment and BCP_mpi_id, which manage the communications between computational modules and the process ids, respectively. In particular, we have added new functions to the old system, to initialize the MPI environment and to determine the number of processes started by mpirun. The MPI interface differs from the PVM one, since it includes no spawn functionality to start dynamically new processes. If the number of started processes is not equal to the number of processes requested by the user, an exception is raised.

BCP-G takes trace of all processes started by mpirun and assigns a type to each of them. It assigns to the master TM a pid equal to zero; higher pids are assigned to the slaves. For example, if the user wants to start 3 LP, 2 VG and 3 CG processes, the system will give type LP to the processes with pid from 1 to 3, type VG to the pid from 4 to 5 and type CG to the pid from 6 to 8. With this new interface, which is now integrated in the COIN-OR framework, the solver can run in a Globus grid environment using the grid-enabled implementation of MPI, MPICH-G2. The user has simply to write a Globus rsl script and, through the globusrun command, he can start the solver execution [14].

3.2 Meta-PBC

Meta-PBC is a parallel solver for solving optimization problems using the Branch and Cut algorithm. It is not the porting of existing software, but it has been developed from scratch for this research project. Our idea in designing this library was to create a parallel implementation, which would take advantage of the best sequential B&C solvers, such as commercial solvers ILOG CPLEX or Dashoptimization Xpress-MP. These sequential solvers are therefore executed on a purposely-developed parallel layer, which manages their workload. Meta-PBC consists of three modules: *manager, worker* and *tree monitor* [9]. The *manager* is the master process. It is responsible for the initialization of the problem and I/O, and manages the message handling between the workers. The *worker* is a sequential solver of Branch and Cut, with some additional functionality to communicate in the parallel layer. The workers communicate with each other through the parallel API to know the state the overall solution process. The parallel interaction between modules is achieved by a separate communication

API. In particular, an abstract message environment is used, which can be implemented on the top of any communication protocol supporting basic message passing functions. The current version is implemented in MPI. In this way, the processes can be executed on the Grid with MPICH-G2. The *tree monitor* collects information about the search tree, storing it in a format that can be handled by a program that can display it in graphical form (GraphViz).

3.3 SWI-Portal

SWI-Portal is the interface to our system (Fig. 2). Users interact with the portal, and, hence, with the solvers and the grid, through this interface. This allows them to submit a new job and, hence, to solve an optimization problem, to monitor their job, to view their output and to download the results. SWI-Portal is implemented using the Java Server Pages technology (JSP). It consists of an user interface, and a set of Java classes, wrapping of the most important and useful Globus functions. Furthermore, it interacts with a database collecting information on users, job and resources.

SWI-Portal is composed of four subsystems. The first is the account subsystem, responsible for managing user access in conjunction with the users DB. This subsystem allows a user to register in the system, and to enter in the portal giving his login and password. The second one is the scheduling subsystem. SWI-Portal currently supports explicit scheduling; the user has to specify the hosts on which he wishes to run his jobs. He must insert a set of parameters describing his problem, and the scheduling system invokes the Globus system to start the run. The subsystem also records information about the runs in the database. It creates automatically the parameter file necessary for the solver, using the information supplied by the user, and creates a Globus rsl script describing the running job. The Grid layer is responsible for the transfer of the files to all the hosts selected to execute the job. From the pages of the Monitoring subsystem, an user can check the status of the search, and consult any other information about all the started processes (such as output, error, rsl, and search tree). Users can download through the Download Subsystem all information regarding his jobs and/or cancel them from the server.

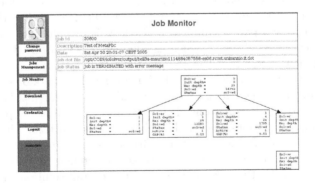

Fig. 2. A screenshot of the SWI-Portal

4   Case Studies

To test the developed software we have firstly built a test environment using Globus and MPICH-G2. This test environment is not, strictly speaking, a Grid, i.e., a geographically-distributed set of high performance computing facilities, but rather a cluster of Globus nodes on a LAN. This was chosen purposely to stress the developed framework by adopting a fairly low-grain decomposition and to compare its performance results to the MPI-only version, where the grid middleware overheads are absent. In particular, we used 9 workstations equipped with Pentium Xeon, 2.8 GHz CPU and 1 GB of RAM over 100 GigaEthernet, and as LP solver, Ilog Cplex version 9.0. The test performed are relative to the solution of two well-known problems, p-median and MIP, briefly described in the following.

The P-median problem is a classic NP-hard problem introduced by K. Hakimi in 1979 and widely studied in the literature. The p-median problem can be easily formulated as a mixed integer linear programming. For solving this problem, we implemented a Branch, Cut and Price algorithm with a simple procedure choosing the core problem, a preprocessing procedure fixing some variables, a procedure of column and rows generation solving the LP relaxed problem, and a procedure of separation of valid cuts violated by the current solution. In the computational experiments of BCP-G with the p-median problem, we used several instances of the OR-Library, a collection of instances for a large variety of problems of OR [15]. To test Meta-PBC, we implemented a generic MIP solver. In particular we take advantage from the MIPLIB library [16] that, since its introduction, has become a standard test set, used to compare the performance of mixed integer optimizers.

The primary objective of our tests was not to obtain absolute performance measurements of the solvers, but to detect possible performance losses due to the use of the grid environment. The interest reader is referred for COIN/BCP and Meta-PBC absolute performance figures to [13] and to [9], respectively.

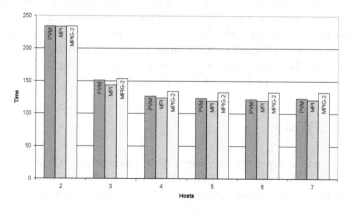

Fig. 3. Comparison between PVM, MPICH and MPICH-G2 response times for the *pmed26* problem of the OR-Library, using BCP-G on a variable number of hosts

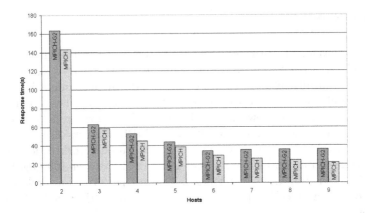

Fig. 4. Comparison between MPICH and MPICH-G2 response times for the *misc07*, problem of the MIPLIB Library, using Meta-PBC on a variable number of hosts

Fig. 3 shows that the performance of the MPICH porting of BCP-G is slightly better than that of the "original" PVM implementation. The MPICH-G2 version, being affected by the grid overheads, performs instead in a very similar way to the latter. Fig. 4 shows a similar behavior using the new Meta-PBC solver.

In fact, our tests have shown that the grid layer introduces a reasonable performance penalty, which is higher than 10 % only for very small-scale problems and becomes negligible as problem size increases. However, this is not sufficient to deduce that the use of a grid environment, particularly on a geographical and/or loaded network, is always satisfactory as far as performance figures are concerned. The topic is too wide to be dealt with in the scope of this paper, and should be suitably supported by extensive testing. However, just to alert the reader on this issue, Fig. 5 compares the response times of COIN/BCP for the pmed26 and pmed39 problem in a LAN and a geographical grid environment under heavy network load, using 1, 2 and 3 host nodes. The bar diagrams show that, unlike what happens on a LAN, where response times decrease with the number of hosts, the exploitation of parallelism is not necessarily convenient in a geographical LAN. In fact, the best response times on a geographical LAN under heavy load are obtained using a single processor. Fortunately, this is just a limit case, and the use of grid environments remains nevertheless appealing for more coarse-grained problems and on fast networks.

5   Related Work

In the last years, many software packages implementing parallel branch and bound have been developed. SYMPHONY [6] is a parallel framework, similar to COIN/BCP, for solving mixed integer linear programs. COIN/BCP and SYMPHONY are combined in a new solver under development, ALPS [17]. Some other parallel solver are PUBB [18], PPBB-Lib [19] and PICO [20]. PARINO [21]

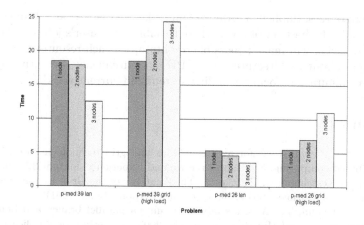

Fig. 5. Response times of the pmed26 and pmed39 problem in a LAN and a geographical grid environment for 1, 2 and 3 grid nodes

and FATCOP [22, 4] are generic parallel MIP solvers, and the second one is designed for Grid systems.

The grid portals allows users to access grid facilities by a standard Web Browser. A grid portal offers an user interface to the functionalities of a Grid system. Currently there are many grid portals, including the Alliance Portal [23], the NEESgrid Portal [24], the Genius Portal [25] and the IeSE Portal [26]. Also, there are many Grid portal development frameworks including the Gridsphere Portal [27], the GridPort Portal [28] and the Grid Portal Development Toolkit (GPDK) [29].

6   Conclusions and Future Work

In this paper, we have described a grid-enabled system for solving large-scale optimization problems made up of two solvers and a portal interface. The two case studies used to test it in a small LAN-based grid led to satisfactory results, in that the measured response times are comparable with the MPI version. The overhead due to grid middleware is higher than 10 % only for small scale problems. However, this does not guarantee reasonable performance in a geographical grid, especially under heavy network load. In these cases, the characteristics of the problem and of the host environment have to be suitably taken into account, to avoid disappointing performance losses.

Currently, in Meta-PBC a checkpointing technique is used to recover the computation if some system component fails. Our intention for future research is to implement also dynamic process generation and task reassignment. Another issue is to implement a global cut pool, studying different ways of sharing the cutting planes between the solvers. As regards the SWI-Portal, we want to extend its functionalities adding implicit scheduling, and a system for monitoring available resources. Implicit scheduling will make the system capable of choosing automatically the best set of hosts where to submit the user's job. On the other hand, monitoring will make it possible to obtain, for each resource in the Grid, information about architecture type, CPU clock frequency, operating system, memory configuration, local scheduling system and current workload.

References

1. Baker, M., Buyya, R., Laforenza, D.: Grids and grid technologies for wide-area distributed computing. Software: Practice and Experience Journal **32** (2002)
2. Foster, I., Kesselman, C., Tuecke, S.: The Anatomy of the Grid. Enabling Scalable Virtual Organizations. Intl J. Supercomputer Applications (2001)
3. Aida, K., Osumi, T.: A case study in running a parallel branch and bound application on the grid. In: Proc. of the The 2005 Symposium on Applications and the Internet (SAINT'05), Washington, DC, USA, IEEE Computer Society (2005) 164–173
4. Chen, Q., Ferris, M., Linderoth, J.: Fatcop 2.0: Advanced features in an opportunistic mixed integer programming solver. Annals of Op. Res. (2001) 17–32
5. Drummond, L.M., Uchoa, E., Goncalves, A.D., Silva, J.M., Santos, M.C., de Castro, M.C.S.: A grid-enabled distributed branch-and-bound algorithm with application on the steiner problem in graph. Technical report, Universidade Federal Fluminense, Instituto de Computacao (2004) http://www.ic.uff.br/PosGrad/RelatTec/Download/rt_02-05.pdf.gz.
6. Ralphs, T., Ladanyi, L., Saltzman, M.: Parallel Branch, Cut, and Price for Large-Scale Discrete Optmization. Mathematical Programming **98** (2003) 253–280
7. Ferreira, L., Jacob, B., Slevin, S., Brown, M., Sundararajan, S., Lepesant, J., Bank, J.: Globus Toolkit 3.0 Quick Start. IBM. (2003)
8. Karonis, N., Toonen, B., Foster, I.: MPICH-G2: A Grid-Enabled Implementation of the Message Passing Interface. J. of Parallel and Dist. Comp. **63** (2003) 551–563
9. Vasil'ev, I., Avella, P.: PBC: A parallel branch-and-cut framework. In: Proc. of 35th Conference of the Italian Operations Res. Society, Lecce, Italy (2004) 138
10. Avella, P., Sassano, A.: On the p-median polytope. Mathematical Programming (2001) 395–411
11. Margot, F.: BAC: A BCP Based Branch-and-Cut Example. (2003)
12. Cordiery, C., Marchandz, H., Laundyx, R., Wolsey, L.: bc-opt: a Branch-and-Cut Code for Mixed Integer Programs. Mathematical Programming (1999) 335–354
13. Ralphs, T., Ladanyi, L.: COIN/BCP User's Manual. (2001) http://www.coin-or.org/Presentations/bcp-man.pdf.
14. Globus Alliance: WS GRAM: Developer's Guide. (2005) http://www-unix.globus.org/toolkit/docs/3.2/gram/ws/developer.
15. Beasley, J.: OR-Library: distributing test problems by electronic mail. Journal of the Operational Research Society **41** (1990) 1069–1072
16. Bixby, R.E., Ceria, S., McZeal, C.M., Savelsbergh, M.W.P.: An updated mixed integer programming library MIPLIB 3.0. Optima (1998) 12–15
17. Ralphs, T.K., Ladanyi, L., Saltzman, M.J.: A library hierarchy for implementing scalable parallel search algorithms. J. Supercomput. **28** (2004) 215–234
18. Shinano, Y., Higaki, M., Hirabayashi, R.: Control schemas in a generalized utility for parallel branch and bound. In: Proc. of the 1997 Eleventh International Parallel Processing Symposium, Los Alamitos, CA, IEEE Computer Society Press (1997)

19. Tschoke, S., Polzer, T.: Portable Parallel Branch-And-Bound Library PPBB-Lib User Manual. Department of computer science Univ. of Paderborn. (1996)
20. Eckstein, J., Phillips, C., Hart, W.: Pico: An object-oriented framework for parallel branch and bound. Technical report, Rutgers University, Piscataway, NJ (2000)
21. Linderoth, J.: Topics in Parallel Integer Optimization. PhD thesis, School of Industrial and Systems Engineering, Georgia Inst. of Tech., Atlanta, GA (1998)
22. Chen, Q., Ferris, M.C.: Fatcop: A fault tolerant condor-pvm mixed integer programming solver. Technical report, University of Wisconsin CS Department Technical Report 99-05, Madison, WI (1999)
23. Alliance Portal Project: Scientific Portals. Argonne National Labs. (2002) http://www.extreme.indiana.edu/alliance/docandpres/SC2002PortalTalk.pdf.
24. Brown, G.E.: Towards a Vision for the NEES Collaboratory. NEES Consortium   Development   Project.   (2002)   http://www.curee.org/projects/NEES/docs/outreach/VisionWhitePaperV3.pdf.
25. Barbera, R., Falzone, A., Rodolico, A.: The genius grid portal. In: Proc. of Computing in High Energy and Nuclear Physics, La Jolla, California (2003) 24–28 https://genius.ct.infn.it.
26. Kleese van Dam, K., Sufi, S., Drinkwater, G., Blanshard, L., Manandhar, A., Tyer, R., Allan, R., O'Neill, K., Doherty, M., Williams, M., Woolf, A., Sastry, L.: An integrated e-science environment for environmental science. In: Proc. of Tenth ECMWF Workshop, Reading, England (2002) 175–188
27. Novotny, J., Russell, M., Wehrens:, O.: Gridsphere: An advanced portal framework. In: Proc. of 30th EUROMICRO Conf., Rennes, Fr., IEEE (2004) 412–419 http://www.gridsphere.org/gridsphere/wp-4/Documents/France/gridsphere.pdf.
28. Thomas, M., Mock, S., Boisseau, J., Dahan, M., Mueller, K., Sutton., D.: The gridport toolkit architecture for building grid portals. In: Proc. of the 10th IEEE Intl. Symp. on High Perf. Dist. Comp. (2001) http://gridport.net.
29. Novotny, J.: The grid portal development kit. Grid Computing (2003) 657–673 http://doesciencegrid.org/projects/GPDK.

An Optimal Broadcast Algorithm Adapted to SMP Clusters

Jesper Larsson Träff and Andreas Ripke

C&C Research Laboratories, NEC Europe Ltd.,
Rathausallee 10, D-53757 Sankt Augustin, Germany
{traff, ripke}@ccrl-nece.de

Abstract. We describe and and evaluate the adaption of a new, optimal broadcast algorithm for "flat", fully connected networks to clusters of SMP nodes. The optimal broadcast algorithm improves over other commonly used broadcast algorithms (pipelined binary trees, recursive halving) by up to a factor of two for the non-hierarchical (non-SMP) case. The algorithm is well suited for clusters of SMP nodes, since intra-node broadcast of relatively small blocks can take place concurrently with inter-node communication over the network. This new algorithm has been incorporated into a state-of-the art MPI library. On a 32-node dual-processor AMD cluster with Myrinet interconnect, improvements of a factor of 1.5 over for instance a pipelined binary tree algorithm has been achieved, both for the case with one and with two MPI processes per node.

1 Introduction

Broadcast is a frequently used collective operation of MPI, the *Message Passing Interface* [8], and there is therefore good reason to pursue the most efficient algorithms and best possible implementations. Recently, there has been much interest in broadcast algorithms and implementations for different systems and MPI libraries [2,5,6,9,10], but none of these achieve the theoretical lower bound for their respective models. An exception is the *LogP* algorithm in [7], but no implementation results were given. A quite different, theoretically optimal algorithm for single-ported, fully connected networks was developed by the authors in [11]. This algorithm has the potential of being up to a factor two faster than the best currently implemented broadcast algorithms based on pipelined binary trees, or on recursive halving as recently implemented in mpich2 [9] and elsewhere [5,10]. These algorithms were developed on the assumption of a "flat", homogeneous, fully connected communication system, and will not perform optimally on systems with a hierarchical communication system like clusters of SMP nodes. Pipelined binary tree algorithms can naturally be adapted to the SMP case [3], whereas the many of the other algorithms will entail a significant overhead.

In this paper we present the ideas behind the new, optimal broadcast algorithm; technical details, however, must be found in [11]. We describe how the

B. Di Martino et al. (Eds.): EuroPVM/MPI 2005, LNCS 3666, pp. 48–56, 2005.

implementation has been extended to clusters of SMP nodes, and compare the performance of the algorithm to a pipelined binary tree algorithm [3] on a 32-node dual-processor AMD based SMP cluster with Myrinet interconnect. Very worthwhile performance improvements of more than a factor of 1.5 are achieved.

2 The Broadcast Algorithm

We first give a high-level description of the optimal broadcast algorithm for "flat", homogeneous, fully connected systems. We assume a linear communication cost model, in which sending m units of data takes time $\alpha + \beta m$, and each processor can both send and receive a message at the same time, possibly from different processors. The p processors are numbered from 0 to $p - 1$ as in MPI, and we let $n = \lceil \log p \rceil$. Without loss of generality we assume that broadcast is from *root* processor 0. Assuming further that the m data is sent as N blocks of m/N units, the number of communication rounds required (for any algorithm) to broadcast the N blocks is $n - 1 + N$, for a time of $(n - 1 + N)(\alpha + \beta m/N) = (n - 1)\alpha + (n - 1)\beta m/N + N\alpha + \beta m$. By balancing the terms $(n - 1)\beta m/N$ and αN, the optimal running time can be found as

$$T_{\text{opt}}(m) = (n - 1)\alpha + 2\sqrt{(n - 1)\alpha}\sqrt{\beta m} + \beta m = (\sqrt{(n - 1)\alpha} + \sqrt{\beta m})^2 \quad (1)$$

Proofs of this lower bound can be found in e.g. [4,6].

The optimal broadcast algorithm is pipelined in the sense that all processors are (after an initial fill phase of n rounds) both sending and receiving blocks at the same time. For sending data each processor acts as if it is a root of a(n incomplete, when p is not a power of 2) binomial tree. Each non-root processor has n different parents from which it receives blocks. To initiate the broadcast, the root first sends n successive blocks to its children. The root continues in this way sending blocks successively to its children in a round-robin fashion. The non-root processors receive their first block from their parent in the binomial tree rooted at processor 0. The non-roots pass this block on to their children in this tree. After this initial *fill phase*, each processor now has a block, and the broadcast enters a *steady state*, in which in each round each processor (except the root) receives a new block from a parent, and sends a previously received block to a child.

A more formal description of the algorithm is given in Figure 1. The buffer containing the data being broadcast is divided into N blocks of roughly m/N units, and the ith block is denoted buffer[i] for $0 \le i < N$.

As can be seen each processor receives N blocks of data. That indeed N different blocks are received and sent is determined by the recvblock(i, r) and sendblock(i, r) functions which specify the block to be received and sent in round i for processor r (see Section 2.2). The functions next and prev determine the *communication pattern* (see Section 2.1). In each phase the same pattern is used, and the n parent and child processors of processor r are next(j, r) and prev(j, r) for $j = 0, \ldots n - 1$. The parent of processor r for the fill phase is first(r), and the first round for processor r is likewise first(r). With these provisions we have:

Root processor 0:

```
/* fill */
for i ← 0, n − 1 do
    send(buffer[sendblock(i, 0)], next(i, 0))
/* steady state */
for i ← 1, N do
    j ← (i − 1) mod n
    send(buffer[sendblock(n − 1 + i, 0)], next(j, 0))
```

Non-root processor r:

```
/* fill */
i ← first(r)
recv(buffer[recvblock(i, r)], prev(i, r))
for i ← first(r) + 1, n − 1
    send(buffer[sendblock(i, r)], next(i, r))
/* first block received, steady state */
for i ← 1, N
    j ← (i − 1) mod n
    if next(j, r) ≠ 0 then /* no sending to root */
        send(buffer[sendblock(n − 1 + i, r)], next(j, r))
    ‖ /* send and receive simultaneously */
    recv(buffer[recvblock(n − 1 + i, r)], prev(j, r))
```

Fig. 1. The optimal broadcast algorithm

Theorem 1. *In the fully-connected, one-ported, bidirectional, linear cost communication model, N blocks of data can be broadcast in $n-1+N$ rounds reaching the optimal running time (1).*

The algorithm is further simplified by the following observations. First, the block to send in round i is obviously

$$\text{sendblock}(i, r) = \text{recvblock}(i, \text{next}(i, r))$$

so it will suffice to determine a suitable recvblock function. Actually, we can determine the recvblock function such that for any processor $r \neq 0$ it holds that

$$\{\text{recvblock}(0, r), \text{recvblock}(1, r), \dots, \text{recvblock}(n − 1, r)\} = \{0, 1, \dots, n − 1\}$$

that is the recvblock for a phase consisting of rounds $0, \dots n-1$ is a permutation of $\{0, \dots, n − 1\}$. For such functions we can take for $i \geq n$

$$\text{recvblock}(i, r) = \text{recvblock}(i \bmod n, r) + n(\lfloor i/n \rfloor − 1 + \delta_{\text{first}(r)}(i \bmod n))$$

where $\delta_j(i) = 1$ if $i = j$ and 0 otherwise. Thus in rounds $i + n, i + 2n, i + 3n, \dots$ for $0 \leq i < n$, processor r receives blocks $\text{recvblock}(i, r), \text{recvblock}(i, r) + n, \text{recvblock}(i, r) + 2n, \dots$ (plus n if $i = \text{first}(r)$). We call such a recvblock function a *full block schedule*, and to make the broadcast algorithm work we need to show that a full block schedule always exists, and how it can be computed. Existence is proved in [11], while the construction is outlined in the following sections.

When $N = 1$ the algorithm is just an ordinary binomial tree broadcast, which is optimal for small m. The number of blocks N can be chosen freely, e.g. to minimize the broadcast time under the linear cost model, or, which is relevant for some systems, to limit the amount of communication buffer space.

2.1   The Communication Pattern

When p is a power of two the communication pattern of the allgather algorithm in [1] can be used. In round j for $j = 0, \ldots, n-1$ processor r receives a block from processor $(r - 2^j) \bmod p$ and sends a block to processor $(r + 2^j) \bmod p$, so for that case we take

$$\mathsf{next}(j, r) = (r + 2^j) \bmod p$$
$$\mathsf{prev}(j, r) = (r - 2^j) \bmod p$$

With this pattern the root successively sends n blocks to processors $1, 2, 4, \ldots, 2^j$ for $j = 0, \ldots n-1$. The subtree of child processor $r = 2^j$ consists of the processors $(r+2^k) \bmod p$, $k = j+1, \ldots, n-1$. We say that processors $2^j, \ldots 2^{j+1}-1$ together form *group* j, since these processors will all receive their first block in round j. The *group start* of group j is 2^j, and the *group size* is likewise 2^j. Note that $\mathsf{first}(r) = j$ for a processor in group j.

For the general case where p is not a power of two, the processors are divided into groups of size approximately 2^j. To guarantee existence of the full block schedule, the communication pattern must satisfy that the total number of processors in groups $0, 1, \ldots, j-1$ plus the root processor must be at least the number of processors in group j, so that all processors in group j can receive their first block in round j. Likewise, the size of the last group $n-1$ must be at least the size of groups $0, 1, \ldots, n-2$ for the processors of the last group to deliver a block to all previous processors in round $n-1$. To achieve this we define for $0 \le j < n$

$$\mathrm{groupsize}(j, p) = \begin{cases} \mathrm{groupsize}(j, \lceil p/2 \rceil) & \text{if } j < \lceil \log p \rceil - 1 \\ \lfloor p/2 \rfloor & \text{if } j = \lceil \log p \rceil - 1 \end{cases}$$

and

$$\mathrm{groupstart}(j, p) = 1 + \sum_{i=0}^{j-1} \mathrm{groupsize}(j, p)$$

When p is a power of two this definition coincides with the above definition, eg. $\mathrm{groupsize}(j, p) = 2^j$.

We now define the next and prev functions analogously to the powers-of-two case:

$$\mathsf{next}(j, r) = (r + \mathrm{groupstart}(j, p)) \bmod p$$
$$\mathsf{prev}(j, r) = (r - \mathrm{groupstart}(j, p)) \bmod p$$

We note that this communication pattern leads to an exception for the fill phase of the algorithm in Figure 1, since it may happen that $\text{next}(j, r) = \text{groupstart}(j + 1, p) = r'$ and $\text{prev}(\text{first}(r'), r') = 0 \neq \text{next}(j, r)$. Such a **send** has no corresponding **recv**, and shall not be performed.

2.2 Computing the Full Block Schedule

The key to the algorithm is the existence and construction of the *full block schedule* given the communication pattern described in the previous section. Existence and correctness of the construction is discussed in [11]. For the construction itself a greedy algorithm almost suffices. Based on what we call the *first block schedule* which determines the first block to be received by each processor r, the construction is as follows.

1. Construct the *first block schedule* schedule:
   set $\text{schedule}[\text{groupstart}(j, p)] = j$, and $\text{schedule}[\text{groupstart}(j, p) + i] = \text{schedule}[i]$ for $i = 1, \ldots, \text{groupstart}(j, p) - 1$.
2. Scan the first block schedule in descending order $i = r - 1, r - 2, \ldots 0$. Record in $\text{block}[j]$ the first block $\text{schedule}[i]$ different from $\text{block}[j - 1], \text{block}[j - 2], \ldots \text{block}[0]$, and in $\text{found}[j]$ the index i at which $\text{block}[j]$ was found.
3. If $\text{prev}(j, r) < \text{found}[j]$ either
   - if $\text{block}[j] > \text{block}[j - 1]$ then swap the two blocks,
   - else mark $\text{block}[j]$ as unseen,
   and continue scanning in Step 2.
4. Set $\text{block}[\text{first}(r)] = \text{schedule}[r]$
5. Find the remainder blocks by scanning the first block schedule in the order $i = p - 1, p - 2, \ldots r + 1$, and swap as in Step 3.

For each r take
$$\text{recvblock}(i, r) = \text{block}[i]$$
with block as computed above.

Space for the full block schedule is $O(p \log p)$, and as described above the construction takes $O(p^2)$ time. However, by a different formulation of the algorithm, the computation time can be reduced to $O(p \log p)$ steps [11]. The correctness proof can also be found in [11]; as anecdotal evidence we mention that we computed and verified all schedules up to 100 000 processors on a 2 GHz AMD Athlon PC. Construction time for the largest schedule was about 225 ms. This is of course prohibitive for on-line construction of the schedule at each MPI_Bcast operation. Instead, a corresponding schedule must be precomputed for each MPI communicator. In applications with many communicators the space consumption of $O(p \log p)$ can become a problem. Fortunately, it is possible to store the full block schedule in a distributed fashion with only $O(\log p)$ space for each processor: essentially each processor i needs only its own $\text{recvblock}(i, j)$ and $\text{sendblock}(i, j)$ functions, assuming that process 0 is the broadcast root (if this is not the case, the trick is that some other processor sends the needed schedule to processor 0, so each processor has to store schedules for two processors which is still only $O(\log p)$ space. The sending of the schedule to processor 0 can be hidden behind the already started broadcast operation and thus does not cost extra time).

2.3 Properties

We summarize the main properties of the broadcast algorithm as described above for flat systems.

1. The algorithm broadcasts N blocks in $n - 1 + N$ communication rounds, which is best possible.
2. The number of blocks can be chosen freely. In the linear cost communication model, the best block size is $\sqrt{(m\alpha)/((n - 1)\beta)}$ resulting in optimal running time (1).
3. The number of rounds can also be chosen such that a given maximum block size, eg. internal communication buffer, is not exceeded.
4. Small messages should be broadcast in $N = 1$ rounds, in which case the algorithm is simply a binomial tree algorithm.
5. The required *full block schedule* can be computed in $O(p \log p)$ time.
6. Space for the full block schedule is likewise $O(p \log p)$ which can be stored in a distributed fashion with $O(\log p)$ space per processor.

2.4 Adaption to Clusters of SMP Nodes

The flat broadcast algorithm can be adapted to systems with a hierarchical communication structure like clusters of SMP nodes as follows. For each node a local root processor is chosen. The flat broadcast algorithm is run over the set of local root processors. In each communication round each root processor receives new block which is broadcast locally over the node. For SMP clusters a simple shared memory algorithm can be used. Furthermore, the steady-state loop of the algorithm in Figure 1 can easily be rewritten such that in each round a) a new block is received, b) an already received block is sent, and c) the block received in the previous round is broadcast locally. This makes it possible – for communication networks supporting this – to hide the local, shared memory broadcast completely behind the sending/receiving of new blocks. Only the node local broadcast of the last block cannot be hidden in this fashion, which adds time proportional to the block size $\sqrt{(m\alpha)/((n - 1)\beta)}$ to the total broadcast time. For broadcast algorithms based on recursive halving [2,9,10], the size of the "last block" received may be proportional to $m/2$ causing a much longer delay.

3 Performance

Both the flat and the SMP cluster broadcast algorithms have been implemented in a state-of-the art MPI implementation for PC clusters. We give results for a dual-processor AMD cluster with Myrinet interconnect.

Figure 2 compares the performance of various broadcast algorithms for the flat case with one MPI process per SMP node. The new, optimal algorithm is compared to an algorithm based on recursive halving [10], a pipelined binary tree algorithm (with a binomial tree for short messages), and a binomial tree algorithm. The theoretical bandwidth improvement over both the recursive halving

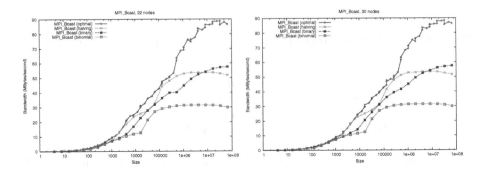

Fig. 2. Bandwidth of 4 different "flat" broadcast algorithms for fixed number of processors $p = 22$ (left) and $p = 30$ (right), one MPI process per node, and data size m up to 64MBytes

and the pipelined binary tree algorithm is a factor of 2, over the binomial tree algorithm a factor $\lceil \log_2 p \rceil$. The actual improvement is more than a factor 1.5 even for messages of only a few K bytes. It should be noted that to obtain the performance and relatively smooth bandwidth growth shown here, the simple, linear cost model is not sufficient. Instead a piecewise linear model with up to 4 different values of α and β is used for estimating the best block size for a given message size m.

Figure 3 compares the new SMP-adapted broadcast algorithm to a pipelined binary tree algorithm likewise adapted to SMP clusters with two MPI processes per node. For both algorithms, lower bandwidth than in the one process/node case is achieved, but also in the SMP case the new broadcast algorithm achieves about a factor 1.4 higher bandwidth than the pipelined binary tree.

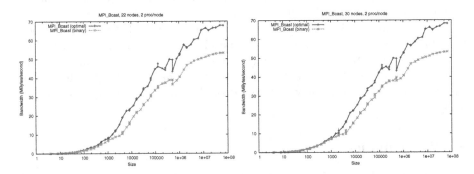

Fig. 3. Bandwidth of optimal and pipelined binary tree with two MPI processes per node, $p = 22$ (left) and $p = 30$ (right), and data size m up to 64MBytes

Finally, Figure 4 illustrates the SMP-overhead of the current implementation. The flat version of the algorithm is compared to the SMP-algorithm with one process/node and to the SMP-algorithm with two processes/node. Even with one process/node, the SMP-adapted algorithm (in the current implementation) has

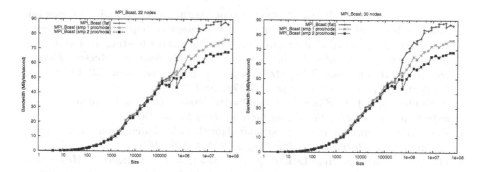

Fig. 4. New broadcast algorithm for the flat case vs. the SMP algorithm with one process/node vs. the SMP-algorithm with two processes/node, $p = 22$ (left) and $p = 30$ (right), and data size m up to 64MBytes. Note that the SMP algorithm even for the one process/node case performs an extra memory copy compared to the flat algorithm.

to perform an extra memory copy compared to the flat algorithm, and Figure 4 estimates the cost of this. Up to about 100KBytes the performance of the three versions is similar, after that the cost of the extra copying and node-internal, shared memory broadcast becomes visible, and degrades the performance. However, improvements in the implementation are still possible to overlap intra-node communication and node-internal broadcast as described in Section 2.4.

4   Conclusion

We described the main ideas behind a new, theoretically optimal broadcast algorithm for "flat", homogeneous, fully connected networks, and discussed an easy adaption to hierarchical systems like clusters of SMP nodes. On a small Myrinet cluster significant bandwidth improvements over other, commonly used broadcast algorithm were demonstrated, both for the "flat" case with one MPI process/node, and for the case with more than one process per node. Further implementation improvements to better overlap network communication and intra-node communication are still possible, and will be pursued in the future.

References

1. J. Bruck, C.-T. Ho, S. Kipnis, E. Upfal, and D. Weathersby. Efficient algorithms for all-to-all communications in multiport message-passing systems. *IEEE Transactions on Parallel and Distributed Systems*, 8(11):1143–1156, 1997.
2. E. W. Chan, M. F. Heimlich, A. Purkayastha, and R. A. van de Geijn. On optimizing collective communication. In *Cluster 2004*, 2004.
3. M. Gołebiewski, H. Ritzdorf, J. L. Träff, and F. Zimmermann. The MPI/SX implementation of MPI for NEC's SX-6 and other NEC platforms. *NEC Research & Development*, 44(1):69–74, 2003.

4. S. L. Johnsson and C.-T. Ho. Optimum broadcasting and personalized communication in hypercubes. *IEEE Transactions on Computers*, 38(9):1249–1268, 1989.
5. S. Juhász and F. Kovács. Asynchronous distributed broadcasting in cluster environment. In *Recent Advances in Parallel Virtual Machine and Message Passing Interface. 11th European PVM/MPI Users' Group Meeting*, volume 3241 of *Lecture Notes in Computer Science*, pages 164–172, 2004.
6. P. Sanders and J. F. Sibeyn. A bandwidth latency tradeoff for broadcast and reduction. *Information Processing Letters*, 86(1):33–38, 2003.
7. E. E. Santos. Optimal and near-optimal algorithms for k-item broadcast. *Journal of Parallel and Distributed Computing*, 57(2):121–139, 1999.
8. M. Snir, S. Otto, S. Huss-Lederman, D. Walker, and J. Dongarra. *MPI – The Complete Reference*, volume 1, The MPI Core. MIT Press, second edition, 1998.
9. R. Thakur, W. D. Gropp, and R. Rabenseifner. Improving the performance of collective operations in MPICH. *International Journal on High Performance Computing Applications*, 19:49–66, 2004.
10. J. L. Träff. A simple work-optimal broadcast algorithm for message passing parallel systems. In *Recent Advances in Parallel Virtual Machine and Message Passing Interface. 11th European PVM/MPI Users' Group Meeting*, volume 3241 of *Lecture Notes in Computer Science*, pages 173–180, 2004.
11. J. L. Träff and A. Ripke. Optimal broadcast for fully connected networks. In *High Performance Computing and Communications (HPCC'05)*, Lecture Notes in Computer Science, 2005.

Efficient Implementation of Allreduce on BlueGene/L Collective Network

George Almási[1], Gábor Dózsa[1], C. Chris Erway[2],
and Burkhardt Steinmacher-Burow[3]

[1] IBM T. J. Watson Research Center, Yorktown Heights, NY 10598
{gheorghe, gdozsa}@us.ibm.com
[2] Dept. of Comp. Sci, Brown University Providence, RI 02912
cce@cs.brown.edu
[3] IBM Germany
Boeblingen 71032, Germany
steinmac@de.ibm.com

Abstract. BlueGene/L is currently in the pole position on the Top500 list [4]. In its full configuration the system will leverage 65,536 compute nodes. Application scalability is a crucial issue for a system of such size. On BlueGene/L scalability is made possible through the efficient exploitation of special communication. The BlueGene/L system software provides its own optimized version for collective communication routines in addition to the general purpose MPICH2 implementation. The *collective network* is a natural platform for reduction operations due to its built-in arithmetic units. Unfortunately ALUs of the collective network can handle only fixed point operands. Therefore efficient exploitation of that network for the purpose of floating point reductions is a challenging task. In this paper we present our experiences with implementing an efficient collective network algorithm for `Allreduce` sums of floating point numbers.

1 Introduction

The BlueGene/L supercomputer is a new massively parallel system being developed by IBM in partnership with Lawrence Livermore National Laboratory (LLNL). BlueGene/L uses system-on-a-chip integration [3] and a highly scalable architecture [1] to assemble a machine with 65,536 dual-processor compute nodes.

BlueGene/L compute nodes can address only its local memory, making message passing the natural programming model. The machine has 5 different networks, 3 of which are used by message passing software. In this paper we concentrate one the *collective network* and its use for reduction operations in message passing. We describe an optimized algorithm for performing floating point `Allreduce` operations of small- to mid-size data buffers, minimizing latency and maximizing bandwidth; we describe how we used the dual-processor layout of BlueGene/L to our advantage, almost doubling the algorithm's throughput. We

B. Di Martino et al. (Eds.): EuroPVM/MPI 2005, LNCS 3666, pp. 57–66, 2005.

present performance figures collected by running MPI micro-benchmarks. We are in the process of integrating and deploying our new algorithm into the Blue-Gene/L MPI library.

The rest of this paper is organized as follows. Section 2 presents an overview of the hardware and software architecture of BlueGene/L. Section 3 discusses the implementation of reduction operations using the collective network. Section 4 presents our network-optimized dual-processor floating-point `Allreduce` sum algorithm. Section 5 is a discussion of our experimental results; finally Section 6 talks about conclusions and plans for future work.

2   BlueGene/L Overview

The BlueGene/L hardware [1] and system software [2] have been exhaustively described in other papers. In this section we present a cross-section of the hardware and software that is relevant to our particular pursuit in this paper, i.e. the optimizing the use of the collective network.

The 65,536 compute nodes of BlueGene/L feature dual PowerPC 440 processors running at 700 MHz, each with their own L1 cache but sharing a small 2 kB L2 prefetch buffer, a 4 MB L3 cache and 512 MBytes of DDR memory. Because the PowerPC 440 architecture is not designed for multiprocessor applications, the L1 caches are not coherent; software has to provide coherency for correct operation of the processors. To aid this process the hardware provides specialized SRAM memory, called the *lockbox*, that supports atomic test-and-set primitives and even inter-processor barrier operations. The hardware also provides a *blind device*, which can be used to force-clean a processor's cache by evicting dirty lines. Our optimized algorithm will be using the blind device for inter-processor cache coherence.

Each processor has a dual SIMD floating point unit with 16 Byte wide registers. This allows the processor to access the networks with 128 bit reads and writes, resulting in increased bandwidth.

The machine has five different networks, three of which are used for userspace communication. The network of interest to us is the collective network, a configurable network for high performance broadcast and reduction operations spanning all compute nodes. It provides a latency of 2.5 microseconds for a 65,536-node system. It also provides point-to-point capabilities that can be used for system file I/O; however, in this paper we will be concentrating on the collective capabilities of the network.

The collective network features two virtual channels. Channel 0 is used by system software for I/O; channel 1 is available to user programs, including MPI. Network packets cannot cross virtual channels, ensuring the safety of the system channel.

User application processes run on compute nodes under the supervision of a custom Compute Node Kernel (CNK). The CNK is a simple, minimalist runtime system written in approximately 5000 lines of C++ that supports a single application running by a single user in each BlueGene/L node, reminescent of

PUMA [5]. It provides exactly two threads, one on each PowerPC 440 processor. The CNK does not require or provide scheduling and context switching. Physical memory is statically mapped, protecting kernel regions from user applications. Porting scientific applications to run on this new kernel has been a straightforward process because we provide a standard `glibc` runtime system with most of the Posix system calls.

To alleviate problems with the lack of hardware cache coherency, the BlueGene/L system software provides a special cache inhibited memory region called *scratchpad* for data exchange between processors on the same node.

3 Allreduce Operation on the Collective Network

Reductions play a central role in many distributed applications, used in many cases in the result accumulation phase. The reduction has to scale up as the machine size grows; on machines like BlueGene/L with tens of thousands of nodes, old-style reduction algorithms require a significant computational effort. BlueGene/L addresses this performance bottleneck with the collective communication network, which provides limited hardware support for reduction operations.

The collective network spans all compute nodes. It appears to each processor as a set of memory-mapped FIFOs. The processors read and write fixed-size packets (256 bytes per packet). Each packet has an attached 32 bit descriptor that determines the reduction operation to be performed on the packet (allowable operations include fixed-point sum, the MAX operation and various bitwise operations). The operation's data width can be varied in 16 bit increments.

The collective network is designed to be used as follows: software breaks up the user's input data buffer into 256 byte packets and feeds the packets to the network on every compute node. The network combines the packets and broadcasts the result to every node, where software has to read out the result and store it into the output buffer.

Latency: The collective network's latency is around 2.5 usecs. Software more than doubles this latency to 6.5 usecs for short `MPI_Allreduce` operations.

Bandwidth: The collective network is synchronized with the CPU clock, delivering 4 bits of data per CPU cycle, or approximatively 350 MBytes/s for the BlueGene/L core CPU frequency of 700 MHz. A network packet with 256 bytes of payload is approximately 265 bytes on the wire; therefore a fully utilized collective network both reads and delivers 256 bytes of payload every 530 or so cycles. Software has this much time to store a packet into the network and to read a reply packet.

The collective network's FIFOs are fed and emptied by special 128 bit ("quad") loads and stores that were originally designed for the SIMD floating point unit. The quad loads and stores allow for faster access to the network FIFOs, but restrict the range of addresses: every load and store must be to a 16 byte aligned address. Quad loads and stores also limit the granularity of access, rather complicating access to the network.

Complex Reduction Operations: Unfortunately the collective network does not support a number of reduction operations that are part of the MPI standard. For example, while the MAX operation is supported in hardware there is no corresponding MIN operation. The obvious solution is to negate (using a 2's complement operation) the input data before network injection and to again negate the results before storage in the result buffer. Here the granularity of network access comes into play; since we cannot perform 32 bit negations on the SIMD floating point registers, we are forced to store the results extracted from the network, load them back into fixed-point registers and then store the negated values into memory. To do this both on the outgoing and on the incoming packet takes about 825 cycles even when the multiple store and load streams are interleaved. Under these circumstances the CPU becomes a bottleneck, reducing bandwidth to $\frac{256\ Bytes}{825\ cycles} = 0.3$ Bytes/cycle, or 210 Mbytes/s. To achieve full bandwidth on MIN operations we need to deploy both CPUs in the compute node, one feeding the network and the other reading out results.

The MIN operation is a good introductory example for the rest of this paper, which deals with a truly complicated reduction algorithm: floating point `Allreduce` using the fixed-point operations available in the collective network.

4   Floating Point `Allreduce`

We implement floating point `Allreduce` sums in terms of the fixed point operations available on the network. In order to describe the `Allreduce` algorithm we need to introduce some notation. We assume that each processor $P^{(j)} \mid 0 \leq j < Nprocs$ has a buffer of N floating point numbers $D^{(j)} = \{d_i^{(j)} \mid 0 \leq i < N\}$. An `Allreduce` SUM operation calculates the following array:

$$R = \{r_i \mid 0 \leq i < N, r_i = \sum_j d_i^{(j)}\}$$

on each processor. Using the notation we just introduced, the summation of IEEE single- or double-precision floating point numbers can be broken down into a multi-stage algorithm as follows:

- **Find a common exponent:** The participating nodes extract exponents $E^{(j)} = \{e_i^{(j)} \mid 0 \leq i < N\}$ from their respective input buffers and insert them into the network. The collective network performs a fixed-point `Allreduce` MAX on the exponents, returning an array of maximum exponents $E^{max} = \{e_i^{max} \mid 0 \leq i < N\}$ to each node.

- **Line up the mantissas:** The nodes extract mantissas $M^{(j)} = \{m_i^{(j)} \mid 0 \leq i < N\}$ from the input and shift them to correspond to the maximum exponents E^{max}. At the end of this stage the mantissa arrays $M^{(j)}$ are element-wise compatible with each other and can be added with a fixed-point operation.

- **Mantissa Allreduce:** The collective network performs the fixed point `Allreduce` sum on the aligned mantissas. This operation results in a new array of summed mantissas $M^{sum} = \{m_i^{sum} \mid 0 \leq i < N\}$ on all participating nodes.

- **Re-normalize result:** The nodes IEEE-normalize incoming mantissas M^{sum} and exponents E^{max} to produce the final result. IEEE normalization entails multiple steps, including the calculation of 2's complements if the mantissa is negative and handling de-normalized IEEE numbers.

Unlike the simple cases of `Allreduce` where the processors' roles is simply to feed and read the collective network, the algorithm outlined above is fairly compute intensive. An earlier implementation of the algorithm resulted in a relatively low performance of 0.06 Bytes/cycle or 42 Mbytes/s, indicating that the processor spends $\frac{256}{0.06}$ = 4266 cycles processing each packet. Our main purpose in this paper is to improve this result by (a) pipelining the computational phases of the algorithm and (b) by using both processors in the compute nodes to help with the computation.

A symmetric data parallel algorithm would have been the most obvious solution for deploying both processors in a node. However, the BlueGene/L hardware/software architecture poses substantial challenges to this approach. There is only one collective network channel available for use by user code, making the symmetric use of network hardware cumbersome and expensive. Fine grain synchronization of this channel would have been prohibitively expensive; to avoid this problem we carefully distributed the algorithm's tasks between the processors, putting one CPU in charge of creating and sending packets and letting the other read out and post-process all results. We deal with the remaining synchronization issues by using buffers of non-cached memory.

4.1   An Interleaved Dual-Processor `Allreduce`

The algorithm we are about to describe interleaves the two kinds of fixed point allreduce operations: maximum selection on the exponents and summing up of the mantissas. Figures 1 and 2 depict the hardware resources involved in the process and the flow of data.

In each figure the compute node is represented as a large rectangle. The processors are represented by circles; various memory buffers are represented by rectangles. Rounded rectangles represent the collective network's memory-mapped FIFOs. The arrows represent load and store operations and they are numbered to show the logical order of data movement.

Processing Exponents: Figure 1 shows the movement of exponent information during the algorithm. Processor #1 extracts the exponents of the input values from the user supplied input buffer *SendBuf* and stores them as 16 bit values in the *ExpBuf* buffer. This buffer is then loaded into the SIMD floating point unit's registers with quad load instructions and stored in the network FIFOs with quad stores.

The collective network processes the exponent packets from all compute nodes, and returns a sequence of maximum exponents to the network receive FIFO on each node. Processor #2 removes these packets from the FIFO and stores two copies of the quad values into buffers called *MaxExpBuf* and `MaxExpCopy` (the reasons for this seeming redundancy will become apparent later).

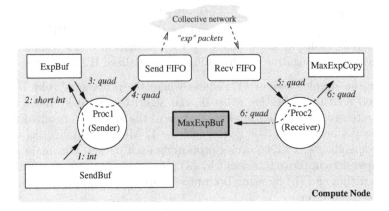

Fig. 1. Floating-point `Allreduce`: exponent processing

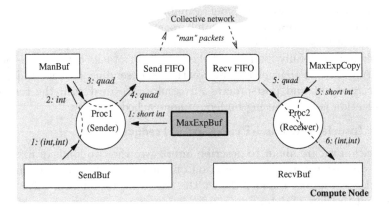

Fig. 2. Floating-point `Allreduce`: mantissa processing

Processing Mantissas: Figure 2 shows the processing of mantissa data. Processor #1 reads the input values again from *SendBuf* this time extracting the mantissa part and the sign information. Mantissas are right-shifted to match the maximum exponent value from the *MaxExpBuf* buffer. The mantissa's implicit leading "1" bit is restored; if the input value was negative it is replaced by its 2's complement. The result is stored in a temporary buffer called `ManBuf`; next, the processor reads this buffer and stores in the collective network's send FIFO using quad loads.

The collective network processes mantissa packets by adding them up as fixed-point numbers.

Processor #2 reads arriving packets of summed mantissas from the network FIFOs into a temporary buffer. The mantissas are then re-loaded into fixed-point registers together with corresponding exponent values from the `MaxExp` buffer. Each mantissa-exponent pair is *normalized* to an IEEE compliant floating point/double precision number. This last operation is fairly complicated, as it has to pay attention to negative and de-normalized IEEE numbers. The results are stored in memory.

Synchronizing the Processors: Using one processor to feed the network and another to read it creates a simple producer-consumer relationship that would seem to make other synchronization devices superfluous. Unfortunately in our algorithm the mantissa processing phase depends on results obtained from the exponent processing phase, making it necessary to synchronize the processors with a *reverse* producer-consumer relationship from processor #2 to processor #1. This relationship is mediated by `MaxExpBuf`.

Because there is no inter-processor cache coherency on BlueGene/L, we use the non-L1-cached (and therefore expensive to access) *scratchpad* memory area for the purpose of storing `MaxExpBuf`. However, maximum exponents are needed by *both* processors for further processing, and it is cheaper for processor #2 to make two copies of it than to read back data from non-cached memory - hence the redundant store to `MaxExpCopy` in the exponent processing phase.

`MaxExpBuf` synchronization is achieved with two counters. One of the counters is allocated in non-cached memory and updated by Processor #2 to keep track of the number of packets inserted into `MaxExpBuf`; the second is private to Processor #1 and counts packets extracted from `MaxExpBuf`. Processor #2 must ensure that the second counter's value never exceeds the first one. Because the first counter's value increases monotonically through the process additional synchronization devices (e.g. atomic access to the counters) are not necessary.

Interleaving algorithm phases is important for two reasons. First, the non-cached memory area on BlueGene/L (and hence `MaxExpBuf`) is of limited size and cannot be allowed to grow indefinitely. Thus, processing of mantissas has to start before the processing of all exponents is done.

The second argument for interleaving is that processing mantissas is more processor intensive than processing exponents. Since the algorithm's performance is limited by CPU processing power rather than the network bandwidth it makes sense to distribute mantissa processing evenly throughout the algorithm.

Thus the `Allreduce` algorithm we deployed starts with sending a number of exponent packets, filling up `MaxExpBuf`, and then proceeds to alternate mantissa and exponent packets at the correct ratio.

Network Efficiency: An 8-byte double precision IEEE number consists of 11 bits of exponent and 53 bits of mantissa. Hence a fixed point sum on 2^{16} nodes requires $53+16=69$ bits of precision. Given the granularity of network operations we have the choice of using 64 bits for the mantissa (at a potential loss of 5 bits of precision), 80 bits or 96 bits. We use 16 bits for each exponent. We end up with $64+16=80$ or $96+16=112$ bits on the network for every 64 bits of payload, resulting in a net efficiency of 57% to 80%. Payload bandwidth decreases accordingly from 350 MBytes/s to 280 and 200 MBytes/s respectively.

5   Performance Analysis

We obtained our performance figures by running a very simple micro-benchmark that calls several implementations of the double precision floating point MPI_Allreduce sum operation. We tested three implementations, the default MPICH2 implementation and two versions of our collective network optimized floating point algorithm, with 64 bit and 96 bit mantissas respectively.

We ran measurements on a variety of BlueGene/L systems ranging from 32 to 1024 nodes. Due to the extremely good scaling properties of the collective network performance variations are below the noise level. Therefore only 1024 node measurements are presented in this paper.

Figure 4 shows the measured bandwidth of the three implementations for message sizes ranging from 1 to 10^7 doubles. While the MPICH2 implementation (using the torus network) reaches less than 40 MBytes/s, the collective network assisted implementation reach 100 to 140 MBytes/s, handily beating the default as well as our non-overlapped old collective assisted implementation (not shown here).

The 100 and 140 MBytes/s figures are essentially half of the collective network's bandwidth limit (200 and 280 MBytes/s respectively). It is thus clear that the collective network based algorithm are CPU limited rather than network limited.

Figure 3 shows the cycles taken by several phases of the algorithm. The main obstacle to bandwidth turns out to be the mantissa post-processing phase of the algorithm, which takes more than 1100 cycles per mantissa packet. It is evident that the two processors do unequal amounts of work, resulting in load imbalance that degrades performance.

On Figure 4 the message size corresponding to 50% of the maximum bandwidth is around 1000 doubles or 8 Kbytes. On BlueGene/L $\frac{N}{2}$ values of more than 1 Kbyte usually indicate a latency problem. Figure 5 confirms this by comparing the small-message latencies of the single- and dual-performance optimized Allreduce implementations. MPICH2 latencies are also shown for comparison. The latency of the dual processor algorithm is around 30 microseconds, almost double the latency of the single-processor algorithm. The latency increase is caused by software cache coherence code executed at the beginning and end of the function and inter-processor communication in the setup phase of the algorithm.

	quad mem	memory	compute	Total
exp pre	110	603	-	637
exp post	241	271	-	269
man pre	124	339	672	815
man post	198	370	945	1141

Fig. 3. Cycle breakdown of the algorithm's phases: exponent and mantissa pre- and postprocessing. The first column lists the cycles spent processing quad loads and stores. The second column lists all cycles spent accessing memory or network. The third column lists all instructions other than loads and stores. The last column lists the grand total of cycles spent per packet in each phase. The memory and non-memory cycles do not add up because there is a certain amount of overlap between them.

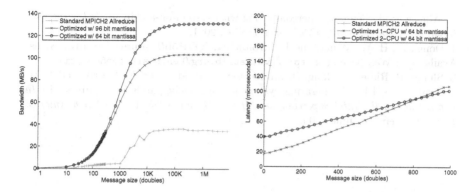

Fig. 4. Bandwidth of `Allreduce` algorithms

Fig. 5. Latency comparison of single- and dual-processor optimized algorithms

The cross-over point for the two algorithms is at the message size of approximatively 850 doubles. At this stage the better bandwidth of the dual-processor algorithm overcomes the higher initial latency.

6   Conclusions and Future Work

The development of this collective network algorithm was made harder by aspects of the code development not discussed in this paper. E.g. the mantissa pre- and postprocessing sequences were written in PowerPC assembler. They are 50 and 76 lines long respectively, and together they cost more than a man-month of sustained work to develop and debug. Another aspect is the development of parallel code on a non-cache-coherent platform.

With careful coding of the assembler functions, software cache coherence and interprocessor communication we were able to more than double the bandwidth of the pre-existing collective network-assisted implementation, at the cost of additional latency. For very short messages a single-core implementation is preferable; for medium size messages the dual-core interleaved implementation is more than twice faster.

In the near future we intend to deploy these algorithms as part of the standard MPI library that is shipped with BlueGene/L.

References

1. N. R. Adiga et al. An overview of the BlueGene/L supercomputer. In *SC2002 – High Performance Networking and Computing*, Baltimore, MD, November 2002.
2. G. Almási, R. Bellofatto, J. Brunheroto, C. Caşcaval, J. G. C. nos, L. Ceze, P. Crumley, C. Erway, J. Gagliano, D. Lieber, X. Martorell, J. E. Moreira, A. Sanomiya, and K. Strauss. An overview of the BlueGene/L system software organization. In *Proceedings of Euro-Par 2003 Conference*, Lecture Notes in Computer Science, Klagenfurt, Austria, August 2003. Springer-Verlag.

3. G. Almasi et al. Cellular supercomputing with system-on-a-chip. In *IEEE International Solid-state Circuits Conference ISSCC*, 2001.
4. J. Dongarra, H.-W. Meuer, and E. Strohmaier. **TOP500 Supercomputer Sites**. Available in Web page at: http://www.netlib.org/benchmark/top500.html.
5. L. Shuler, R. Riesen, C. Jong, D. van Dresser, A. B. Maccabe, L. A. Fisk, and T. M. Stallcup. The PUMA operating system for massively parallel computers. In *In Proceedings of the Intel Supercomputer Users' Group. 1995 Annual North America Users' Conference*, June 1995.

Scalable Fault Tolerant MPI: Extending the Recovery Algorithm

Graham E. Fagg, Thara Angskun, George Bosilca,
Jelena Pjesivac-Grbovic, and Jack J. Dongarra

Dept. of Computer Science, 1122 Volunteer Blvd., Suite 413,
The University of Tennessee, Knoxville, TN 37996-3450, USA

Abstract. Fault Tolerant MPI (FT-MPI)[6] was designed as a solution
to allow applications different methods to handle process failures beyond
simple check-point restart schemes. The initial implementation of FT-
MPI included a robust heavy weight system state recovery algorithm
that was designed to manage the membership of MPI communicators
during multiple failures. The algorithm and its implementation although
robust, was very conservative and this effected its scalability on both
very large clusters as well as on distributed systems. This paper details
the FT-MPI recovery algorithm and our initial experiments with new
recovery algorithms that are aimed at being both scalable and latency
tolerant. Our conclusions shows that the use of both topology aware
collective communication and distributed consensus algorithms together
produce the best results.

1   Introduction

Application developers and end-users of high performance computing systems
have today access to larger machines and more processors than ever before. Additionally, not only the individual machines are getting bigger, but with the
recently increased network capacities, users have access to higher number of
machines and computing resources. Concurrently using several computing resources, often referred to as Grid- or Metacomputing, further increases the number of processors used in each single job as well as the overall number of jobs,
which a user can launch.

With increasing number of processors however, the probability, that an application is facing a node or link failure is also increasing. The current de-facto
means of programming scientific applications for such large and distributed systems is via the message passing paradigm using an implementation of the Message Passing Interface (MPI) standard [10,11]. Implementations of MPI such as
FT-MPI [6] are designed to give users a choice on how to handle failures when
they occur depending on the applications current state.

The internal algorithms used within FT-MPI during failure handling and
recovery are also subject to the same scaling and performance issues that the rest
of the MPI library and applications face. Generally speaking, failure is assumed
to be such a *rare* event that the performance of the recovery algorithm was

B. Di Martino et al. (Eds.): EuroPVM/MPI 2005, LNCS 3666, pp. 67–75, 2005.
© Springer-Verlag Berlin Heidelberg 2005

considered secondary to its robustness. The system was designed originally for use on LAN based Clusters where some linear performance was acceptable at up to several hundred nodes, its scaling is however become an issue when FT-MPI is used on both larger MPP which are becoming more common and when running applications in a Meta/Grid environment.

This paper describes current work on FT-MPI to make its internal algorithms both scalable on single network (MPP) systems as well as more scalable when running applications across the wide area on potentially very high latency links.

This paper is ordered as follows: Section 2 detailed related work in fault tolerent MPIs, collective communication and distributed algorithms, section 3 details HARNESS/FT-MPIs architecture and the current recovery algorithm, section 4 the new methods togther with some initial experiment results (including transatlantic runs) and section 5 the conclusions and future work.

2   Related Work

Work on making MPI implementations both fault tolerant and scalable can be split into the different categories based on the overall goals, either usually fault tolerance or scalablity [8] but rarely both. On the scalability front, related work includes both collective communication tuning and the development of distributed consensus algorithms though various schemes.

Most other fault tolerant MPI implementations support checkpoint and restart models, with or with various levels of message logging to improve performance. Coordinated checkpoint restart versions of MPI include: Co-Check MPI [12], LAM/MPI[14]. MPICH-V [4] uses a mix of uncoordinated check-pointing and distributed message logging. More relevant work includes: Starfish MPI [1] which uses low level atomic communications to maintain state and MPI/FT [2] which provides fault-tolerance by introducing a central co-ordinator and/or replicating MPI processes. Using these techniques, the library can detect erroneous messages by introducing a voting algorithm among the replicas and can survive process-failures. The drawback however is increased resource requirements and partially performance degradation. Finally, the Implicit Fault Tolerance MPI project MPI-FT [9] supports several master-slave models where all communicators are built from grids that contain 'spare' processes. These spare processes are utilized when there is a failure.

Starfish and MPI-FT are interesting in that they use classical distributed system solutions such as atomic communications and replication [17] to solve underlying state management problems.

3   Current Algorithm and Architecture of Harness and FT-MPI

FT-MPI was built from the ground up as an independent MPI implementation as part of the Department of Energy Heterogeneous Adaptable Reconfigurable Networked SyStems (HARNESS) project [3]. HARNESS provides a dynamic framework for adding new capabilities by using runtime plug-ins. FT-MPI is one such plug-in. A number of HARNESS services are required both during startup, failure-recovery and shutdown. This services are built in the form of plug-ins that can also be compiled as standalone daemons. The ones relevant to this work are:

- Spawn and Notify service. This service is provided by a plug-in which allows remote processes to be initiated and then monitored. The service notifies other interested processes when a failure or exit of the invoked process occurs. The notify message is either sent directly to all other MPI tasks directly or more commonly via the Notifier daemon which can provide additional diagnostic information if required.
- Naming services. These allocate unique identifiers in the distributed environment for tasks, daemons and services (which are uniquely addressable). The name service also provides temporary internal system (not application) state storage for use during MPI application startup and recovery, via a comprehensive record facility.

An important point to note is that the Spawn and Notify Service together with the Notifier daemons are responsible for delivering Failure/Death events. When the notifier daemon is used it forces an ordering on the delivery of the death event messages, but it does not impose a time bounding other than that provided by the underlying communication system SNIPE [7]. i.e. it is best effort, with multiple retries. Processes can be assumed to be dead when either their Spawn service detects their death (SIGCHLD etc), another MPI process cannot contact them or their Spawn service is unreachable.

It is useful to know what exactly the meaning of *state* is. The state in the context of FT-MPI is which MPI processes make up the MPI Communicator MPI_COMM_WORLD. In addition, the state also contains the process connection information, i.e. IP host addresses and port numbers etc. (FT-MPI allows processes the right to migrate during the recovery, thus the connection information can change and needs to be recollected). The contact information for all processes is stored in the Name Service, but during the recovery each process *can* receive a complete copy of all other processes contact information, reducing accesses to the Name Service at the cost of local storage within the MPI runtime library.

Current Recovery Algorithm. The current recovery algorithm is a multistage algorithm that can be viewed as a type of conditional ALL2ALL communication based on who failed and who recovered. The aim of the algorithm is to build a new consistent state of the system after a failure is detected. The algorithm itself must also be able to handle failures during recovery (i.e. recursive failures). The overall design is quite simple, first we detect who failed and who survived, then we recover processes (if needed), verify that the recovery proceeded correctly, build a new consistent state, disseminate this new state and check that the new state has been accepted by all processes. The following is a simple outline:

- State Discovery (initial)
- Recovery Phase
- State Discovery (verification if starting new processes or migration)
- New State Proposal
- New State Acceptance
- Continue State if accepted, or restart if not accepted

The current implementation contains the notion of two classes of participating processes within the recovery; *Leaders* and *Peons*. The leader tasks are responsible for synchronization, initiating the Recovery Phase, building and disseminating the new state atomically. The peons just follow orders from the leaders. In the event that a peon dies during the recovery, the leaders will restart the recovery algorithm. If the leader dies, the peons will enter an election controlled by the name service using an atomic test_and_set primitive. A new leader will be elected, and this will restart the recovery algorithm. This process will continue until either the algorithm succeeds or until everyone has died.

As mentioned previously the delivery of the death events is ordered but not time bounded. This is the reason why the *Initial* and *verification* State Discovery and New State Acceptance phases exist. The leader processes cannot rely on only the death events to know the current state of the system. In the case of bust failures, the death events may not all arrive at the same time. A consequence of this could be that the leader recovers only a single failed process and either completes the algorithm only to immediately have to restart it, or it discovers at the end of a recovery that the one of processes in the final state has also failed. The Acceptance phase prevents some processes from receiving the New State and continuing, while other processes receive a late death event and then restart the recovery process. This is essential as the recovery is *collective* and hence synchronizing across MPI_COMM_WORLD and must therefore be consistent.

Implementation of Current Algorithm. Each of the phases in the recovery algorithm are implemented internally by the Leaders and Peons as a state machine. Both classes of processes migrate from one state to another by sending or receiving messages (i.e. a death event is receiving a message). The Leader processes store a copy of the state in the Name Service. This assists new processes in finding out what to do in the case that they were started after a particular state has already been initialized.

The State Discovery phase is implemented by the Leader telling all Peons that they need to send him an acknowledgment message (ACK). The Peons reply back to the Leader by sending a message which contains their current contact information, thus combining two messages into one slightly larger message. The Leader then waits for the number of replies plus the number of failures (m) plus one (for themselves) to equal the number of total processes (n) in the original MPI_COMM_WORLD. As the recovery system does not have any explicit timeouts, it relies on the conservation of messages, i.e. no ACK or death event messages are lost.

The Recovery phase involves the leaders using their copy of the current state and then building a new version of MPI_COMM_WORLD. Depending on the

FT-MPI communicator mode [6] used this could involve rearranging processes ranks or spawning new processes. The phase starts with the Leaders telling their existing Peons to WAIT via a short message broadcast. (This is an artifact left over from an earlier version that used polling of the Name Service). If new processes are started, they discover from the Name Service that they need to ACK the Leaders, without the Leaders needing to send the ACK requests to the new processes directly. After this the Leaders again perform a State Discovery to ensure that any of the new (and current) processes have not since failed. If no new processes are required, the Leaders build the new state and then move to the New State Proposal phase.

The New State Proposal phase is where the Leaders propose the new state to all other processes. During the Acceptance phase, all processes get the chance to reject the new state. This currently only occurs if they detect the death of a processes that is included in the new state otherwise they automatically accept it. The Leader(s) collect the votes and if ALL voted YES it sends out a final STATE OK message. Once this message has started to be transmitted, any further failures are IGNORED until a complete global restart of the algorithm by the Leader entering State Discovery phase again. This is implemented by associating each recovery with a unique value (*epoch*) assigned atomically by the Name Service. A Peon may detect the failure of a process, but will follow instructions from the Leaders, in this case STATE OK. The failure will however still be stored and not lost. This behavior prevents some processes from exiting the recovery while other processes continue to attempt to recover again.

Cost of Current Algorithm. The current algorithm can be broken down into a number of linear and collective communication patterns. This then allows us to both model and then predict the performance of the algorithm for predetermined conditions such as a burst of m failures.

- Discovery phases can be viewed and as a small message broadcast (request ACK) followed by a larger message gather (receive ACK).
- Recovery phase is first a small message broadcast (WAIT) followed by a series of sequential message transfers between the Leader, Spawn & Notify service, Name Server etc to start any required new processes.
- New State Proposal phase is a broadcast of the complete state (larger message).
- New State Acceptance phase is a small message reduce (not a gather).
- OK State phase is a small message broadcast.

Assuming a time per operation of $Top(n)$ where n is the participants (including the root), the current algorithm would take:

$Ttotal = Tbcast\_ack(n-m) + Tgather\_ack(n-m) + Tbcast\_wait(n-m) + Tspawn(m) + Tbcast\_ack(n) + Tgather\_ack(n) + Tbcast\_state(n) + Treduce\_accept(n) + Tbcast\_ok(n)$

As some message sizes are identical for a number of operations we can replace them with an operator based solely on their message size. Also if we assume that $n > m$ (which typically is true as we have single failures) we can simplify the cost to:

$$T total = 4 \; T bcast\_small(n) + 2 \; T gather\_large(n) + T spawn(m) + T bcast\_large(n)$$
$$+ \; T reduce\_small(n)$$

The initial implementation for LAN based systems used simple linear fault tolerant algorithms, as any collective communication had to be fault tolerant itself. Take for example the initial ACK broadcast. The Leader only potentially knows who some of the failed tasks might be, but the broadcast cannot deadlock if unexpected tasks are missing.

Assuming the cost of spawning is constant, and that all operations can be performed using approximately n messages then the current algorithm could be further simplified if we consider the time for sending only small $T small$ or larger messages $T large$ to:

$$T total = 5n \; T small + 3n \; T large + T spawn \text{ or } O(8n) + T spawn.$$

4   New Scalable Methods and Experimental Results

The aim of any new algorithm and implementation is to reduce the total cost for the recovery operation on both large LAN based systems as well for Grid and Metacomputing environments where the FT-MPI application is split across multiple clusters (from racks to continents).

Fault Tolerant Tuned Collectives. The original algorithm is implemented by a number of broadcast, gather and reduce operations. The first obvious solution is to replace the linear operations by tuned collective operations using a variety of topologies etc. This we have done using both binary and binomial trees. This work was not trivial for either the LAN or wide area cases simply due to the fault tolerant requirements placed on the operations being able to handle recursive failures (i.e. nodes in any topology disappearing unexpectedly). This has been achieved by the development of self healing tree/mesh topologies. Under normal

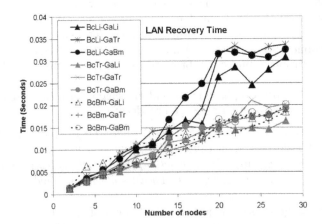

Fig. 1. Recovery time of the original algorithm when using various combinations of tuned collective operations

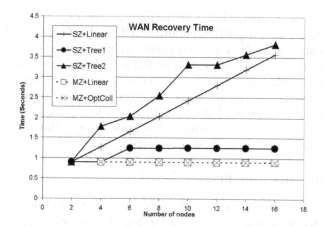

Fig. 2. Recovery time of various algorithms operating across a very high latency link

conditions they operate by passing message as expected in static topologies. In the case of a failures they use neighboring nodes to reroute the messages.

Figure 1 shows how the original algorithm performs when using various combinations of linear and tuned FT-collectives. As can be expected the tree based (Binary Tree (Tr) / Binomial Tree (Bm)) implementations out performed the Linear (Li) versions. The results also clearly show that the implementation is more sensitive to the performance of the broadcast operations (Bc) than the gather/reduce (Ga) operations. This is to be expected as the previous section showed that algorithm contains five broadcasts verses three gather/reduce operations. The best performing implementation used a binary tree broadcast and a linear gather, although we know this not to be true for larger process counts [13].

Multi-zone Algorithms. For the wide area we have taken two approaches. The first is to arrange the process topologies so that they minimize the wide area communication, much the same as Magpie [16].

The second is to develop a multi-zone algorithm. This is where there is a leader process per zone. The lowest MPI ranked Leader becomes the master leader for the whole system. This leader synchronizes with the other zone leaders, who in turn execute the recovery algorithm across their zones. Unfortunately this algorithm does not benefit us much more than the latency sensitive topology algorithms due to the synchronizing nature of the New State Proposal and New State Acceptance phases, unless there are additional failures during the recovery.

Figure 2 shows how the original algorithm and the new multi-zone algorithm performs when executed between two remote clusters of various sizes. One cluster is located at the University of Tennessee USA, and the other is located at the University of Strasbourg France. A size of eight refers to four nodes at each site. The labels *SZ* indicate Single-Zone and *MZ* indicates Multi-Zone algorithms. Single-Zone *Linear* is the normal algorithm without modification, which performed badly as expected. Single-Zone *Tree1* is the single leader algorithm but using a binary tree where the layout of the process topology reducing the number of wide area hops. This algorithm performs well. Single-Zone *Tree2* is the single leader algorithm using a binary tree topology where no changes have been made to the layout to avoid wide area communication. Multi-Zone *Linear* is a multi leader algorithm using linear collectives per zone, and Multi-Zone *OpColl* uses the best FT-tuned collective per zone. The two multi-zone algorithms perform best, although the node count is so low that it hides any advantage of the internal tree collectives within each individual zone. Overall the multi-zone tuned collective method perform the best as expected.

5   Conclusions and Future Work

The current FT-MPI recovery algorithm is robust but also heavyweight at approximately $O(8n)$ messages. Although FT-MPI has successfully executed fault tolerant applications on medium sized IBM SP systems of up to six hundred processes its recovery algorithm is not scalable or latency tolerant. By using a combination of fault tolerant tree topology collective communications and a more distributed multi-coordinator (leader) based recovery algorithm, these scalability issues have been overcome.

Work is continuing on finding better distributed coordination algorithms and reducing the amount of state exchanged at the final stages of recovery. This is the first stage in moving FT-MPIs process fault tolerant model into the ultra scale arena. A latter stage will involve taking FT-MPIs recovery algorithm and adding it to the community Open MPI implementation [15].

Acknowledgments. This material is based upon work supported by the Department of Energy under Contract No. DE-FG02-02ER25536 and 8612-001-0449 through a subcontract from Rice University No. R7A827-792000. The NSF CISE Research Infrastructure program EIA-9972889 supported the infrastructure used in this work.

References

1. A. Agbaria and R. Friedman. Starfish: Fault-tolerant dynamic mpi programs on clusters of workstations. In *In 8th IEEE International Symposium on High Performance Distributed Computing*, 1999.
2. R. Batchu, J. Neelamegam, Z. Cui, M. Beddhua, A. Skjellum, Y. Dandass, and M. Apte. Mpi/ft^{TM}: Architecture and taxonomies for fault-tolerant, message-passing middleware for performance-portable parallel computing. In *In Proceedings of the 1st IEEE International Symposium of Cluster Computing and the Grid held in Melbourne, Australia.*, 2001.
3. Beck, Dongarra, Fagg, Geist, Gray, Kohl, Migliardi, K. Moore, T. Moore, Papadopoulous, Scott, and Sunderam. HARNESS:a next generation distributed virtual machine. *Future Generation Computer Systems*, 15, 1999.
4. G. Bosilca, A. Bouteiller, F. Cappello, S. Djilali, G. Fédak, C. Germain, T. Hérault, P. Lemarinier, O. Lodygensky, F. Magniette, V. Néri, and A. Selikhov. MPICH-v: Toward a scalable fault tolerant MPI for volatile nodes. In *SuperComputing*, Baltimore USA, November 2002.

5. G. Burns and R. Daoud. Robust MPI message delivery through guaranteed resources. In *MPI Developers Conference*, June 1995.
6. G. E. Fagg, A. Bukovsky, and J. J. Dongarra. HARNESS and fault tolerant MPI. *Parallel Computing*, 27:1479–1496, 2001.
7. G. E. Fagg, K. Moore, and J. J. Dongarra. Scalable networked information processing environment (SNIPE). *Future Generation Computing Systems*, 15:571–582, 1999.
8. R. L. Graham, S.-E. Choi, D. J. Daniel, N. N. Desai, R. G. Minnich, C. E. Rasmussen, L. D. Risinger, and M. W. Sukalski. A network-failure-tolerant message-passing system for terascale clusters. In *ICS*. New York, USA, June. 22-26 2002.
9. S. Louca, N. Neophytou, A. Lachanas, and P. Evripidou. Mpi-ft: Portable fault tolerance scheme for MPI. In *Parallel Processing Letters, Vol. 10, No. 4, 371-382, World Scientific Publishing Company.*, 2000.
10. Message Passing Interface Forum. *MPI: A Message Passing Interface Standard*, June 1995. http://www.mpi-forum.org/.
11. Message Passing Interface Forum. *MPI-2: Extensions to the Message Passing Interface*, July 1997. http://www.mpi-forum.org/.
12. G. Stellner. Cocheck: Checkpointing and process migration for MPI. In *Proceedings of the 10th International Parallel Processing Symposium (IPPS '96)*, Honolulu, Hawaii, 1996.
13. S. S. Vadhiyar, G. E. Fagg, and J. J. Dongarra. Performance modeling for self-adapting collective communications for MPI. In *LACSI Symposium*. Springer, Eldorado Hotel, Santa Fe, NM, Oct. 15-18 2001.
14. Sriram Sankaran and Jeffrey M. Squyres and Brian Barrett and Andrew Lumsdaine and Jason Duell and Paul Hargrove and Eric Roman. The LAM/MPI Checkpoint/Restart Framework: System-Initiated Checkpointing. In *LACSI Symposium*. Santa Fe, NM. October 2003.
15. E. Gabriel and G.E. Fagg and G. Bosilica and T. Angskun and J. J. Dongarra J.M. Squyres and V. Sahay and P. Kambadur and B. Barrett and A. Lumsdaine and R.H. Castain and D.J. Daniel and R.L. Graham and T.S. Woodall. Open MPI: Goals, Concept, and Design of a Next Generation MPI Implementation. In *Proceedings 11th European PVM/MPI Users' Group Meeting*, Budapest, Hungry, 2004.
16. Thilo Kielmann and Rutger F.H. Hofman and Henri E. Bal and Aske Plaat and Raoul A. F. Bhoedjang. MagPIe: MPI's collective communication operations for clustered wide area systems. In *ACM SIGPLAN Symposium on Principles and Practice of Parallel Programming (PPoPP'99)*, 34(8), pp131–140, May 1999.
17. Andrew S. Tanenbaum and Maarten van Steen. Distributed Systems: Principles and Paradigms, Prentice Hall, 2002.

Hash Functions for Datatype Signatures in MPI

Julien Langou, George Bosilca, Graham Fagg, and Jack Dongarra

Dept. of Computer Science, The University of Tennessee, Knoxville, TN 37996

Abstract. Detecting misuse of datatypes in an application code is a desirable feature for an MPI library. To support this goal we investigate the class of hash functions based on checksums to encode the type signatures of MPI datatype. The quality of these hash functions is assessed in terms of hashing, timing and comparing to other functions published for this particular problem (Gropp, 7th European PVM/MPI Users' Group Meeting, 2000) or for other applications (CRCs). In particular hash functions based on Galois Field enables good hashing, computation of the signature of unidatatype in $\mathcal{O}(1)$ and computation of the concatenation of two datatypes in $\mathcal{O}(1)$ additionally.

1 Introduction

MPI datatypes are error prone. Detecting such errors in the user application code is a desirable feature for an MPI library and can potentially provide an interesting feedback to the user. Our goal is to detect when the type signature from the sender and the receiver do not respect the MPI standard. The cost of doing so should be negligible with respect to the cost of the communication.

The idea was previously mentioned by Gropp [3] and we merely agree with his solution and specifications that we recall in Section 2. In Section 3, we give a more general framework to his study that enables us to rederive his solution but also create new solutions. In particular, our hash functions have the property of being $\mathcal{O}(1)$ time to solution for computing the signature of an unidatatype. We conclude with some experimental results that assess the quality of our hash functions in term of hashing and timing. The codes to reproduce the experiments are available online [7].

2 Specifications of the Problem

The MPI standard [1,2] provides a full set of functions to manipulate datatypes. Using the basic datatypes predefined by the standard, these functions allow the programmer to describe most of the datatypes used in parallel applications from regular distributions in memory (i.e. contiguous, vector, indexed) to more complex patterns (i.e. structures and distributed arrays). The type signature of a datatype is defined as the (unique) decomposition of a datatype in a succession of basic datatypes. From the MPI standard point of view, a communication is correct if and only if the type signature of the sender exactly matches the

B. Di Martino et al. (Eds.): EuroPVM/MPI 2005, LNCS 3666, pp. 76–83, 2005.

beginning of the type signature of the receiver. For the remaining of this paper we consider that the type signature of the sent datatype exactly matches the type signature of the received datatype (in its whole). (The case of a type signature of the datatype of the receiver longer than the one of the sender is dealt in the last paragraph of the paper.)

In the framework developed by Gropp [3], a hash function is used to create a signature for the type signature of the datatype of the sender, the receiver then checks whether the signature of the sender matches the signature of the type signature of its datatype. Note that some errors might still be unnoticed when two type signatures have the same hash value.

At this point of the specification, any hash functions existing in the literature would be adequate. However, the fact that MPI datatypes can be combined together implies that we would like to be able to efficiently determine the signature of the concatenatenation of two datatypes. (Note that this is not only a desirable feature but also mandatory since we want to check the datatype when the count of the sender and the receiver mismatch.) The datatype signature function σ shall be such that the signature of the concatenation of two datatypes can be obtained from the signatures of the two datatypes, we therefore need an operator $\odot$ such that:

$$\sigma([\alpha_1, \alpha_2]) = \sigma(\alpha_1) \odot \sigma(\alpha_2).$$

If there exists such a $\odot$ for σ, we call σ assossiative.

We call unidatatype a derived datatype made of just one datatype (derived or not). This type or class of datatype is fairly frequent in user application codes and we therefore also would like to be able to efficiently compute their signature.

3   Some Associative Hash Functions

3.1   Properties of Checksum-Based Hash Functions

Considering a set of elements, E, and two binary operations in E, $\oplus$ (the addition, required to be associative) and $\otimes$ (the multiplication), the checksum of $X = (x_i)_{i=1,...,n} \in E^n$ is defined as

$$f(X) = f((x_i)_{i=1,...,n}) = \bigoplus_{i=1}^{n} (x_i \otimes \alpha_i), \tag{1}$$

where α_i are predefined constants in E.

Let us define the type signature of the datatype X, $\sigma(X)$, as the tuple

$$\sigma(X) = (f(X), n). \tag{2}$$

In this case, we can state the following theorem.

Theorem 1. *Given $(E, \oplus, \otimes)$ and an $\alpha \in E$, defining the checksum function, f, as in Equation (1), and the type signature σ as in Equation (2) and providing that*

$$\oplus \text{ and } \otimes \text{ are associative} , \tag{3}$$

$$\text{the } \alpha_i \text{ are chosen such that } \alpha_i = \bigotimes_{j=1,\ldots,i} \alpha = \alpha^i, \tag{4}$$

$$\text{for any } x,y \in E, \quad (x \otimes \alpha) \oplus (y \otimes \alpha) = (x \oplus y) \otimes \alpha, \tag{5}$$

then the concatenation datatype operator, $\odot$, is defined as

$$\sigma(X) \odot \sigma(Y) = ((f(X) \otimes \alpha^m) \oplus f(Y), n+m), \tag{6}$$

and satisfies

$$\sigma([X,Y]) = \sigma(X) \odot \sigma(Y). \tag{7}$$

for any $X \in E^n$ and $Y \in E^m$.

Proof. The proof is as follows, let us define $X = (x_i)_{i=1,\ldots,n}$, $Y = (y_i)_{i=1,\ldots,m}$ and $Z = [X,Y] = (z_i)_{i=1,\ldots,n+m}$, then

$$
\begin{aligned}
f(Z) &= \bigoplus_{i=1}^{n+m} (z_i \otimes \alpha^i) = (\bigoplus_{i=1}^{n} x_i \otimes \alpha^{m+i}) \oplus (\bigoplus_{i=1}^{m} y_i \otimes \alpha^i) \\
&= (\bigoplus_{i=1}^{n} ((x_i \otimes \alpha^i) \otimes \alpha^m)) \oplus (\bigoplus_{i=1}^{m} y_i \otimes \alpha^i) \\
&= (\left(\bigoplus_{i=1}^{n} (x_i \otimes \alpha^i) \right) \otimes \alpha^m) \oplus (\bigoplus_{i=1}^{m} y_i \otimes \alpha^i) \\
&= (f(X) \otimes \alpha^m) \oplus f(Y)
\end{aligned}
$$

The equality of the first line is the consequence of the associativity of $\oplus$ (3), the second line is the consequence of the associativity of $\otimes$ (3) and the definition of the α_i (4), the third line is the consequence of the distributivity of $\oplus$ versus $\otimes$ (5).

In our context, E is included in the set of the integers ranging from 0 to $2^w - 1$ (i.e. E represents a subset of the integers that we can be encoded with w bits). In the next three sections we give operators, $\oplus$ and $\otimes$ that verifies (3), (4) and (5) over E.

Any binary operations over E, $\oplus$ and $\otimes$, such that $(E, \oplus, \otimes)$ is a ring verifies the necessary properties (3), (4) and (5), therefore our study will focus on rings over E.

3.2   Checksum mmm-bs1

Given any integer a, $(E, \oplus, \otimes)$ where E is the set of integer modulo a, $\oplus$ the integer addition modulo a and $\otimes$ the integer multiplication modulo a defines a ring. (Even a field iff a is prime.)

A natural choice for α and a is $\alpha = 2$ and $a = 2^w - 1$ which defines the signature mmm-bs1.

The multiplication of integers in E by power of 2 modulo $2^w - 1$ corresponds to a circular leftshift over w bits, $\ll_{c,w}$. This operation as well as the modulo $2^w - 1$ operation can both be efficiently implemented on a modern CPU.

Remark 1. Note that the type signature is encoded on w bits but only $2^w - 1$ values are taken.

Remark 2. The computation of the signature of a unidatatype, X, that is composed of n identical datatypes x, costs as much as a single evaluation thanks to the formula:

$$f(X) = f([x, .., x]) = (x \otimes 2^0) \oplus ... \oplus (x \otimes 2^{n-1}) = x \otimes (2^n \ominus 1), \qquad (8)$$

where $\ominus$ denotes the substraction operation in the group $(E, \oplus)$. The evaluation of 2^n in formula (6) is efficiently computed thanks to $1 \ll_{c,w} n$.

Remark 3. In [3], $\oplus$ is set to the integer addition modulo 2^w and $\otimes$ is set to the circular leftshift over w bits (that is the integer multiplication by power of 2 modulo $2^w - 1$ for numbers between 0 to $2^w - 2$ and the identity for $2^w - 1$). This mix of the moduli breaks the ring property of $(E, \oplus, \otimes)$, (the distributivity relation (5) is not anymore true,) and consequently the Equation (6) is not true. We definitely do not recommend this choice since it fails to meet the concatenation requirement.

3.3 Checksum xor-bs1

Gropp [3] proposed to use the xor operation, $\wedge$, for the addition and a circular leftshift over w bits by one, $\ll_{c,w}$, for the multiplication operation. E represents here the integers modulo 2^w. The condition of the Theorem 1 holds and thus this represents a valid choice for $(E, \oplus, \otimes)$.

Note that in this case, we can not evaluate in $\mathcal{O}(1)$ time the checksum of unidatatype datatype. (See [3, §3.1], for a $\mathcal{O}(log(n))$ solution.)

3.4 Checksum gfd

Another ring to consider on the integers modulo 2^w is the Galois field $\mathrm{GF}(2^w)$. (A comprehensive introduction to Galois field can be found in [4].)

The addition in $\mathrm{GF}(2^w)$ is xor. The multiplication in $\mathrm{GF}(2^w)$ is performed thanks to two tables corresponding to the logarithms, gflog, and the inverse logarithms, gfilog in the Galois Field thus

$$a \otimes b = \mathtt{gfilog}(\mathtt{gflog}[a] + \mathtt{gflog}[b]).$$

where $+$ is the addition modulo 2^w. This requires to store the two tables gflog and gfilog of 2^w words of length w bits.

Since $(E, \oplus, \otimes)$ is a ring, the Theorem1 applies and thus the signature of the concatenation of two datatypes can be computed thanks to formula (6). Note that the value 2^n in the formula (6) is directly accessed from the table of the inverse logarithms since $2^n = gfilog[n-1]$ (see [4]).

Finally since we have a field, we can compute type signatures of unidatatype in $\mathcal{O}(1)$ time via

$$x \otimes 2^0 \oplus \ldots \oplus x \otimes 2^n = x \otimes (2^{n+1} \ominus 1) \oslash (2 \ominus 1) = x \otimes (2^{n+1} \ominus 1) \oslash (3), \quad (9)$$

where $\oslash$ denotes the division operation in the field $(E, \oplus, \otimes)$.

gfd needs to store two tables of 2^w word each of length w therefore we do not want w to be large. (A typical value would be $w = 8$.) In this case to encode the derived datatype on 32 bits, we would use $m = 4$ checksums in $\mathrm{GF}(2^{w=8})$.

3.5   Cyclic Redundancy Check crc

Since we are considering hash functions, we also want to compare to at least one class of hash functions that are known to be efficient in term of computation time and quality. We choose to compare with the cyclic redundancy check.

Stating briefly if we want to have a w-bit CRC of the datatype X, we interpret X as a binary polynomial and choose another binary polynomial C of degree exactly equal to w (thus representing a number between 2^w and $2^{w+1} - 1$). The CRC, R, is the polynomial $X \times x^w$ modulo C, that is to say:

$$X.x^w = Q.C + R.$$

For a more detailed explanations we refer to [5]. Various choices for C are given in the literature, we consider in this paper some standard value, for more about the choice of C, we referred to [6].

The concatenation operation is possible with CRC signatures. If $X = (x_i)_{i=1,\ldots,n}$ and $Y = (y_i)_{i=1,\ldots,m}$ are two datatypes and we have computed their signatures $\sigma(X) = (R_X, n)$ $\sigma(Y) = (R_Y, m)$, then

$$\sigma([X, Y]) = (R_Z, n + m)$$

where R_Z is $R_X x^m + R_Y$ modulo Q.

Even though it is possible given the signatures of two datatypes to compute the signature of the concatenation, the cost of this operation is proportional to the size of the second data-type (in our case m). This is an important drawback in comparison with the checksum where the cost of a concatenation is $\mathcal{O}(1)$. Although, it is not possible to compute quickly the signatures of a datatype composed of all the same datatype.

4   Experimental Validation of the Hash Functions to Encode MPI Datatype

4.1   Software Available

Our software is available on the web [7]. We believe it is high quality in the sense of efficiency, robustness and ease of use. It is provided with a comprehensive set of testing routines, timing routines and hash function quality assessment routines. The experimental results presented thereafter are based on this software and thus are meant to be easily reproducible. We can also verify the correctness of the theoretical results in part 3 through the software.

4.2   Quality of the Hash Functions

To evaluate the quality of a hash function we consider a sample of datatypes and check how often collisions appear. (Note, that since the number of values taken by the hash function is finite (encoded on 16 or 32 bits) and the number of datatypes is infinite, collisions in the value of the hash functions are unavoidable.) Considering the fact that we are interested in the quality of the hash function when applied to MPI datatypes, it makes sense to have a sample that reflects what an MPI library might encounter in an application code. In this respect, we follow the experimental method of Gropp [3]. In our experiments, we consider 6994 datatypes that are made of $wg = 13$ different pre-defined datatypes. (We refer to [7] for the exact description of the datatypes.)

The results of the hash functions are given in Table 1. Two quantities are given: the percentage of collisions and the percentage of duplicates. A collision is when a type signature has its hash value already taken by another type signature in the sample. A duplicate is a hash value that has strictly more than one type signature that maps to it in the sample.

Since we are mapping 6994 words on 2^{16} value, it is possible to give a *perfect hash function* for our sample example. (That is to say a function where no collision happens.) However this is not the goal. We recall the fact that if we apply a random mapping to m out of n possible states. Then we expect that this mapping will produce about

$$n/(n/m + (1/2) + \mathcal{O}(m/n)) \tag{10}$$

distinct results providing $\sqrt{n} < m$. Thus, a random mapping from $m = 6994$ to $n = 2^{16}$ shall give about 5.07% collisions. This number is representative of a good hash functions.

We considered three different CRCs (namely 16-bit CRC/CCITT,XMODEM and ARC) but only report the one of CRC/CCITT that is best suited to the considered panel.

Table 1. Percentage of collisions and duplications for some 16-bit and 32-bit hash functions. We have used the panel of 6994 type-signatures described in [7].

16 bit type signature			32 bit type signature		
	Collisions	Duplicates		Collisions	Duplicates
16-bit CRC/CCITT	4.76 %	4.88 %	CRC-04C11DB7	0.00 %	0.00 %
xor-bs1(w=16,m= 1)	61.10 %	26.75 %	xor-bs1(w=16,m= 2)	37.32 %	15.76 %
mmm-bs1(w=16,m= 1)	52.03 %	19.67 %	mmm-bs1(w=16,m= 2)	23.29 %	11.48 %
gfd (w=16,m= 1)	12.65 %	8.48 %	gfd (w=16,m= 2)	0.00 %	0.00 %
xor-bs1(w= 8,m= 2)	64.40 %	23.04 %	xor-bs1(w= 8,m= 4)	60.71 %	13.65 %
mmm-bs1(w= 8,m= 2)	51.44 %	27.47 %	mmm-bs1(w= 8,m= 4)	30.08 %	14.91 %
gfd (w= 8,m= 2)	3.73 %	3.77 %	gfd (w= 8,m= 4)	0.00 %	0.00 %

From Table 1, gfd performs fairly well, it is as good as 16-bit CRC-CCITT or a random mapping (5.07%). To have a better hash function, one can simply increase the number of bits on which the data is encoded. In Table 1, we also give the percentage of collisions and duplications for 32-bit signatures, we obtain a perfect hash function for gfd and CRC-04C11DB7. In conclusion, gfd has the same quality of hashing as some well known CRCs on our sample and is much better than xor-bs1 and mmm-bs1.

4.3   Timing Results for the Hash Functions

We present timing results of optimized routine for crc and gfd. The code is compiled with GNU C compiler with -O3 flag and run on two architectures: torc5.cs.utk.edu a Pentium III 547 MHz and earth.cs.utk.edu an Intel Xeon 2.3 GHz, both machines are running Linux OS. Results are presented in Table 2. signature represents the time to encode a word of length nx, concatsignature represents the time to encode the concatenation of two words of size $nx/2$, unisignature represents the time to encode a word of length nx with all the same datatype (basic or derived). Either for crc or for gfd, the computation of a signature lasts $\mathcal{O}(n)$ time. The $\mathcal{O}(1)$ time for the concatsignature and unisignature is an obvious advantage of gfd over crc. Note that in Table 2, the encoding with gfd is done by a pipeline algorithm ($\mathcal{O}(n)$ work) whereas encoding with a tree algorithm by using concatsignature. Such an algorithm should have given us better performance for gfd signature (see Gropp [3, §]).

Table 2. Time in μs to compute the datatype signatures of different MPI datatype type signatures. for $gfd(w = 8, m = 2)$ and $w = 16$ for $crc(w = 16)$.

Pentium III (547 MHz)						
$nx = 100$			$nx = 1000$			
	gfd	crc		gfd	crc
signature	$17.80\mu s$	$11.68\mu s$	signature	$176.88\mu s$	$114.87\mu s$
concatsignature	$0.30\mu s$	$5.95\mu s$	concatsignature	$0.34\mu s$	$57.48\mu s$
unisignature	$0.36\mu s$	$11.68\mu s$	unisignature	$0.37\mu s$	$114.87\mu s$
$nx = 100$ Intel Xeon 2.392 GHz						
$nx = 100$			$nx = 1000$			
	gfd	crc		gfd	crc
signature	$4.60\mu s$	$2.04\mu s$	signature	$48.26\mu s$	$20.21\mu s$
concatsignature	$0.09\mu s$	$1.12\mu s$	concatsignature	$0.10\mu s$	$10.07\mu s$
unisignature	$0.09\mu s$	$2.04\mu s$	unisignature	$0.10\mu s$	$20.21\mu s$

5   Conclusions

In this paper, we assess that checksums based on Galois field provide a quality of hashing comparable to CRCs (on our test problem). The big advantage of gfd is that the computation of signatures of unidatatypes is $\mathcal{O}(1)$ work and the computation of concatenations of two datatypes is $\mathcal{O}(1)$ work (CRCs are $\mathcal{O}(n)$ for those two special cases). In a typical application, having concatsignature and unisignature in $\mathcal{O}(1)$ leads to the computation of most signatures in $\mathcal{O}(1)$ times as well.

Notes and Comments. The MPI standard requires the type signature (datatype, count) of the sender to fit in the first elements of the type signature (datatype, count) of the receiver. When the number of basic datatypes in the (datatype, count) of the receiver is longer than the number of basic datatypes in the (datatype, count) of the sender, the receiver needs to look inside the structure of its datatype to find the point when the number of basic datatypes is the same as the one sent. The MPI correctness of the communication can then be assessed by checking if the signature of this part of the datatype matches the signature of the sender.

Special care has to be taken for the datatypes MPI_PACKED and MPI_BYTE (see Gropp [3]).

More information theory could have been exploited, for example, $\mathtt{gfd}(w = 8, m = 2)$ guarantees that any swap between two basic datatypes is detected as long as there is less than 253 basic datatypes in the two derived datatypes considered. These considerations are out of the scope of this paper.

References

1. Message Passing Interface Forum: MPI: A message-passing interface standard. http://www.mpi-forum.org
2. Message Passing Interface Forum: MPI: A message-passing interface standard. International Journal of Supercomputer Applications **8** (1994) 165–414
3. Gropp, W.D.: Runtime checking of datatype signatures in MPI. In Dongarra, J., Kacsuk, P., Podhorszki, N., eds.: Recent Advances in Parallel Virtual Machine and Message Passing Interface. Number 1908 in Springer Lecture Notes in Computer Science (2000) 160–167. 7th European PVM/MPI Users' Group Meeting, http://www-unix.mcs.anl.gov/~gropp/bib/papers/2000/datatype.ps
4. Plank, J.S.: A tutorial on Reed-Solomon coding for fault-tolerance in RAID-like systems. Software – Practice & Experience **27** (1997) 995–1012. http://www.cs.utk.edu/~plank/plank/papers/CS-96-332.html
5. Knuth, D.E.: The Art of Computer Programming, 2nd Ed. (Addison-Wesley Series in Computer Science and Information). Addison-Wesley Longman Publishing Co., Inc. (1978)
6. Koopman, P., Chakravarty, T.: Cyclic redundancy code (CRC) polynomial selection for embedded networks. IEEE Conference Proceeding (2004) 145–154. 2004 International Conference on Dependable Systems and Networks (DSN'04)
7. Langou, J., Bosilca, G., Fagg, G., Dongarra, J.: TGZ for hash functions of MPI datatypes (2004) http://www.cs.utk.edu/~langou/articles/LBFD:05/2005-LBFD.html

Implementing MPI-IO Shared File Pointers Without File System Support

Robert Latham, Robert Ross, Rajeev Thakur, and Brian Toonen

Mathematics and Computer Science Division,
Argonne National Laboratory,
Argonne, IL 60439, USA
{robl, rross, thakur, toonen}@mcs.anl.gov

Abstract. The ROMIO implementation of the MPI-IO standard provides a portable infrastructure for use on top of any number of different underlying storage targets. These targets vary widely in their capabilities, and in some cases additional effort is needed within ROMIO to support all MPI-IO semantics. The MPI-2 standard defines a class of file access routines that use a *shared file pointer*. These routines require communication internal to the MPI-IO implementation in order to allow processes to atomically update this shared value. We discuss a technique that leverages MPI-2 one-sided operations and can be used to implement this concept without requiring any features from the underlying file system. We then demonstrate through a simulation that our algorithm adds reasonable overhead for independent accesses and very small overhead for collective accesses.

1 Introduction

MPI-IO [1] provides a standard interface for MPI programs to access storage in a coordinated manner. Implementations of MPI-IO, such as the portable ROMIO implementation [2] and the implementation for AIX GPFS [3], have aided in the widespread availability of MPI-IO. These implementations include a collection of optimizations [4,3,5] that leverage MPI-IO features to obtain higher performance than would be possible with the less capable POSIX interface [6].

One feature that the MPI-IO interface provides is *shared file pointers*. A shared file pointer is an offset that is updated by any process accessing the file in this mode. This feature organizes accesses to a file on behalf of the application in such a way that subsequent accesses do not overwrite previous ones. This is particularly useful for logging purposes: it eliminates the need for the application to coordinate access to a log file.

Obviously coordination must still occur; it just happens implicitly within the I/O software rather than explicitly in the application. Only a few historical file systems have implemented shared file pointers natively (Vesta [7], PFS [8], CFS [9], SPIFFI [10]) and they are not supported by parallel file systems being deployed today. Thus, today shared file pointer access must be provided by the MPI-IO implementation.

B. Di Martino et al. (Eds.): EuroPVM/MPI 2005, LNCS 3666, pp. 84–93, 2005.
© Springer-Verlag Berlin Heidelberg 2005

This paper discusses a novel method for supporting shared file pointer access within a MPI-IO implementation. This method relies only on MPI-1 and MPI-2 communication functionality and not on any storage system features, making it portable across any underlying storage. Section 2 discusses the MPI-IO interface standard, the portions of this related to shared file pointers, and the way shared file pointer operations are supported in the ROMIO MPI-IO implementation. Section 3 describes our new approach to supporting shared file pointer operations within an MPI-IO implementation. Two algorithms are used, one for independent operations and another for collective calls. Section 4 evaluates the performance of these two algorithms on synthetic benchmarks. Section 5 concludes and points to future work in this area.

2   Background

The MPI-IO interface standard provides three options for referencing the location in the file at which I/O is to be performed: explicit offsets, individual file pointers, and shared file pointers. In the explicit offset calls the process provides an offset that is to be used for that call only. In the individual file pointer calls each process uses its own internally stored value to denote where I/O should start; this value is referred to as a file pointer. In the shared file pointer calls each process in the group that opened the file performs I/O starting at a single, shared file pointer.

Each of these three ways of referencing locations have both independent (non-collective) and collective versions of read and write calls. In the shared file pointer case the independent calls have the _shared suffix (e.g., MPI_File_read_shared), while the collective calls have the _ordered suffix (e.g., MPI_File_read_ordered). The collective calls also guarantee that accesses will be ordered by rank of the processes. We will refer to the independent calls as the *shared mode* accesses and the collective calls as the *ordered mode* accesses.

2.1   Synchronization of Shared File Pointers in ROMIO

The fundamental problem in supporting shared file pointers at the MPI-IO layer is that the implementation never knows when some process is going to perform a shared mode access. This information is important because the implementation must keep a single shared file pointer value somewhere, and it must access and update that value whenever a shared mode access is made by any process.

When ROMIO was first developed in 1997, most MPI implementations provided only MPI-1 functionality (point-to-point and collective communication), and these implementations were not thread safe. Thread safety makes it easier to implement algorithms that rely on nondeterministic communication, such as shared-mode accesses, because a separate thread can be used to wait for communication related to shared file pointer accesses. Without this capability, a process desiring to update a shared file pointer stored on a remote process could stall indefinitely waiting for the remote process to respond. The reason is that the implementation could check for shared mode communication only when an

MPI-IO operation was called. These constraints led the ROMIO developers to look for other methods of communicating shared file pointer changes.

Processes in ROMIO use a second hidden file containing the current value for the shared file pointer offset. A process reads or writes the value of the shared file pointer into this file before carrying out I/O routines. The hidden file acts as a communication channel among all the processes. File system locks serialize access and prevent simultaneous updates to the hidden file. This approach works well as long as the file system meets two conditions:

1. The file system must support file locks
2. The file system locks must prevent access from other processes, and not just from other file accesses in the same program.

Unfortunately, several common file systems do not provide file system locks (e.g., PVFS, PVFS2, GridFTP) and the NFS file system provides advisory lock routines but makes no guarantees that locks will be honored across processes. On file systems such as these, ROMIO cannot correctly implement shared file pointers using the hidden file approach and hence disables support for shared file pointers. For this reason a portable mechanism for synchronizing access to a shared file pointer is needed that does not rely on any underlying storage characteristics.

3   Synchronization with One-Sided Operations

The MPI-2 specification adds a new set of communication primitives, called the one-sided or remote memory access (RMA) functions, that allow one process to modify the contents of remote memory without the remote process intervening. These passive target operations provide the basis on which to build a portable synchronization method within an MPI-IO implementation. This general approach has been used in a portable atomic mode algorithm [11]. Here we extend that approach to manage a shared file pointer and additionally to address efficient ordered mode support.

MPI-2 one-sided operations do not provide a way to atomically read and modify a remote memory region. We can, however, construct an algorithm based on existing MPI-2 one-sided operations that lets a process perform an atomic modification. In this case, we want to serialize access to the shared file pointer value.

Before performing one-sided transfers, a collection of processes must first define a *window object*. This object contains a collection of memory *windows*, each associated with the rank of the process on which the memory resides. After defining the window object, MPI processes can then perform put, get, and accumulate operations into the memory windows of the other processes.

MPI passive target operations are organized into *access epochs* that are bracketed by MPI_Win_lock and MPI_Win_unlock calls. Clever MPI implementations [12] will combine all the data movement operations (puts, gets, and accumulates) into one network transaction that occurs at the unlock. The MPI-2

standard allows implementations to optimize RMA communication by carrying
out operations in any order at the end of an epoch. Implementations take advantage of this fact to achieve much higher performance [12]. Thus, within one
epoch a process cannot read a byte, modify that value, and write it back because the standard makes no guarantee about the order of the read-modify-write
steps. This aspect of the standard complicates, but does not prevent, the use of
one-sided to build our shared file pointer support.

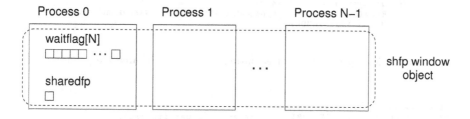

Fig. 1. Creation of MPI windows

Our algorithms operate by using the following data structure. We define a
window object with an N-byte `waitflag` array and an MPI_Offset-sized
`sharedfp`, both residing on a single process (Figure 1). In our discussion we
will assume that this data structure is stored on process 0, but for multiple files
being accessed in shared file pointer mode, these structures could be distributed
among different processes. This data structure is used differently for shared mode
than for ordered mode access. We will discuss each in turn.

3.1 Shared Mode Synchronization

The MPI-2 standard makes no promises as to the order of concurrent shared
mode accesses. Additionally, the implementation does not need to serialize access
to the file system, only the value of the shared file pointer. After a process updates
the value of the file pointer, it can carry out I/O while the remaining processes
attempt to gain access to the shared file pointer. This approach minimizes the
time during which any one process has exclusive access to the shared file pointer.

In our shared mode approach, we use the `waitflag` array to synchronize
access to the shared file pointer. Figure 2 gives pseudocode for acquiring the
shared file pointer, and Figure 3 demonstrates how we update the shared file
pointer value.

Each byte in the `waitflag` array corresponds to a process. A process will
request a lock by putting a 1 in the byte corresponding to its rank in the communicator used to open the file. Doing so effectively adds it to the list of processes
that want to access the shared file pointer. In the same access epoch the process
gets the remaining N-1 bytes of `waitflag` and the `sharedfp` value. This combination effectively implements a test and set. If a search of `waitflag` finds no
other 1 values, then the process has permission to access the shared file pointer,
and it already knows what that value is without another access epoch.

```
val = 1; /* add self to waitlist */
MPI_Win_lock(MPI_LOCK_EXCLUSIVE, homerank, 0, waitlistwin);
MPI_Get(waitlistcopy, nprocs-1, MPI_BYTE, homerank, FP_SIZE, 1,
        waitlisttype, waitlistwin);
MPI_Put(&val, 1, MPI_BYTE, homerank, FP_SIZE + myrank, 1, MPI_BYTE,
        waitlistwin);
MPI_Get(fpcopy, 1, fptype, homerank, 0, 0, fptype, waitlistwin);
MPI_Win_unlock(homerank, waitlistwin);

/* check to see if lock is already held */
for (i=0; i < nprocs-1 && waitlistcopy[i] == 0; i++);
if (i < nprocs - 1) {
    /* wait for notification */
    MPI_Recv(&fpcopy, 1, fptype, MPI_ANY_SOURCE, WAKEUPTAG, comm,
        MPI_STATUS_IGNORE);
}
```

Fig. 2. MPI pseudocode for acquiring access to the shared file pointer

```
val=0; /* remove self from waitlist */
MPI_Win_lock(MPI_LOCK_EXCLUSIVE, homerank, 0, waitlistwin);
MPI_Get(waitlistcopy, nprocs-1, MPI_BYTE, homerank, FP_SIZE, 1,
        waitlisttype, waitlistwin);
MPI_Put(&val, 1, MPI_BYTE, homerank, FP_SIZE + myrank, 1,
        MPI_BYTE, waitlistwin);
MPI_PUT(&fpcopy, 1, fptype, homerank, 0, 1, fptype, waitlistwin);
MPI_Win_unlock(homerank, waitlistwin);

for (i=0; i < nprocs-1 && waitlistcopy[i] == 0; i++);
if (i < nprocs - 1) {
    int nextrank = myrank;

    /* find the next rank waiting for the lock */
    while (nextrank < nprocs-1 && waitlistcopy[nextrank] == 0) nextrank++;
    if (nextrank < nprocs - 1) {
        nextrank++; /* nextrank is off by one */
    }
    else {
        nextrank = 0;
        while (nextrank < myrank && waitlistcopy[nextrank] == 0) nextrank++;
    }
    /* notify next rank with new value of shared file pointer */
    MPI_Send(&fpcopy, 1, fptype, nextrank, WAKEUPTAG, comm);
}
```

Fig. 3. MPI pseudocode for updating shared file pointer and (if needed) waking up the next process

In this case the process saves the current shared file pointer value locally for subsequent use in I/O. It then immediately performs a second access epoch (Figure 3). In this epoch the process updates **sharedfp**, puts a zero in its corresponding **waitflag** location, and gets the remainder of the **waitflag** array. Following the access epoch the process searches the remainder of **waitflag**. If all the values are zero, then no processes are waiting for access. If there is a 1 in the array, then some other process is waiting. For fairness the first rank after the current process's rank is selected to be awakened, and a point-to-point send (**MPI_Send**) is used to notify the process that it may now access the shared file pointer. The contents of the send is the updated shared file pointer value; this optimization eliminates the need for the new process to reread `sharedfp`. Once the process has released the shared file pointer in this way, it performs I/O using the original, locally-stored shared file pointer value. Again, by moving I/O after the shared file pointer update, we minimize the length of time the shared file pointer is held by any one process.

If during the first access epoch a process finds a 1 in any other byte, some other process has already acquired access to the shared file pointer. The requesting process then calls `MPI_Recv` with `MPI_ANY_SOURCE` to block until the process holding the shared file pointer notifies it that it now has permission to update the pointer and passes along the current value. It is preferable to use point-to-point operations for this notification step, because they allow the underlying implementation to best manage making progress. We know, in the case of the sender, that the process we are sending to has posted, or will very soon post, a corresponding receive. Likewise, the process calling receive knows that very soon some other process will release the shared file pointer and pass it to another process. The alternative, polling using one-sided operations, has been shown less effective [11].

3.2   Ordered Mode Synchronization

Ordered mode accesses are collective; in other words, all processes participate in them. The MPI-IO specification guarantees that accesses in ordered mode will be ordered by rank for these calls: the I/O from a process with rank N will appear in the file after the I/O from all processes with a lower rank (in the write case). However, the actual I/O need not be carried out sequentially. The implementation can instead compute *a priori* where each process will access the file and then carry out the I/O for all processes in parallel.

MPI places several restrictions on collective I/O. The most important one, with regard to ordered mode, is that the application ensure all outstanding independent I/O (e.g. shared mode) routines have completed before initiating collective I/O (e.g. ordered mode) ones. This restriction simplifies the implementation of the ordered mode routines. However, the standard also states that

> in order to prevent subsequent shared offset accesses by the same processes from interfering with this collective access, the call might return only after all the processes within the group have initiated their accesses. When the call returns, the shared file pointer points to the next etype accessible.

This statement indicates that the implementation should guarantee that changes to the shared file pointer have completed before allowing the MPI-IO routine to return.

Figure 4 outlines our algorithm for ordered mode. Process 0 uses a single access epoch to get the value of the shared file pointer. Since the value is stored locally, the operation should complete with particularly low latency. It does not need to access `waitlist` at all, because the MPI specification leaves it to the application not to be performing shared mode accesses at the same time. All

Process 0	Process 1 through (N minus 2)	Process (N minus 1)
Lock		
MPI_Get		
Unlock		
MPI_Scan	MPI_Scan	MPI_Scan
		Lock
		MPI_Put
		Unlock
MPI_Bcast	MPI_Bcast	MPI_Bcast
perform collective I/O	*perform collective I/O*	*perform collective I/O*

Fig. 4. Synchronizing in the ordered mode case. Process 0 acquires the current value for the shared file pointer. After the call to MPI_Scan, process $(N-1)$ knows the final value for the shared file pointer after the I/O completes and can MPI_Put the new value into the window. Collective I/O can then be carried out in parallel with all processes knowing their appropriate offset into the file. An MPI_Bcast ensures that the shared file pointer value is updated before any process exits the call, and imposes slightly less overhead than an MPI_Barrier.

processes can determine, based on their local datatype and count parameters, how much I/O they will carry out. In the call to MPI_Scan, each process adds this amount of work to the ones before it. After this call completes, each process knows its effective offset for subsequent I/O. The $(N-1)$th process can compute the new value for the shared file pointer by adding the size of its access to the offset it obtained during the MPI_Scan. It performs a one-sided access epoch to put this new value into sharedfp, again ignoring the waitlist.

To ensure that a process doesn't race ahead of the others and start doing I/O before the shared file pointer has been updated, the $(N-1)$th process performs a MPI_Bcast of one byte after updating the shared file pointer. All other processes wait for this MPI_Bcast, after which they may all safely carry out collective I/O and then exit the call. If we used an MPI_Barrier instead of an MPI_Bcast, the $(N-1)$th process would block longer than is strictly necessary.

4   Performance Evaluation

We simulated both algorithms (independent and collective) with a test program that implemented just the atomic update of the shared file pointer value. We ran tests on a subset of Jazz, a 350-node Linux cluster at Argonne National Laboratory, using the cluster's Myrinet interconnect.

Earlier in the paper we laid out the requirements for the hidden file approach to shared file pointers. On Jazz, none of the available clusterwide file systems meet those requirements. In fact, the NFS volume on Jazz has locking routines that not only fail to enforce sequential access to the shared file pointer but fail silently. Thus, we were unable to compare our approach with the hidden file technique.

This silent failure demonstrates another benefit of the RMA approach: if an MPI-I/O implementation tests for the existence of RMA routines, it can assume

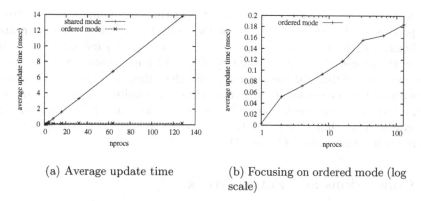

(a) Average update time

(b) Focusing on ordered mode (log scale)

Fig. 5. Average time to perform one update of the shared file pointer

they work (otherwise the MPI-2 implementation is buggy). Tests for file locks, especially testing how well they prevent concurrent access, are more difficult, because such tests would have to ensure not just that the locking routines exist, but that they perform as advertised across multiple nodes.

4.1   Shared Mode Synchronization

In our simulation, processes repeatedly perform an atomic update of the shared file pointer in a loop. We measured how long processes spent updating the shared file pointer and then computed the average time for one process to carry out one shared file pointer update. Figure 5(a) shows how the number of processes impacts the average update time and gives an indication of the scalability of our algorithm.

All the independent I/O routines leave little room for the implementation to coordinate processes, so when N processes attempt to update the shared file pointer, we have to carry out $O(N)$ one-sided operations. Only one process can lock the RMA window at a time. This serialization point will have more of an impact as the number of processes — and the number of processes blocking — increases. The shared mode graph in Figure 5(a) confirms linear growth in time to update the shared file pointer value.

When considering performance in the independent shared file pointer case, one must bear in mind the nature of independent access. As with all independent MPI routines, the implementation does not have enough information to optimize accesses from multiple processes. Also, the simulation provides a worst-case scenario, with multiple processes repeatedly updating the shared file pointer as quickly as possible. In a real application, processes would perform some I/O before attempting to update the shared file pointer again.

4.2   Ordered Mode Synchronization

Implementations have more options for optimizing collective I/O, especially when performing collective I/O with a shared file pointer. As outlined in Section 3.2, N processes need only perform two access epochs — one for reading the current shared file pointer value, one for writing the updated value — from two processes. Our algorithm does use two collective routines (MPI_Scan and MPI_Bcast), so we would expect to see roughly $O(\log(N))$ increase in time to update the shared file pointer using a quality MPI implementation. Figure 5(a) compares update time for the ordered mode algorithm with that of the shared mode algorithm. The ordered mode algorithm scales quite well and ends up being hard to see on the graph. Figure 5(b) shows the average update time for just the ordered algorithm. The graph has a log scale X axis, emphasizing that the ordered mode algorithm is $O(\log(N))$.

5   Conclusions and Future Work

We have outlined two algorithms based on MPI-2 one-sided operations that an MPI-IO implementation could use to implement the shared mode and ordered mode routines. Our algorithms rely solely on MPI communication, using one-sided, point-to-point, and collective routines as appropriate. This removes any dependency on file system features and makes shared file pointer operations an option for all file systems. Performance in the shared mode case scales as well as can be expected, while performance in the ordered mode case scales very well.

We designed the algorithms in this paper with an eye toward integration into ROMIO. At this time, one-sided operations make progress only when the target process hosting the RMA window is also performing MPI communication. We will have to add a progress thread to ROMIO before we can implement the shared file pointer algorithms. In addition to their use in ROMIO, the primitives used could be made into a library implementing test-and-set or fetch-and-increment for other applications and libraries.

The simulations in this paper focused on relatively small numbers of processors (128 or less). As the number of processors increases to thousands, we might need to adjust this algorithm to make use of a tree. Leaf nodes would synchronize with their parents before acquiring the shared file pointer. Such an approach reduces contention on the process holding the memory windows, but it also introduces additional complexity.

Our synchronization routines have been used for MPI-IO atomic mode as well as MPI-IO shared file pointers. In future efforts we will look at using these routines to implement extent-based locking and other more sophisticated synchronization methods.

Acknowledgments

This work was supported by the Mathematical, Information, and Computational Sciences Division subprogram of the Office of Advanced Scientific Computing Research, Office of Science, U.S. Department of Energy, under Contract W-31-109-Eng-38.

References

1. The MPI Forum: MPI-2: Extensions to the Message-Passing Interface (1997)
2. Thakur, R., Gropp, W., Lusk, E.: On implementing MPI-IO portably and with high performance. In: Proceedings of the Sixth Workshop on Input/Output in Parallel and Distributed Systems. (1999) 23–32
3. Prost, J.P., Treumann, R., Hedges, R., Jia, B., Koniges, A.: MPI-IO/GPFS, an optimized implementation of MPI-IO on top of GPFS. In: Proceedings of SC2001. (2001)
4. Thakur, R., Gropp, W., Lusk, E.: A case for using MPI's derived datatypes to improve I/O performance. In: Proceedings of SC98: High Performance Networking and Computing, ACM Press (1998)
5. Latham, R., Ross, R., Thakur, R.: The impact of file systems on MPI-IO scalability. In: Proceedings of EuroPVM/MPI 2004. (2004)
6. IEEE/ANSI Std. 1003.1: Portable operating system interface (POSIX)–Part 1: System application program interface (API) [C language] (1996 edition)
7. Corbett, P.F., Feitelson, D.G.: Design and implementation of the Vesta parallel file system. In: Proceedings of the Scalable High-Performance Computing Conference. (1994) 63–70
8. Intel Supercomputing Division: Paragon System User's Guide. (1993)
9. Pierce, P.: A concurrent file system for a highly parallel mass storage system. In: Proceedings of the Fourth Conference on Hypercube Concurrent Computers and Applications, Monterey, CA, Golden Gate Enterprises, Los Altos, CA (1989) 155–160
10. Freedman, C.S., Burger, J., Dewitt, D.J.: SPIFFI — a scalable parallel file system for the Intel Paragon. IEEE Transactions on Parallel and Distributed Systems **7** (1996) 1185–1200
11. Ross, R., Latham, R., Gropp, W., Thakur, R., Toonen, B.: Implementing MPI-IO atomic mode without file system support. In: Proceedings of CCGrid 2005. (2005)
12. Thakur, R., Gropp, W., Toonen, B.: Minimizing synchronization overhead in the implementation of MPI one-sided communication. In: Proceedings of the 11th European PVM/MPI Users' Group Meeting (Euro PVM/MPI 2004). (2004) 57–67

An Efficient Parallel File System
for Cluster Grids

Franco Frattolillo and Salvatore D'Onofrio

Research Centre on Software Technology,
Department of Engineering, University of Sannio, Italy
frattolillo@unisannio.it

Abstract. ePVM is an extension of the well known PVM programming system, whose main goal is to enable PVM applications to run on
computing nodes belonging to non-routable private networks, but connected to the Internet through publicly addressable IP front-end nodes.
However, in order to enable large-scale PVM applications to effectively
manage the enormous volumes of data they usually generate, ePVM
needs a parallel file system. This paper presents ePIOUS, the optimized
porting of the PIOUS parallel file system under ePVM. ePIOUS has
been designed so as to take into account the two-levels physical network
topology characterizing the "cluster grids" normally built by ePVM. To
this end, ePIOUS exploits the ePVM architecture to implement a file
caching service that is able to speed up file accesses across clusters.

1   Introduction and Motivations

Cluster grids [1,2] are nowadays considered a promising alternative to both grid
computing systems [3,4,5] and traditional supercomputing systems and clusters
of workstations exploited as a unique, coherent, high performance computing
resource. ePVM [6,7] is an extension of PVM [8], whose main goal is to enable
PVM applications to run on cluster grids made up by computing nodes belonging
to non-routable private networks, but connected to the Internet through publicly
addressable IP front-end nodes. In fact, ePVM enables the building of "extended
virtual machines" (EVMs) made up by sets of clusters. Each cluster can be a
set of interconnected computing nodes provided with private IP addresses and
hidden behind a publicly addressable IP front-end node. During computation,
it is managed as a normal PVM virtual machine where a master *pvmd* daemon
is started on the front-end node, while slave *pvmd*s are started on all the other
nodes of the cluster. However, the front-end node is also provided with a specific
ePVM daemon, called *epvmd*, which allows the cluster's nodes to interact with
the nodes of all other clusters of the EVM, thus creating a same communication
space not restricted to the scope of the PVM daemons belonging to a single
cluster, but extended to all the tasks and daemons running within the EVM. In
fact, due to ePVM, both publicly addressable IP nodes and those ones hidden
behind publicly addressable IP front-end nodes of the clusters in the EVM can
be directly referred to as hosts. This means that all the hosts belonging to an

B. Di Martino et al. (Eds.): EuroPVM/MPI 2005, LNCS 3666, pp. 94–101, 2005.

EVM, even though interconnected by a two-levels physical network, can run PVM tasks as in a single, flat distributed computing platform.

However, in order to enable large-scale PVM applications to efficiently run on such cluster grids and to effectively manage the enormous volumes of data they usually generate, ePVM needs a parallel file system (PFS) [9]. To this end, it is worth noting that, in the past, a PFS, called PIOUS (Parallel Input/OUtput System) [10], was specifically designed to incorporate parallel I/O into existing parallel programming systems and has also been widely used as a normal parallel application within PVM. Therefore, PIOUS could be directly exploited within ePVM. However, its execution results in being penalized by the two-levels network topology characterizing the cluster grids normally built by ePVM, and this ends up also penalizing the applications exploiting ePVM and PIOUS.

This paper presents ePIOUS, the optimized porting of PIOUS under ePVM. The porting has been carried out taking into account the two-levels physical network topology characterizing the cluster grids built by ePVM. To this end, the porting exploits the basic ePVM ideas and architecture to provide ePIOUS with a file caching service that is able to speed up file accesses across clusters.

The outline of the paper is as follows. Section 2 describes the software architecture of ePIOUS. Section 3 describes the file operations supported by ePIOUS, while Section 4 describes the implemented file caching service. Section 5 reports on some preliminary experimental results. In Section 6 a conclusion is available.

2   The Software Architecture of ePIOUS

ePIOUS preserves as much as possible the original behavior, semantics and software architecture of PIOUS. It supports a coordinated access to *parafile* objects. In particular, the access is characterized by a sequential consistency semantics and a dynamically-selectable fault tolerance level, while *parafile* objects are logically single files composed of one or more physically disjoint segments, which are data units stored by hosts.

The software architecture of ePIOUS (Figure 1), replicated on each cluster taking part in an EVM, consists of a set of parallel data servers (PDSs), a service coordinator (PSC), a data cache server (CPDS), and a software library, called *libepious*, linked to tasks and comprising all the routines implemented by PIOUS. As in ePVM, PDSs, CPDS and PSC are all implemented as normal PVM tasks,

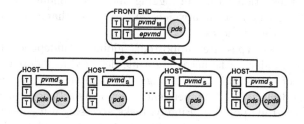

Fig. 1. The software architecture of ePIOUS

and so they can communicate by exploiting the standard PVM communication routines, such as pvm_send and pvm_receive.

One PDS has to reside on each host over which files are "declustered", i.e. over hosts that store physical segments of files. It provides transaction-based access to its local portions of files. On the contrary, a PSC is unique within each cluster of an EVM. It participates in the high level operations involving files, such as open and close operations, while it does not participate in general file access. In fact, PSCs are contacted by tasks upon file openings to obtain file descriptors and to ensure the correct semantics in file access.

A CPDS is unique within each cluster of an EVM. It implements a file caching service in order to speed up file accesses that require inter-cluster operations, thus taking into account the particular network topology characterizing the cluster grids built by ePVM. Therefore, a CPDS is always involved in "remote" file accesses, i.e. whenever a task belonging to a cluster wants to access a file declustered within a different cluster. To this end, ePIOUS assumes that files created by tasks belonging to a cluster may be stored only onto hosts within the cluster, even though they can be opened and written by tasks running on hosts belonging to any cluster composing the EVM. To implement this, file descriptors have been modified: each descriptor now includes the "cluster identifier" (CID) [7] specifying the cluster whose hosts store the file's physical segments. Thus, if a task within a cluster wants to open a file created within a different cluster, it has to specify the pair (CID, pathname), which becomes the "filename" in the open routine, so as to unambiguously identify the file within the EVM. However, the task, once obtained the file descriptor, can exploit it to transparently access the file, without having to specify the CID or any other information about the file.

3   Operations on Files

Parallel operations on files are performed within the context of transactions, transparently to the user, to provide sequential consistency of access and tolerance of system failures. In particular, the *libepious* routines behave as in PIOUS and act as transaction managers for coordinator and data servers participating in a distributed transaction across the EVM clusters satisfying a user request.

Two user-selectable transaction types are supported: "stable" and "volatile". Both transaction types ensure serializability of access. However, the former is a "traditional" transaction based on logging and synchronous disk write operations, while the latter is a "lightweight" transaction that does not guarantee fault-tolerance to the file access.

Three file access types are supported: "global", "independent" and "segmented". In particular, tasks accessing a file declared as "global" view it as a liner sequence of bytes and share a single file pointer. On the contrary, tasks accessing a file declared as "independent" view it as a liner sequence of bytes, but maintain local file pointers. Finally, tasks accessing a file declared as "segmented" view the actual segmented file structure exposed by the associated *parafile* object and can access segments via local file pointers.

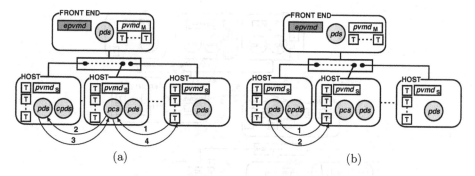

Fig. 2. The scheme of local PSC-PDSs (a) and PDSs (b) interactions

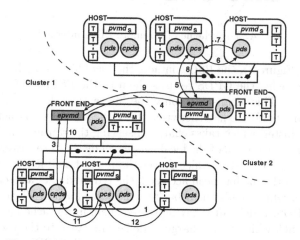

Fig. 3. The scheme of a remote PSC-PDSs interaction

Two types of file operations are supported: those involving PSC and PDSs, and those solely involving PDSs. The former includes the open and close operations as well as the operations that change file attributes. The latter includes the read and write operations. In fact, both types of operations involve CPDSs, when they require inter-cluster actions to be performed.

The scheme of PSC-PDSs operations is depicted in Figures 2(a) and 3. In particular, when a task allocated onto a host of a cluster invokes an ePIOUS routine to perform a PSC-PDSs operation, the *libepious* library contacts the PSC running within the cluster, which decides whether the actions to be performed are "local" or "remote", i.e. if the actions can be confined to the cluster or involve other clusters of the EVM. In fact, such a decision can be taken by examining the file descriptor, which specifies the CID of the cluster within which the file has been created. If the required actions are "local", the routine is served as in PIOUS (Figure 2(a)), i.e. the task wanting to access the file contacts the PSC running within its cluster, which takes charge of completing the routine service by accessing the PDSs managing the file. On the contrary, if the required actions

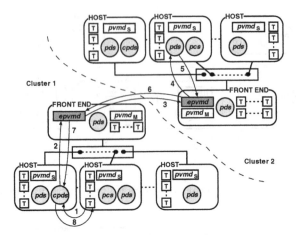

Fig. 4. The scheme of a remote PDSs interaction

are "remote" (Figure 3), the PSC contacts the local CPDS. If the required data are already in cache, the CPDS returns them to the PSC, which delivers them to the user task. This means that only the interactions 1, 2, 11 and 12 shown in Figure 3 are performed. Otherwise, the CPDS has to contact the *epvmd* managing its cluster, which takes charge of serving the routine within the EVM. This means that the *epvmd* of the local cluster sends the file access request to the *epvmd* of the remote cluster, which serves the request as if it were issued within its cluster. Once the file has been accessed, the required data are returned to the remote *epvmd*, which sends them back to the local *epvmd*. Then, the data are sent to the CPDS, which, after having stored them in its cache memory, returns them to the PSC, which takes charge of delivering them to the user task.

The PDSs operations follow a scheme similar to PSC-PDSs operations (Figures 2(b) and 4). In particular, when the file access requires "remote" interactions (Figure 4), the task wanting to access the file first contacts the CPDS, which can directly return the required data if they are available in its cache. In this case, only the interactions 1 and 8 are performed. Otherwise, the CPDS has to start the "remote" interaction scheme by involving the local and remote *epvmd*s. In this case, the remote *epvmd* behaves as the task wanting to access the file, thus interacting with the PDSs local to its cluster, as depicted in Figure 2(b).

4   The File Caching Service

ePIOUS supplies a file caching service able to speed up file accesses involving different clusters of an EVM. The service is implemented by the CPDS servers, each of which may be allocated on any host of a cluster.

A CPDS implements a "segmented least-recently used" (SLRU) caching policy. However, it has been designed so as to be able to exploit different caching algorithms that can be selected by users at ePIOUS start-up. In fact, a CPDS behaves as a normal PDS, even though it takes charge of managing the interac-

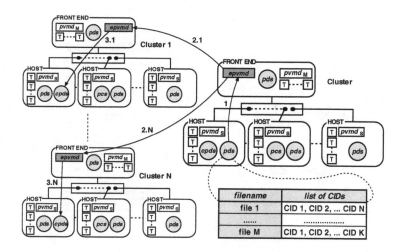

Fig. 5. The multicast messages invalidating the CPDSs' cache memories

tions solely involving remote files. To this end, it has to update its cache memory whenever: (1) a cache miss is generated in a remote file access (Figures 3 and 4); (2) a file, declustered within a cluster but whose segments have been cached by the CPDSs of other clusters of an EVM, is updated. In fact, to easily invalidating the cache memories of the CPDSs in the latter case, each PDS manages a table in which each entry is associated to a file locally declustered and refers to the list of the clusters whose CPDSs have cached segments of the file. Thus, when a file included in the table is updated, a specific invalidating message is multicasted to all the CPDSs belonging to the clusters reported in the list (Figure 5).

5   Experimental Results

Preliminary tests have been conducted on two PC clusters connected by a Fast Ethernet network. The first cluster is composed of 4 PCs interconnected by a Fast Ethernet hub and equipped with Intel Pentium IV 3 GHz, hard disk EIDE 60 GB, and 1 GB of RAM. The second cluster is composed of 4 PCs connected by a Fast Ethernet switch and equipped with Intel Xeon 2.8 GHz, hard disk EIDE 80 GB, and 1 GB of RAM. All the PCs run Red Hat Linux rel. 9.

The benchmark applications implement two parallel file access patterns, defined as "partitioned" and "self-scheduled" [11]. The former divides a file into contiguous blocks, with each block accessed sequentially by a different process. The latter results when a linear file is accessed sequentially by a group of tasks via a shared file pointer. In fact, partitioned and self-scheduled access correspond to the "independent" and "global" file access types, respectively. Furthermore, the files used in the tests are characterized by a number of segments equal to the number of the PDSs employed, while the number of the tasks performing the read and write file operations may vary from 4 to 12. The tasks are all allocated on the hosts belonging to the first cluster according to a round-robin strategy,

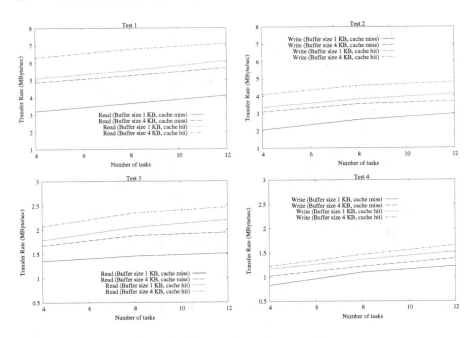

Fig. 6. Some preliminary tests conducted on ePIOUS

while the PDSs accessed during the tests are allocated on the hosts belonging to the second cluster. Finally, files are always accessed in "volatile" mode.

Test 1 in Figure 6 depicts read performance achieved by a test application that accesses a file according to a "partitioned" access pattern. The test has been conducted with two sizes of the transfer buffer used by each task: 1 KB and 4 KB. The file segment size is 1 KB. Despite both the high latency induced by the PVM transport layer in the intra-cluster communications and the overhead induced by the inter-cluster operations managed by *epvmd*s, Test 1 shows a good performance of the ePIOUS implementation. This is due both to the excess of parallelism developed by the application tasks and to the internal structure of the *epvmd*s, which makes them able to serve in parallel many remote file access requests. Furthermore, the caching service significantly improves the observed bandwidth, thus diminishing the effects of the two-levels network topology characterizing the cluster grid. However, performance does not scale well as the number of tasks is increased, and this is due to an increased number of message interrupts at the server transport layer and to the not particularly optimized original implementation of PIOUS, which is largely preserved in ePIOUS.

Test 2 in Figure 6 depicts write performance obtained by the test application implementing a "partitioned" access pattern. In particular, data caching results in being effective for write access among different clusters, even though aggregate performance is not much sensitive to a continued increase in the number of tasks.

Finally, Test 3 and 4 in Figure 6 depict the read and write performance obtained by the test application that accesses a file according to a "self-scheduled" access pattern. The achieved results show that this access pattern does not scale in performance beyond the number of PDSs, and this because the use of a shared file pointer forces a serialized file access.

6   Conclusions

ePIOUS is the PIOUS porting under ePVM, a PVM extension that enables PVM applications to run on cluster grids. The limited modifications to the PIOUS library and the addition of a caching service able to speed up file accesses involving different clusters have enabled ePIOUS to achieve a good performance in all the executed tests, and this demonstrates that existing large-scale PVM applications can exploit ePVM together with ePIOUS to run on cluster grids built by harnessing high performance/cost ratio computing platforms widely available within localized network environments and departmental organizations without having to be rearranged or penalized by the use of complex software systems for grid computing. In particular, the preliminary tests have been performed for a caching service implementing an SLRU policy. However, further tests will be performed under different cache algorithms and several file access patterns.

References

1. Springer, P.L.: PVM support for clusters. In: Procs of the 3rd IEEE Int Conf. on Cluster Computing, Newport Beach, CA, USA (2001)
2. Stefán, P.: The hungarian clustergrid project. In: Procs of the MIPRO'2003, Opatija, Croatia (2003)
3. Berman, F., Fox, G., Hey, T., eds.: Grid Computing: Making the Global Infrastructure a Reality. Wiley & Sons (2003)
4. Foster, I., Kesselman, C., eds.: The Grid: Blueprint for a New Computing Infrastructure. 2nd edn. Morgan Kaufmann (2004)
5. Joseph, J., Fellenstein, C.: Grid Computing. Prentice Hall PTR (2003)
6. Frattolillo, F.: A PVM extension to exploit cluster grids. In: Procs of the 11th EuroPVM/MPI Conference. Volume 3241 of Lecture Notes in Computer Science., Budapest, Hungary (2004) 362–369
7. Frattolillo, F.: Running large-scale applications on cluster grids. Intl Journal of High Performance Computing Applications 19 (2005) 157–172
8. Geist, A., Beguelin, A., et al.: PVM: Parallel Virtual Machine. A Users' Guide and Tutorial for Networked Parallel Computing. MIT Press (1994)
9. Frattolillo, F., D'Onofrio, S.: Providing PVM with a parallel file system for cluster grids. In: Procs of the 9th World Multi-Conference on Systemics, Cybernetics and Informatics, Orlando, Florida, USA (2005)
10. Moyer, S.A., Sunderam, V.S.: PIOUS: a scalable parallel I/O system for distributed computing environments. In: Procs of the Scalable High-Performance Computing Conference. (1994) 71–78
11. Crockett, T.W.: File concepts for parallel I/O. In: Procs of the ACM/IEEE Conference on Supercomputing, Reno, Nevada, USA (1989) 574–579

Cooperative Write-Behind Data Buffering for MPI I/O

Wei-keng Liao[1], Kenin Coloma[1], Alok Choudhary[1], and Lee Ward[2]

[1] Electrical and Computer Engineering Department, Northwestern University
[2] Scalable Computing Systems Department, Sandia National Laboratories

Abstract. Many large-scale production parallel programs often run for a very long time and require data checkpoint periodically to save the state of the computation for program restart and/or tracing the progress. Such a write-only pattern has become a dominant part of an application's I/O workload and implies the importance of its optimization. Existing approaches for write-behind data buffering at both file system and MPI I/O levels have been proposed, but challenges still exist for efficient design to maintain data consistency among distributed buffers. To address this problem, we propose a buffering scheme that coordinates the compute processes to achieve the consistency control. Different from other earlier work, our design can be applied to files opened in read-write mode and handle the patterns with mixed MPI collective and independent I/O calls. Performance evaluation using BTIO and FLASH IO benchmarks is presented, which shows a significant improvement over the method without buffering.

Keywords: Write behind, MPI I/O, file consistency, data buffering, I/O thread.

1 Introduction

Periodical checkpoint write operations are commonly seen in today's long-running production applications. Checkpoint data typically are snapshot of the current computation status to be used for progress tracking and/or program restart. Once written, files created by checkpointing are usually not touched for the rest of the run. In many large-scale applications, such write-once-never-read patterns are observed to dominate the overall I/O workload and, hence, designing efficient techniques for such operations becomes very important. Write-behind data buffering has been known to operating system designers as a way to speed up sequential writes [1]. Write-behind buffering accumulates multiple writes into large contiguous file requests in order to better utilize the I/O bandwidth. However, implementing the write-behind strategy requires the support of client-side caching which often complicates the file system design due to the cache coherence issues. System-level implementations for client-side caching often hold dedicated servers or agents to be responsible for maintaining the coherence. In the parallel environment, this problem gets even obvious since processes running the same parallel application tends to operate their I/O on shared files concurrently. User-level implementation of write-behind data buffering has been proposed in [2] which demonstrated a significant performance improvement when the buffering scheme is embedded in ROMIO [3], an I/O library implementation for Message Passing Interface [4]. However, due to the possible file consistency problem, it is limited to MPI collective write operations with the file opened in write-only mode.

B. Di Martino et al. (Eds.): EuroPVM/MPI 2005, LNCS 3666, pp. 102–109, 2005.

We propose *cooperative write-behind buffering*, a scheme that can benefit both MPI collective and independent I/O operations. To handle the consistency issue, we keep tracking the buffering status at file block level and at most one copy of the file data can be buffered among processes. The status metadata is cyclically distributed among the application processes such that processes cooperate with each other to maintain the data consistency. To handle MPI independent I/O, each process must respond to the queries, remote and local, to the status assigned without explicitly stopping the program's main thread. Thus, we create an I/O thread in each process to handle the requests to the status as well as the buffered data. Our implementation requires every process first look up the buffering status for file blocks covered by the read/write request to determine whether the request should create a new buffer or overwrite the existing buffers (locally or remotely). To ensure I/O atomicity, a status-locking facility is implemented and locks must be granted prior to the read/write calls. We evaluate the cooperative buffering on the IBM SP at San Diego supercomputing center using its GPFS file system. Two sets of I/O benchmarks are presented: BTIO, and FLASH I/O. Compared with the native I/O method without data buffering, cooperative buffering shows a significant performance enhancement for both benchmarks.

The rest of the paper is organized as follows. Section 2 discusses the background information and related works. The design and implementation for cooperative buffering is presented in section 3. Performance results are given in section 4 and the paper is concluded in section 5.

2   Background

Message passing interface (MPI) standard defines two types of I/O functions: collective and independent calls [4]. Collective I/O must be called by all processes that together opened the file. Many collective I/O optimizations take advantage of this synchronization requirement to exchange information among processes such that I/O requests can be analyzed and reconstructed for better performance. However, independent I/O does not require process synchronization, which makes existing optimizations difficult to apply.

2.1   Active Buffering and ENWRICH Write Caching

Active buffering is considered an optimization for MPI collective write operations [2]. It buffers output data locally and uses an I/O thread to perform write requests at background. Using I/O threads allows to dynamically adjust the size of local buffer based on available memory space. Active buffering creates only one thread for the entire run of a program, which alleviates the overhead of spawning a new thread every time a collective I/O call is made. For each write request, the main thread allocates a buffer, copies the data over, and appends this buffer into a queue. The I/O thread, running at background, later retrieves the buffers from the head of the queue, issues write calls to the file system, and releases the buffer space. Although write behind enhances parallel write performance, active buffering is applicable if the I/O patterns only consist of collective writes. Lacking of consistency control, active buffering could not handle the operations mixed with reads and writes as well as independent and collective calls.

Similar limitations are observed in ENWRICH write caching scheme which is a system-level optimization [5]. The client-side write caching proposed in ENWRICH can only handle the files that are opened in write-only mode. For I/O patterns mixed with reads and writes, caching is performed at I/O servers due to the consistency concern. Similar to active buffering, ENWRICH also appends each write into a write queue and all files share the same cache space. During the flushing phase, the I/O threads on all processes coordinate the actual file operations.

2.2   I/O Thread in GPFS

IBM GPFS parallel file system performs the client-side file caching and adopts a strategy called data shipping for file consistency control [6,7]. Data shipping binds each GPFS file block to a unique I/O agent which is responsible for all the accesses to this block. The file block assignment is made in round-robin striping scheme. Any I/O operations on GPFS must go through the I/O agents which will ship the requested data to appropriate processes. To avoid incoherent cache data, a distributed file locking is used to minimize the possibility of I/O serialization that can be caused by lock contention. I/O agents are multi-threaded residing in each process and are responsible for combining I/O requests in collective operations. I/O thread also performs advanced caching strategies at background, such as read ahead and write behind.

3   Design and Implementation

The idea of cooperative buffering is to let application processes cooperate with each other to manage a consistent buffering scheme. Our goals are first to design a write-behind data buffering as an MPI I/O optimization at client side without adding overhead to I/O servers. Secondly, we would like to support read-write operations in arbitrary orders, which implies the incorporation of consistency control. Thirdly, the buffering scheme would benefit both MPI collective and independent I/O.

3.1   Buffering Status Management and Consistency Control

We logically divide a file into blocks of the same size and buffering status of these blocks is assigned in a round-robin fashion across the MPI processes that together open the file. Our consistency control is achieved by tracking the buffering status of each block and keeps at most one copy of file data in the buffers globally. As illustrated in Figure 1(a), the status for block i is held by the process of rank (i mod $nproc$), where $nproc$ is the number of processes in the MPI communicator supplied at file open. The status indicates if the block is buffered, its file offset, current owner process id, a dirty flag, byte range of the dirty data, and the locking mode. Note that buffering is performed for all opened files, but buffering status is unique to each file. An I/O request begins with checking the status of the blocks covered by the request. If the requested blocks have not been buffered by any process, the requesting process will buffer them locally and update the status. Otherwise, the request will be forwarded to the owner(s) of the blocks and appropriate reads or over-writes from/to the remote buffer are performed. Unlike active buffering and ENWRICH appending all writes into a queue, cooperative

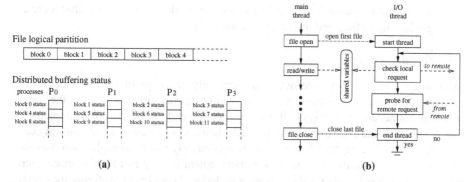

Fig. 1. (a) The buffering status is statically distributed among processes in a round-robin fashion. (b) Design of the I/O thread and its interactions with the main thread and remote requests.

buffering writes to existing buffers whenever possible. Since at most one copy of the file data can be buffered in the process memory at any time, data consistency is maintained. In addition, a status locking facility is implemented, in which locks to the file blocks covered by an I/O request must be all granted before proceeding with any operation on the blocks. To enforce MPI sequential consistency and atomicity, the locks to multiple blocks are granted in an increasing order. For example, if an I/O request covers file blocks from i to j, where $i \leq j$, lock request for block k, $i \leq k \leq j$, will not be issued until lock to block $(k-1)$ is granted. This design is similar to the two-phase locking method [8] used to serialize multiple overlapping I/O requests to guarantee the I/O atomicity.

3.2 I/O Thread

Since buffered data and buffering status are distributed among processes, each process must be able to respond to remote requests for accessing to the status and buffered data stored locally. For MPI collective I/O, remote queries can be fulfilled through inter-process communication during the process synchronization. However, the fact that MPI independent I/O is asynchronous makes it difficult for one process to explicitly receive remote requests. Our design employs an I/O thread in each process to handle remote requests without interrupting the execution of the main thread. To increase the portability, our implementation uses the POSIX standard thread library [9]. Figure 1(b) illustrates the I/O thread design from the viewpoint of a single process. Details of the I/O thread design are described as follows.

- The I/O thread is created when the application opens the first file and destroyed when the last file is closed. Each process can have multiple files opened, but only one thread is created.
- The I/O thread performs an infinite loop to serve the local and remote I/O requests.
- All I/O and communication operations are carried out by the I/O thread only.
- A conditional variable protected by a mutual exclusion lock is used to communicate the two threads.
- To serve remote requests, the I/O thread keeps probing for incoming I/O requests from any process in the MPI communicator group. Since each opened file is associated with a communicator, the probe will check for all the opened files.

– The I/O thread manages the local memory space allocation which includes creating/releasing buffers and resizing the status data.

3.3 Flushing Policy

Since our file consistency control ensures that only one copy of the file data can be buffered globally among processes and any read/write operations must check the status first, it is not necessary to flush buffered data prior to end of the run. In our implementation, buffered data can be explicitly flushed when file is closed or the file flushing call is made. Otherwise, implicit data flushing is needed only when the application runs out of memory, in which the memory space management facility handles the space overflow. Our design principle for data flushing includes: 1) declining buffering for overly large requests exceeding one-forth of entire memory size (direct read/write calls are made for such requests); 2) least-recent-used buffered data is flushed first (an accessing time stamp is associated with each data buffer); and 3) when flushing, all buffers are examined if any two buffers can be coalesced to reduce the number of write calls. For file systems that do not provide consistency automatically, we mimic the approach used in ROMIO that wraps byte-range file locking around each read/write call to disable client-side caching [10].

3.4 Incorporate into ROMIO

We place cooperative buffering at the ADIO layer of ROMIO to catch every read/write system call and determines whether the request should access the existing buffers or create a new buffer. ADIO is an abstract-device interface providing uniform and portable I/O interfaces for parallel I/O libraries [11]. This design preserves the existing optimizations used by ROMIO, such as two-phase I/O and data sieving, both implemented above ADIO [3,12]. In fact, cooperative buffering need not know if the I/O operation is collective or independent, since it only deals with system read/write calls.

4   Experimental Results

The evaluation of cooperative buffering implementation was performed using BTIO benchmark and FLASH I/O benchmark on the IBM SP machine at San Diego Supercomputing Center. The IBM SP contains 144 Symmetric Multiprocessing (SMP) compute nodes and each node is an eight-processor shared-memory machine. We use the IBM GPFS file system to store the files. The peak performance of the GPFS is 2.1 GBytes per second for reads and 1 GBytes per second for writes. The I/O will approximately max out at about 20 compute nodes. In order to simulate a distributed-memory environment, we ran the tests using one processor per compute node.

4.1   BTIO Benchmark

BTIO is the I/O benchmark from NASA Advanced Supercomputing (NAS) parallel benchmark suite (NPB 2.4) [13]. BTIO uses a block-tridiagonal (BT) partitioning

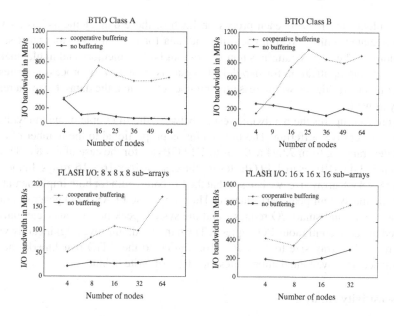

Fig. 2. I/O bandwidth results for BTIO and FLASH I/O benchmarks

pattern on a three-dimensional array across a square number of compute nodes. Each processor is responsible for multiple Cartesian subsets of the entire data set, whose number increases as the square root of the number of processors participating in the computation. BTIO provides four types of evaluations, each with different I/O implementations, including MPI collective I/O, MPI independent I/O, Fortran I/O, and separate-file I/O. In this paper, we only present the performance results for the type of using MPI collective I/O, since collective I/O generally results in the best performance [14]. The benchmark performs 40 collective MPI writes followed by 40 collective reads. We evaluated two I/O sizes: classes A and B, which generate I/O amount of 800 MBytes and 3.16 GBytes, respectively. Figure 2 compares the bandwidth results of using cooperative buffering with the native approach (without buffering.) We observe that cooperative buffering out-performs the native approach in most of the cases. Especially, when the number of compute nodes becomes large, cooperative buffering can achieve bandwidth near the system peak performance. The main contribution to this performance improvement is due to the effect of write behind and read from buffered data.

4.2 FLASH I/O Benchmark

FLASH is an AMR application that solves fully compressible, reactive hydrodynamic equations, developed mainly for the study of nuclear flashes on neutron stars and white dwarfs [15]. The FLASH I/O benchmark [16] uses HDF5 for writing checkpoints, but underneath is using MPI I/O for performing parallel reads and writes. The in-memory data structures are 3D sub-arrays of size $8 \times 8 \times 8$ or $16 \times 16 \times 16$ with a perimeter of four guard cells that are left out of the data written to files. In the simulation, 80 of these blocks are held by each processor. Each of these data elements has 24 variables associated with it. Within each file, the data for the same variable must stored contiguously. The access pattern is non-contiguous both in memory and in file, making it a challenging application for parallel I/O systems. Since every processor writes 80 FLASH blocks to file, as we increase the number of clients, the dataset size increases linearly as well.

Figure 2 compares the bandwidth results between the I/O implementation with and without cooperative buffering. The I/O amount is proportional to the number of compute nodes, ranging from 72.94 MBytes to 1.14 GBytes for the case of $8 \times 8 \times 8$ arrays and from 573.45 MBytes to 4.49 GBytes for the case of $16 \times 16 \times 16$ arrays. In the case of using $8 \times 8 \times 8$ array size, we can see that the I/O bandwidth for both implementations is far from the system peak performance. This is because FLASH I/O generates many non-contiguous and small I/O requests and the system peak performance can only be achieved by large contiguous I/O requests. The bandwidth improves significantly when we increase the array size to $16 \times 16 \times 16$. Similar to the BTIO benchmark, the I/O performance improvement demonstrates the effect of write behind.

4.3 Sensitivity Analysis

Due to the consistency control, cooperative buffering bears the cost of remotely and.or locally buffering status inquiry each time an I/O request is made. For the environment with a relative slow communication network or the I/O patterns with many small requests, this overhead may become significant. Another parameter that may affect the I/O performance is the file block size. The granularity of file block size determines the number of local/remote accesses generated from an I/O request. For different file access patterns, one file block size cannot always deliver the same performance enhancement. For the patterns with frequent and small amounts of I/O, using a large file block can cause contention when multiple requests access to the same buffers. On the other hand, if small file block size is used when the access pattern is less frequent and with large amount of I/O, a single large request can result in many remote data accesses. In most cases, such parameters can only be fine-tuned by the application users. In MPI, such user inputs usually are implemented through MPI_Info objects which, in our case, can also be used to activate or disable cooperative buffering.

5   Conclusions

Write-behind data buffering is known to be able to improve I/O performance by combining multiple small writes into large writes to be executed later. However, the overhead for write behind is the cost of maintaining file consistency. The cooperative buffering proposed in this paper addresses the consistency issue by coordinating application processes to manage buffering status data at file block level. This buffering scheme can benefit both MPI collective and independent I/O while the file open mode is no longer limited to write-only. The experimental results have shown a great improvement for two I/O benchmarks. In the future, we plan to investigate in depth the effect of the file block size and study irregular access patterns from scientific applications.

Acknowledgments

This work was supported in part by Sandia National Laboratories and DOE under Contract number 28264, DOE's SCiDAC program (Scientific Data Management Center), award number DE-FC02-01ER25485, NSF's NGS program under grant CNS-0406341, NSF/DARPA ST-HEC program under grant CCF-0444405, and NSF through the SDSC under grant ASC980038 using IBM DataStar.

References

1. Callaghan, B.: NFS Illustrated. Addison-Wesley (2000)
2. Ma, X., Winslett, M., Lee, J., Yu, S.: Improving MPI-IO Output Performance with Active Buffering Plus Threads. In: the International Parallel and Distributed Processing Symposium (IPDPS). (2003)
3. Thakur, R., Gropp, W., Lusk, E.: Users Guide for ROMIO: A High-Performance, Portable MPI-IO Implementation. Technical Report ANL/MCS-TM-234, Mathematics and Computer Science Division, Argonne National Laboratory. (1997)
4. Message Passing Interface Forum: MPI-2: Extensions to the Message Passing Interface. (1997) http :// www.mpi-forum.org / docs / docs.html.
5. Purakayastha, A., Ellis, C.S., Kotz, D.: ENWRICH: A Compute-Processor Write Caching Scheme for Parallel File Systems. In: the Fourth Workshop on Input/Output in Parallel and Distributed Systems (IOPADS). (1996)
6. Prost, J., Treumann, R., Hedges, R., Jia, B., Koniges, A.: MPI-IO/GPFS, an Optimized Implementation of MPI-IO on top of GPFS. In: Supercomputing. (2001)
7. Schmuck, F., Haskin, R.: GPFS: A Shared-Disk File System for Large Computing Clusters. In: the Conference on File and Storage Technologies (FAST'02). (2002) 231–244
8. Bernstein, P., Hadzilacos, V., Goodman, N.: Concurrency Control and Recovery in Database Systems. Addison-Wesley (1987)
9. IEEE/ANSI Std. 1003.1: Portable Operating System Interface (POSIX)-Part 1: System Application Program Interface (API) [C Language]. (1996)
10. Thakur, R., Gropp, W., Lusk, E.: On Implementing MPI-IO Portably and with High Performance. In: the Sixth Workshop on I/O in Parallel and Distributed Systems. (1999) 23–32
11. Thakur, R., Gropp, W., Lusk, E.: An Abstract-Device Interface for Implementing Portable Parallel-I/O Interfaces. In: the 6th Symposium on the Frontiers of Massively Parallel Computation. (1996)
12. Thakur, R., Gropp, W., Lusk, E.: Data Sieving and Collective I/O in ROMIO. In: the 7th Symposium on the Frontiers of Massively Parallel Computation. (1999)
13. Wong, P., der Wijngaart, R.: NAS Parallel Benchmarks I/O Version 2.4. Technical Report NAS-03-002, NASA Ames Research Center, Moffet Field, CA (2003)
14. Fineberg, S., Wong, P., Nitzberg, B., Kuszmaul, C.: PMPIO - A Portable Implementation of MPI-IO. In: the 6th Symposium on the Frontiers of Massively Parallel Computation. (1996)
15. Fryxell, B., Olson, K., Ricker, P., Timmes, F.X., Zingale, M., Lamb, D.Q., MacNeice, P., Rosner, R., Tufo, H.: FLASH: An Adaptive Mesh Hydrodynamics Code for Modelling Astrophysical Thermonuclear Flashes. Astrophysical Journal Suppliment (2000) 131–273
16. Zingale, M.: FLASH I/O Benchmark Routine – Parallel HDF 5 (2001) http://flash.uchicago.edu/~zingale/flash_benchmark_io.

Hint Controlled Distribution with Parallel File Systems

Hipolito Vasquez Lucas and Thomas Ludwig

Parallele und Verteilte Systeme, Institut für Informatik,
Ruprecht-Karls-Universität Heidelberg, 69120 Heidelberg, Germany
{hipolito.vasquez, thomas.ludwig}@informatik.uni-heidelberg.de

Abstract. The performance of scientific parallel programs with high file-I/O-activity running on top of cluster computers strongly depends on the qualitative and quantitative characteristics of the requested I/O-accesses. It also depends on the corresponding mechanisms and policies being used at the parallel file system level. This paper presents the motivation and design of a set of MPI-IO-hints. These hints are used to select the distribution function with which a parallel file system manipulates an opened file. The implementation of a new physical distribution function called `varstrip_dist` is also presented in this article. This function is proposed based upon spatial characteristics presented by I/O-access patterns observed at the application level.

1   Introduction

Hard disks offer a cost effective solution for secondary storage, but mainly due to mechanical reasons their access time has not kept pace with the speed development of processors. Disk and microprocessor performance have evolved at different rates [1]. This difference of development at the hardware level is one of the main causes of the so-called I/O-bottleneck problem [3] in disk-based computing systems.

The performance of I/O intensive scientific applications, which convey huge amounts of data between primary and secondary storage, suffers heavily due to this bottleneck. The performance of such an application depends on the I/O-subsystem architecture and on the corresponding usage of it, which is inherent to the application's nature.

In order to design computing systems with the cost effective advantages of hard disks and at the same time favor I/O intensive scientific applications, which run on top of such systems, the parallel I/O approach [4] has been adopted. This consists in arranging a set of disks over which files are striped or declustered [2]. By applying this mechanism, the applications take advantage of the resulting aggregated throughput.

A Beowulf cluster computer [5] in which many nodes have their own hard disk device inherently constitutes an appropriate hardware testbed for supporting parallel I/O. In order to make this parallelism, at the hardware level, visible to the applications, corresponding parallel I/O operations at the file system and

B. Di Martino et al. (Eds.): EuroPVM/MPI 2005, LNCS 3666, pp. 110–118, 2005.

middleware level must be supported. Two implementations which fullfill these tasks are the PVFS2 [6] parallel file system and the ROMIO [7] library, an implementation of MPI-2[16]. ROMIO accepts so-called hints that are communicated via the `info` argument in the functions `MPI_File_open`, `MPI_File_set_view`, and `MPI_File_set_info`. Their purpose is mainly to communicate information, which may improve the I/O-subsystem's performance. A hint is represented by a key-value pair mainly concerning parameters for striping, collective I/O, and access patterns.

In this work we propose a set of hints, which we call *distribution hints*. This set gives the user the opportunity to choose the type of physical distribution function [20] to be applied by the PVFS2 parallel file system for the manipulation of an opened file. After choosing the type of distribution function, the user can set its corresponding parameters. Assigning a value to the `strip_size`, for example, requires information on the type of distribution function to which this parameter belongs. To augment the set of distribution functions, which can be manipulated via distribution hints, we also propose the new `varstrip_dist` distribution for PVFS2. We propose this distribution function taking into consideration the characteristics of spatial I/O access patterns generated from scientific parallel applications. Through the usage of the varstrip distribution the programers can control the *throughput* or the *load balancing degree* in a PVFS2-ROMIO-based I/O subsytem, thus influencing the performance of their MPI-IO-based application.

2   Parallel I/O Access Patterns

2.1   Introduction

Our objective in this section is to present an abstract set of spatial I/O access patterns at the application level and their parameters. These patterns represent the assignation of storage areas of a logical file to monothreaded processes of a parallel program. This assignation is known as logical file partitioning [8]. The logical file can be interpreted as a one dimensional array of data blocks, whose smallest granularity is one byte. We use this set of patterns as a reference model, in order to propose distribution functions for the PVFS2 parallel file system.

We have summarized these characteristics based upon studies, which have been done on I/O intensive parallel scientific applications running mainly on multiprocessor systems [10], [12], [14]. These patterns depend on the application's nature [15], but they are also conditioned by the kind of application programming interface being used and furthermore by the way this interface is used. ROMIO's interface, for example, offers four different levels to communicate a request pattern to the I/O subsystem. Each of these levels might have different performance implications for the application [13].

2.2   Parameters

We use the following parameters to characterize a spatial I/O access pattern: *request size*, *type of operation*, and *sequentiality*.

Table 1. Relative Sizes of R

Condition	Relative Size
$R < M_{size} * 0.5$	Small
$M_{size} * 0.5 < R < M_{size}$	Medium
$R > M_{size}$	Big

We differentiate between *absolute* and *relative request sizes*. An absolute request size, R, is the requested number of bytes from the perspective of each involved process within a parallel program. R can be *uniform* or *variable* across processes. In order to express the *relative size* of R, we define M_{size} as the main memory size of the compute node, where the accessing process runs. Taking M_{size} as reference we distinguish the types of relative sizes shown in Table 1. Requests are also characterized by the type of operations they make. In this work we consider basically *read* and *write* operations.

The main criterion that we use to characterize the set of spatial access patterns used in this work is the *sequentiality* from the program's perspective. We consider especially two types: *partitioned* and *interleaved* [11]. A partitioned sequentiality appears when the processes collectively access the entire file in disjoint sequential segments. There is no common area in the file being used by two processes. The *interleaved* sequentiality appears when the accesses of every process are strided, or noncontiguous, to form a global sequential pattern.

2.3 Spatial Patterns

Figure 1 shows snapshots of five spatial patterns. The circles represent processes running within a common program that are accessing a common logical file, and the arrows mean any type of operation.

Pattern 0 represents a non-MPI parallel I/O to multiple files where every process is sequential with respect to I/O. This pattern has drawbacks such as a non-one logical view of the entire data set, a difficulty to manage the number of files, and a dependency on the number of original processes. Since it can be generated using language I/O [17], it will often be applied.

Patterns 1 through 4 are MPI parallel I/O variants. Their main advantage consists in offering the user a one logical view of the file. Patterns 1 and 3 fall into the category of global partitioned sequentiality, whereas 2 and 4 are variants of interleaved global sequentiality. Pattern 4 appears when each process accesses the file in a noncontiguous manner. This happens when parallel scientific applications access multidimensional data structures. It can be generated through calling the `darray` or the `subarray` function of the MPI-2 interface. We call Pattern 4 *irregular* because it is the result of irregularly distributed arrays. In such a pattern each process has a data array and a map array, which indicates the position in the file of the corresponding data in the data array. Such a pattern can be expressed using the MPI-2 interface through the `MPI_Type_create_indexed_block`. It can also unknowingly be generated by using `darray` in the cases where the size of

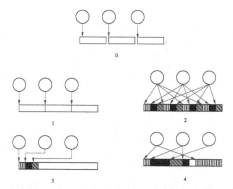

Fig. 1. Parallel I/O Application Patterns

the array in any dimension is not evenly divisible by the number of processes in that dimension. For this kind of access load balancing is an issue.

3 Distribution Functions in PVFS2

File distribution, physical distribution or simply distribution, is a set of methods describing a mapping from a logical sequence of bytes to a physical layout of bytes on PVFS2 I/O servers, which we here simply call I/O nodes. These functions are similar to declustering or striping methods used to scatter data across many disks such as in RAID 0 systems [18]. One of these functions is the round robin scheme, which is implemented in PVFS2.

In the context of PVFS2, the logical file consists of a set of *strip sizes*, *ss*, which are stored in a contiguous manner on I/O servers [9]. These strips are stored in *datafiles* [19] on I/O nodes through a distribution function.

4 A Set of Distribution Hints

To ease our discussion in this section we define an *I/O cluster* as a Beowulf cluster computer where every physical node has a secondary storage device.

The default distribution function in PVFS2 is the so called `simple_stripe`, which is a round robin mechanism, that uses a fixed value of 64KB for *ss*.

Suppose that PVFS2 is configured on an I/O cluster such that each node is a compute and I/O node at the same time and on top of this configuration an application generates pattern 1. Under these circumstances the simple stripe might penalize some strips by sending them over the network, thus slowing down I/O operations.

In this work we propose the *varstrip distribution*. Our approach consists in reproducing pattern 1 at each level of the software stack down to the raw hardware, thus the varstrip distribution does not scatter strips over I/O nodes in a RAID 0 manner, but instead it guarantees that each compute node accesses

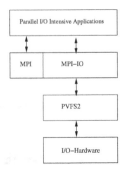

Fig. 2. Software Stack Environment for Distribution Hints

only its own local hard disk. Furthermore the strip size to be stored or retrieved on an I/O node can be defined. The varstrip distribution allows the definition of flexible strip sizes that can be assigned to a defined datafile number, thus influencing the load balancing degree among the different I/O servers.

In order to control the parameters of any distribution function from an MPI-Program, running on a similar software stack as that shown in figure 2, we introduce *distribution hints*. The purpose of such a hint is to select not only a type of distribution function, but also its parameters. The hint-key must have the following format: `<distribution name>:<parameter type>:<parameter name>`.

At the moment the user can choose, using this format, the following functions: `basic_dist`, `simple_stripe`, and `varstrip_dist`. By choosing the first one, the user saves the data on one single I/O node. The second applies the round robin mechanism with a strip size of 64 KB. These functions are already part of the standard set of distributions in PVFS2. By selecting our proposed `varstrip_dist` function the user can influence the throughput or the amount of data to be assign to the I/O nodes when manipulating an opened file.

In the hint-key the parameter name must be given with its type, in order for ROMIO and PVFS2 to manipulate it. Currently the `strip_size`, type `int64`, parameter for the `simple_stripe` is supported. The parameter `strips` is supported for `varstrip_dist`. This parameter represents the assignation between datafile numbers and strip sizes. The following piece of code shows the usage of `varstrip_dist`.

```
MPI_Info_set(theinfo,
            ''distribution_name'', ''varstrip_dist'')

/*Throughput */
MPI_Info_set(theinfo,
             ''varstrip_dist:string:strips'',''0:1000;1:1000'')

/*Load Balancing*/
MPI_Info_set(theinfo,
             ''varstrip_dist:string:strips'', ''0:8000;1:1000'')
```

5   Experiments

5.1   Testbed

The hardware testbed used for the implementation and tests was an I/O cluster consisting of 5 SMP nodes (master1, node01..node04). Each node had two Xeon hyper-threaded processors running at 2 Ghz, a main memory size of 1 GB, and an 80 GB hard disk. These nodes were networked using a store-and-forward Gigabit Ethernet switch.

The used operating system was linux with kernel 2.6.8. On top of this operating system we installed version 1.0.1 of PVFS2 and MPICH2. PVFS2 was running on top of an ext3 file system and every node was configured both as client and server. The node called master1 was configured as the metadata server.

5.2   Objective

The purpose of the measurements was to compare the bandwidth observed at the nodes when using the varstrip distribution with the bandwidth observed when using pattern 0 or two variants of the round robin PVFS2 distribution: the default distribution function with a strip size of 64KB and a variant which we called `simple_stripe` with *fitted* strip size. This variant resulted from setting the same value for R, ss, and datafile. When using the fitted `simple_stripe` a compute node did not necessarily access its own secondary storage device.

5.3   Measurements

Figures 3, 4, and 5 show the bandwidths, y-axes, calculated from the measured times before and after `MPI_File_write` or `MPI_File_read` operations. One single process was started per node. Each process made small, medium, read, and write R requests following pattern 1. The requests ($R < 1GB$) are shown on the x-axes.

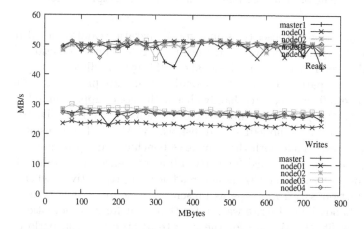

Fig. 3. Measured Bandwidth: Pattern 1, `varstrip_dist`

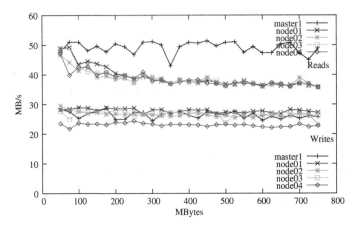

Fig. 4. Measured Bandwidth: Pattern 1, `simple_stripe`, fitted strip size

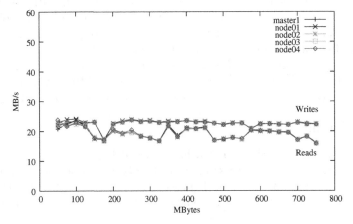

Fig. 5. Measured Bandwidth: Pattern 1, `simple_stripe`

For comparison purposes the same type of operations and values of R were requested at the application level using the unix `write` and `read` functions. The data was saved or retrieved to/from the local ext3 file system directly on the involved nodes following pattern 0. The corresponding values are presented in Figure 6.

For pattern 0 the measured bandwidth at the nodes approximately was of 50 MB/s and 40 MB/s for read and write operations respectively. The bandwidth for write operations of node01's hard disk was 30 MB/s. These results correlate with similar tests made with the `bonnie++` benchmarking program.

Using the values obtained for pattern 0 as reference, we obtained only 55% and 40% of performance for write and read accesses respectively when using the default function `simple_stripe` as presented in figure 5. It was the only case where the bandwidht of write was better than that for read operations.

With the fitted strip size for the `simple_stripe` function performances of approximately 75% and 80% were measured for write and read operations re-

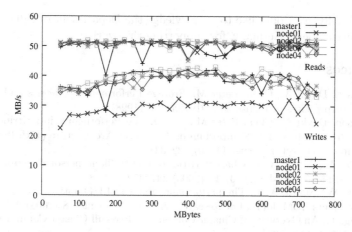

Fig. 6. Bandwidth obtained using the UNIX interface

spectively. Since the compute nodes were not necessarily using their own local hard disks, node04 accessed the hard disk of node01 during reading operations as shown in figure 4. The node master1, also the metadata server, used its own local disk during read operations.

Figure 3 presents the performance observed when using our proposed varstrip distribution. The bandwidth reached 80% and 100% of the reference bandwidth for write and read operations respectively.

6   Conclusion and Future Work

In this paper we have described a set of MPI-IO-hints, which the user can choose to select a certain distribution function of the PVFS2 parallel file system and its corresponding parameters. We have also described the varstrip distribution function. This function is proposed taking into consideration pattern 1, a parallel I/O spatial pattern, which appears at the application level. For this type of workload the varstrip distribution performs better than the other distribution functions, as shown through the experiments. Furthermore, by selecting varstrip the user can manipulate the load balancing degree among the I/O servers.

Our future work consists in implementing other distribution functions and constructing a matrix with pattern-distribution pairs, which will provide information about the functions best suited for particular application patterns. During this process we shall find out for which pattern and configuration the simple_stripe performs best and how well varstrip_dist performs with some other patterns as workload.

Acknowledgment

We thank Tobias Eberle and Frederik Grüll for the implementations, and Sven Marnach, our cluster administrator.

Additionally, we would like to acknowledge the Department of Education of Baden Württemberg, Germany, for supporting this work.

References

1. Patterson, David A., Chen, Peter M.: Storage Performance - Metrics and Benchmarks. http://citeseer.ist.psu.edu/91919.html. (1998)
2. Patterson, David A., Chen, Peter M.: Maximizing Performance in a Striped Disk Array. Proc. 17th Annual Symposium on Computer Architecture (17th ISCA'90), Computer Architecture News. (1990) 322–331
3. Hsu, W. W., Smith, A. J.: Characteristics of I/O traffic in personal computer and server workloads. IBM Syst. J. **42** (2003) 347–372
4. Hsu, W. W., Smith, A. J.: The performance impact of I/O optimizations and disk improvements. IBM Journal of Research and Development. **48** (2004) 255–289
5. Sterling, T.: An Overview of Cluster Computing. Beowulf Cluster Computing with Linux. (2002) 015–029
6. PVFS2 URL: http://www.pvfs.org/pvfs2/
7. ROMIO URL: http://www-unix.mcs.anl.gov/romio/
8. Ligon, W.B., Ross, R.B.: Implementation and Performance of a Parallel File System for High Performance Distributed Applications. Proceedings of the Fifth IEEE International Symposium on High Performance Distributed Computing. (1996) 471–480
9. Ross, Robert B., Carns, Philip H., Ligon III, Walter B., Latham, Robert: Using the Parallel Virtual File System. http://www.parl.clemson.edu/pvfs/user-guide.html (2002)
10. Madhyastha, Tara M.: Automatic Classification of Input/Output Access Patterns. PhD Thesis. (1997)
11. Madhyastha, Tara M., Reed, Daniel A.: Exploiting Global Input/Output Access Pattern Classification. Proceedings of SC97: High Performance Networking and Computing. (1997)
12. Thakur, Rajeev, Gropp, William, Lusk, Ewing: On implementing MPI-IO portably and with high performance. Proceedings of the 6th Workshop on I/O in Parallel and Distributed Systems (IOPADS-99). (1999) 23–32
13. Thakur, Rajeev S., Gropp, William, Lusk, Ewing: A Case for ung MPI's derived datatypes to improve I/O Performance. Proceedings of Supercomputing'98 (CD-ROM)". (1998)
14. Rabenseifner, Rolf, Koniges, Alice E., Prost, Jean-Pierre, Hedges, Richard: The Parallel Effective I/O Bandwidth Benchmark: b_eff_io. Parallel I/O for Cluster Computing. (2004) 107–132
15. Miller, Ethan L., Katz, Randy H.: Input/output behavior of supercomputing applications . SC. (1991) 567–576
16. MPI-2 URL: http://www.mpi-forum.org
17. Gropp, William, Lusk, Ewing, Thakur Rajeev: Using MPI-2: Advanced Features of the Message-Passing Interface. (1999) 15–16
18. Patterson, David, Gibson, Garth, Katz Randy: A case for redundant arrays of inexpensive disks (RAID). Proceedings of the ACM SIGMOD International Conference on Management of Data. (1988) 109–116
19. PVFS Development Team: PVFS 2 Concepts: the new guy's guide to PVFS. PVFS 2 Documentation (2004)
20. PVFS Development Team: PVFS 2 Distribution Design Notes. PVFS 2 Documentation. (2004)

Implementing Byte-Range Locks Using MPI One-Sided Communication

Rajeev Thakur, Robert Ross, and Robert Latham

Mathematics and Computer Science Division,
Argonne National Laboratory, Argonne, IL 60439, USA
{thakur, rross, robl}@mcs.anl.gov

Abstract. We present an algorithm for implementing byte-range locks using MPI passive-target one-sided communication. This algorithm is useful in any scenario in which multiple processes of a parallel program need to acquire exclusive access to a range of bytes. One application of this algorithm is for implementing MPI-IO's atomic-access mode in the absence of atomicity guarantees from the underlying file system. Another application is for implementing data sieving, a technique for optimizing noncontiguous writes by doing an atomic read-modify-write of a large, contiguous block of data. This byte-range locking algorithm can be used instead of POSIX `fcntl` file locks on file systems that do not support `fcntl` locks, on file systems where `fcntl` locks are unreliable, and on file systems where `fcntl` locks perform poorly. Our performance results demonstrate that the algorithm has low overhead and significantly outperforms `fcntl` locks on NFS file systems on a Linux cluster and on a Sun SMP.

1 Introduction

Often, processes must acquire exclusive access to a range of bytes. One application of byte-range locks is to implement the atomic mode of access defined in MPI-IO, the I/O interface that is part of MPI-2 [7]. MPI-IO, by default, supports weak consistency semantics in which the outcome of concurrent overlapping writes from multiple processes to a common file is undefined. The user, however, can optionally select stronger consistency semantics on a per file basis by calling the function `MPI_File_set_atomicity` with `flag=true`. In this mode, called the atomic mode, if two processes associated with the same open file write concurrently to overlapping regions of the file, the result is the data written by either one process or the other, and nothing in between.

In order to implement the atomic mode, either the underlying file system must provide functions that guarantee atomicity, or the MPI-IO implementation must ensure that a process has exclusive access to the portion of the file it needs to access [13]. Many POSIX-compatible file systems support atomicity for contiguous reads and writes, such as those issued by a single `read` or `write` function call, but some high-performance parallel file systems, such as PVFS [1] and PVFS2 [9], do not. MPI-IO's atomic mode supports atomicity even for noncontiguous file accesses that are made with a single MPI function call by using

B. Di Martino et al. (Eds.): EuroPVM/MPI 2005, LNCS 3666, pp. 119–128, 2005.

noncontiguous file views. No file system supports atomicity for noncontiguous reads and writes. For such accesses, the MPI-IO implementation must explicitly acquire exclusive access to the byte range being read or written by a process.

Another use of byte-range locks is to implement data sieving [14]. Data sieving is a technique for optimizing noncontiguous accesses. For reading, it involves reading a large chunk of data and extracting the necessary pieces from it. For writing, the process must read the large chunk of data from the file into a temporary buffer, copy the necessary pieces into the buffer, and then write it back. This read-modify-write must be done atomically to prevent other processes from writing to the same region of the file while the buffer is being modified in memory. Therefore, the process must acquire exclusive access to the range of bytes before doing the read-modify-write.

POSIX defines a function `fcntl` by which processes can acquire byte-range locks on an open file [5]. However, many file systems, such as PVFS [1], PVFS2 [9], some installations of NFS, and various research file systems [2,4,8], do not support `fcntl` locks. On some file systems, for example, some installations of NFS, `fcntl` locks are not reliable. In addition, on some file systems, the performance of `fcntl` locks is poor. Therefore, one cannot rely solely on `fcntl` for file locking.

In this paper, we present an algorithm for implementing byte-range locks that can be used instead of `fcntl`. This algorithm extends an algorithm we described in an earlier work [12] for acquiring exclusive access to an entire file (not a range of bytes). Both algorithms have some similarities with the MCS lock [6], an algorithm devised for efficient mutex locks in shared-memory systems, but differ from it in that they use MPI one-sided communication (which does not have atomic read-modify-write operations) and can be used on both distributed- and shared-memory systems. Byte-range locks add significant complications to the algorithm for exclusive access to an entire file [12], which is essentially just a mutex. Byte-range locks are an important improvement because they enable multiple processes to perform I/O concurrently to nonoverlapping regions of the file, a feature whole-file locks preclude.

The rest of this paper is organized as follows. In Section 2, we give a brief overview of MPI one-sided communication, particularly those aspects used in our algorithm. In Section 3, we describe the byte-range locking algorithm. In Section 4, we present performance results. In Section 5, we conclude with a brief discussion of future work.

2   One-Sided Communication in MPI

To enable one-sided communication in MPI, a process must first specify a contiguous memory region, called a *window*, that it wishes to expose to other processes for direct one-sided access. Each process in the communicator must call the function `MPI_Win_create` with the starting address of the local memory window, which could be `NULL` if the process has no memory to expose to one-sided communication. `MPI_Win_create` returns an opaque object, called a *window object*, which is used in subsequent one-sided communication functions.

Process 0	Process 1	Process 2
MPI_Win_create(&win)	MPI_Win_create(&win)	MPI_Win_create(&win)
MPI_Win_lock(shared,1)		MPI_Win_lock(shared,1)
MPI_Put(1)		MPI_Put(1)
MPI_Get(1)		MPI_Get(1)
MPI_Win_unlock(1)		MPI_Win_unlock(1)
MPI_Win_free(&win)	MPI_Win_free(&win)	MPI_Win_free(&win)

Fig. 1. An example of MPI one-sided communication with passive-target synchronization. Processes 0 and 2 perform one-sided communication on the window memory of process 1 by requesting shared access to the window. The numerical arguments indicate the target rank.

Three one-sided data-transfer functions are provided: MPI_Put (remote write), MPI_Get (remote read), and MPI_Accumulate (remote update). In addition, some mechanism is needed for a process to indicate when its window is ready to be accessed by other processes and to specify when one-sided communication has completed. For this purpose, MPI defines three synchronization mechanisms. The first two synchronization mechanisms require both the origin and target processes to call synchronization functions and are therefore called *active-target synchronization*. The third mechanism requires no participation from the target and is therefore called *passive-target synchronization*. We use this method in our byte-range locking algorithm because a process must be able to acquire a lock independent of any other process.

2.1 Passive-Target Synchronization

In passive-target synchronization, the origin process begins a synchronization epoch by calling MPI_Win_lock with the rank of the target process and indicating whether it wants shared or exclusive access to the window on the target. After issuing the one-sided operations, it calls MPI_Win_unlock, which ends the synchronization epoch. The target does not make any synchronization call. When MPI_Win_unlock returns, the one-sided operations are guaranteed to be completed at the origin and the target. Figure 1 shows an example of one-sided communication with passive-target synchronization.

An implementation is allowed to restrict the use of this synchronization method to window memory allocated with MPI_Alloc_mem. MPI_Win_lock is *not* required to block until the lock is acquired, except when the origin and target are one and the same process. In other words, MPI_Win_lock does not establish a critical section of code; it ensures only that the one-sided operations issued between the lock and unlock will be executed on the target window in a shared or exclusive manner (as requested) with respect to the one-sided operations from other processes.

2.2 Completion and Ordering

MPI puts, gets, and accumulates are nonblocking operations, and an implementation is allowed to reorder them within a synchronization epoch. They are guaranteed to be completed, both locally and remotely, only when the synchronization epoch has ended. In other words, a get operation is not guaranteed to see the data that was written by a put issued before it in the same synchronization epoch. Consequently, it is difficult to implement an atomic read-modify-write operation by using MPI one-sided communication [3]. One cannot simply do a lock-get-modify-put-unlock because the data from the get is not available until after the unlock. In fact, the MPI Standard defines such an operation to be erroneous (doing a put and a get to the same location in the window in the same synchronization epoch). One also cannot do a lock-get-unlock, modify the data, and then do a lock-put-unlock because the read-modify-write is no longer atomic. This feature of MPI complicates the design of a byte-range locking algorithm.

3   Byte-Range Locking Algorithm

In this section, we describe the design of the byte-range locking algorithm together with snippets of the code for acquiring and releasing a lock.

3.1   Window Layout

The window memory for the byte-range locking algorithm is allocated on any one process—in our prototype implementation, on rank 0. Other processes pass NULL to MPI_Win_create. All processes needing to acquire locks access this window by using passive-target one-sided communication. The window comprises three values for each process, ordered by process rank, as shown in Figure 2. The three values are a flag, the start offset for the byte-range lock, and the end offset. In our implementation, for simplicity, all three values are represented as integers. The window size, therefore, is 3 * sizeof(int) * nprocs. In practice, the flag could be a single byte, and the start and end offsets may each need to be eight bytes to support large file sizes.

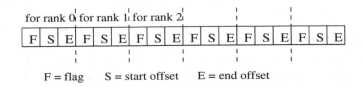

for rank 0 for rank 1 for rank 2

F = flag   S = start offset   E = end offset

Fig. 2. Window layout for the byte-range locking algorithm

3.2   Acquiring the Lock

The algorithm for acquiring a lock is as follows. The pseudocode is shown in Figure 3. The process wanting to acquire a lock calls MPI_Win_lock with the lock_type as MPI_LOCK_EXCLUSIVE, followed by an MPI_Put, an MPI_Get, and then MPI_Win_unlock. With the MPI_Put, the process sets its own three values in the window: It sets the flag to 1 and the start and end offsets to those needed

```
Lock_acquire(int start, int end)
{
  val[0] = 1; /* flag */  val[1] = start; val[2] = end;

  while (1) {
      /* add self to locklist */
      MPI_Win_lock(MPI_LOCK_EXCLUSIVE, homerank, 0, lockwin);
      MPI_Put(&val, 3, MPI_INT, homerank, 3*(myrank), 3, MPI_INT, lockwin);
      MPI_Get(locklistcopy, 3*(nprocs-1), MPI_INT, homerank, 0, 1, locktype1,
              lockwin);
      MPI_Win_unlock(homerank, lockwin);

      /* check to see if lock is already held */
      conflict = 0;
      for (i=0; i < (nprocs - 1); i++) {
          if ((flag == 1) && (byte ranges conflict with lock request)) {
              conflict = 1; break;
          }
      }

      if (conflict == 1) {
          /* reset flag to 0, wait for notification, and then retry the lock */
          MPI_Win_lock(MPI_LOCK_EXCLUSIVE, homerank, 0, lockwin);
          val[0] = 0;
          MPI_Put(val, 1, MPI_INT, homerank, 3*(myrank), 1, MPI_INT, lockwin);
          MPI_Win_unlock(homerank, lockwin);
          /* wait for notification from some other process */
      MPI_Recv(NULL, 0, MPI_BYTE, MPI_ANY_SOURCE, WAKEUP, comm,
                  MPI_STATUS_IGNORE);
          /* retry the lock */
      }
      else {
          /* lock is acquired */
          break;
      }
  }
}
```

Fig. 3. Pseudocode for obtaining a byte-range lock. The derived datatype `locktype1` is created at lock-creation time and cached in the implementation.

for the lock. With the MPI_Get, it gets the three values for all other processes (excluding its own values) by using a suitably constructed derived datatype, for example, an indexed type with two blocks. After MPI_Win_unlock returns, the process goes through the list of values returned by MPI_Get. For all other processes, it first checks whether the flag is 1 and, if so, checks whether there is a conflict between that process's byte-range lock and the lock it wants to acquire. If there is no such conflict with any other process, it considers the lock acquired. If a conflict (flag and byte range) exists with any process, it considers the lock as not acquired.

If the lock is not acquired, the process resets its flag in the window to 0 by doing an MPI_Win_lock–MPI_Put–MPI_Win_unlock and leaves its start and end offsets in the window unchanged. It then calls a zero-byte MPI_Recv with MPI_ANY_SOURCE as the source and blocks until it receives such a message from any other process (that currently has a lock; see the lock-release algorithm below). After receiving the message, it tries again to acquire the lock by using the above algorithm (further explained below).

```
Lock_release(int start, int end)
{
  val[0] = 0; val[1] = -1; val[2] = -1;

  /*set start and end offsets to -1, flag to 0, and get everyone else's status*/
  MPI_Win_lock(MPI_LOCK_EXCLUSIVE, homerank, 0, lockwin);
  MPI_Put(val, 3, MPI_INT, homerank, 3*(myrank), 3, MPI_INT, lockwin);
  MPI_Get(locklistcopy, 3*(nprocs-1), MPI_INT, homerank, 0, 1, locktype2,
          lockwin);
  MPI_Win_unlock(homerank, lockwin);

  /* check if anyone is waiting for a conflicting lock. If so, send them a
     0-byte message, in response to which they will retry the lock. For
     fairness, we start with the rank after ours and look in order. */

  i = myrank;  /* ranks are off by 1 because of the derived datatype */
  while (i < (nprocs - 1)) {
    /* the flag doesn't matter here. check only the byte ranges */
    if (byte ranges conflict) MPI_Send(NULL, 0, MPI_BYTE, i+1, WAKEUP, comm);
    i++;
  }
  i = 0;
  while (i < myrank) {
    if (byte ranges conflict) MPI_Send(NULL, 0, MPI_BYTE, i, WAKEUP, comm);
    i++;
  }
}
```

Fig. 4. Pseudocode for releasing a byte-range lock. The derived datatype `locktype2` is created at lock-creation time and cached in the implementation.

3.3 Releasing the Lock

The algorithm for releasing a lock is as follows. The pseudocode is shown in Figure 4. The process wanting to release a lock calls `MPI_Win_lock` with the lock_type as `MPI_LOCK_EXCLUSIVE`, followed by an `MPI_Put`, an `MPI_Get`, and then `MPI_Win_unlock`. With the `MPI_Put`, the process resets its own three values in the window: It resets its flag to 0 and the start and end offsets to −1. With the `MPI_Get`, it gets the start and end offsets for all other processes (excluding its own values) by using a derived datatype. This derived datatype could be different from the one used for acquiring the lock because the flags are not needed. After `MPI_Win_unlock` returns, the process goes through the list of values returned by `MPI_Get`. For all other processes, it checks whether there is a conflict between the byte range set for that process and the lock it is releasing. The flag is ignored in this comparison. For fairness, it starts with the next higher rank after its own, wrapping back to rank 0 as necessary. If there is a conflict with the byte range set by another process—meaning that process is waiting to acquire a conflicting lock—it sends a 0-byte message to that process, in response to which that process will retry the lock. After it has gone through the entire list of values and sent 0-byte messages to all other processes waiting for a lock that conflicts with its own, the process returns.

3.4 Discussion

In order to acquire a lock, a process opportunistically sets its flag to 1, before knowing whether it has got the lock. If it determines that it does not have the lock, it resets its flag to 0 with a separate synchronization epoch. Had we chosen the opposite approach, that is, set the flag to 0 initially and then set it to 1 after determining that the lock has been acquired, there could have been a race condition because another process could attempt the same operation between the two distinct synchronization epochs. The lack of an atomic read-modify-write operation in MPI necessitates the approach we use.

When a process releases a byte-range lock, multiple processes waiting on a conflicting lock may now be able to acquire their lock, depending on the byte range being released and the byte ranges those processes are waiting for. In the lock-release algorithm, we use the conservative method of making the processes waiting on a conflicting lock retry their lock instead of having the releasing process hand the lock to the appropriate processes directly. The latter approach can get fairly complicated for byte-range locks, and in Section 5 we describe some optimizations that we plan to explore.

If processes are multithreaded, the current design of the algorithm requires that either the user must ensure that only one thread calls the lock acquisition and release functions (similar to MPI_THREAD_SERIALIZED), or the lock acquisition and release functions themselves must acquire and release a thread mutex lock. We plan to extend the algorithm to allow multiple threads of a process to acquire and release nonconflicting locks concurrently.

The performance of this algorithm depends on the quality of the implementation of passive-target one-sided communication in the MPI implementation. In particular, it depends on the ability of the implementation to make progress on passive-target one-sided communication without requiring the target process to call MPI functions for progress. On distributed-memory environments, it is also useful if the implementation can cache derived datatypes at the target, so that the derived datatypes need not be communicated to the target each time.

4   Performance Evaluation

To measure the performance of our algorithm, we wrote two test programs: one in which all processes try to acquire a conflicting lock (same byte range) and another in which all processes try to acquire nonconflicting locks (different byte ranges). In both tests, each process acquires and releases the lock in a loop several times. We measured the time taken by all processes to complete acquiring and releasing all their locks and divided this time by the number of processes times the number of iterations. This measurement gave the average time taken by a single process for acquiring and releasing a single lock. We compared the performance of our algorithm with that using fcntl locks. We ran the tests on a Myrinet-connected Linux cluster at Argonne and on a 24-CPU Sun SMP at the University of Aachen in Germany.

On the Linux cluster, we used a beta version of MPICH2 1.0.2 with the GASNET channel running over GM. To measure the performance of fcntl locks, we used an NFS file system that mounted a GFS [10] backend. This is the only way to use fcntl locks on this cluster; the parallel file system on the cluster, PVFS, does not support fcntl locks. On the Sun SMP, we could not use Sun

MPI because of a bug in the implementation that caused one of our tests to hang when run with more than four processes. We instead used a beta version of MPICH2 1.0.2 with the sshm (scalable shared-memory) channel. For fcntl locks, we used an NFS file system. When we ran our test for conflicting locks with more than 16 processes, fcntl returned an error with errno set to "no record locks available." This is an example of the unreliability of fcntl locks with NFS, mentioned in Section 1.

On the Linux cluster, this version of MPICH2 has two limitations that can affect performance of the byte-range locking algorithm. One limitation is that MPICH2 requires the target process to call MPI functions in order to make progress on passive-target one-sided communication. This restriction did not affect our test programs because the target (rank 0) also tried to acquire byte-range locks and therefore made MPI function calls. Furthermore, all processes did an MPI_Barrier at the end, which also guaranteed progress at the target. The other limitation is that MPICH2 does not cache derived datatypes at the target process, so they need to be communicated each time. Both these limitations will be fixed in a future release of MPICH2. These limitations do not exist on the Sun SMP because when the window is allocated with MPI_Alloc_mem, the sshm channel in MPICH2 allocates the window in shared memory and implements puts and gets by directly copying data to/from the shared-memory window.

Figure 5 shows the average time taken by a single process to acquire and release a single lock on the Linux cluster and the Sun SMP. On the Linux cluster, for nonconflicting locks, our algorithm is on average about twice as fast as fcntl, and the time taken does not increase with the number of processes. For conflicting locks, the time taken by our algorithm increases with the number of processes because of the overhead induced by lock contention. In Section 5, we describe some optimizations we plan to incorporate that will reduce communication traffic in the case of conflicting locks and therefore improve performance and scalability. The graph for conflicting locks with fcntl on NFS on the Linux

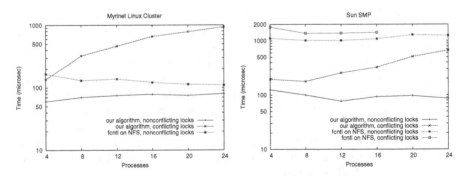

Fig. 5. Average time for acquiring and releasing a single lock on a Myrinet-connected Linux cluster (left) and a Sun SMP (right). The graph for conflicting locks with fcntl on the Linux cluster is not shown because the time was three orders of magnitude higher than the other results. On the Sun SMP, when run on more than 16 processes, fcntl on NFS for conflicting locks failed.

cluster is not shown because the time taken was on the order of seconds—about three orders of magnitude higher than any of the other results!

On the Sun SMP, for nonconflicting locks, our algorithm is about 10 times faster than `fcntl`, and the time taken does not increase with the number of processes. For conflicting locks, our algorithm is 5–9 times faster than `fcntl`. As mentioned above, `fcntl` on NFS for conflicting locks failed when run on more than 16 processes on the Sun SMP.

5   Conclusions and Future Work

We have presented an efficient algorithm for implementing byte-range locks using MPI one-sided communication. We have shown that our algorithm has low overhead and outperforms NFS `fcntl` file locks on the two environments we tested. We plan to use this algorithm for byte-range locking in our implementation of MPI-IO, called ROMIO [11].

This algorithm requires that the MPI implementation handle passive-target one-sided communication efficiently, which is not the case in many MPI implementations today. For example, with IBM MPI on the IBM SP at the San Diego Supercomputer Center, we observed wide fluctuations in the performance of our algorithm. We hope that such algorithms that demonstrate the usefulness of passive-target one-sided communication will spur MPI implementers to optimize their implementations.

While we have focused on making the algorithm correct and efficient, it can be improved in several ways. For example, in our current implementation of the lock-release algorithm, we make the processes waiting for conflicting locks retry their lock, instead of having the releasing process grant the locks directly. We chose this approach because, in general, the analysis required to determine which processes can be granted their locks is tricky. For special cases, however, such as processes waiting for a byte range that is a subset of the byte range being held by the releasing process, it is possible for the releasing process to grant the lock directly to the other process. With such an approach, the releasing process could grant the lock to processes for which it can easily determine that the lock can be granted, deny it to processes for which it can determine that the lock cannot be granted, and have others retry their lock. Preventing too many processes from retrying conflicting locks will improve the performance and scalability of the algorithm significantly. Another optimization for scalability is to replace the linear list of values with a tree-based structure, so that a process does not have to fetch and check the values of all other processes. We plan to explore such optimizations to the algorithm.

Acknowledgments

This work was supported by the Mathematical, Information, and Computational Sciences Division subprogram of the Office of Advanced Scientific Computing Research, Office of Science, U.S. Department of Energy, under Contract

W-31-109-ENG-38. We thank Chris Bischof for giving us access to the Sun SMP machines at the University of Aachen.

References

1. Philip H. Carns, Walter B. Ligon III, Robert B. Ross, and Rajeev Thakur. PVFS: A parallel file system for Linux clusters. In *Proceedings of the 4th Annual Linux Showcase and Conference, Atlanta*, pages 317–327, October 2000.
2. Peter F. Corbett and Dror G. Feitelson. The Vesta parallel file system. *ACM Transactions on Computer Systems*, 14(3):225–264, August 1996.
3. William Gropp, Ewing Lusk, and Rajeev Thakur. *Using MPI-2: Advanced Features of the Message-Passing Interface*. MIT Press, Cambridge, MA, 1999.
4. Jay Huber, Christopher L. Elford, Daniel A. Reed, Andrew A. Chien, and David S. Blumenthal. PPFS: A high performance portable parallel file system. In *Proceedings of the 9th ACM International Conference on Supercomputing*, pages 385–394. ACM Press, July 1995.
5. IEEE/ANSI Std. 1003.1. Portable Operating System Interface (POSIX)–Part 1: System Application Program Interface (API) [C Language], 1996 edition.
6. J. M. Mellor-Crummey and M. L. Scott. Algorithms for scalable synchronization on shared-memory multiprocessors. *ACM Transactions on Computer Systems*, 1991.
7. Message Passing Interface Forum. MPI-2: Extensions to the Message-Passing Interface, July 1997. http://www.mpi-forum.org/docs/docs.html.
8. Nils Nieuwejaar and David Kotz. The Galley parallel file system. *Parallel Computing*, 23(4):447–476, June 1997.
9. PVFS2: Parallel virtual file system. http://www.pvfs.org/pvfs2/.
10. Red Hat Global File System. http://www.redhat.com/software/rha/gfs.
11. ROMIO: A high-performance, portable MPI-IO implementation. http://www.mcs.anl.gov/romio.
12. Robert Ross, Robert Latham, William Gropp, Rajeev Thakur, and Brian Toonen. Implementing MPI-IO atomic mode without file system support. In *Proceedings of CCGrid 2005*, May 2005.
13. Rajeev Thakur, William Gropp, and Ewing Lusk. On implementing MPI-IO portably and with high performance. In *Proceedings of the 6th Workshop on I/O in Parallel and Distributed Systems*, pages 23–32. ACM Press, May 1999.
14. Rajeev Thakur, William Gropp, and Ewing Lusk. Optimizing noncontiguous accesses in MPI-IO. *Parallel Computing*, 28(1):83–105, January 2002.

An Improved Algorithm for (Non-commutative) Reduce-Scatter with an Application

Jesper Larsson Träff

C&C Research Laboratories, NEC Europe Ltd,
Rathausallee 10, D-53757 Sankt Augustin, Germany
traff@ccrl-nece.de

Abstract. The collective *reduce-scatter operation* in MPI performs an element-wise reduction using a given associative (and possibly commutative) binary operation of a sequence of m-element vectors, and distributes the result in m_i sized blocks over the participating processors. For the case where the number of processors is a power of two, the binary operation is commutative, and all resulting blocks have the same size, efficient, butterfly-like algorithms are well-known and implemented in good MPI libraries.

The contributions of this paper are threefold. First, we give a simple trick for extending the butterfly algorithm also to the case of non-commutative operations (which is advantageous also for the commutative case). Second, combining this with previous work, we give improved algorithms for the case where the number of processors is not a power of two. Third, we extend the algorithms also to the *irregular* case where the size of the resulting blocks may differ extremely.

For p processors the algorithm requires $\lceil \log_2 p \rceil + (\lceil \log_2 p \rceil - \lfloor \log_2 p \rfloor)$ communication rounds for the regular case, which may double for the irregular case (depending on the amount of irregularity). For vectors of size m with $m = \sum_{i=0}^{p-1} m_i$ the total running time is $O(\log p + m)$, irrespective of whether the m_i blocks are equal or not. The algorithm has been implemented, and on a small Myrinet cluster gives substantial improvements (up to a factor of 3 in the experiments reported) over other often used implementations. The reduce-scatter operation is a building block in the *fence* one-sided communication synchronization primitive, and for this application we also document worthwhile improvements over a previous implementation.

1 Introduction

MPI_Reduce_scatter is the slightly exotic collective reduction operation of the MPI standard [8]. Each involved MPI process contributes an m-element vector which is element-wise reduced using a given, associative (and possibly commutative) binary reduction operation with the result scattered in blocks of given (possibly different) sizes of m_i elements across the processes. The involved p processes are numbered from 0 to $p-1$, and process i receives the ith m_i-element block of the m-element result vector with $m = \sum_{i=0}^{p-1} m_i$. The associative, binary

B. Di Martino et al. (Eds.): EuroPVM/MPI 2005, LNCS 3666, pp. 129–137, 2005.

reduction operation can be either an MPI predefined operation (like MPI_SUM), in which case it is (mathematically) commutative, or a user-defined operation, in which case the user must inform the MPI library whether it is commutative or only associative. In an MPI_Reduce_scatter call each process must specify the same reduction operation.

In order to ensure consistent results with finite-precision arithmetic MPI poses certain requirements to the collective reduction operations which in turn have consequences for the algorithms that can be used. A canonical reduction order is defined for associative operations, and preferred even for commutative operations [8, p. 228]. A natural (but not explicitly stated) requirement is that all elements of the result vector are computed in the same way, that is with the same order and bracketing. The algorithm presented in this paper meets both these requirements (as do the reduction algorithms in [7]), and in fact does not distinguish between commutative and non-commutative operations. Since it is at least as good as all previously published results (in the same cost model), we claim that commutativity gives no advantage to the implementer of the MPI reduction operations. The reduce-scatter algorithm has the following properties:

- It has the same complexity for commutative and non-commutative (associative) binary reduction operations, and reduction is performed in canonical order with the same bracketing for each result vector element.
- It provides an efficient solution also for the case where the number of processes is not a power of two.
- As the first, it solves the reduce-scatter problem efficiently also for the *irregular* case where the size of the result blocks differs.

The algorithm requires $\lceil \log_2 p \rceil + (\lceil \log_2 p \rceil - \lfloor \log_2 p \rfloor)$ communication rounds for the regular case, where the $(\lceil \log_2 p \rceil - \lfloor \log_2 p \rfloor)$ term accounts for an extra round needed to send the result to processes in excess of $2^{\lfloor \log_2 p \rfloor}$ (for the non-power of two case). Each process sends and receives less than m vector elements, and the total computation effort is $O(m)$. For the irregular case where the m_is are different, the number of communication rounds may double, depending on the amount of irregularity, and another m elements may have to be sent and received. Overall, the time taken by the reduce-scatter algorithm is $O(\log p + m)$, assuming a linear communication cost model in which sending and receiving m elements takes time $\alpha + \beta m$, as well as linear time for performing a binary reduction of two m-element vectors.

Recent work [9] on implementing MPI_Reduce_scatter for mpich2 in the same cost model uses a "recursive halving" algorithm which works only for commutative operations. The irregular case is not addressed – in that case the running time becomes $O(m \log p)$ – and a less efficient solution for the non-power of two case is used. An algorithm that is optimal in the more refined *LogGP* cost model for the regular case with commutative reduction operations is developed, analyzed and implemented in [1,5]. For non-commutative operations a (theoretically) significantly less efficient algorithm is provided, and the irregular case is not addressed at all. It is by the way curious that the MPI standard defines only an irregular version of the reduce-scatter collective, and not a regular *and* an irregular version as for most other collectives [8, Chapter 4].

The reduce-scatter operation is convenient for computing vector-matrix products for distributed matrices and vectors. The reduce-scatter operation is also used for MPI internal purposes, for instance for *fence* synchronization in the MPI-2 one-sided communication model [3, Chapter 4], see [10,12] and Section 4. Another recent application is generation of file recovery data as described in [4]. In both of these applications a regular reduce-scatter computation is called for.

2   The Algorithm

We first describe a standard butterfly algorithm for the reduce-scatter collective for the case where p is a power of two, assuming furthermore that each result block m_i has the same size. We denote m-element vectors divided into p *blocks* of m_i elements by $B = (B_0, B_1, \ldots, B_{p-1})$. Each process maintains a reduction buffer B which during any stage of the algorithm will contain a partial result for some of the p blocks. Initially, each process copies its input vector into its reduction buffer. The algorithm runs in $\log_2 p$ rounds. In round k, $k = 0, \ldots, p-1$ each process i exchanges data with process $j = i \oplus 2^k$, where $\oplus$ denotes the *exclusive-or* operation. If bit k of i is 0 (and bit k of j is therefore 1), process i sends the upper half of its reduction buffer to process j, and receives the lower half of process js reduction buffer. After the exchange each process invokes the binary reduction operation on the upper/lower half of its reduction buffer; to ensure canonical order, reduction is performed as $f(B, R)$ if $i < j$, and as $f(R, B)$ if $i > j$, where R are the elements received. Thus, if process i had a partial result of $p/2^k$ blocks *before* round k, it has a partial result of $p/2^{k+1}$ blocks *after* round k, and thus exactly *one* block of m_i elements after the $\log_2 p$ rounds, which is the result for process i.

Let the blocks of process i before round k be $(B_a, B_{a+1}, \ldots, B_{a+p/2^k-1})$. After round k, process i has the lower blocks $(B_a, B_{a+1}, \ldots, B_{a+p/2^{k+1}-1})$ of the partial result if bit k of i is 0, and the upper blocks $(B_{a+p/2^{k+1}}, \ldots, B_{a+p/2^k-1})$ if bit k of i is 1. As can be seen, process i receives upper/lower blocks in the reverse order of the bits of i. Thus, after the $\log_2 p$ rounds the block of process i is $B_{\sigma^{\log_2 p}(i)}$ where $\sigma^n(i)$ denotes the reverse (or *mirror*) permutation of an n-bit number i: bit 0 (least significant bit) of i becomes bit $n-1$, bit 1 of i becomes bit $n-2$ and so on. Note that the mirror permutation is idempotent, eg. $\sigma^n(\sigma^n(i)) = i$. It is also used in eg. FFT [6].

An example showing the result after 3 rounds for $p = 8$ is given in Figure 1. Each process ends up with one block of $m/2^{\log_2 p} = m_i$ elements of the result, each computed in canonical order and with the same bracketing, but for some of the processes the resulting block is not the block intended for that process: instead of block B_i for process i, process i has block $B_{\sigma^3(i)}$.

To remedy this situation and get the correct result the blocks that are at their incorrect positions have to be permuted. In Figure 1, processes 1 and 4, and processes 3 and 6 must exchange their blocks. The permutation to be used is σ^3. This costs (for some of the processes) an extra communication round.

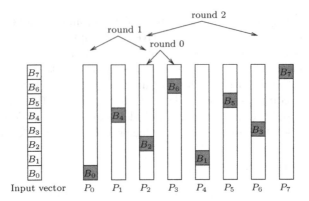

Fig. 1. Result of the butterfly reduce-scatter algorithm for $p = 8$ without permutation. Double arrows show the exchanges performed by process 2. The result blocks B_4 and B_1, and B_6 and B_3 are at incorrect processes.

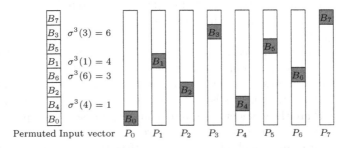

Fig. 2. Result of the butterfly reduce-scatter algorithm for $p = 8$ with mirror permutation before reduction. All result blocks end at their correct processes.

It is now easy to see that the following trick resolves the situation *without* extra communication. Instead of doing the mirror permutation *after* the reduction, the blocks can be permuted into their right order *before* the reduction starts: each processor puts block k in position $\sigma^{\log_2 p}(k)$. So instead of running the butterfly algorithm on input buffers $B = (B_0, B_1, \ldots, B_{p-1})$, the algorithm is run on the permuted buffer $B' = (B_{\sigma(0)}, B_{\sigma(1)}, B_{\sigma(2)}, \ldots, B_{\sigma(p-1)})$. This is illustrated for $p = 8$ in Figure 2.

To avoid having to compute $\sigma^{\log_2 p}$ repeatedly, a permutation table can be precomputed whenever a new MPI communicator is created. For the regular reduce-scatter problem where the m_is are (roughly) equal, the running time is $O(\log p + m)$.

2.1 The Non-power of Two Case

The butterfly exchange scheme above assumes that p is a power of two. A simple way of handling the non-powers of two case is to first eliminate processes in excess of $2^{\lfloor \log_2 p \rfloor}$ by letting such processes send their input to neighboring processes. The $2^{\lfloor \log_2 p \rfloor}$ remaining processes then perform the butterfly algorithm, and at the end some send a result back to the eliminated processes. This simple solution has the drawback that the eliminated processes stay idle for most of the reduce-scatter computation, and that some processes reduce an extra full m element vector. Furthermore, the size of the blocks B_i of processes that are neighbors to eliminated processes double. Thus, from a regular reduce-scatter problem, an irregular problem arises when p is not a power of two!

By using the reduction order and handling of excess processes as described in [7], the amount of extra reduction work can be reduced by at least a factor of 2, and the excess processes can be kept busy with useful work much longer. Finally, the increase in block size can be reduced from a factor of two to a factor of $3/2$. Due to space limitations, the exact details are not given here.

The number of rounds for the improved algorithm is $\lceil \log_2 p \rceil + (\lceil \log_2 p \rceil - \lfloor \log_2 p \rfloor)$, and the total number of elements to be reduced by any process is $< 3/2m$ (down from $2m$ for the naive solution) when p is not a power of two, and m for the power-of-two case.

2.2   The Irregular Case

The mirror permutation idea can also be used for the irregular reduce-scatter problem where the block sizes m_i differ to ensure that each process ends up with its correct result block without extra communication rounds. However, since the blocks B_i are treated as atomic entities that are not divided, the running time can be severely affected. In the extreme case where one only process gets a result block ($m_i = 0$ for all except one process) the running time of the algorithm is $O(m \log p)$, which is not acceptable for large p. We are striving for an improvement from $O(\max(\log_2 p + m, \max_i(m_i \log_2 p)))$ to $O(\log_2 p + m)$.

To achieve this, a new, *regular problem B'* is created with m'_i roughly equal to m/p. To get as many elements to end up at their correct processes (without extra communication) the elements of the blocks of the original problem $B = (B_0, B_1, \ldots, B_{p-1})$ are assigned to the blocks of the regular problem as far as possible: for blocks with $m_i > m'_i$ the excess elements will be assigned later to other blocks B'_j for which $m_j < m'_j$, and for these excess elements some rerouting at the end of the butterfly algorithm will be necessary.

After filling the blocks B'_i of the regular problem, these blocks are permuted according to the mirror permutation σ as described in Section 2. The excess elements are assigned to the permuted input vector $B' = (B'_{\sigma(0)}, B'_{\sigma(1)}, \ldots, B_{\sigma(p-1)})$ with the kth excess element put into the kth free slot. This is illustrated in Figure 3. The butterfly algorithm is then executed to bring all result elements in blocks with $m_i \leq m/p$ to their correct processes. To reroute the excess elements, the butterfly communication pattern is used in reverse. In step k with $k = \lceil \log_2 p \rceil - 1, \ldots, 0$, a process j which has excess elements decides how to route as follows: if excess element x belongs to block B_i, and the kth bit of i equals the kth bit of $j \oplus 2^k$, then x is sent to the process $j \oplus 2^k$ with which j communicates in step k (the reverse butterfly pattern). The rerouting for the irregular

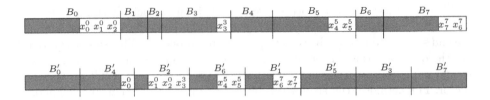

Fig. 3. Permuted blocks and excess elements of the new regular problem $B' = (B_0', B_4', B_2', B_6', B_1', B_5', B_3', B_7')$ (bottom) created to solve the original, irregular problem $B = (B_0, B_1, B_2, B_3, B_4, B_5, B_6, B_7)$ (top) with minimal rerouting. Excess element k belonging to block B_i is denoted by x_k^i.

	P_0	P_1	P_2	P_3	P_4	P_5	P_6	P_7
After butterfly		x_6^7, x_7^7	x_1^0, x_2^0, x_3^3		x_0^0		x_4^5, x_5^5	
After step 2	$\mathbf{x_0^0}$					x_6^7, x_7^7		
After step 1	$\mathbf{x_1^0, x_2^0}$				x_4^5, x_5^5			$\mathbf{x_6^7, x_7^7}$
After step 0				$\mathbf{x_3^3}$		$\mathbf{x_4^5, x_5^5}$		

Fig. 4. Rerouting of excess elements of the irregular reduce-scatter problem of Figure 3. When excess elements have reached their final destination they are shown in bold.

problem of Figure 3 is shown in Figure 4. It is clear that at most $\lfloor \log_2 \rfloor$ extra communication rounds are needed, and that in total $O(m)$ data are exchanged. It can furthermore be proved that in each communication step a process either sends or receives excess elements (or does nothing), that the excess elements end up at their correct processes at the end of the rerouting phase, and that excess elements are received in consecutive segments, so that no further reorganization is needed at any step of the rerouting.

3   Performance

The new reduce-scatter algorithm, including the handling of non-power of two number of processors and the possibly extra routing for the irregular case has been implemented in a state-of-the-art MPI implementation [2]. Since rerouting is expensive, this takes effect only from a certain message threshold, and only from a certain degree of irregularity.

Experiments have been carried out on a 32-node AMD Athlon based cluster with Myrinet interconnect. We compare three implementations of MPI_Reduce_scatter:

1. The first, trivial implementation is an MPI_Reduce (to intermediate root 0) followed by an MPI_Scatterv, using a simple binomial tree algorithm for reduce and a linear MPI_Scatterv algorithm.
2. The second implementation is also reduce followed scatter, but with optimized algorithms for both MPI_Reduce [7] and MPI_Scatterv [11].
3. The third algorithm is the direct algorithm described in this paper.

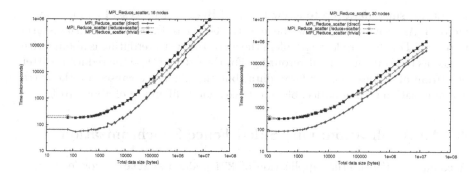

Fig. 5. Performance of the three reduce-scatter algorithms on a regular problem, 16 nodes (left) and 30 nodes (right). Plot is doubly logarithmic.

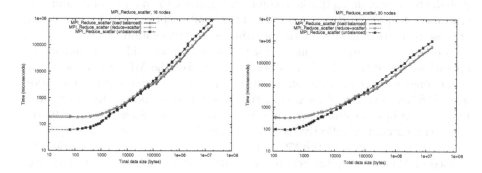

Fig. 6. Irregular problem with whole result at one process, 16 nodes (left) and 30 nodes (right). Plot is doubly logarithmic.

In Figure 5 the performance of the three algorithms for *regular* reduce-scatter problems, ie. equal block sizes m_i, is shown. Running time is given as a function of the total problem size $m = \sum_{i=0}^{p-1} m_i$ for $p = 16$ and $p = 30$ processes with one process per node. For the non-power of two case $p = 30$ the new algorithm is a factor 3 faster than the trivial solution for both small (latency) and large data (bandwidth). Compared to the second implementation, the new algorithm is almost a factor 1.5 faster for large data, and again a factor 3 faster for small data. The results for $p = 16$ are similar.

To test the performance in the irregular case we consider for total problem size m an extreme case with $m_{p/2} = m$ and all other $m_i = 0$. The results are shown in Figure 6 again for $p = 16$ and $p = 30$. We compare the new algorithm with rerouting to the new algorithm without rerouting, and to the second algorithm listed above. For the extreme case where one process (different from 0) receives all data, this algorithm requires only one more communication round than the new algorithm, namely for sending the resulting m-vector from process 0 to process $m_{p/2}$. As can clearly be seen, rerouting is too expensive for small data: up to about $m = 60$KBytes the new algorithm without rerouting is fastest, for small data (up to a few hundred Bytes per process) by more than a factor 3. As problem size grows beyond the 60KByte threshold rerouting starts to pay off. For the very large problem the algorithm with rerouting is about twice as fast as the version without rerouting. For the extreme case the reduce+scatter algorithm also does well for large data, but the difference caused by the extra communication step is noticeable and makes for a difference of about 10%.

4   An Application: One-Sided Fence Synchronization

A common, MPI internal application of MPI_Reduce_scatter is for fence synchronization in the MPI-2 one-sided communication model [3, Chapter 4]. The MPI_Win_fence synchronization primitive must ensure for each process that all one-sided communication operations with that process as target has been completed. One way of doing this is to count, per process, the number of pending operations to be processed at the MPI_Win_fence call, and then wait for completion of these requests. For the counting a reduce-scatter operation can be used with each process supplying as input vector the number of one-sided communication operations issued to each of the other processes. For the MPI/SX implementation of one-sided communication [12] a special-purpose reduce-scatter operation is used, which is now implemented by the algorithm described in this paper. Also in the recent implementation of one-sided communication for mpich2 in [10], the MPI_Reduce_scatter collective is used for completion (message) counting.

In Table 1 the cost of the MPI_Win_fence call with the implementation described in [12] and using the new reduce-scatter algorithm is given. An improvement of up to 20% over the previous MPI_Win_fence implementation using a different reduce-scatter implementation is achieved.

Table 1. Time for MPI_Win_fence operations opening and closing an epoch for varying number of processes

Processes:	4	8	12	16	20	24	28	30
Old	65.95	104.08	120.89	136.89	164.91	165.93	180.01	182.98
New	49.11	78.05	110.99	110.11	141.95	142.01	148.97	150.08

5   Concluding Remarks

We gave a new, scalable, more efficient algorithm for the collective reduce-scatter operation for both commutative and *non-commutative* (associative) binary reduction operations. The algorithm is the first to address efficiently the irregular case in which the size of the resulting blocks differ (substantially). The trick underlying the algorithm was a reordering of the blocks to fit the butterfly communication pattern. An open question is whether this trick can also be of use for the *LogGP* algorithms in [5].

References

1. M. Bernaschi, G. Iannello, and M. Lauria. Efficient implementation of reduce-scatter in MPI. Technical report, University of Napoli, 1997.
2. M. Gołebiewski, H. Ritzdorf, J. L. Träff, and F. Zimmermann. The MPI/SX implementation of MPI for NEC's SX-6 and other NEC platforms. *NEC Research & Development*, 44(1):69–74, 2003.
3. W. Gropp, S. Huss-Lederman, A. Lumsdaine, E. Lusk, B. Nitzberg, W. Saphir, and M. Snir. *MPI – The Complete Reference*, volume 2, The MPI Extensions. MIT Press, 1998.
4. W. D. Gropp, R. Ross, and N. Miller. Providing efficient I/O redundancy in MPI environments. In *11th European PVM/MPI Users' Group Meeting*, volume 3241 of *Lecture Notes in Computer Science*, pages 77–86, 2004.
5. G. Iannello. Efficient algorithms for the reduce-scatter operation in LogGP. *IEEE Transactions on Parallel and Distributed Systems*, 8(9):970–982, 1997.
6. F. T. Leighton. *Introduction to Parallel Algorithms and Architechtures: Arrays, Trees, Hypercubes*. Morgan Kaufmann Publishers, 1992.
7. R. Rabenseifner and J. L. Träff. More efficient reduction algorithms for message-passing parallel systems. In *11th European PVM/MPI Users' Group Meeting*, volume 3241 of *Lecture Notes in Computer Science*, pages 36–46, 2004.
8. M. Snir, S. Otto, S. Huss-Lederman, D. Walker, and J. Dongarra. *MPI – The Complete Reference*, volume 1, The MPI Core. MIT Press, second edition, 1998.
9. R. Thakur, W. D. Gropp, and R. Rabenseifner. Improving the performance of collective operations in MPICH. *International Journal on High Performance Computing Applications*, 19:49–66, 2004.
10. R. Thakur, W. D. Gropp, and B. Toonen. Minimizing synchronization overhead in the implementation of MPI one-sided communication. In *11th European PVM/MPI Users' Group Meeting*, volume 3241 of *Lecture Notes in Computer Science*, pages 57–67, 2004.
11. J. L. Träff. Hierarchical gather/scatter algorithms with graceful degradation. In *International Parallel and Distributed Processing Symposium (IPDPS 2004)*, 2004.
12. J. L. Träff, H. Ritzdorf, and R. Hempel. The implementation of MPI-2 one-sided communication for the NEC SX-5. In *Supercomputing*, 2000. http://www.sc2000.org/proceedings/techpapr/index.htm#01.

Collective Error Detection for MPI Collective Operations*

Chris Falzone[1], Anthony Chan[2], Ewing Lusk[2], and William Gropp[2]

[1] University of Pennsylvania at Edinboro, Edinboro, Pennsylvania 16444
[2] Mathematics and Computer Science Division,
Argonne National Laboratory, Argonne, Illinois 60439

Abstract. An MPI profiling library is a standard mechanism for intercepting MPI calls by applications. Profiling libraries are so named because they are commonly used to gather performance data on MPI programs. Here we present a profiling library whose purpose is to detect user errors in the use of MPI's collective operations. While some errors can be detected locally (by a single process), other errors involving the consistency of arguments passed to MPI collective functions must be tested for in a collective fashion. While the idea of using such a profiling library does not originate here, we take the idea further than it has been taken before (we detect more errors) and offer an open-source library that can be used with any MPI implementation. We describe the tests carried out, provide some details of the implementation, illustrate the usage of the library, and present performance tests.

Keywords: MPI, collective, errors, datatype, hashing.

1 Introduction

Detection and reporting of user errors are important components of any software system. All high-quality implementations of the Message Passing Interface (MPI) Standard [6,2] provide for runtime checking of arguments passed to MPI functions to ensure that they are appropriate and will not cause the function to behave unexpectedly or even cause the application to crash. The MPI collective operations, however, present a special problem: they are called in a coordinated way by multiple processes, and the Standard mandates (and common sense requires) that the arguments passed on each process be consistent with the arguments passed on the other processes. Perhaps the simplest example is the case of MPI_Bcast:

```
MPI_Bcast(buff, count, datatype, root, communicator)
```

in which each process must pass the same value for root. In this case, "consistent" means "identical," but more complex types of consistency exist. No

---

* This work was supported by the Mathematical, Information, and Computational Sciences Division subprogram of the Office of Advanced Scientific Computing Research, Office of Science, U.S. Department of Energy, SciDAC Program, under Contract W-31-109-ENG-38.

B. Di Martino et al. (Eds.): EuroPVM/MPI 2005, LNCS 3666, pp. 138–147, 2005.

single process by itself can detect inconsistency; the error check itself must be a collective operation.

Fortunately, the MPI profiling interface allows one to intercept MPI calls and carry out such a collective check before carrying out the "real" collective operation specified by the application. In the case of an error, the error can be reported in the way specified by the MPI Standard, still independently of the underlying MPI implementation, and without access to its source code.

The profiling library we describe here is freely available as part of the MPICH2 MPI-2 implementation [4]. Since the library is implemented entirely as an MPI profiling library, however, it can be used with any MPI implementation. For example, we have tested it with the IBM MPI implementation for Blue Gene/L [1].

The idea of using the MPI profiling library for this purpose was first presented by Jesper Träff and Joachim Worringen in [7], where they describe the error-checking approach taken in the NEC MPI implementation, in which even local checks are done in the profiling library, some collective checks are done portably in a profiling library as we describe here, and some are done by making NEC-specific calls into the proprietary MPI implementation layer. The datatype consistency check in [7] is only partial, however; the sizes of communication buffers are checked, but not the details of the datatype arguments, where there is considerable room for user error. Moreover, the consistency requirements are not on the datatypes themselves, but on the datatype *signatures*; we say more about this in Section 3.1.

To address this area, we use a "datatype signature hashing" mechanism, devised by William Gropp in [3]. He describes there a family of algorithms that can be used to assign a small amount of data to an MPI datatype signature in such a way that only small messages need to be sent in order to catch most user errors involving datatype arguments to MPI collective functions. In this paper we describe a specific implementation of datatype signature hashing and present an MPI profiling library that uses datatype signature hashing to carry out more thorough error checking than is done in [7]. Since extra work (to calculate the hash) is involved, we also present some simple performance measurements, although one can of course use this profiling library just during application development and remove it for production use.

In Section 2 we describe the nature and scope of the error checks we carry out and compare our approach with that in [7]. Section 3 lays out details of our implementation, including our implementation of the hashing algorithm given in [3]; we also present example output. In Section 4 we present some performance measurements. Section 5 summarizes the work and describes future directions.

2   Scope of Checks

In this section we describe the error checking carried out by our profiling library. We give definitions of each check and provide a table associating the checks made on the arguments of each collective MPI function with that function. We also compare our collective error checking with that described in [7].

2.1   Definitions of Checks

The error checks for each MPI(-2) collective function are shown in Table 1. The following checks are made:

call checks that all processes in the communicator have called the same collective function in a given event, thus guarding against the error of calling MPI_Reduce on some processes, for example, and MPI_Allreduce on others.

root means that the same argument was passed for the root argument on all processes.

datatype refers to datatype signature consistency. This is explained further in Section 3.1.

MPI_IN_PLACE means that every processes either did or did not provide MPI_IN_PLACE instead of a buffer.

op checks operation consistency, for collective operations that include computations. For example, each process in a call to MPI_Reduce must provide the same operation.

local leader and tag test consistency of the local_leader and tag arguments. They are used only for MPI_Intercomm_create.

high/low tests consistency of the high argument. It is used only for MPI_Intercomm_merge.

dims checks for dims consistency across the communicator.

graph tests the consistency of the graph supplied by the arguments to MPI_Graph_create and MPI_Graph_map.

amode tests for amode consistency across the communicator for the function MPI_File_open.

size, datarep, and flag verify consistency on these arguments, respectively.

etype is an additional datatype signature check for MPI file operations.

order checks for the collective file read and write functions, therefore ensuring the proper order of the operations. According to the MPI Standard [2], a begin operation must follow an end operation, with no other collective file functions in between.

2.2   Comparison with Previous Work

This work can be viewed as an extension of the NEC implementation of collective error checking via a profiling library presented in [7]. The largest difference between that work and this is that we incorporate the datatype signature hashing mechanism described in Section 3, which makes this paper also an extension of [3], where the hashing mechanism is described but not implemented. In the NEC implementation, only message lengths, rather than datatype signatures, are checked. We do not check length consistency since it would be incorrect to do so in a heterogeneous environment. We also implement our library as a pure profiling library. This precludes us from doing some MPI-implementation-dependent checks that are provided in the NEC implementation, but allows our library to be used with any MPI implementation. In this paper we also present some performance tests, showing that the overhead, even of our unoptimized version, is acceptable. Finally, the library described here is freely available.

Table 1. Checks performed on MPI functions

MPI_Barrier	call
MPI_Bcast	call, root, datatype
MPI_Gather	call, root, datatype
MPI_Gatherv	call, root, datatype
MPI_Scatter	call, root, datatype
MPI_Scatterv	call, root, datatype
MPI_Allgather	call, datatype, MPI_IN_PLACE
MPI_Allgatherv	call, datatype, MPI_IN_PLACE
MPI_Alltoall	call, datatype
MPI_Alltoallw	call, datatype
MPI_Alltoallv	call, datatype
MPI_Reduce	call, datatype, op
MPI_AllReduce	call, datatype, op, MPI_IN_PLACE
MPI_Reduce_scatter	call, datatype, op, MPI_IN_PLACE
MPI_Scan	call, datatype, op
MPI_Exscan	call, datatype, op
MPI_Comm_dup	call
MPI_Comm_create	call
MPI_Comm_split	call
MPI_Intercomm_create	call, local leader, tag
MPI_Intercomm_merge	call, high/low
MPI_Carte_create	call, dims
MPI_Carte_map	call, dims
MPI_Graph_create	call, graph
MPI_Graph_map	call, graph
MPI_Comm_spawn	call, root
MPI_Comm_spawn_multiple	call, root
MPI_Comm_connect	call, root
MPI_Comm_disconnect	call
MPI_Win_create	call
MPI_Win_fence	call
MPI_File_open	call, amode
MPI_File_set_size	call, size
MPI_File_set_view	call, datarep, etype
MPI_File_set_automicity	call, flag
MPI_File_preallocate	call, size
MPI_File_seek_shared	call, order
MPI_File_read_all_begin	call, order
MPI_File_read_all	call, order
MPI_File_read_all_end	call, order
MPI_File_read_at_all_begin	call, order
MPI_File_read_at_all	call, order
MPI_File_read_at_all_end	call, order
MPI_File_read_ordered_begin	call, order
MPI_File_read_ordered	call, order
MPI_File_read_ordered_end	call, order
MPI_File_write_all_begin	call, order
MPI_File_write_all	call, order
MPI_File_write_all_end	call, order
MPI_File_write_at_all_begin	call, order
MPI_File_write_at_all	call, order
MPI_File_write_at_all_end	call, order
MPI_File_write_ordered_begin	call, order

3   Implementation

In this section we describe our implementation of the datatype signature matching presented in [3]. We also show how we use datatype signatures in coordination with other checks on collective operation arguments.

3.1   Datatype Signature Matching

An MPI datatype signature for n different datatypes $type_i$ is defined to ignore the relative displacement among the datatypes as follows [6]:

$$Typesig = \{type_1, type_2, \ldots, type_n\}. \tag{1}$$

A datatype hashing mechanism was proposed in [3] to allow efficient comparison of datatype signature over any MPI collective call. Essentially, it involves comparison of a tuple (α, n), where α is the hash value and n is the total number of basic predefined datatypes contained in it. A tuple of form $(\alpha, 1)$ is assigned for each basic MPI predefined datatype (e.g. MPI_INT), where α is some chosen hash value. The tuple for an MPI derived datatype consisting of n basic predefined datatypes $(\alpha, 1)$ becomes (α, n). The combined tuple of any two MPI derived datatypes, (α, n) and (β, m), is computed based on the hashing function:

$$(\alpha, n) \oplus (\beta, m) \equiv (\alpha \wedge (\beta \lhd n), n + m), \tag{2}$$

where $\wedge$ is the bitwise exclusive or (xor) operator, $\lhd$ is the circular left shift operator, and $+$ is the integer addition operator. The noncommutative nature of the operator $\oplus$ in equation (2) guarantees the ordered requirement in datatype signature definition [1].

One of the obvious potential hash collisions is caused by the $\lhd$ operator's circular shift by 1 bit. Let us say there are four basic predefined datatypes identified by tuples $(\alpha, 1)$, $(\beta, 1)$, $(\gamma, 1)$, and $(\lambda, 1)$ and that $\alpha = \lambda \lhd 1$ and $\gamma = \beta \lhd 1$. For $n = m = 1$ in equation (2), we have

$$\begin{aligned}
(\alpha, 1) \oplus (\beta, 1) &\equiv (\alpha \wedge (\beta \lhd 1), 2) \\
&\equiv ((\beta \lhd 1) \wedge \alpha, 2) \\
&\equiv (\gamma \wedge (\lambda \lhd 1), 2) \\
&\equiv (\gamma, 1) \oplus (\lambda, 1),
\end{aligned} \tag{3}$$

If the hash values for all basic predefined datatypes are assigned consecutive integers, there will be roughly a 25 percent collision rate as indicated by equation (3). The simplest solution for avoiding this problem is to choose consecutive odd integers for all the basic predefined datatypes. Also, there are composite predefined datatypes in the MPI standard (e.g., MPI_FLOAT_INT), whose hash values are chosen according to equation (2) such that

$$MPI\_FLOAT\_INT = MPI\_FLOAT \oplus MPI\_INT.$$

The tuples for MPI_UB and MPI_LB are assigned $(0, 0)$, so they are essentially ignored. MPI_PACKED is a special case, as described in [3].

More complicated derived datatypes are decoded by using
MPI_Type_get_envelope() and MPI_Type_get_content() and their hashed tuple
computed during the process.

3.2   Collective Datatype Checking

Because of the different comunication patterns and the different specifications of
the send and receive datatypes in various MPI collective calls, a uniform method
of collective datatype checking is not attainable. Hence five different procedures
are used to validate the datatype consistency of the collectives. The goal here is
to provide error messages at the process where the erroneous argument has been
passed. To achieve that goal, we tailor each procedure to match the communication pattern of the profiled collective call. For convenience, each procedure is
named by one of the MPI collective routines being profiled.

Collective Scatter Check

1. At the root, compute the sender's datatype hash tuple.
2. Use PMPI_Bcast() to broadcast the hash tuple from the root to other
   processes.
3. At each process, compute the receiver's datatype hash tuple locally and
   compare it to the hash tuple received from the root.

A special case of the collective scatter check is when the sender's datatype signa-
ture is the same as the receiver's. This special case can be refered to as a collec-
tive bcast check. It is used in the profiled version of MPI_Bcast(), MPI_Reduce(),
MPI_Allreduce(), MPI_Reduce_scatter(), MPI_Scan(), and MPI_Exscan().
   The general collective scatter check is used in the profiled version of
MPI_Gather() and MPI_Scatter().

Collective Scatterv Check

1. At the root, compute the vector of the sender's datatype hash tuples.
2. Use PMPI_Scatter() to broadcast the vector of hash tuples from the root to
   the corresponding process in the communicator.
3. At each process, compute the receiver's datatype hash tuple locally and
   compare it to the hash tuple received from the root.

The collective scatterv check is used in the profiled version of MPI_Gatherv()
and MPI_Scatterv().

Collective Allgather Check

1. At each process, compute the sender's datatype hash tuple.
2. Use PMPI_Allgather() to gather other senders' datatype hash tuples as a
   local hash tuple vector.
3. At each process, compute the receiver's datatype hash tuple locally, and
   compare it to each element of the hash tuple vector received.

The collective allgather check is used in the profiled version of MPI_Allgather() and MPI_Alltoall().

Collective Allgatherv Check

1. At each process, compute the sender's datatype hash tuple.
2. Use PMPI_Allgather() to gather other senders' datatype hash tuples as a local hash tuple vector.
3. At each process, compute the vector of the receiver's datatype hash tuples locally, and compare this local hash tuple vector to the hash tuple vector received element by element.

The collective allgatherv check is used in the profiled version of MPI_Allgatherv().

Collective Alltoallv/Alltoallw Check

1. At each process, compute the vector of the sender's datatype hash tuples.
2. Use PMPI_Alltoall() to gather other senders' datatype hash tuples as a local hash tuple vector.
3. At each process, compute the vector of the receiver's datatype hash tuples locally, and compare this local hash tuple vector to the hash tuple vector received element by element.

The difference between collective alltoallv and collective alltoallw checks is that alltoallw is more general than alltoallv; in other words, alltoallw accepts a vector of MPI_Datatype in both the sender and receiver.

The collective alltoallv check is used in the profiled version of MPI_Alltoallv(), and the collective alltoallw check is used in the profiled version of MPI_Alltoallw().

3.3 Example Output

In this section we illustrate what the user sees (on stderr) when a collective call is invoked incorrectly.

Example 1. In this example, run with five processes, all but the last process call MPI_Bcast; the last process calls MPI_Barrier.

```
aborting job:
Fatal error in MPI_Comm_call_errhandler:

VALIDATE BARRIER (Rank 4) --> Collective call (BARRIER) is Inconsistent with Rank 0's (BCAST).

rank 4 in job 204  ilsig.mcs.anl.gov_32779   caused collective abort of all ranks
  exit status of rank 4: return code 13
```

Example 2. In this example, run with five processes, all but the last process give MPI_CHAR; but the last process gives MPI_INT.

```
aborting job:
Fatal error in MPI_Comm_call_errhandler:

VALIDATE BCAST (Rank 4) --> Datatype Signature used is Inconsistent with Rank 0s.

rank 4 in job 205  ilsig.mcs.anl.gov_32779   caused collective abort of all ranks
  exit status of rank 4: return code 13
```

Example 3. In this example, run with five processes, all but the last process use 0 as the `root` parameter; the last process uses its rank.

```
aborting job:
Fatal error in MPI_Comm_call_errhandler:

VALIDATE BCAST (Rank 4) --> Root Parameter (4) is inconsistent with rank 0 (0)

rank 4 in job 207  ilsig.mcs.anl.gov_32779    caused collective abort of all ranks
  exit status of rank 4: return code 13
```

4 Experiences

Here we describe our experiences with the collective error checking profiling library in the areas of usage, porting, and performance.

After preliminary debugging tests gave us some confidence that the library was functioning correctly, we applied it to the collective part of the MPICH2 test suite. This set of tests consists of approximately 70 programs, many of which carry out multiple tests, that test the MPI-1 and MPI-2 Standard compliance for MPICH2. We were surprised (and strangely satisfied, although simultaneously embarrassed) to find an error in one of our test programs. One case in one test expected a datatype of one `MPI_INT` to match a vector of `sizeof(int)` `MPI_BYTE`s. This is incorrect, although MPICH2 allowed the program to execute.

To test a real application, we linked FLASH [5], a large astrophysics application utilizing many collective operations, with the profiling library and ran one of its model problems. In this case no errors were found.

A profiling library should be automatically portable among MPI implementations. The library we describe here was developed under MPICH2. To check for portability and to obtain separate performance measurements, we also used it in conjunction with IBM's MPI for BlueGene/L [1], without encountering any problems.

Table 2. The maximum time taken (in seconds) among all the processes in a 32-process job on BlueGene/L. Count is the number of MPI_Doubles in the datatype, and N_{itr} refers to the number of times the MPI collective routine was called in the test. The underlined digits indicates that the corresponding digit could be less in one of the processes involved.

Test Name	count×N_{itr}	No CollChk	With CollChk
MPI_Bcast	1×10	0.000269	0.002880
MPI_Bcast	1K×10	0.000505	0.003861
MPI_Bcast	128K×10	0.031426	0.135138
MPI_Allreduce	1×1	0.000039	0.000318
MPI_Allreduce	1K×1	0.000233	0.000586
MPI_Allreduce	128K×1	0.022263	0.032420
MPI_Alltoallv	1×1	0.000043	0.000252
MPI_Alltoallv	1K×1	0.000168	0.000540
MPI_Alltoallv	128K×1	0.015357	0.035828

We carried out performance tests on two platforms. On BlueGene/L, the collective and datatype checking library and the test codes were compiled with xlc with -O3 and linked with IBM's MPI implementation available on BlueGene/L.

The performance of the collective and datatype checking library of a 32-process job is listed in Table 2, where each test case is linked with and without the collective and datatype checking library.

Similarly on a IA32 Linux cluster, the collective and datatype checking library and the test codes were compiled with gcc with -O3 and linked with MPICH2-1.0.1. The performance results of the library are tabulated in Table 3.

Table 3. The maximum time taken (in seconds) on a 8-process job on Jazz, an IA32 Linux cluster. Count is the number of MPI_Double in the datatype, and N_{itr} refers to the number of times the MPI collective routine has been called in the test.

Test Name	count$\times N_{itr}$	No CollChk	With CollChk
MPI_Bcast	1×10	0.034332	0.093795
MPI_Bcast	1K×10	0.022218	0.069825
MPI_Bcast	128K×10	1.704995	1.730708
MPI_Allreduce	1×1	0.000423	0.006863
MPI_Allreduce	1K×1	0.003896	0.005795
MPI_Allreduce	128K×1	0.233541	0.236214
MPI_Alltoallv	1×1	0.000320	0.009682
MPI_Alltoallv	1K×1	0.002415	0.003593
MPI_Alltoallv	128K×1	0.271676	0.355068

Both Tables 2 and 3 indicate that the cost of the collective and datatype checking library diminishes as the size of the datatype increases. The cost of collective checking can be significant when the datatype size is small. One would like the performance of such a library to be good enough that it is convenient to use and does not affect the general behavior of the application it is being applied to. On the other hand, performance is not absolutely critical, since it is basically a debug-time tool and is likely not to be used when the application is in production. Our implementation at this stage does still present a number of opportunities for optimization, but we have found it highly usable.

5   Summary

We have presented an effective technique to safeguard users from making easy-to-make but hard-to-find mistakes that often lead to deadlock, incorrect results, or worse. The technique is portable and available for all MPI implementations. We have also presented a method for checking datatype signatures in collective operations. We intend to extend the datatype hashing mechanism to point-to-point operations as well.

This profiling library is freely available as part of MPICH2 [4] in the MPE subdirectory, along with other profiling libraries.

References

1. G. Almási, C. Archer, J. G. Casta nos, M. Gupta, X. Martorell, J. E. Moreira, W. D. Gropp, S. Rus, and B. Toonen. MPI on BlueGene/L: Designing an efficient general purpose messaging solution for a large cellular system. In Jack Dongarra, Domenico Laforenza, and Salvatore Orlando, editors, *Recent Advances in Parallel Virtual Machine and Message Passing Interface*, number LNCS2840 in Lecture Notes in Computer Science, pages 352-361. Springer Verlag, 2003.
2. William Gropp, Steven Huss-Lederman, Andrew Lumsdaine, Ewing Lusk, Bill Nitzberg, William Saphir, and Marc Snir. *MPIThe Complete Reference: Volume 2, The MPI-2 Extensions*. MIT Press, Cambridge, MA, 1998.
3. William D. Gropp. Runtime checking of datatype signatures in MPI. In Jack Dongarra, Peter Kacsuk, and Norbert Podhorszki, editors, *Recent Advances in Parallel Virutal Machine and Message Passing Interface*, number 1908 in Springer Lecture Notes in Computer Science, pages 160-167, September 2000.
4. MPICH2 Web page. http://www.mcs.anl.gov/mpi/mpich2.
5. R. Rosner, A. Calder, J. Dursi, B. Fryxell, D. Q. Lamb, J. C. Niemeyer, K. Olson, P. Ricker, F. X. Timmes, J. W. Truran, H. Tufo, Y. Young, M. Zingale, E. Lusk, and R. Stevens. Flash code: Studying astrophysical thermonuclear flashes. *Computing in Science and Engineering*, 2(2):33, 2000.
6. Marc Snir, Steve W. Otto, Steven Huss-Lederman, David W. Walker, and Jack Dongarra. *MPIThe Complete Reference: Volume 1, The MPI Core*, 2nd edition. MIT Press, Cambridge, MA, 1998.
7. Jesper Larsson Träff and Joachim Worringen. Verifying collective MPI calls. In Dieter Kranslmuller, Peter Kacsuk, and Jack Dongarra, editors, *Recent Advances in Parallel Virutal Machine and Message Passing Interface*, number 3241 in Springer Lecture Notes in Computer Science, pages 18-27, 2004.

Implementing OpenMP for Clusters on Top of MPI[*]

Antonio J. Dorta[1], José M. Badía[2], Enrique S. Quintana[2],
and Francisco de Sande[1]

[1] Depto. de Estadística, Investigación Operativa y Computación,
Universidad de La Laguna, 38271, La Laguna, Spain
{ajdorta, fsande}@ull.es
[2] Depto. de Ingeniería y Ciencia de Computadores,
Universidad Jaume I, 12.071, Castellón, Spain
{badia, quintana}@icc.uji.es

Abstract. llc is a language designed to extend OpenMP to distributed memory systems. Work in progress on the implementation of a compiler that translates llc code and targets distributed memory platforms is presented. Our approach generates code for communications directly on top of MPI. We present computational results for two different benchmark applications on a PC-cluster platform. The results reflect similar performances for the llc compiled version and an *ad-hoc* MPI implementation, even for applications with fine-grain parallelism.

Keywords: MPI, OpenMP, cluster computing, distributed memory, OpenMP compiler.

1 Introduction

The lack of general purpose high level parallel languages is a major drawback that limits the spread of High Performance Computing (HPC). There is a division between the users who have the needs of HPC techniques and the experts that design and develop the languages as, in general, the users do not have the skills necessary to exploit the tools involved in the development of the parallel applications. Any effort to narrow the gap between users and tools by providing higher level programming languages and increasing their simplicity of use is thus welcome.

MPI [1] is currently the most successful tool to develop parallel applications, due in part to its portability (to both shared and distributed memory architectures) and high performance. As an alternative to MPI, OpenMP [2] has emerged in the last years as the industry standard for shared memory programming. The OpenMP Application Programming Interface is based on a small set of compiler directives together with some library routines and environment variables.

---

[*] This work has been partially supported by the Canary Islands government, contract PI2003/113, and also by the EC (FEDER) and the Spanish MCyT (Plan Nacional de I+D+I, contracts TIC2002-04498-C05-05 and TIC2002-04400-C03-03).

B. Di Martino et al. (Eds.): EuroPVM/MPI 2005, LNCS 3666, pp. 148–155, 2005.

One of the main drawbacks of MPI is that the development of parallel applications is highly time consuming as major code modifications are generally required. In other words, parallelizing a sequential application in MPI requires a considerable effort and expertise. In a sense, we could say that MPI represents the assembler language of parallel computing: you can obtain the best performance but at the cost of quite a high development investment.

On the other hand, the fast expansion of OpenMP comes mainly from its simplicity. A first rough parallel version is easily built as no significative changes are required in the sequential implementation of the application. However, obtaining the best performance from an OpenMP program requires some specialized effort in tuning.

The increasing popularity of commodity clusters, justified by their better price/performance ratio, is at the source of the recent efforts to translate OpenMP codes to distributed memory (DM) architectures, even if that implies a minor loss of performance. Most of the projects to implement OpenMP in DM environments employ software distributed shared memory systems; see, e.g., [3,4,5]. Different approaches are developed by Huang *et al.* [6], who base their strategy for the translation in the use of Global Arrays, and Yonezawa *et al.* [7], using an *array section descriptor* called *quad* to implement their translation.

In this paper we present the language llc, designed to extend OpenMP to DM systems, and the llc compiler, llCoMP, which has been built on top of MPI. Our own approach is to generate code for communications directly on top of MPI, and therefore does not rely on a coherent shared memory mechanism. Through the use of two code examples, we show the experimental results obtained in a preliminary implementation of new constructs which have been incorporated into the language in order to deal with irregular memory access patterns. Compared to others, the main benefit of our approach is its simplicity. The use of direct generation of MPI code for the translation of the OpenMP directives conjugates the simplicity with a reasonable performance.

The remainder of the paper is organized as follows. In Section 2 we outline the computational model underlying our strategy. We next discuss some implementation details of the llc compiler in Section 3. Case studies and the experimental evaluation of a preliminary implementation of the new constructs are given in Section 4. Finally, we summarize a few concluding remarks and future work in Section 5.

2   The llc Computational Model

The llCoMP compiler is a source-to-source compiler implemented on top of MPI. It translates code annotated with llc directives into C code with explicit calls to MPI routines. The resulting program is then compiled using the native back-end compiler, and properly linked with the MPI library. Whenever possible, the llc pragmas are compatible with their counterparts in OpenMP, so that minor changes are required to obtain a sequential, an MPI, or an OpenMP binary from the same source code.

The llc language follows the *One Thread is One Set of Processors* (*OTOSP*) computational model. Although the reader can refer to [8] for detailed information, a few comments are given next on the semantics of the model and the behaviour of the compiler.

The *OTOSP* model is a DM computational model where all the memory locations are private to each processor. A key concept of the model is that of *processor set*. At the beginning of the program (and also in the sequential parts of it), all processors available in the system belong to the same unique set. The processor sets follow a fork-join model of computation: the sets divide (fork) into subsets as a consequence of the execution of a *parallel construct*, and they join back together at the end of the execution of the construct. At any point of the code, all the processors belonging to the same set replicate the same computation; that is, they behave as a single thread of execution.

When different processors (sub-)sets join into a single set at the end of a *parallel construct*, *partner processors* exchange the contents of the memory areas they have modified inside the *parallel construct*. The replication of computations performed by processors in the same set, together with the communication of modified memory areas at the end of the *parallel construct*, are the mechanisms used in *OTOSP* to guarantee a coherent image of the memory.

Although there are different kinds of *parallel constructs* implemented in the language, in this paper we focus on the **parallel for** construct.

3   The llCoMP Compiler

The simplicity of the *OTOSP* model greatly eases its implementation on DM systems. In this section we expose the translation of parallel loops performed by llCoMP. For each of the codes in the NAS Parallel Benchmark [9] (columns of the table), Table 1 indicates the number of occurrences of the directive in the corresponding row. All the directives in Table 1 can be assigned a semantic under the premises of the *OTOSP* model. Using this semantic, the directives and also the data scope attribute clauses associated with them can be implemented using MPI on a DM system. For example, let us consider the implementation

Table 1. Directives in the NAS Parallel Benchmark codes

	BT	CG	EP	FT	IS	LU	MG	SP
parallel	2	2	1	2	2	3	5	2
for	54	21	1	6	1	29	11	70
parallel for		3				1		
master	2	2	1	10	4	2	1	2
single		12		5		2	10	
critical			1	1	1	1	1	
barrier				1	2	3	1	3
flush						6		
threadprivate			1					

of a `parallel` directive: since all the processors are running in parallel at the beginning of a computation, in our model the `parallel` directive requires no translation.

Shared memory variables (in the OpenMP sense) need special care, though. Specifically, any shared variable in the left-hand side of an assignment statement inside a `parallel loop` should be annotated with an `llc result` or `nc_result` clause. Both clauses are employed to notify the compiler of a region of the memory that is potentially modifiable by the set of processors which execute the loop. Their syntax is similar: the first parameter is a pointer to the memory region (`addr`), the second one is the size of that region (`size`), and the third parameter, only present in `nc_result`, is the name of the variable holding that memory region. Directive `result` is used when all the memory addresses in the range [`addr, addr+size`] are (potentially) modified by the processor set. This is the case, for example, when adjacent positions in a vector are modified. If there are write accesses to non-contiguous memory regions inside the parallel loop, these should be notified with the `nc_result` clause.

llCoMP uses *Memory Descriptors* (MD) to guarantee the consistency of memory at the end of the execution of a *parallel construct*. MD are data structures based on queues which hold the necessary information about memory regions modified by a processor set. The basic information holded in MD are pairs (*address, size*) that characterize a memory region. Prior to their communication to other processor sets, these memory regions (pairs) are compacted in order to minimize the communication overhead. In most of the cases, the communication pattern involved in the translation of a `result` or `nc_result` is an *all-to-all* pattern. The post-processing performed by a processor receiving a MD is straightforward: it writes the bytes received in the address annotated in the MD. In section 4 we present an experiment that has been designed to evaluate the overhead introduced in the management of MDs.

In [8] we presented several examples of code with the `result` directive. In this paper we focus in the implementation of non-contiguous memory access patterns, the most recent feature incorporated into llCoMP.

```
1   #pragma omp parallel for private(ptr, temp, k, j )
2   for (i=0; i<Blks->size1; i++) {
3     ptr = Blks->ptr[i];
4     temp = 0.0;
5     k = index1_coordinate(ptr); // First element in i-th row
6     for (j=0; j<elements_in_vector_coordinate(Blks, i); j++) {
7       temp += value_coordinate(ptr) *
8                 x[index2_coordinate(ptr)*incx];
9       inc_coordinate(ptr);
10    }
11    #pragma llc nc_result(&y[k*incy], 1, y)
12    y[k*incy] += alpha * temp;
13  }
```

Listing 1. A parallelization of the USMV operation

In particular, consider the code in Listing 1, which shows the parallelization using llc of the main loop of the *sparse matrix-vector product* operation $y = y + \alpha Ax$, where x and y are both vectors and A is a sparse matrix (this is known as operation USMV in the Level-2 sparse BLAS). Matrix elements are stored using a rowwise coordinate format, but we also store pointers to the first element on each row in vector ptr. In the code, each iteration of the external loop in line 2 performs a dot product between a row of the sparse matrix and vector x, producing one element of the solution vector y. The code uses three C macros (index1_coordinate(ptr), index2_coordinate(ptr) and value_coordinate(ptr)) in order to access to the row index, column index and value of an element of the sparse matrix pointed by ptr. A fourth macro, namely inc_coordinate, moves the pointer to the next element in the same row. Values incx and incy allow the code to access to vectors x and y with strides different from 1.

A direct parallelization of the code can be obtained having into account that different dot products are fully independent. Therefore, a parallel for directive is used in line 1 to indicate that the set of processors executing the loop in line 2 has to fork to execute the loop. The llc specific directive nc_result in line 11 indicates to the compiler that the value of the *y[k*incy]* element has to be "annotated".

4   Experimental Results

The experiments reported in this section were obtained on a cluster composed of 32 Intel Pentium Xeon processors running at 2.4 GHz, with 1 GByte of RAM memory each, and connected through a Myrinet switch. The operating system of the cluster was Debian Sarge Testing Linux. We used the MPICH [10] implementation on top of the vendor's communication library GM-1.6.3.

In order to evaluate the performance of the llCoMP translation we have used two benchmarks: the *sparse matrix-vector product* USMV, and a Molecular Dynamics (MD) simulation code [11]. These benchmarks were selected because they are composed of irregular, non-contiguous accesses to memory, and also because they are simple codes representative for a much larger class. Besides, the USMV operation is a common operation in sparse linear algebra, extremely useful in a a vast amount of applications arising, among many others, in VLSI design, structural mechanics and engineering, computational chemistry, and electromagnetics.

Using MPI and llCoMP we developed two parallel versions of the USMV code. Consider first the parallelization using MPI. For simplicity, we assume vectors x and y to be both replicated. With A distributed by rows, the matrix-vector product is performed as a series of inner products which can proceed in parallel. An *all-to-all* communication is required at the final stage to replicate the result y onto all nodes. On the other hand, using llc to implement the product, we parallelize the external for loop, so that each thread deals with a group of inner products (see Listing 1). As all threads share vectors x and y, it is not necessary to perform any additional gathering of partial results in this case.

Table 2. Execution time (in secs.) for the *ad hoc* MPI vs. `llCoMP` parallel versions of the USMV code. Problem sizes of 30000 and 40000 are employed with sparsity degrees of 1% and 2%.

| | *ad hoc* MPI | | | | llCoMP | | | |
| | 30000 | | 40000 | | 30000 | | 40000 | |
#Proc.	1%	2%	1%	2%	1%	2%	1%	2%
SEQ	11.09	21.55	26.67	45.16	11.09	21.55	26.67	45.16
2	5.85	11.99	11.59	30.11	8.24	15.15	21.70	34.08
4	3.64	6.65	8.26	16.72	4.85	8.18	10.34	21.34
8	2.23	3.33	4.44	11.39	2.93	4.23	6.42	10.57
16	1.62	2.13	2.86	6.76	1.86	2.68	3.48	6.57
24	1.45	1.82	2.43	6.15	1.80	2.32	3.29	6.23
32	1.27	1.55	2.00	5.25	1.94	2.40	3.02	4.35

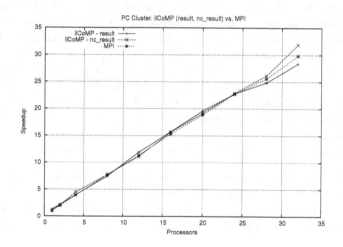

Fig. 1. MPI and `llCoMP` (`result` and `nc_result`) speedups for the MD code

Table 2 compares the accumulated execution time of 100 runs for the USMV code using an *ad hoc* MPI implementation and the translation produced by `llCoMP`. The executions correspond to square sparse random matrices of dimensions 30000, 40000 and sparsity factors of 1% and 2%.

The fine-grain parallelism present in the USMV code and the small amount of computations performed by this operation are at the source of the limited speed-up reported in the experiment. Not surprisingly, coarse-grain algorithms are the best scenario to achieve high performance for the translations provided by `llCoMP`. Nevertheless, if we compare the results obtained from an *ad hoc* program using MPI with those produced by the `llc` variant, we can expect this overhead to be compensated in some situations by the much smaller effort invested in the development of the parallel code.

The source code for the MD code written in OpenMP can be obtained from the *OpenMP Source Code Repository* [12,13]; translation of this code using llc is straight-forward. With this experiment our aim is to evaluate the overhead introduced by llCoMP in the management of non-regular memory access patterns. The MD code exhibits a regular memory access pattern, but as the nc_result clause is a general case of result, we have implemented it using both clauses. Figure 1 shows the speedup achieved by the MD code for three different implementations: and *ad hoc* MPI implementation and two different llc implementations using result and nc_result. We observe an almost linear behaviour for all the implementations. For this particular code, no relevant differences are appreciated when using regular and non-regular memory access patterns.

5   Conclusions and Future Work

We believe that preserving the sequential semantics of the programs is a major key to achieve the objective of alleviating the difficulties in the development of parallel applications. Surely the extension of the OpenMP programming paradigm to the DM case is a desirable goal. At the present time the technology and the ideas are not mature enough as to show a clear path to the solution of the problem and, in this line, our own approach does not intend to compete with other authors' work. We show that a compiler working under the premises of the *OTOSP* computational model and using direct generation of MPI code for communications can produce acceptable results, even in the case of fine-grain parallel algorithms.

Work in progress in our project includes the following issues:

- To unburden the final user of the specification of memory regions to be communicated (using the result clauses).
- To explore the potential sources of improvement for the compiler prototype.
- To collect OpenMP applications suitable to be targeted by the llCoMP compiler.

Acknowledgments

We wish to thank the anonymous reviewers for their suggestions on how to improve the paper.

References

1. Message Passing Interface Forum, MPI: A Message-Passing Interface Standard, University of Tennessee, Knoxville, TN, 1995 http://www.mpi-forum.org/.
2. OpenMP Architecture Review Board, OpenMP Application Program Interface v. 2.5, electronically available at
   http://www.openmp.org/drupal/mp-documents/spec25.pdf (May 2005).

3. S.-J. Min, A. Basumallik, R. Eigenmann, Supporting realistic OpenMP applications on a commodity cluster of workstations, in: Proc. of of WOMPAT 2003, Workshop on OpenMP Applications and Tools, Toronto, Canada, 2003, pp. 170–179.

4. M. Sato, H. Harada, A. Hasegawa, Cluster-enabled OpenMP: An OpenMP compiler for the SCASH software distributed shared memory system., Scientific Programming, Special Issue: OpenMP 9 (2-3) (2001) 123–130.

5. Y. C. Hu, H. Lu, A. L. Cox, W. Zwaenepoel, OpenMP for Networks of SMPs, Journal of Parallel and Distributed Computing 60 (12) (2000) 1512–1530.

6. L. Huang, B. Chapman, Z. Liu, Towards a more efficient implementation of openmp for clusters via translation to global arrays, Tech. Rep. UH-CS-04-05, Department of Computer Science, Univeristy of Houston, electronically available at http://www.cs.uh.edu/docs/preprint/2004_11_15.pdf (dec 2004).

7. N. Yonezawa, K. Wada, T. Ogura, Quaver: OpenMP compiler for clusters based on array section descriptor, in: Proc. of the 23rd IASTED International Multi-Conference Parallel and Distributed Computing and Networks, IASTED /Acta Press, Innsbruck, Austria, 2005, pp. 234–239, electronically available at http://www.actapress.com/Abstract.aspx?paperId=6530.

8. A. J. Dorta, J. A. González, C. Rodríguez, F. de Sande, llc: A parallel skeletal language, Parallel Processing Letters 13 (3) (2003) 437–448.

9. D. H. Bailey et al., The NAS parallel benchmarks, Technical Report RNR-94-007, NASA Ames Research Center, Moffett Field, CA, USA, electronically available at http://www.nas.nasa.gov/News/Techreports/1994/PDF/RNR-94-007.pdf (Oct. 1994).

10. W. Gropp, E. Lusk, N. Doss, A. Skjellum, A high-performance, portable implementation of the message passing interface standard, Parallel Computing 22 (6) (1996) 789–828.

11. W. C. Swope, H. C. Andersen, P. H. Berens, K. R. Wilson, A computer simulation method for the calculation of equilibrium constants for the formation of physical clusters of molecules: Application to small water clusters, Journal of Chemical Physics 76 (1982) 637–649.

12. OmpSCR OpenMP Source Code Repository http://www.pcg.ull.es/ompscr/ and http://ompscr.sf.net.

13. A. J. Dorta, A. González-Escribano, C. Rodríguez, F. de Sande, The OpenMP source code repository, in: Proc. of the 13th Euromicro Conference on Parallel, Distributed and Network-based Processing (PDP 2005), Lugano, Switzerland, 2005, pp. 244–250.

Designing a Common Communication Subsystem*

Darius Buntinas and William Gropp

Mathematics and Computer Science Division,
Argonne National Laboratory
{buntinas, gropp}@mcs.anl.gov

Abstract. Communication subsystems are used in high-performance parallel computing systems to abstract the lower network layer. By using a communication subsystem, an upper middleware library or runtime system can be more easily ported to different interconnects. By abstracting the network layer, however, the designer typically makes the communication subsystem more specialized for that particular middleware library, making it ineffective for supporting middleware for other programming models. In previous work we analyzed the requirements of various programming-model middleware and the communication subsystems that support such requirements. We found that although there are no mutually exclusive requirements, none of the existing communication subsystems can efficiently support the programming model middleware we considered. In this paper, we describe our design of a common communication subsystem, called CCS, that can efficiently support various programming model middleware.

1 Introduction

Communication subsystems are used in high-performance parallel computing systems to abstract the lower network layer. By using a communication subsystem, an upper middleware library or runtime system can be ported more easily to different interconnects. By abstracting the network layer, however, the designer typically makes the communication subsystem less general and more specialized for that particular middleware library. For example, a communication subsystem for a message-passing middleware might have been optimized for transferring data located anywhere in a process's address space, whereas a communication subsystem for a global address space (GAS) language might have been better optimized for transferring small data objects located in a specially allocated region of memory. Thus, the communication subsystem designed for a GAS language cannot efficiently support the message-passing middleware because, for example, it cannot efficiently transfer data that is located on the stack or in dynamically allocated memory.

* This work was supported by the Mathematical, Information, and Computational Sciences Division subprogram of the Office of Advanced Scientific Computing Research, Office of Science, U.S. Department of Energy, under Contract W-31-109-ENG-38.

B. Di Martino et al. (Eds.): EuroPVM/MPI 2005, LNCS 3666, pp. 156–166, 2005.

Despite their differences, communication subsystems have many common features, such as bootstrapping and remote memory access (RMA) operations. In [1] we analyzed the requirements of various programming model middleware and the communication subsystems that support them. We found that although there are no mutually exclusive requirements, none of the existing communication subsystems can efficiently support the programming-model middleware we considered. In this paper, we describe our design of a common communication subsystem, called CCS, that can efficiently support various programming-model middleware. We specifically targeted CCS to efficiently support the requirements of MPICH2 [2,3], the Global Arrays (GA) toolkit [4,5], and the Berkeley UPC runtime [6,7]; however, we believe that CCS is general enough to efficiently support any message-passing, global address space, or remote-memory middleware.

The rest of this paper is organized as follows. In Section 2 we briefly describe the critical design issues necessary to support the various programming models. In Section 3 we present our design for a common communication subsystem. In Section 4 we show performance results from our preliminary implementation of CCS. In Section 5 we conclude and present future work.

2   Design Issues for Communication Subsystems

In this section, we briefly describe the important issues for designing a common communication subsystem. These design issues are covered in more detail in [1]. We divide the design issues into required features and desired features. A required feature is a feature that, if lacking, would prevent the communication subsystem from effectively supporting a particular programming model. Desired features are features that, when implementing a programming model on top of the communication subsystem, make the implementation simpler or more efficient.

2.1   Required Features

Remote Memory Access Operations. RMA operations allow a process to transfer data between its local memory and the local memory of remote process without active participation of the remote process. RMA operations are important for global address space and remote-memory programming models, as well as for message-passing applications that have irregular communication patterns.

In order to allow better overlap of communication and computation, non-blocking RMA operations should be provided. A mechanism is then needed to check whether the operation has completed.

MPI-2 RMA Support. In order to support MPI-2 [8] active-mode RMA operations, the communication subsystem must be able to perform RMA operations between any memory location in the process's address space. In order to support passive-mode RMA operations, the communication subsystem need only be able to perform RMA operations on memory that has been dynamically allocated using a special allocation function.

GAS Language and Remote-Memory Model Support. GAS language and remote-memory model runtime systems need to be able to perform concurrent conflicting RMA operations to the same memory region. Similarly, they require the ability to perform local load/store operations concurrently with RMA operations, possibly to the same memory location. While the result of such conflicting operations may be undefined, the communication subsystem must not consider it an error to perform them. RMA operations also must be very lightweight, since typical RMA operations in these programming models are single-word operations.

Efficient Transfer of Large MPI Two-Sided Messages. MPI and other message-passing interfaces provide two-sided message passing, where the sending process specifies the source buffer, and the receiving process specifies the destination buffer. Typically, in message-passing middleware, large data is transferred by using a rendezvous protocol, where one process sends the address of its buffer to the other process, so that one process has the location of both the source and destination buffers. Once one process has the location of both buffers, it can use RMA operations to transfer the data. In MPI, the source and destination buffers can be located anywhere in the process's address space. In order to support transferring large two-sided messages in this way, the communication subsystem must be able to perform RMA operations on any memory location in the process's address space.

2.2   Desired Features

Active Messages. Active messages [9] allow the sender to specify a handler function that is executed at the receiver when the message is received. This function can be used, for example, to match an incoming message with a pre-posted receive in MPI or to perform an accumulate operation in Global Arrays.

In order to support multiple middleware libraries at the same time, active messages from one middleware library must not interfere with those of another middleware library. One solution is to ensure that each library uniquely specifies its own handlers.

In-Order Message Delivery. In-order message delivery is a requirement for many message-passing programming models. If the communication subsystem provides this feature, the middleware doesn't have to deal with reordering messages. However, in other programming models such as GAS languages, message ordering is not required, and in some cases performance can be improved by reordering or coalescing messages. A common communication subsystem should be able to provide FIFO ordering when it is required, and allow messages to be reordered otherwise.

Noncontiguous Data. Programming model instances such as MPI and Global Arrays have operations for specifying the transfer of noncontiguous data. Furthermore, modern interconnects such as InfiniBand (IBA) [10], support noncontiguous data transfer. Hence, a common communication subsystem needs to support the transfer of noncontiguous data in order to take advantage of such functionality.

Table 1. Feature summary of the communication subsystems

	RMA operations	MPI-2 active-mode RMA	MPI-2 passive-mode RMA	GAS language support	Transfer of large MPI messages	Active messages	In-order message delivery	Noncontiguous data*	Portability
ARMCI	•			•				V, S	•
GASNet	•		•	•		•			•
LAPI	•	•	•	•	•	•		V	
Portals	•	•	•	•	•		•		•
MPI-2	•	•	•		•		•	V, S, B	•

* V = I/O vector; S = strided; B = blockindexed

2.3 Feature Support by Current Communication Subsystems

In [1] we examined several communication subsystems and evaluated how well each addresses the features described above. Table 1 summarizes the results. We evaluated ARMCI [11], GASNet [12], LAPI [13], Portals [14], and MPI-2 [8] as communication subsystems. We can see from this table that none of the communication subsystems we studied supports all of the features necessary for message-passing, remote-memory, and GAS language programming models.

3 Proposed Communication Subsystem

In this section we describe our design for a common communication subsystem, called CCS, that addresses the issues identified in the previous section. The CCS communication subsystem is based on nonblocking RMA operations, with active messages used to provide for control and remote invocation of operations. Active messages could be used to implement an operation to deliver the message in message passing middleware or to perform an accumulate operation in remote memory middleware.

3.1 Remote Memory Access Operations

CCS provides nonblocking RMA operations. It is intended that RMA operations be implemented by using the interconnect's native RMA operations in order to maximize performance. If an interconnect does not natively provide all or some of the required RMA operations, active messages can be used to implement the missing RMA operations. For example, if an interconnect has a native Put operation but not a Get operation, the Get can be implemented with an active message in which the handler performs a Put operation.

CCS uses a *callback* mechanism to indicate the completion of RMA operations. A callback function pointer is specified by the upper layer as a parameter to the RMA function. Then, when an RMA operation completes remotely, CCS calls the callback function. This can be used to implement fence and global fence operations.

Because the user-level communication libraries of most interconnects require memory to be registered before RMA operations can be performed on that memory, CCS also requires memory registration. The upper layer is responsible for ensuring that any dynamically allocated memory is deregistered before it is freed. The current design is to limit the amount of memory that a process can register to the amount that can be registered with the interconnect. A future design is to lift this restriction. If the upper layer registers more memory with CCS than the interconnect can register, CCS would handle deregistering and reregistering memory with the interconnect as needed. A mechanism similar to the *firehose* [15] mechanism used in GASNet could be employed.

Registering and deregistering memory with a network library usually involve a system call, which makes them costly operations. In order to reduce the overhead of registering and deregistering memory, CCS implements a registration cache and uses lazy deregistration. CCS keeps track of which pages have already been registered, to avoid registering pages twice. CCS also does not immediately deregister memory when the upper layer calls the CCS deregistration function. Instead, CCS simply decrements the usage count and deregisters pages once the number of unused pages reaches a certain threshold. This scheme reduces the number of network library registration and deregistration calls.

CCS RMA operations can access all of the process's memory and have no restrictions on concurrent access to memory. While this feature simplifies implementing upper layers on CCS, it can impact performance on machines that are not cache coherent and on interconnects that do not have byte granularity for their RMA operations. In these cases, CCS will have to handle the RMA operations in software taking care of cache coherence and data transfer.

3.2   Efficient Transfer of Large MPI Two-Sided Messages

CCS nonblocking RMA operations are to be used for transferring large messages. CCS RMA operations are intended to be implemented by using native interconnect RMA operations to maximize throughput. Because the operations are nonblocking, the communication can be overlapped with computation.

As described in the previous section, large MPI messages are typically transferred by using a rendezvous operation. In CCS, the rendezvous operation can be performed with active messages. Once the exchange of buffer locations has been done, the data can be transferred with RMA operations. When the RMA operations have completed, another active message would be sent to notify the other side of completion.

A future design is to implement a *large data active message* operation, which would function similar to the LAPI active messages using the header handler. The large data active message would be nonblocking. The sender would specify an active message handler and a local completion handler. The active message handler would be executed at the receiver before any data has been transferred.

The handler would specify the receive buffer and its local completion handler. Once the data had been transferred, the completion handlers on the sender and receiver would be called. A mechanism would be needed for the receiver to abort or delay the operation in the active message handler if it was not ready to receive the data yet.

3.3   Active Messages

We are including active messages in CCS because of the flexibility they provide to upper-layer developers. In our design, active messages are intended to be used for small message sizes, so the implementation should be optimized for low latency.

When an active message is received and the handler is executed, the handler gets a pointer to a temporary buffer where the received data resides. The handlers are responsible for copying the data out of the buffer. Noncontiguous source data will be packed contiguously into the temporary buffer. If the final data layout is to be noncontiguous, the message handler will have to unpack the data.

Depending on the implementation, the active message handlers will be called either asynchronously or from within another CCS function. CCS provides locks that are appropriate to be called from within the handler and includes a mechanism to prevent a handler from interrupting a thread.

To allow multiple upper layer libraries to use CCS at the same time, we introduce the notion of a *context*. Each separate upper layer library, or module, allocates its own context. Active message handlers are registered with a particular context. When an active message handler is registered, the upper layer provides the handler function pointer along with an ID number and the context. The ID number must be unique within that context. When an active message is sent, the context is specified along with the handler ID to uniquely identify the active message handler at the remote side.

3.4   In-Order Message Delivery

In order to support the message-passing programming model, CCS guarantees in-order delivery of active messages. However, RMA operations are not guaranteed to be completed in order. This approach allows CCS, or the underlying interconnect, to reorder messages in order to improve performance.

3.5   Noncontiguous Data

CCS supports noncontiguous data in active messages and RMA operations. CCS uses *datadescs* to describe the layout of noncontiguous data. Datadescs are similar to MPI datatypes and are, in fact, implemented by using the same mechanism that MPICH2 uses for datatypes [16]. Datadescs are defined recursively like MPI datatypes; however, datadescs do not currently store information about the native datatype (e.g., `double` or `int`) of the data. Because datadescs do not keep track of native datatypes, datadescs CCS cannot be used on heterogeneous systems, where byte-order translation would need to be done. We will address this situation in future work.

While datadescs are defined recursively, they need not be implemented recursively. In the implementation the datadesc can be unrolled into a set of component loops, rather than use recursive procedure calls that would affect performance. These unrolled representations can be used to efficiently and concisely describe common data layouts such as ARMCI strided layouts.

MPI datatypes can be implemented by using datadescs having the upper layer keep track of the native datatypes of the data. I/O vector and strided data layouts in LAPI and ARMCI can also be represented with datadescs. An implementation optimization would be to include specialized operations to create datadescs quickly from the commonly used I/O vector and strided representations in LAPI and ARMCI.

3.6   Summary of Proposed Communication Subsystem

Our proposed communication subsystem addresses all of the issues raised in the previous section. Active messages can be used by GAS language and remote-memory copy middleware for remote-memory allocation and locking operations and by message-passing middleware for message matching. Because CCS supports multiple contexts for active messages, it can be used for hybrid programming models, for example, where an application uses both MPI-2 and UPC.

CCS provides RMA operations that are compatible with MPI-2 RMA operations, as well as GAS language and remote memory copy RMA operations. CCS has primitives that can be used to implement fence and global fence operations. With the addition of a symmetric allocation function, GAS language and remote memory copy RMA support can be implemented very efficiently. The CCS RMA operations can also be used for transferring large messages in message-passing middleware.

CCS also provides in-order message delivery for active messages but does not force RMA operations to be in order. This feature allows active messages to be used for MPI-2 message-passing, while allowing RMA operations to be reordered for efficiency.

CCS supports transfer of noncontiguous data. The data layout is described in a recursive manner but can be internally represented compactly and efficiently. CCS's *datadescs* are compatible with MPI-2 datatypes. Strided and IOV data descriptions used in Global Arrays can also be efficiently represented with datadescs.

Our design of using RMA operations with active messages was inspired by LAPI and GASNet. But, as we showed in [1], LAPI and GASNet do not support all of the key features necessary to efficiently support all of the programming models we targeted. LAPI does not guarantee in-order message delivery, supports only I/O vector style of noncontiguous data, and is not portable. GASNet does not support MPI-2 active-mode RMA operations, the efficient transfer of large MPI messages, in-order message delivery, or noncontiguous data.

We note that the lack of some of the features we described does not necessarily mean that a middleware cannot be implemented over a particular communication subsystem. In fact, MPI has been implemented over LAPI [17], UPC has been

```
void get_callback (void *arg) {
    ++gets_completed;
}
#define NEW_MSG_HANDLER_ID 0
void new_msg_handler (CCS_token_t token, void *buffer, unsigned buf_len,
                      void *remote_buf, int remote_buflen) {
    int sender;
    CCS_sender_rank (token, &sender);
    CCS_get (sender, remote_buf, remote_buflen, CCS_DATA8, buf, buflen,
             CCS_DATA8, get_callback, 0 /* callback argument */);
}
int main (int argc, char **argv) {
    CCS_init();
    CCS_new_context (&context);
    CCS_register_handler (context, NEW_MSG_HANDLER_ID, new_msg_handler);
    buf = malloc (buflen);
    CCS_register_mem (buf, buflen);
    CCS_barrier();
    ...
    CCS_amrequest (context, other_node, NULL, 0, CCS_DATA_NULL,
                   NEW_MSG_HANDLER_ID, 2, buf, buflen);
    ...
    CCS_finalize();
}
```

Fig. 1. Sample CCS code

implemented over MPI [7], and MPI-2 has been implemented over GASNet [2]. But the lack of these features makes these implementations less efficient and more difficult. By implementing all of the key features, CCS can efficiently support all of the programming-model middleware.

Figure 1 shows some sample code using CCS. The code sends an active message, using CCS_amrequest(), to another node with no data, but with the pointer and length to its local buffer as parameters to the message handler. The message handler on the receiving side calls CCS_get() to get the data stored in the buffer specified by the sender. When the Get operation completes CCS will call the callback function get_callback() specified in the call to CCS_get().

4   Preliminary Performance Results

In this section we present performance results for our preliminary implementation of CCS over GM2 [18]. We performed latency and bandwidth tests on two dual 2 GHz Xeon nodes running Linux 2.4.18 and connected with a Myrinet2000 network [19] using Myricom M3F-PCI64C-2 NICs through a 16-port switch.

Figure 2 shows the latency results for CCS as well as for GASNet 1.3 and ARMCI 1.2B. For CCS and GASNet, we performed the test using active messages. Because ARMCI supports only RMA operations, we performed the test using Put. The results are averaged over 1,000 iterations. The 4-byte latency for GASNet is 8.8 μs, for CCS is 9.6 μs, and for ARMCI is 10.8 μs. We see from these numbers and Figure 2 that CCS performs better than ARMCI but not as well as GASNet.

Figure 3 shows the bandwidth results. We used nonblocking Put operations to perform the test. In this test, for each message size, we performed 10,000 Put operations, then waited for the operations to complete. We see that CCS

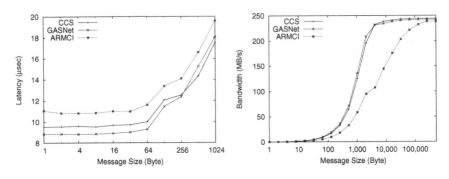

Fig. 2. Latency   **Fig. 3.** Bandwidth

performs better than ARMCI for all message sizes and better than GASNet for messages larger than 4 KB. For messages smaller than 4 KB, CCS performs only slightly worse than GASNet. The maximum bandwidth for CCS was 244 MBps, for GASNet was 242 MBps, and for ARMCI was 238 MBps.

The performance of CCS is comparable to the other communication subsystems. The additional functionality of CCS does not have a large impact on performance. We have not yet tuned the CCS source code for performance and expect that that with some performance tuning, we can further improve the performance of CCS.

We note that ARMCI over GM is implemented by using a server thread on each node. In ARMCI, RMA operations from remote processes are performed by the server thread. While using a server thread may affect performance compared to CCS and GASNet, it does allow RMA operation to complete asynchronously, independent of what the application thread is doing. We intend to evaluate such functionality for CCS.

5   Discussion and Future Work

In this paper we have presented our design for a common communication subsystem, CCS. CCS is designed to support the middleware libraries and runtime systems of various programming models efficiently by taking advantage of the high-performance features of modern interconnects. We evaluated a preliminary implementation of CCS and found the performance to be comparable to that of ARMCI and GASNet.

In the future, we intend to address thread safety, RMA Accumulate operations, and collective communication operations. We also intend to implement atomic remote memory operations, such as compare-and-swap and fetch-and-add, as well as more complex operations like an *indexed Put*, where the address of a Put operation is specified by a pointer at the destination process, and the pointer is incremented after the Put completes. Such operations can be used to efficiently implement remote queues on shared memory architectures.

We are also investigating using CCS to support a hybrid UPC/MPI programming model. In such a hybrid programming environment, a process can perform both UPC and MPI operations. By porting both the Berkeley UPC runtime and MPICH2 over CCS, CCS would perform the communication operations for both paradigms, allowing both paradigms to benefit from CCS's high-performance implementation.

In remote-memory model and GAS language middleware, when a process accesses a remote portion of a shared object distributed across different processes, the virtual address of that portion at the remote process needs to be computed. This operation can be simplified by allocating shared memory regions *symmetrically* across all processes; that is, the address of the local portion of the shared object is the same at each process. This also improves the scalability of the operation because less information needs to be kept for each remote memory region. We intend to address this issue, perhaps by including a special symmetric allocation function.

Acknowledgments

We thank Rusty Lusk, Rajeev Thakur, Rob Ross, Brian Toonen, and Guillaume Mercier for their valuable comments and suggestions while we were designing and implementing CCS.

References

1. Buntinas, D., Gropp, W.: Understanding the requirements imposed by programming model middleware on a common communication subsystem. Technical Report ANL/MCS-TM-284, Argonne National Laboratory (2005)
2. Argonne National Laboratory: MPICH2. (http://www.mcs.anl.gov/mpi/mpich2)
3. Gropp, W., Lusk, E., Doss, N., Skjellum, A.: A high-performance, portable implementation of the MPI message passing interface standard. Parallel Computing **22** (1996) 789–828
4. Nieplocha, J., Harrison, R.J., Littlefield, R.L.: Global Arrays: A portable shared memory programming model for distributed memory computers. In: Supercomputing 94. (1994)
5. Pacific Northwest National Laboratory: Global arrays toolkit. (http://www.emsl.pnl.gov/docs/global/ga.html)
6. Carlson, W.W., Draper, J.M., Culler, D.E., Yelick, K., Brooks, E., Warren, K.: Introduction to UPC and language specification. Technical Report CCS-TR-99-157, Center for Computing Sciences, IDA, Bowie, MD (1999)
7. Lawrence Berkeley National Laboratory and University of California, Berkeley: Berkeley UPC runtime. (http://upc.lbl.gov)
8. Message Passing Interface Forum: MPI-2: Extensions to the Message-Passing Interface. (1997)
9. von Eicken, T., Culler, D.E., Goldstein, S.C., Schauser, K.E.: Active messages: A mechanism for integrated communication and computation. In: Proceedings of the 19th International Symposium on Computer Architecture. (1992) 256–266
10. InfiniBand Trade Association: (InfiniBand Architecture Specification) http://www.infinibandta.com.

11. Nieplocha, J., Carpenter, B.: ARMCI: A portable remote memory copy library for distributed array libraries and compiler run-time systems. 3rd Workshop on Runtime Systems for Parallel Programming (RTSPP) of International Parallel Processing Symposium IPPS/SPDP '99 (1999)

12. Bonachea, D.: GASNet specification, v1.1. Technical Report CSD-02-1207, University of California, Berkeley (2002)

13. International Business Machines: RSCT for AIX 5L LAPI Programming Guide. Second edn. (2004) SA22-7936-01.

14. Brightwell, R., Riesen, R., Lawry, B., Maccabe, A.B.: Portals 3.0: Protocol building blocks for low overhead communication. In: Proceedings of the 2002 Workshop on Communication Architecture for Clusters (CAC). (2002)

15. Bell, C., Bonachea, D.: A new DMA registration strategy for pinning-based high performance networks. In: Workshop on Communication Architecture for Clusters (CAC03) of IPDPS'03. (2003)

16. Ross, R., Miller, N., Gropp, W.D.: Implementing fast and reusable datatype processing. In Dongarra, J., Laforenza, D., Orlando, S., eds.: Recent Advances in Parallel Virtual Machine and Message Passing Interface. Number LNCS2840 in Lecture Notes in Computer Science, Springer Verlag (2003) 404–413

17. Banikazemi, M., Govindaraju, R.K., Blackmore, R., Panda, D.K.: Implementing efficient MPI on LAPI for IBM RS/6000 SP systems: Experiences and performance evaluation. In: Proceedings of the 13th International Parallel Processing Symposium. (1999) 183–190

18. Myricom: The GM-2 message passing system – The reference guide to the GM-2 API. (http://www.myri.com/scs/GM-2/doc/refman.pdf)

19. Boden, N.J., Cohen, D., Felderman, R.E., Kulawik, A.E., Seitz, C.L., Seizovic, J., Su, W.: Myrinet - A gigabit per second local area network. In: IEEE Micro. (1995) 29–36

Dynamic Interoperable Message Passing

Michal Kouril[1] and Jerome L. Paul[2]

Department of ECECS, University of Cincinnati,
Cincinnati, OH 45221-0030, USA
{mkouril, jpaul}@ececs.uc.edu

Abstract. In this paper we present two solutions to dynamic interoperable MPI communication. These solutions use the Inter-Cluster Interface library that we have developed. The first solution relies on the MPI-2 Standard; specifically on general requests, threads and "external32" encoding. The second solution discusses adjustments to the first solution that allow its implementation in environments where some parts of the MPI-2 Standard are not implemented, and can even work independently of MPI. We have successfully implemented these solutions in a number of scenarios, including parallelizing SAT solvers, with good speedup results.

1 Introduction

When connecting multiple clusters there are two main paradigms and concomitant programming environments, *static* versus *dynamic*. In the static environment, all participating processes are captured at the outset. In the dynamic environment, processes can join and leave during the execution of the computation. Using the Inter-Cluster Interface (ICI) library [13] that we have developed, we have applied this dynamic environment to computationally intensive, hard problems (such as NP complete problems) with good speedup results. In particular, we have parallelized some SAT solvers, including SBSAT [15, 17] and zChaff [3], as well as our own special solver. The latter solver was targeted to determine exact values or improved lower bounds for van der Waerden numbers $W(k,n)$ [16]. Amongst other results, we have almost doubled the previously known lower bound on $W(2,6)$ from 696 to 1132. ICI together with our backtracking framework BkFr [9] is being used in an attempt to establish that this bound is sharp. The computation is estimated to take a several months, but it is critical that we use all available computing resources, including using heterogeneous clusters dynamically.

There is a large problem domain for which it is extremely advantageous to have both interoperability and dynamic cluster participation available to the programmer. Grid computing and the heterogeneity of the current grid make yet another case for having an interoperable way to dynamically connect running MPI programs. Current MPI standards do not explicitly support *interoperability* (communication between two different MPI implementations). On the other hand, the IMPI Standard, while it supports interoperability, it is limited to the static environment.

---

[1] Partially supported by DoD contract MDA-904-02-C-1162 and a fellowship grant of the Ohio Board of Regents.
[2] Partially supported by NSF Grant No. 9871345.

B. Di Martino et al. (Eds.): EuroPVM/MPI 2005, LNCS 3666, pp. 167–174, 2005.
© Springer-Verlag Berlin Heidelberg 2005

In the next section we summarize the existing implementations available for static and dynamic message passing programming environments. We then introduce our methods of solving the interoperability issue in the dynamic setting and discuss the performance data that we obtained.

2   Current Static and Dynamic Message Passing Environments

In our scenario we assume that the applications we wish to connect are started independently, as opposed to being spawned from the currently running task. In other words, there is no previous connection between the running tasks. We also assume that the applications are compiled by different MPI implementations, therefore making it impossible to directly use the existing functions introduced in MPI-2. In Table 1 we list the dynamic/interoperability status of the current standards. Although using compatible MPI implementations on both ends of intercluster communication is an option, usually every cluster has an optimized MPI implementation installed and compatible implementations might not be available. The user should not be expected to recompile MPI implementation on each cluster. Our intercluster interface library is small and provides the bare minimum for the intercluster communication, so that applications can take advantage of the optimized intracluster communication. ICI does not include collective communication functions at this time, since many applications, including those discussed in this paper, do not require this functionality.

Table 1. Dynamicity and interoperability supported by the current MPI Standards

Standard	Participating nodes	Interoperability status
IMPI	Static[3]	Interoperable
MPI-1	Static	Non-interoperable
MPI-2	Dynamic	Non-interoperable
Not currently available	Dynamic	Interoperable

Spreading static MPI-1 computations among multiple clusters without changing the source code has been addressed by PACX-MPI [6], PLUS [18], MPI-Glue [20] and MPICH-Madeleine [1]. The number of nodes for such computation is statically preallocated and no new dynamic connections are possible. The project StaMPI [19] also allows the spawning of new processes, but does not allow connecting independent processes. Additional projects touching on the subjects are MagPIe [12] and MPICH-G2 [11]. None of these projects cover dynamic connections of independent MPI implementations.

Interoperable MPI (IMPI) [10] addressed the problem of interconnecting different MPI implementations, but it is done statically on a group of preallocated nodes without the possibility of connecting additional nodes or clusters using the IMPI protocol.

Relative to the dynamic environment, MPI_Connect [4] allowed dynamic connections between clusters by forming inter-communicators. This project had similar goals

---

[3] Number of nodes participating over IMPI connections is static.

as our project, but it is no longer supported. MPI_Connect implementation was based on MetaComputing system SNIPE. MPI_Connect was succeeded by Harness/FT-MPI [2] which is a light-weight MPI implementation with a focus on fault tolerance, which retained the dynamic communication capabilities from MPI_Connect. Harness/FT-MPI has its own MPI implementation, and therefore is not interoperable.

3   The MPI Standards Needed for Our Solutions

We now briefly describe the parts of the MPI Standards necessary to implement our solutions given in section 4 (see the MPI Standards [14, 7] for additional details). MPI-2 extends the existing MPI-1 Standard by adding a number of new functions. In particular, communication functions were added between two sets of MPI processes that do not share a communicator, but no standard was mandated for interoperable communication. However, the new functions for establishing communication, such as `MPI_Comm_connect` and `MPI_Comm_join`, have a few associated specifics that are not interoperable, such as port names, `MPI_Info` parameters, and communication protocols. Thus, implementers must decide how to communicate (no protocol specification is given), thereby effectively making any communication implementation specific, i.e., not interoperable.

The MPI-2 Standard defines `MPI_Init_thread` (to be used instead of `MPI_Init`) specifically targeted towards threaded applications. This function returns the highest level of thread safety the implementation can provide.

Three functions (`MPI_Pack_external`, `MPI_Pack_external_size` and `MPI_Unpack_external`) were added that allow conversion of the data stored in the internal format to a portable format "external32". This format ensures that all MPI implementations will interpret the data the same way. For example "external32" uses big-endian for storing integral values, big-endian IEEE format for floating point values, and so forth. The MPI-2 Standard mentions that this could be used to send typed data in a portable way from one MPI implementation to another. In both of our solutions, ICI allows the utilization of these functions when they are available.

Finally, MPI-2 added a group of functions called *generalized requests* allowing for the definition of additional non-blocking operations and their integration into MPI. These functions are `MPI_Grequest_start` and `MPI_Grequest_complete`. `MPI_Grequest_start` returns a handle that can be used in MPI functions such as `MPI_Test` and `MPI_Wait`. The completion of the operation is signaled to MPI using `MPI_Grequest_complete`. Typically the non-blocking operation runs in a separate thread, so that it is only usable in a sufficiently thread-safe environment.

While MPI-2 also added functions for spawning new processes, they are not considered in this paper, since in our problem domain parallel applications are typically started independently.

4   Two Solutions for Dynamic Interoperable Communication

We will describe two solutions to the problem of establishing interoperable dynamic connections, where the first solution assumes the relevant functions from the MPI-2 Standard are available, and the second is suitable for environments in which some of the relevant functions in this standard are missing. In fact, the second solution creates a communication layer that is independent of MPI and is usable with other communication libraries. In the both solutions, a user could choose to call ICI functions directly and thereby explicitly recognize when the communication is with a process of another MPI implementation. On the other hand, a user could use the profiling interface where interoperable communication is hidden from the user.

4.1 A Solution That Assumes a Thread-Safe Implementation of MPI-2

Together with ICI, the following set of functions from the MPI-2 Standard can be used to create dynamic interoperable MPI communication:

1. Canonical `MPI_Pack_external` and `MPI_Unpack_external` for interoperability.
2. Generalized requests for integration with MPI.
3. Full thread-safeness.
4. Profiling interface for transparent integration with MPI (optional).

The ICI library has to be used on both ends of the communication. The library uses the existing MPI implementations and ICI low level functions (such as `recv_data`, `send_data` and so on). The ICI functions use the Berkeley TCP/IP API interface, which is available on the majority of platforms.

In Table 2 we list the basic ICI communication functions required to support dynamic interoperable communication. These functions are counterparts to MPI functions having the same arguments, with the possibility of using the profiling interface. In Table 2 we only show a sample set of MPI functions and their ICI counterparts. The full set would contain all MPI functions that allow communication using point-to-point communicators.

Table 2. A sample set of ICI functions

MPI	ICI counterpart	Description
MPI_Send	ICI_Send	Send data, blocking
MPI_Recv	ICI_Recv	Receive data, blocking
MPI_Isend	ICI_Isend	Send data, non-blocking
MPI_Irecv	ICI_Irecv	Receive data, non-blocking
MPI_Probe	ICI_Probe	Blocking wait for message
MPI_Comm_connect	ICI_Connect	Initiate connection, blocking
-	ICI_Iconnect	Initiate connection, non-blocking

4.2 A Solution for Implementations Missing Some Functionality of the MPI-2 Standard

For interoperability, we encode the data using the MPI-2 "external32" format by `MPI_Pack_external` and `MPI_Unpack_external`. This encoding incurs overhead during sending and receiving data, but guaranties interoperability not only among implementations, but also among platforms and used high-level languages.

Generalized requests are used to integrate inter-cluster communication into the existing MPI code. Non-blocking `ICI_Isend` and `ICI_Irecv` calls setup a request handle using `MPI_Grequest_start`, execute in a separate thread and call `MPI_Grequest_complete` once the operation is complete. It is therefore possible to use ICI just like MPI for point-to-point communication; that is, analogous to communication using an inter-communicator. Since the ICI communication happens in a separate thread, `MPI_Grequest_complete` is called in a separate thread and therefore the MPI implementation has to be fully thread-safe (see Section 5).

We chose not to use the MPI profiling interface to create a uniform API interface for intra-cluster and inter-cluster communication, since its use would negatively impact performance. However, it could be added in future ICI implementations.

In addition to dynamic interoperability, the ICI library allows implementing properties such as encryption, authentication, tunneling, and so forth.

Although some of the MPI implementations comply with the MPI-2 Standard, at the time of writing this paper the implementation level and thread safety is low (see Section 5). We now discuss how the solution in Section 4.1 can be adjusted for the current state of the implementations when one or more of the following functionalities are missing. In the special case where MPI is not even available, the missing MPI functionality can be easily obtained by substituting appropriate ICI functions.

1. Implementations missing the canonical `MPI_Pack_external` and `MPI_Unpack_external` functions. These functions are not only platform dependent but also language dependent. However, it is feasible to implement them using `MPI_Type_get_envelope` and `MPI_Type_get_contents` which are generally available.
2. Implementations missing generalized requests. Instead of integrating ICI communication with MPI, we substituted blocking functions that potentially use both MPI and ICI communication, such as using `MPI_WaitAny` with a polling loop of non-blocking tests of ICI and MPI communication. In our applications the communication is not the critical component, and therefore the polling overhead is acceptable.
3. Implementations missing thread–safety. This can be overcome by implementing not only blocking ICI communication within ICI function calls, but also enabling progress checks on non-blocking ICI communication within all ICI function calls. In our pseudocode this is called the `do_background_ops` function.

This solution was implemented as part of the BkFr project [9], and tested with MPI-1 implementations MPICH [8] and LAM [5] with good results.

5   Feasibility of Each Solution in Current Implementations

The first solution places the most requirements on the implementation level of the MPI-2 Standard. Moreover, this solution requires full thread-safety, which renders it unusable for the widely available MPI implementations shown in Table 3.

Table 3. MPI-2 implementation level in various MPI implementations

	Thread safety	**MPI_Grequests**	**MPI_Pack_external**
MPICH 1.2.7	FUNNELED	N	N^4
MPICH2 1.0.2	FUNNELED	Y	Y
MPICH G2 1.2.5.3	SINGLE	N	N^4
LAM 7.1.1	SERIALIZED	N	N^4
SUN HPC 5.0	SINGLE	Y	Y

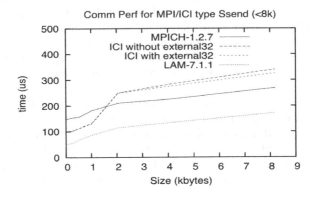

Fig. 1. Performance data for small and midsize messages

Our tests of the second solution show a comparison of MPICH 1.2.7, LAM 7.1.1 with our implementation of the ICI communication with and without encoding the messages into an interoperable format "external32." The data were measured on AMD MP-1800+ cluster connected by 1000Base-T adapters using modified perftest 1.3a (performance testing tool included with MPICH). MPICH and LAM data were measured using perftest as a communication performance between two nodes of the cluster during intracluster communication. ICI data we measured also using perftest as a communication performance between the same two computers as with LAM and MPICH but the ICI library as described in the Section 4.2 was used. In the prototype the exchanged messages were MPI_INT type typed arrays.

Figure 1 shows the performance for small and midsize messages up to 8K. LAM outperforms both ICI and MPICaH and ICI with and without "external32" outperforms MPICH for small messages (size < 1k). Figure 2 shows performance for long messages. The overhead of "external32" encoding is obvious and ICI with the "external32" encoding is outperformed by LAM and MPICH. Although the performance was not our primary goal for small and large messages (size<1000 and size>32K), the ICI implementation without "external32" encoding outperforms MPICH 1.2.7. For small messages the "external32" encoding poses only negligible overhead, which increases with the increasing message size up to 58%, after which it remains constant (see Figure 3).

---

[4] MPI_Type_get_contents and MPI_Type_get_envelope are available.

Table 4. Performance comparisons of typical instances run on the BkFr with and without ICI

	One node	**One cluster**	**2 clusters[5] connected via ICI**
Sum of subsets	1:13:27	9:04	3:36
BkFr SAT SBSAT	1:07:13	7:14	5:06
BkFr SAT zChaff	50:59	4:13	2:35

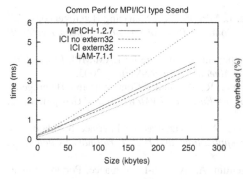

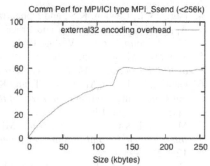

Fig. 2. Performance data for long messages   **Fig. 3.** Perf. data for encoding overhead

The performance of the BkFr[9] project utilizing ICI is shown in the Table 4 and further described in [9].

6 Conclusion

There are many MPI applications where it would be very advantageous to dynamically add clusters, with possibly varying MPI implementations, to an ongoing computation. Using resources dynamically as they become available is currently possible, but the MPI-2 Standard lacks interoperability. We presented two solutions to the dynamic interoperable problem, both of which isolate inter-cluster communication within a separate library layer (ICI), thereby possibly introducing stronger fault tolerant capabilities for inter-cluster communication. The first solution utilizes MPI-2 general requests, threads and "external32" encoding. The second solution provides adjustments to the first solution assuming that some parts of the MPI-2 Standard are not implemented, including the absence of thread safeness. We have implemented a prototype version of the second solution, and successfully utilized it in a general backtracking framework BkFr [9] over multiple clusters, as well as verifying its feasibility in various environments and implementations. The results show that ICI efficiently handles inter-cluster communication, and that BkFr running SAT instances that scale well within the cluster continue to scale well into multiple clusters. We hope that dynamic interoperable communication will eventually become part of the MPI Standard.

---

[5] 16 node AMD 1800+ and 32 node PIII-450 clusters.

References

1. Aumage, O., Mercier, G. and Namyst, R., MPICH/Madeleine: a True Multi-Protocol MPI for High Performance Networks. *IPDPS*, 2001.
2. Dongarra, J., Fagg, G.E., Geist, G.A., Kohl, J.A., Papadopoulos, P.M., Scott, S.L., Sunderam, V. and Magliardi, M. HARNESS: Heterogeneous Adaptable Reconfigurable NEtworked SystemS. *HPDC*, 1998.
3. Moskewicz, M.W., C. Madigan, Zhao, Y., Zhang, L., and Malik, S., Engineering an Efficient SAT Solver. *Proceedings of the 38th ACM/IEEE Design Automation Conference* (2001).
4. Fagg, G. E, London, K.S., and Dongarra, J., MPI_Connect Managing Heterogeneous MPI Applications Interoperation and Process Control. *Proc. of the 5th European PVM/MPI Users' Group*, Springer-Verlag, 1998.
5. Squyres, J.M. and Lumsdaine, A., A Component Architecture for LAM/MPI. *Proceedings of 10th European PVM/MPI Users' Group Meeting*, Springer-Verlag, 2003.
6. Gabriel, E., Resch, M., Beisel, T. and Keller, R., Distributed Computing in a Heterogeneous Computing Environment. *Proc. of the 5th European PVM/MPI Users' Group Meeting*, Springer-Verlag, 1998.
7. Message Passing Interface Forum. MPI-2: A Message-Passing Interface Standard. *The International Journal of Supercomputer Applications and High Performance Computing*, 12(1-2), 1998.
8. Gropp, W., Lusk, E. Doss, N., and Skjellum, A., A High-Performance, Portable Implementation of the MPI Message-Passing Interface Standard. *Parallel Computing,* 1996.
9. Kouril, M. and Paul, J.L., A Parallel Backtracking Framework (BkFr) for Single and Multiple Clusters. *Conf. Computing Frontiers*, ACM Press, 2004.
10. IMPI Steering Committee: IMPI - Interoperable Message-Passing Interface, 1998. http://impi.nist.gov/IMPI/.
11. Karonis N., Toonen B. and Foster I., MPICH-G2: A Grid-Enabled Implementation of the Message Passing Interface. *Journal of Parallel and Distributed Computing,* 2003.
12. Kielmann, T., Hofman, R.F.H., Bal, H.E., Plaat, A. and Bhoedjang, R.A.F., MagPIe: MPI's Collective Communication Operations for Clustered Wide Area Systems. *ACM SIG-PLAN Notices*, 1999.
13. Kouril, M, Paul, J.L., Dynamic Interoperable Point-to-Point Connection of MPI Implementations. Brief Announcement. *Twenty-Fourth Annual ACM SIGACT-SIGOPS Symposium on Principles of Distributed Computing*, 2005.
14. Message Passing Interface Forum: MPI: A Message-Passing Interface Standard. *International Journal of Supercomputer Applications*, 8(3/4), 1994. Special issue on MPI
15. Franco, J., Kouril, M., Schlipf, J. S., Ward, J., Weaver, S., Dransfield, M., Vanfleet, W. M., SBSAT: a state-based, BDD-based Satisfiability solver. LNCS 2919, Springer, 2004.
16. Kouril, M., Franco, J., Resolution Tunnels for Improved SAT Solver Performance. *Eighth International Conference on Theory and Applications of Satisfiability Testing*, 2005.
17. Franco, J., Kouril, M., Schlipf, J. S., Weaver, S., Dransfield, M., Vanfleet, W. M., Function-complete lookahead in support of efficient SAT search heuristics. *Journal of Universal Computer Science*, Know Center and IICM, Graz University, Austria, 2004.
18. Brune, M., Gehring, J., Reinefeld, A.: A Lightweight Communication Interface for Parallel Programming Environments. *HPCN'97*, Springer-Verlag, 1997.
19. Imamura, T., Tsujita, Y., Koide, H., and Takemiya, H. 2000. An Architecture of Stampi: MPI Library on a Cluster of Parallel Computers. *Proc. of the 7th European PVM/MPI Users' Group Meeting*, 2000.
20. Rabenseifner, R., MPI-GLUE: Interoperable High-Performance MPI Combining Different Vendor's MPI Worlds. *Euro-Par*, 1998.

Analysis of the Component Architecture Overhead in Open MPI

B. Barrett[1], J.M. Squyres[1], A. Lumsdaine[1],
R.L. Graham[2], and G. Bosilca[3]

[1] Open Systems Laboratory, Indiana University
{brbarret, jsquyres, lums}@osl.iu.edu
[2] Los Alamos National Lab
rlgraham@lanl.gov
[3] Innovative Computing Laboratory, University of Tennessee
bosilca@cs.utk.edu

Abstract. Component architectures provide a useful framework for developing an extensible and maintainable code base upon which large-scale software projects can be built. Component methodologies have only recently been incorporated into applications by the High Performance Computing community, in part because of the perception that component architectures necessarily incur an unacceptable performance penalty. The Open MPI project is creating a new implementation of the Message Passing Interface standard, based on a custom component architecture – the Modular Component Architecture (MCA) – to enable straightforward customization of a high-performance MPI implementation. This paper reports on a detailed analysis of the performance overhead in Open MPI introduced by the MCA. We compare the MCA-based implementation of Open MPI with a modified version that bypasses the component infrastructure. The overhead of the MCA is shown to be low, on the order of 1%, for both latency and bandwidth microbenchmarks as well as for the NAS Parallel Benchmark suite.

1 Introduction

MPI implementations are designed around two competing goals: high performance on a single platform and support for a range of platforms. Vendor optimized MPI implementations must support ever evolving hardware offerings, each with unique performance characteristics. Production quality open source implementations, such as Open MPI [6], LAM/MPI [13], and MPICH [9], face an even wider range of platform support. Open MPI is designed to run efficiently on platforms ranging from networks of workstations to custom built supercomputers with hundreds of thousands of processors and high speed interconnects.

Open MPI meets the requirements of high performance and portability with the Modular Component Architecture (MCA), a component system designed for High Performance Computing (HPC) applications. While component based programming is widely used in industry and many research fields, it is only recently gaining acceptance in the HPC community. Most existing component

B. Di Martino et al. (Eds.): EuroPVM/MPI 2005, LNCS 3666, pp. 175–182, 2005.

architectures do not provide the low overheads necessary for use in HPC applications. Existing architectures are generally designed to provide features such as language interoperability and to support rapid application development, with performance as a secondary concern.

In this paper, we show that Open MPI's MCA design provides a component architecture with minimal performance implications. Section 2 presents similar work for other component architectures. An overview of the Open MPI architecture is presented in Section 3, focusing on the component architecture. Finally, Section 4 presents performance results from our experiments with Open MPI.

2   Related Work

Component architectures have found a large degree of success in commercial and internet applications. Enterprise JavaBeans, Microsoft COM and DCOM, and CORBA provide a rich environment for quickly developing a component based application. These environments focus on industrial applications, providing reasonable performance for such applications. However, either the languages supported or the overheads involved make them unsuitable for the high performance computing community. Literature on the performance of such component infrastructures is sparse, most likely due to the fact that performance is not a concern for the intended uses of these component architectures.

The Common Component Architecture (CCA) [2] is designed to provide a high performance component architecture for scientific applications. Bernholdt et. al. [3] study the overheads involved in the CCA design and found them to be small, on the order of two extra indirect function calls per invocation. CCA components are designed to be large enough that component boundaries are not crossed for inner loops of a computation. Therefore, the overhead of CCA is negligible for most applications. Much of the overhead is due to inter-language data compatibility, an overhead that is not applicable in Open MPI.

3   Open MPI Architecture

Open MPI is a recently developed MPI implementation, tracing its history to the LAM/MPI [13], LA-MPI [8], FT-MPI [5], and PACX-MPI [10] projects. Open MPI provides support for both MPI-1 and MPI-2 [7,12].[1] Open MPI is designed to be scalable, fault tolerant, and provide high performance in a variety of HPC environments. The use of a component architecture allows for a well architected code base that is both easy to test across multiple configurations and easy to integrate into a new platform.

3.1   Component Architecture

The Modular Component Architecture (MCA) is designed to allow users to build a customized version of Open MPI at runtime using components. The high over-

---

[1] One-sided support is scheduled to be added to Open MPI shortly after the first public release.

heads generally associated with CORBA and COM are avoided in the MCA by not supporting inter-process object communication or cross-language support – MCA components provide a C interface and interface calls are local to the MPI process. Components are opened and loaded at runtime on demand, using the GNU Libtool `libltdl` software package for portable dynamic shared object (DSO) handling. Components can also be linked into the MPI library for platforms that lack support from `libltdl` or when a static library is desired. Current MPI level component frameworks include point-to-point messaging, collective communication, MPI-2 I/O, and topology support. The runtime infrastructure for Open MPI include component frameworks for resource discovery, process startup, and standard I/O forwarding, among others.

In order to provide a manageable number of measurements while still measuring the overhead of the MCA design, we focus on the components directly responsible for MPI point-to-point communication for the remainder of this paper. Many common MPI benchmarks are based primarily on point-to-point communication, providing the best opportunities for analyzing the performance impact of the MCA on real applications.

3.2   MPI Point-to-Point Design

Open MPI implements MPI point-to-point functions on top of the Point-to-point Management Layer (PML) and Point-to-point Transport Layer (PTL) frameworks (Fig. 1). The PML fragments messages, schedules fragments across PTLs, and handles incoming message matching. Currently, there is one PML component, TEG [14]. TEG is designed to support message fault tolerance, recovery from corrupted data, and dropped packets.[2] It can also simultaneously use multiple communication channels (PTLs) for a single message. The PTL provides an interface between the PML and underlying network devices.

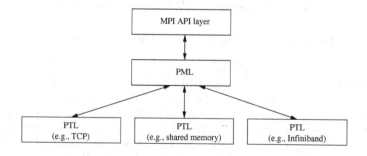

Fig. 1. Open MPI component frameworks for MPI point-to-point messages

---

[2] These features remain under active development and may not be available in the first release of Open MPI.

4   Component Overhead Analysis

The MCA design's primary source of overhead is the use of indirect calls through function pointers for dispatching into a component. There are two designs for calling into components in the MCA, depending on how many component instances (modules) are active within the framework. For frameworks like the PML, where only one module is active per process, a global structure is used to hold the set of function pointers. The address of the global structure is known at link time. In the case of the PTL frameworks, there are multiple components active, so there is not a single global structure of function pointers. Instead, there are multiple tables stored by the caller of the framework, the PML in this case. The PML must compute the address of the function pointer in a PTL structure, load the value of the function pointer, and make the function call.

To measure the overhead of the component architecture in Open MPI, we added the ability to bypass the indirect function call overhead inherent in the MCA design. Calls from the MPI layer into the PML and from PML into the PTL are made directly rather than using the component architecture. The GM PTL, supporting the Myrinet/GM interconnect, was chosen because it offered a low latency, high bandwidth environment best suited for examining the small overheads involved in the MCA. The ability to "hard code" the PML is available as part of Open MPI as a configure time option. Bypassing the PTL component interface is not part of the Open MPI release, as it greatly limits the functionality of the resulting MPI implementation. In particular, bypassing the PTL component interface disables message striping over multiple devices and the ability to send messages to self. For the majority of the tests discussed in this paper, such limitations were not relevant to examining the overheads of the MCA.

Two variables relevant to the overhead of the MCA system for point-to-point communication are how `libmpi` is built and whether the MCA interface is used for point-to-point communication. Table 1 describes the MPI configurations used in testing. Open MPI alpha release r5408 was used for testing Open MPI and was modified to support bypassing the PTL component overhead. MPICH-GM

Table 1. Build configurations used in performance tests

Configuration	Description
MPICH-GM	Myricom MPICH-GM 1.2.6..14a, built using the default build script for Linux with static library
LAM/MPI	LAM/MPI 7.1.1, with GM support and static library
Open MPI shared DSO	Open MPI, `libmpi` shared library. Components dynamically loaded at runtime, using the component interface
Open MPI shared direct	Open MPI, `libmpi` shared library. Point-to-point components part of `libmpi`, bypassing the component interface
Open MPI static DSO	Open MPI, `libmpi` static library. Components part of `libmpi`, using the component interface
Open MPI static direct	Open MPI `libmpi` static library. Point-to-point components part of `libmpi`, bypassing the component interface

1.2.6..14a, the latest version of MPICH available for Myrinet,[3] and LAM/MPI 7.1.1 were used to provide a baseline performance reference.

All MPI tests were performed on a cluster of 8 dual processor machines connected using Myrinet. The machines contain 2.8 GHz Intel Xeon processors with 2 GB of RAM. A Myricom PCIX-D NIC is installed in a 64 bit 133 MHz PCI-X slot. The machines run Red Hat 8.0 with a Linux 2.4.26 based kernel and Myricom's GM 2.0.12. Additional CPU overhead tests were performed on a dual 2.0 GHz AMD Opteron machine with 8 GB of RAM running Gentoo Linux and the 2.6.9 kernel and an Apple Power Mac with dual 2.0 GHz IBM PPC 970 processors and 3.5 GB of memory running Mac OS X 10.4.

4.1   Indirect Function Call Overhead

Fig. 2 presents the overhead, measured as the time to make a call to a function with no body, for different call methods. A tight loop is used to make the calls, so all loads should be satisfied from L1 cache, giving a best case performance. The direct call result is the time to call a function in a static library, a baseline for function call overheads on a particular platform. The function pointer result is the cost for calling the same function, but with a load dependency to determine the address of the function. As expected, the cost is approximately the cost of a load from L1 cache plus the cost of a direct function call. Calling a function in a shared library directly (the shared library call result) requires indirect addressing, as the location of a function is unknown until runtime. There is some additional overhead in a shared library call due to global offset table (GOT) computations, so a direct call into a shared library is generally more expensive than an indirect call into a static library [11]. The unusually high overhead for the PPC 970 when making shared library calls is due to the Mach-O ABI used by Mac OS X, and not the PPC 970 hardware itself [1].

Function calls into a DSO are always made through a function pointer, with the address of the function explicitly determined at runtime using dlsym() (or similar). In modern DSO loader implementations, GOT computations are not required. The cost of calling a function in a DSO is therefore much closer to the cost of an indirect function call into a static library than a direct function call into a shared library. From this result, it should be expected that the performance impact of the component architecture in Open MPI will be more from the use of shared libraries than from the component architecture itself.

4.2   Component Effect on Latency and Bandwidth

MPI latency for zero byte messages using a ping-pong application and bandwidth using NetPIPE are presented in Fig. 3. All builds of Open MPI exhibit performance differences of less than 2%, with most of the performance difference related to whether Open MPI used shared or static libraries. Bypassing the component infrastructure for point-to-point messages shows little impact on

---

[3] At the time of writing, Myricom does not provide an MPICH-2 based implementation of MPICH-GM.

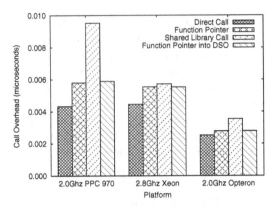

Fig. 2. Time to make a call into an empty function for a number of common architectures

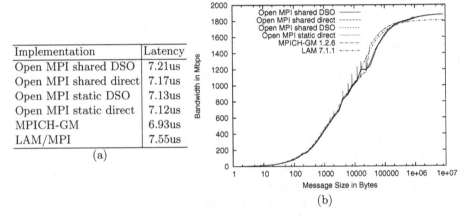

Implementation	Latency
Open MPI shared DSO	7.21us
Open MPI shared direct	7.17us
Open MPI static DSO	7.13us
Open MPI static direct	7.12us
MPICH-GM	6.93us
LAM/MPI	7.55us

(a)

(b)

Fig. 3. Latency of zero byte messages and NetPIPE bandwidth for component and direct call configurations of Open MPI

either latency or bandwidth. In the worst case, the MCA overhead was .04 microseconds, which is a fraction of the end-to-end latency for the GM software stack. The bandwidth results show Open MPI is comparable to MPICH-GM and LAM/MPI for small messages. For large messages, Open MPI is comparable to MPICH-GM and approximately 70 Mbps faster than LAM/MPI. Open MPI suffers a slight performance drop for messages between 32 KB and 256 KB when compared to LAM/MPI and MPICH-GM. The performance drop appears to be caused by our wire protocol, and should be solved through further tuning.

4.3   Component Effect on NAS Parallel Benchmarks

To approximate real application performance impact from Open MPI's use of components, the NAS Parallel Benchmark suite version 2.4 [4] was run with the build configurations described in Table 1. The results in Table 2 are for four processes, using the B sized benchmarks. Each process was executed on a

Table 2. NAS Parallel Benchmark results for Open MPI, MPICH-GM, and LAM/MPI using 4 processors and the B sized tests

Implementation	BT	CG	EP	IS	LU	MG	SP
Open MPI shared DSO	471.16s	95.58s	77.20s	4.37s	297.06s	12.12s	422.43s
Open MPI shared direct	475.91s	95.82s	77.33s	4.34s	298.49s	13.54s	422.16s
Open MPI static DSO	472.48s	95.08s	77.17s	4.35s	297.26s	12.96s	416.76s
Open MPI static direct	477.21s	95.15s	77.21s	4.28s	299.35s	13.50s	421.19s
MPICH-GM	475.63s	96.83s	77.14s	4.22s	296.98s	13.74s	421.95s
LAM/MPI	473.93s	99.54s	75.98s	4.01s	298.14s	13.69s	420.70s

separate node, to prevent use of the shared memory communication channel by configurations that support multiple interconnects. Each test was run five times, with the lowest time given. Variance between runs of each test was under 2%.

The CG and MG tests invoke MPI communication that requires sending a message to self. Due to the design of Open MPI, this requires multiple PTL components be active, which is disabled in the direct call PTL configuration. Therefore, the CG and MG direct call results are with only the PML component interface bypassed. Performance of the Open MPI builds is generally similar, with variations under 3% in most cases. Similar to Section 4.2, the NAS Parallel Benchmarks show that there is very little measurable overhead in utilizing the MCA in Open MPI. Open MPI performance is comparable to both LAM/MPI and MPICH-GM for the entire benchmark suite.

5   Summary

Open MPI provides a high performance implementation of the MPI standard across a variety of platforms through the use of the Modular Component Architecture. We have shown that the component architecture used in Open MPI provides negligible performance impact for a variety of benchmarks. Further, the Open MPI project provides performance comparable to existing MPI implementations, and has only recently begun optimizing performance. The component architecture allows users to customize their MPI implementation for their hardware at run time. Only features that are needed by the application are included, removing the overhead introduced by unused features.

Acknowledgments

This work was supported by a grant from the Lilly Endowment and National Science Foundation grants NSF-0116050, EIA-0202048 and ANI-0330620. Los Alamos National Laboratory is operated by the University of California for the National Nuclear Security Administration of the United States Department of Energy under contract W-7405-ENG-36. This paper was reviewed and approved as LA-UR-05-4576. Project support was provided through ASCI/PSE and the

Los Alamos Computer Science Institute, and the Center for Information Technology Research (CITR) of the University of Tennessee.

References

[1] Apple Computer, Inc. Mach-O Runtime Architecture for Mac OS X version 10.3. Technical report, August 2004.

[2] Rob Armstrong, Dennis Gannon, Al Geist, Katarzyna Keahey, Scott R. Kohn, Lois McInnes, Steve R. Parker, and Brent A. Smolinski. Toward a common component architecture for high-performance scientific computing. In *HPDC*, 1999.

[3] D. E. Bernholdt et al. A component architecture for high-performance scientific computing. *to appear in Intl. J. High-Performance Computing Applications.*

[4] Rob F. Van der Wijngaart. NAS Parallel Benchmarks version 2.4. Technical Report NAS Technical Report NAS-02-007, NASA Advanced Supercomputing Division, NASA Ames Research Center, October 2002.

[5] G. E. Fagg, A. Bukovsky, and J. J. Dongarra. HARNESS and fault tolerant MPI. *Parallel Computing*, 27:1479–1496, 2001.

[6] E. Garbriel et al. Open MPI: Goals, concept, and design of a next generation MPI implementation. In *Proceedings, 11th European PVM/MPI Users' Group Meeting*, 2004.

[7] A. Geist, W. Gropp, S. Huss-Lederman, A. Lumsdaine, E. Lusk, W. Saphir, T. Skjellum, and M. Snir. MPI-2: Extending the Message-Passing Interface. In *Euro-Par '96 Parallel Processing*, pages 128–135. Springer Verlag, 1996.

[8] R. L. Graham, S.-E. Choi, D. J. Daniel, N. N. Desai, R. G. Minnich, C. E. Rasmussen, L. D. Risinger, and M. W. Sukalksi. A network-failure-tolerant message-passing system for terascale clusters. *International Journal of Parallel Programming*, 31(4), August 2003.

[9] W. Gropp, E. Lusk, N. Doss, and A. Skjellum. A high-performance, portable implementation of the MPI message passing interface standard. *Parallel Computing*, 22(6):789–828, September 1996.

[10] Rainer Keller, Edgar Gabriel, Bettina Krammer, Matthias S. Mueller, and Michael M. Resch. Towards efficient execution of parallel applications on the grid: porting and optimization issues. *International Journal of Grid Computing*, 1(2):133–149, 2003.

[11] John R. Levine. *Linkers and Loaders*. Morgan Kaufmann, 2000.

[12] Message Passing Interface Forum. MPI: A Message Passing Interface. In *Proc. of Supercomputing '93*, pages 878–883. IEEE Computer Society Press, November 1993.

[13] J.M. Squyres and A. Lumsdaine. A Component Architecture for LAM/MPI. In *Proceedings, 10th European PVM/MPI Users' Group Meeting*, Lecture Notes in Computer Science, Venice, Italy, September 2003. Springer-Verlag.

[14] T.S. Woodall et al. TEG: A high-performance, scalable, multi-network point-to-point communications methodology. In *Proceedings, 11th European PVM/MPI Users' Group Meeting*, Budapest, Hungary, September 2004.

A Case for New MPI Fortran Bindings

C.E. Rasmussen[1] and J.M. Squyres[2]

[1] Advanced Computing Lab, Los Alamos National Laboratory
crasmussen@lanl.gov
[2] Open Systems Laboratory, Indiana University
jsquyres@open-mpi.org

Abstract. The Fortran language has evolved substantially from the Fortran 77 bindings defined in the MPI-1 (Message Passing Interface) standard. Fortran 90 introduced interface blocks (among other items); subsequently, the MPI-2 standard defined Fortran 90 bindings with explicit Fortran interfaces to MPI routines. In this paper, we describe the Open MPI implementation of these two sets of Fortran bindings and point out particular issues related to them. In particular, we note that strong typing of the Fortran 90 MPI interfaces with user-choice buffers leads to an explosion of interface declarations; each choice buffer must be expanded to all possible combinations of Fortran type, kind, and array dimension. Because of this (and other reasons outlined in this paper), we propose a new set of Fortran MPI bindings that uses the intrinsic ISO_C_BINDING module in Fortran 2003. These new bindings will allow MPI interfaces to be defined in Fortran that directly invoke their corresponding MPI C implementation routines – no additional layer of software to marshall parameters between Fortran and C is required.

1 Introduction

The MPI-1 (Message Passing Interface) standard [5] has been very successful, in part, because it provided MPI bindings in both C and Fortran. Thus, programmers were able to write parallel message passing applications in the language of their choice. Most implementations of MPI are written in C (or C++) and provide a thin translation layer to effect the Fortran bindings.

The MPI-2 standard [4] continued this successful treatment of language interoperability by tracking the Fortran standard as it evolved by defining Fortran 90 bindings using *explicit* interfaces. Similar to the benefits enjoyed by C and C++ programmers, these new Fortran bindings allow the Fortran compiler to fail to compile a program if actual procedure arguments do not conform to the dummy arguments specified by the standard. This level of type safety (at the procedure call) is not possible with the original *implicit* Fortran 77 bindings.

Section 2 provides a brief overview of common implementation techniques and problems associated with the Fortran 77 bindings. Section 3 discusses the Open MPI [3] approach to implementing the Fortran 90 MPI bindings. It also includes details of automatic code generation techniques as well as practical problems that arise from the Fortran 90 MPI bindings specification. In particular, the strong typing of the Fortran 90 explicit interfaces require the specification

B. Di Martino et al. (Eds.): EuroPVM/MPI 2005, LNCS 3666, pp. 183–190, 2005.
© Springer-Verlag Berlin Heidelberg 2005

of a separate interface for each potential type, kind, and array dimension that could be specified by a user for an MPI user-choice buffer argument. This interface explosion for generic MPI procedures with choice buffer arguments is unattractive and can lead to extremely long build times for the MPI library. In response to new advances in the Fortran language standard [2], and to specific problems with the existing Fortran bindings discussed in Section 3, new Fortran MPI bindings are proposed in Section 4.

2   Open MPI Fortran 77 Bindings

Before the 1990 Fortran standard was introduced, Fortran did not provide the ability to explicitly define interfaces describing external procedures (functions and subroutines). In Fortran 77, external interfaces must be *inferred* from the parameters provided in the call to the external procedure. Thus, while the MPI Fortran 77 bindings define standard Fortran interfaces for calling the MPI library, the Fortran compiler does not check to ensure that correct types are supplied to the MPI routines by the programmer.

This lack of type safety actually makes it easier for an MPI implementation to provide a layer of code bridging between user Fortran and an MPI C implementation. Many MPI routines take the address of a data buffer as a parameter and a count, representing the length of the buffer (e.g., MPI_SEND). Since the Fortran convention is to pass arguments by address, virtually any Fortran type can be supplied as the data buffer, including basics scalar types (e.g., real) and arrays of these types.

In most MPI implementations, the Fortran bindings are a thin translation layer that marshals parameters between Fortran and C and invokes corresponding back-end C MPI functions. For scalar types, this only requires dereferencing pointers from Fortran before passing to the C implementation routines; array-valued parameters may be passed directly. MPI handle parameters must also be converted (typically either by pointer dereferencing or table lookup) to the back-end C MPI objects.

2.1   Issues

The primary difficulty in developing MPI Fortran 77 bindings is that Fortran does not define a standard for compiler generated symbols, For example, the symbol for MPI_SEND may be MPI_SEND or mpi_send, followed by one or two underscores (it will likely not be the C symbol MPI_Send). However this uncertainty is relatively easy to overcome and various strategies have evolved over time.

While type safety is still an issue (a programmer may mistakenly supply a real type for a buffer count parameter, for example), the Fortran 77 bindings have been successfully used in practice for many years. However, it should be noted that these bindings are implemented outside of the Fortran language specification and may fail in future compiler versions. For example, an MPI subroutine with choice arguments may be called with different arguments types. This violates the letter of the Fortran standard, although such a violation is common practice [5].

3 Open MPI Fortran 90 Bindings

Several enhancements were made to Fortran in the 1990 standard. MPI-2 defined a Fortran 90 module and support for additional Fortran intrinsic numeric types. MPI interfaces could therefore both be defined for and expressed in Fortran. High-quality MPI implementations are encouraged to provide strong type checking in the MPI module, allowing the compiler to enforce consistency between the parameters supplied by the programmer and those defined in the MPI standard.

While this enhances type safety, there is no Fortran equivalent of the C (void *) data type used by the MPI C standard to declare a generic data buffer. *Every* data type that could conceivably be used as a data buffer must be declared in an *explicit* Fortran interface. The only fallback, for instance for user-defined data types, is for the programmer to resort to the older Fortran 77 *implicit* interfaces.

Not only must interfaces be defined for arrays of each intrinsic data type, but for each array dimension as well. Depending on the compiler, there may be approximately 15 type / size combinations.[1] Each of these combinations can be paired with up to seven array dimensions. With approximately 50 MPI functions that have one choice buffer, this means that 5,250 interface declarations must be specified (i.e., 15 types × 7 dimensions × 50 functions). Note that this does not include the approximately 25 MPI functions with two choice buffers. This leads to an additional 6.8M interface declarations (i.e., $(15 \times 7 \times 25)^2$). Currently, no Fortran 90 compiler can compile a module with this many interface functions.

3.1 Code Generation

Because of the large number of separate interfaces that the MPI standard requires, automatic generation of this code is an attractive option. Chasm [6] was used to accomplish this task.

Chasm is a toolkit providing language interoperability between Fortran 90 and C / C++. It uses static analysis to produce a translation layer between language pairs by first parsing source files to produce an XML representation of existing interfaces and then using XSLT stylesheets to generate the final bridging code. Fig. 1 depicts the code generation process.

There are several different types of files generated by the Chasm XSLT stylesheets. The primary file is the MPI module declaring explicit Fortran interfaces for each MPI function. Similar to a C header file, this file allows the Fortran compiler to check the actual parameter types supplied by user applications to make sure they conform to the interface. The actual Open MPI implementation of these interfaces is the Fortran 77 binding layer (see Section 2). In addition, there are separate files generated to test each MPI function.

MPI functions with choice parameters are handled somewhat differently. They require an additional translation layer to convert Fortran array-valued parameters to C pointers when invoking the corresponding Fortran 77 binding.

---

[1] Assuming the compiler supports CHARACTER, LOGICAL{1,2,4,8}, INTEGER{1,2,4,8}, REAL{4,8,16}, and COMPLEX{8,16,32}.

Fig. 1. Code generation for Fortran 90 MPI bindings in Open MPI

All of the XSLT stylesheets take the XML file `mpi.h.xml` as input. This file was created by the Chasm tools from the Open MPI `mpi.h` header file and subsequently altered slightly by hand to add additional information. An example of the annotations made to `mpi.h.xml` was the addition of the name `ierr` to MPI functions returning an error parameter. This gave notice to the Chasm XSLT stylesheets to create an interface for a subroutine rather than a function, with the `ierr` return value as the last parameter to the procedure (Fortran intent(out)) as defined by the MPI Fortran bindings. Another example of the modifications to `mpi.h.xml` was the "choice" tag, added to specify MPI choice (`void *`) arguments. This allowed the XSLT stylesheets to create explicit interfaces for each possible type provided by the programmer.

3.2 Issues

While the Fortran 90 MPI bindings allow explicit type checking, there are a number of issues with these bindings and with the Open MPI implementation. No explicit interfaces were created for MPI functions with multiple choice parameters (e.g., MPI_ALLREDUCE), because this would have exploded the type system to unmanageable proportions. User application utilizing these functions access the Fortran 77 layer directly with no type checking.

4   Proposed MPI Fortran `BIND(C)` Interfaces

The Fortran 2003 standard [2] contains a welcome addition that vastly improves language interoperability between Fortran and C. These additions are summarized in this section and a new set of Fortran MPI bindings based on this standard is proposed.

4.1 Fortran 2003 C Interoperability Standard

The Fortran 2003 standard includes the ability to declare interfaces to C procedures within Fortran itself. This is done by declaring procedure interfaces as BIND(C) and employing only interoperable arguments. It allows C function names to be given explicitly and removes the mismatch between procedure symbols generated by the C and Fortran compilers. Fortran BIND(C) interoperable types include primitive types, derived types or C structures (if all attributes are interoperable), C pointers, and C function pointers. BIND(C) interfaces are callable from either Fortran or C and may be implemented in either language.

This standard greatly simplifies language interoperability because it places the burden on the Fortran compiler to marshall procedure arguments and to create interoperable symbols for the linker, rather than placing the burden on the programmer. This includes the ability to use pass arguments by value. For MPI, this means that MPI C functions may be called directly from Fortran rather than from an intermediate layer. No additional work is needed other than to declare Fortran interfaces to the MPI C functions.

4.2 MPI C Type Mappings

The intrinsic ISO_C_BINDING module in the Fortran 2003 standard provides mappings between Fortran and C types. This mapping includes Fortran equivalents for the C types commonly used in MPI functions. For example, C integers, null-terminated character strings, and function pointers are declared in Fortran as INTEGER(C_INT), CHARACTER(C_CHAR), and TYPE(C_FUNPTR), respectively.

Most importantly, the ISO_C_BINDING module defines a Fortran equivalent to MPI choice (void *) buffers (TYPE(C_PTR)). This directly solves the interface explosion problem. It also allows interfaces to be declared for MPI functions with multiple choice buffers. In addition, the ISO_C_BINDING module provides functions for converting between Fortran pointers (including pointers associated with arrays) and the C_PTR type.

4.3 MPI_Send Example

An example of the proposed BIND(C) interface for MPI_SEND is shown in Fig. 2.

Note the explicit name attribute given to the BIND(C) declaration in line 3 of Fig. 2. This attribute instructs the Fortran compiler to create the equivalent C symbol of the provided name. The MPI_C_BINDING module name is proposed in line 5 to distinguish it from the Fortran 90 MPI module name. The value attribute used in lines 6-11 instructs the Fortran compiler to use pass-by-value semantics and means that no dereferencing of the arguments need be done on the C side.

The TYPE(C_PTR) declaration in line 6 is the Fortran BIND(C) equivalent of a C (void *) parameter. The usage of this generic C pointer declaration removes the interface explosion for the Fortran 90 MPI_SEND implementation, as described in the previous section. A C_PTR can be obtained from a Fortran

Fig. 2. BIND(C) interface declaration for MPI_SEND

scalar or array variable that has the TARGET attribute via the C_LOC() intrinsic function. The TARGET attribute must be used in Fortran to specify any variable to which a Fortran pointer may be associated.

The Fortran equivalent of MPI handle types are declared in lines 8 and 11. The MPI_HANDLE_KIND attribute must be defined by the MPI implementation and allows flexibility in specifying the size of an MPI handle. At this point it is uncertain if MPI handle types declared in this way will work across all MPI implementations without the need for extra marshalling by the MPI library. The form proposed here should be considered tentative until MPI implementors can consider the consequences of this choice.

While not shown here, there are also interoperable equivalents for null terminated C strings (CHARACTER(C_CHAR), DIMENSION(*)) and C function pointers (C_FUNPTR). In addition, variables defined in the scope of the MPI_C_BINDING module may interoperate with global C variables, further merging the Fortran bindings with the MPI C implementations.

4.4   Issues

Unlike the MPI Fortran 77 and 90 bindings, the proposed bindings describe language interoperability within the Fortran language. Therefore the proposed bindings are guaranteed to work by the Fortran compiler (and companion C compiler) and are not just *expected* to work for a particular Fortran compiler vendor and version. Users will be expected to do some parameter conversions themselves, as noted above in regards to the use of the C_LOC intrinsic function.

In addition, the Fortran 2003 standard is new and vendors are just coming out with ISO_C_BINDING module implementations. Therefore, a period of time will be needed before one can test the proposed features against existing MPI implementations.

5   Conclusions

Because of evolving Fortran language standards and limitations in the MPI Fortran 77 and 90 bindings, we have proposed a new set of Fortran MPI bindings based on the intrinsic ISO_C_BINDING module. These new bindings have several distinct advantages:

1. They solve the interface explosion problem of the Fortran 90 bindings through the use of TYPE(C_PTR). This new type allows a direct mapping to and from the C (void *) choice buffers.
2. They allow direct calls to the MPI C implementation from Fortran. This is more efficient and is less error prone, as the MPI implementor does not need to maintain and test an extra binding layer. The Fortran compiler is responsible for marshalling between C and Fortran data types, not the MPI library.
3. The names of the C functions implementing the MPI procedures can be specified in Fortran. This means that the tricks required to create common symbols between compilers are no longer needed.
4. The proposed bindings are defined entirely within the Fortran language and are guaranteed to work by the Fortran compiler. The proposed bindings are not compiler dependent. While not likely, Fortran 77 bindings are implemented outside of the Fortran language specification and may fail in future compiler versions.

It should be pointed out that as of this writing, only two major compiler vendors support the Fortran 2003 ISO_C_BINDING module (others will likely do so by the fall of 2005 [1]). However, even this support is partial. Thus, there exists a window of opportunity to consider and modify the proposed bindings before widespread adoption.

To this end, we will post the full set of new Fortran bindings and a reference implementation on the Open MPI web site (http://www.open-mpi.org/) and solicit comments and feedback (at mpi-comments@www.mpi-forum.org) from the Fortran HPC community.

Acknowledgments

This work was supported by a grant from the Lilly Endowment and National Science Foundation grants EIA-0202048 and ANI-0330620.

Los Alamos National Laboratory is operated by the University of California for the National Nuclear Security Administration of the United States Department of Energy under contract W-7405-ENG-36.

References

1. Personal communication with compiler vendors. Meeting 168 of the J3 Fortran Standards Committee, August 2004.
2. Fortran 2003 Final Committee Draft, J3/03-007R2. see www.j3-fortran.org.

3. E. Garbriel, G.E. Fagg, G. Bosilica, T. Angskun, J. J. Dongarra J.M. Squyres, V. Sahay, P. Kambadur, B. Barrett, A. Lumsdaine, R.H. Castain, D.J. Daniel, R.L. Graham, and T.S. Woodall. Open mpi: Goals, concept, and design of a next generation mpi implementation. In *Proceedings, 11th European PVM/MPI Users' Group Meeting*, 2004.
4. A. Geist, W. Gropp, S. Huss-Lederman, A. Lumsdaine, E. Lusk, W. Saphir, T. Skjellum, and M. Snir. MPI-2: Extending the Message-Passing Interface. In *Euro-Par '96 Parallel Processing*, pages 128–135. Springer Verlag, 1996.
5. Message Passing Interface Forum. MPI: A Message Passing Interface. In *Proc. of Supercomputing '93*, pages 878–883. IEEE Computer Society Press, November 1993.
6. Craig E Rasmussen, Matthew J. Sottile, Sameer Shende, and Allen D. Malony. Bridging the language gap in scientific computing: The Chasm approach. *Concurrency and Computation: Practice and Experience*, 2005.

Design Alternatives and Performance Trade-Offs for Implementing MPI-2 over InfiniBand*

Wei Huang, Gopalakrishnan Santhanaraman,
Hyun-Wook Jin, and Dhabaleswar K. Panda

Department of Computer Science and Engineering,
The Ohio State University
{huanwei, santhana, jinhy, panda}@cse.ohio-state.edu

Abstract. MPICH2 provides a layered architecture to achieve both portability and performance. For implementations of MPI-2 over Infini-Band, it provides the flexibility for researchers at the RDMA channel, CH3 or ADI3 layer. In this paper we analyze the performance and complexity trade-offs associated with implementations at these layers. We describe our designs and implementations, as well as optimizations at each layer. To show the performance impacts of these design schemes and optimizations, we evaluate our implementations with different micro-benchmarks, HPCC and NAS test suite. Our experiments show that although the ADI3 layers adds complexity in implementation, the benefits achieved through optimizations justify moving to the ADI layer to extract the best performance.

Keywords: MPI-2, InfiniBand, RDMA channel, CH3, ADI3.

1 Introduction

In the last decade, MPI (message passing interface) has become the *de facto* standard for programming parallel applications. MPI-1 standard [12] was proposed by the MPI forum to provide a uniform standard for MPI developers. As a follow-up, MPI-2 [9] standard aims to extend MPI-1 in the areas of one sided communication, I/O and dynamic process management.

MPICH2 [10] from Argonne National Laboratory is one popular implementation of the MPI-2 standard. It aims to combine performance with portability over different interconnects. It tries to achieve this by maximal sharing of platform independent code like MPI datatypes, groups, and communicators, etc., and calls the Abstract Device Interface (ADI3) for platform dependent code. The porting to different interconnects is achieved by having a separate ADI implementation for each interconnect. To further ease the porting, the ADI itself is also layered and can be implemented in terms of lower level interfaces.

---

* This research is supported in part by Department of Energy's Grant #DE-FC02-01ER25506, National Science Foundation's grants #CNS-0204429, and #CUR-0311542, and a grant from Intel.

B. Di Martino et al. (Eds.): EuroPVM/MPI 2005, LNCS 3666, pp. 191–199, 2005.

In the field of High Performance Computing, InfiniBand [5] is emerging as a strong player. InfiniBand supports several advanced hardware features including RDMA capability. To implement MPI-2 on InfiniBand, MPICH2 provides the flexibility to implement the ADI3, or its lower level interfaces like CH3 and RDMA channel layers. Understanding the benefits and limitations of implementing at each layer (ADI3, CH3, and RDMA channel) is very important to come up with an efficient design. The lower layer interfaces are easier to port at the cost of some performance penalties. This is rather expected, but it would be of more interest to quantitatively understand the performance impact and try to come up with different levels of optimizations at each layer. To the best of our knowledge, there is no literature that does an in-depth study on this topic.

In this paper, we attempt to do an detailed analysis of the performance and complexity trade-offs for implementing MPI-2 over InfiniBand at the RDMA channel, CH3, and ADI3 layer. We focus on the point to point and one sided operations in the current work. For fair comparison, we provide our design and implementation at each of these layers. In the rest of the paper, Section 2 introduces the background of our work. Section 3 describes and analyzes our design choices and optimizations. In Section 4 we conduct performance evaluations. Conclusions and future work are presented in Section 5.

2   Background

2.1   InfiniBand Architecture

The InfiniBand Architecture (IBA) [5] defines a System Area Network (SAN) to interconnect processing nodes and I/O nodes. In addition to send/receive semantics it also provides RDMA semantics which can be used to directly access/modify the contents of the remote memory. RDMA operations are one sided and do not incur software overhead on remote side. Further, InfiniBand verbs provide scatter/gather features to handle non contiguity. InfiniBand verbs specification also provides useful features like atomic operations and multi-cast, etc.

2.2   Layered Design of MPICH2 and MVAPICH2

MPICH2 supports both point to point and one sided operations. Figure 1 describes the layered approach provided by MPICH2 for designing MPI-2 over RDMA capable networks like InfiniBand. Implementation of MPI-2 on InfiniBand can be done at one of the three layers in the current MPICH2 stack: RDMA channel, CH3 or ADI3. One of the objectives of such kind of design is to get a better balance between performance and complexity.

RDMA channel is at the bottom most position in the hierarchical structure. All communication operations

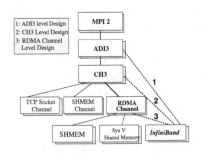

Fig. 1. Layered Design of MPICH2

that MPICH2 supports are mapped to just five functions at this layer. Among them only two (*put* and *read*) are communication functions, thus the porting overhead is minimized. The interface needs to conform to stream semantics. It is especially designed for the architectures with RDMA capabilities, which directly fits with the InfiniBand's RDMA semantics.

The CH3 provides a *channel* device that consists of a dozen functions. It accepts the communication requests from the upper layer and informs the upper layer once the communication has completed. It is responsible to make communication progress, which is the added complexity associated with the implementations at this layer. From a performance perspective, it has more flexibility to improve the performance since it can access more performance oriented features than the RDMA channel layer. Argonne National Lab[10] and the University of Chemnitz[3] have both developed their CH3 devices for Infiniband.

The ADI3 is a full featured, abstract device interface used in MPICH2. It is the highest portable layer in MPICH2 hierarchy. A large number of functions must be implemented to bring out an ADI3 design, but meanwhile it provides flexibility for many optimizations, which are hidden from lower layers.

MVAPICH2 is a high performance implementation of MPI-2 over InfiniBand [6] from the Ohio State University. The latest release version implements the point to point communication at the RDMA channel layer [8] and also optimizes the one sided communication at the ADI3 level [7,4]. The continuous research progress of this project has further motivated us to carry out the proposed in-depth study of design alternatives and performance trade-offs.

3   Designs and Implementations at Different Layers

For an implementation over InfiniBand, MPICH2 design provides choices at three different layers. In this sense, understanding the exact trade-offs of the performance constraints and implementation complexity at each layer will be critical to have an efficient design. To study these issues and to carry out a fair comparison, we have designed and implemented MPI2 over InfiniBand at each of the MPICH2 hierarchy: RDMA channel, CH3 and ADI3, respectively. We present our strategies in the following subsections.

3.1   RDMA Channel Level Design and Implementation

At RDMA channel, as mentioned in Section 2.2, all architecture dependent communication functionalities are encapsulated into a small set of interfaces. The interface needs to provide only stream semantics and the communication progress of MPI messages are left to the upper layers.

Our design is purely based on the RDMA capability of InfiniBand [8]. Fig. 2 illustrates the design issues at this layer. For short messages, the eager protocol is used to achieve good latency. It copies messages to pre-registered buffers and sends them through RDMA write. For large messages, using the eager protocol will introduce high copy overhead, so a zero-copy rendezvous protocol is used.

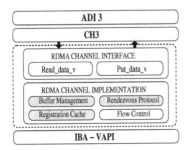
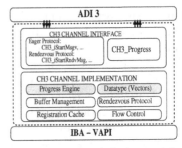

Fig. 2. RDMA Channel level design and implementation

Fig. 3. CH3 level design and implementation

User buffer is registered on the fly and sent through RDMA. Registration cache [13] is implemented to reduce the registration overhead.

The RDMA channel receives the communication requests from the CH3 layer above it. In the current stack, the CH3 layer makes only one outstanding request to the RDMA channel and will not issue the next request until the previous one has completed. This results in the serialization of the communication requests, which causes inefficient utilization of the network. For small messages, an optimization would be to copy the message to the pre-registered buffer and immediately report completion to the CH3 layer. By this early completion method, the CH3 layer can issue the next communication request so that multiple requests can be issued to the RDMA layer. But for large messages which are sent through the rendezvous protocol, we can only report completion after the whole rendezvous process finishes since we need to hold the user buffer for zero-copy send, making it difficult to obtain higher bandwidth for medium-large messages at the RDMA layer. We show this performance impact in Section 4.

3.2   CH3 Level Design and Implementation

Fig. 3 shows the basic function blocks in our CH3 level design. Messages can be sent through eager or rendezvous protocols similar to the approach taken by the RDMA channel implementation. Hence functionalities such as buffer management, registration cache, etc., need to be also implemented at the CH3 layer.

The need to implement progress engine is a significant difference between the CH3 level and the RDMA channel design. The CH3 layer must keep track of all the communication requests coming from the ADI3 layer, finish them and report completions. ADI3 can keep sending requests but the underlying network resources are limited. So we implement a queue to buffer the requests which cannot be finished immediately due to the lack of network resources. Requests in the queue will be retried when resources become available again. The benefit of implementing the progress engine is obvious. We can get access to all the requests at the sender side. Now for large messages we can start multiple rendezvous progresses at the same time so that the throughput is expected to be greatly improved as compared to the RDMA channel level design.

Datatype communication can also be optimized at this layer. ADI3 flattens the datatype and provides the datatype information to the CH3 layer as a vector list. With this information, a CH3 level design can have a global picture of all the buffer vectors that need to be sent in a particular MPI message. So optimizations such as zero-copy datatype [11,14] can be applied at this level.

3.3   ADI3 level Design and Implementation

Compared to the CH3 and RDMA channel, the ADI3 interface is full-featured. This allows the implementations to take opportunities for more efficient communication. To re-implement the whole ADI3 is very complicated, so we decided to reuse most of the CH3 level implementation described in Section 3.2 and perform several optimizations at the ADI3 layer, shown in Fig 4. These optimizations are possible at the ADI3 layer since it is allowed direct access to several global data structures which are abstracted out for the lower layers.

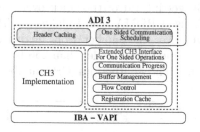

Fig. 4. Optimizations at ADI3 layer

Header caching is an optimization that can be implemented at the ADI3 to reduce the small message latency. The basic idea is to cache some fields of the MPI header for each connection at the receiver side. So that if the next message between this connection has the same header information in those cached fields, we can reduce the size of MPI header being sent. If these fields differ, there is a copy overhead at the receiver for the header caching, which is quite negligible according to our experience. It is be noted that MPI header caching cannot be performed at lower layers since only ADI3 is supposed to know the contents in a MPI header.

Another significant optimization is in the area of one-sided communication. Originally in MPICH2, one sided operations are implemented by the point to point interfaces provided by CH3. Our previous work [7] has shown that by directly using the RDMA features provided by InfiniBand instead of going through the point to point path, the performance for the one-sided operations can be greatly enhanced. We can also schedule one sided operations to achieve much better latency and throughput. The scheduling schemes are described in detail in [4]. These optimizations are done by extending the CH3 interface [7]. At the CH3 layer we cannot distinguish between data for two sided and one sided operations and hence cannot perform such optimizations. The latest MPICH2 also has extended the CH3 one sided interface for shared memory architectures, which reflects the potential to optimize one sided operations at the ADI3 layer.

4   Performance Evaluation

In this section we evaluate our implementations at RDMA channel, CH3 and ADI3 layers by a set of micro-benchmarks, HPC Challenge Benchmark [2] and

NAS test suite [1]. Tests are conducted on on two different clusters. The first cluster (Cluster A) consists of computing nodes with dual Intel Xeon 3.0 GHz processors, 2GB memory, and MT23108 PCI-X HCAs. They are connected by an InfiniScale MTS2400 switch. The second cluster (Cluster B) is equipped with dual Intel Xeon 2.66 GHz processors, 2GB memory, and MT23108 PCI-X HCAs, connected through an InfiniScale MTS14400 switch.

4.1 Point to Point Communication

Point to point communication test are conducted on Cluster A. Fig. 5 shows the uni-directional bandwidth test results for our implementation at RDMA channel, CH3 and ADI3 layers. For medium-large messages, the bandwidth is significantly improved by up to 28% by moving from RDMA channel to CH3 layer, because CH3 handles multiple send requests simultaneously. ADI3 layer shows similar numbers as CH3 layer. It is to be noted that all bandwidth numbers in this paper are reported in MillionBytes/Sec (MB/s).

Fig. 6 shows the ping-pong latency. By getting rid of the stack overhead, the one byte message latency drops from 5.6us at RDMA layer to 5.3us at CH3 layer. By performing header caching at ADI3 layer, the number drops further to 4.9us. Header caching technique is applied to messages smaller than 256 bytes. So for this range the ADI3 level numbers consistently outperform CH3 numbers.

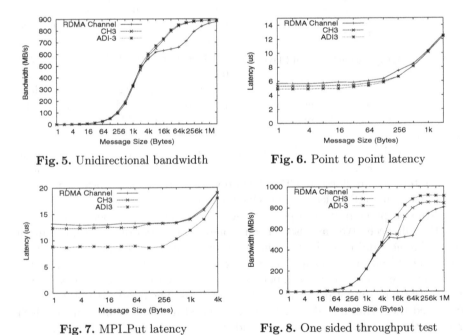

Fig. 5. Unidirectional bandwidth   **Fig. 6.** Point to point latency

Fig. 7. MPI_Put latency   **Fig. 8.** One sided throughput test

4.2 One Sided Operations

Evaluation with one sided operations are also conducted on Cluster A. The results for MPI_Put test are shown in Fig. 7. The test times the ping-pong latency for performing the put operation followed by synchronization. For the CH3 and the RDMA Channel level design, one sided operations are implemented based on point to point communication. Their numbers are similar, with the CH3 level design performing slightly better because point to point communication is optimized. By optimizing one sided operation at ADI3, we observe 30% reduction in MPI_Put latency as compared to the CH3 level design.

We also measure the throughput of one sided communication. Here the origin process issues 16 MPI_Put and 16 MPI_Get operations of the same size. The target process just starts an exposure epoch. We measure the maximum throughput we can achieve (MillionBytes/sec) for multiple iterations of the above sequence. Fig. 8 shows the results. The improvement on point to point bandwidth makes the CH3 level design outperform the RDMA channel level design by up to 49%. And with one sided scheduling at the ADI3 layer, the peak throughput can reach around 920MB/s, which is another 8.1% higher than the CH3 level design numbers.

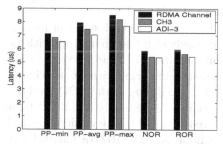
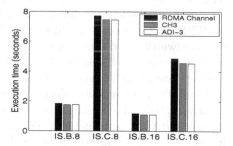

Fig. 9. HPCC 8 bytes latency. PP-min: minimum ping-pong latency; PP-avg: average ping-pong latency; PP-max: maximum ping-pong latency; NOR: Natural ordered ring access latency; ROR: random ordered ring access latency.

Fig. 10. NAS (IS) results on Cluster B

4.3 HPC Challenge (HPCC) Suite

The HPCC suite [2] contains tests which evaluate the latency for different types of communication patterns. It performs the ping-pong tests between all possible different pairs of processors. It also uses two different ring types of communication patterns to evaluate latency: Naturally Ordered Ring and Randomly Ordered Ring. HPCC tests were performed on 16 nodes of Cluster B and we report 8 bytes latency numbers. As shown in Fig. 9, we clearly observe an performance improvement up to 7% for the CH3 level design over the RDMA Channel level design; and the benefit from header caching at the ADI3 layer enhances the performance by up to another 6%.

4.4 NAS Integer Sort

In this section we show the performance evaluation for the NAS-IS benchmark [1]. IS is an integer sort benchmark kernel that stresses the communication aspect of the network. The experiments were conducted for classes B and C on 8 nodes and 16 nodes of Cluster B. The results are shown in Fig. 10. The ADI3 level implementation here shows up to 7% improvement comparing with the RDMA channel level design. The ADI3 level optimizations does not directly help the communication pattern that is observed in NAS-IS. Hence the performance seen at the CH3 level is the same as that of the ADI3 level.

5   Conclusions and Future Work

In this paper we have analyzed the trade-offs associated with implementing MPI-2 over InfiniBand at the RDMA channel, CH3 and ADI3 layer of MPICH2. We have also described the various optimizations that are possible at each level.

With respect to design complexity, a CH3 level design needs to implement the progress engine, which is the main cause of added complexity. A fully featured ADI3 level design is very complicated but optimizations like header caching, one sided communication scheduling can be done at this level.

With respect to performance, the CH3 and ADI3 level design can increase the bandwidth significantly, up to 28% for bandwidth test comparing with the RDMA channel level design. Header caching at the ADI3 can lower the small message latency to 4.9 us, a 12.5% improvement comparing with 5.6 us achieved by the RDMA channel level design. One sided scheduling at the ADI3 level also greatly improves the performance. We see an enhancement up to 30% in MPI_Put latency and 8.1% in throughput test compared with the CH3 level design, which in turn shows 49% improvement on throughput over the RDMA channel level design. Effects of these optimizations also show benefits at the application level evaluation of HPCC and NAS suite. As a conclusion, although the ADI3 layers adds complexity in implementation, the benefits achieved through optimizations justify moving to the ADI layer to extract the best performance.

As a part of future work we would like to come up with a full fledged MPI-2 design over InfiniBand at the ADI3 layer to deliver good performance. We are planning to support communication through shared memory, and optimize collective operations using InfiniBand's RDMA and multi-cast features.

References

1. D. H. Bailey, E. Barszcz, L. Dagum, and H.D. Simon. NAS Parallel Benchmark Results. Technical Report 94-006, RNR, 1994.
2. HPC Challenge Benchmark. http://icl.cs.utk.edu/hpcc/.
3. R. Grabner, F. Mietke, and W. Rehm. An MPICH2 Channel Device Implementation over VAPI on InfiniBand. In *Proceedings of the International Parallel and Distributed Processing Symposium*, 2004.
4. W. Huang, G. Santhanaraman, H. W. Jin, and D. K. Panda. Scheduling of MPI-2 One Sided Operations over InfiniBand. Workshop On Communication Architecture on Clusters (CAC), in conjunction with IPDPS'05, April 2005.
5. InfiniBand Trade Association. InfiniBand Architecture Specification, Release 1.2.
6. Network Based Computing Laboratory. http://nowlab.cis.ohio-state.edu/.

7. J. Liu, W. Jiang, H. W. Jin, D. K. Panda, W. Gropp, and R. Thakur. High Performance MPI-2 One-Sided Communication over InfiniBand. International Symposium on Cluster Computing and the Grid (CCGrid 04), April 2004.
8. J. Liu, W. Jiang, P. Wyckoff, D. K. Panda, D. Ashton, D. Buntinas, W. Gropp, and B. Toonen. Design and Implementation of MPICH2 over InfiniBand with RDMA Support. In *Proceedings of the International Parallel and Distributed Processing Symposium*, 2004.
9. Message Passing Interface Forum. MPI-2: A Message Passing Interface Standard. *High Performance Computing Applications*, 12(1–2):1–299, 1998.
10. MPICH2. http://www-unix.mcs.anl.gov/mpi/mpich2/.
11. G. Santhanaraman, J. Wu, and D. K. Panda. Zero-Copy MPI Derived Datatype Communication over InfiniBand. EuroPVM-MPI 2004, September 2004.
12. M. Snir, S. Otto, S. Huss-Lederman, D. Walker, and J. Dongarra. *MPI–The Complete Reference. Volume 1 - The MPI-1 Core, 2nd edition*. The MIT Press, 1998.
13. H. Tezuka, F. O'Carroll, A. Hori, and Y. Ishikawa. Pin-down cache: A virtual memory management technique for zero-copy communication. In *Proceedings of the 12th International Parallel Processing Symposium*, 1998.
14. J. Wu, P. Wyckoff, and D. K. Panda. High Performance Implementation of MPI Datatype Communication over InfiniBand. In *Proceedings of the International Parallel and Distributed Processing Symposium*, 2004.

Designing a Portable MPI-2 over Modern Interconnects Using uDAPL Interface*

L. Chai, R. Noronha, P. Gupta, G. Brown, and D. K. Panda

Department of Computer Science and Engineering,
The Ohio State University, USA
{chail, noronha, guptapr, browngre, panda}@cse.ohio-state.edu

Abstract. In the high performance computing arena, there exist several implementations of MPI-1 and MPI-2 for different networks. Some implementations allow the developer to work with multiple networks. However, most of them require the implementation of a new device, before they can be deployed on a new networking interconnect. The emerging uDAPL interface provides a network-independent interface to the native transport of different networks. Designing a portable MPI library with uDAPL might allow the user to move quickly from one networking technology to another. In this paper, we have designed the popular MVAPICH2 library to use uDAPL for communication operations. To the best of our knowledge, this is the first open-source MPI-2 compliant implementation over uDAPL. Evaluation with micro-benchmarks and applications on InfiniBand shows that the implementation with uDAPL performs comparably with that of MVAPICH2. Evaluation with micro-benchmarks on Myrinet and Gigabit Ethernet shows that the implementation with uDAPL delivers performance close to that of the underlying uDAPL library.

Keywords: MPI-1, MPI-2, uDAPL, InfiniBand, Myrinet, Gigabit Ethernet, Cluster.

1 Introduction

Message Passing is a popular paradigm for writing parallel applications. Application written with Message Passing Interface (MPI) libraries like MPICH and MPICH2 [2] from Argonne National Labs may be run on a wide variety of architectures and networks. This is easily achieved by relinking the application with the appropriate library for that architecture or network.

Modern interconnection technologies like InfiniBand [5], 10 GigE, Myrinet [9] and Quadrics [11] offer improved performance. This is both in terms of lower latency of the order of a few micro-seconds and higher bandwidth. Applications may also take advantage of the RDMA capabilities of these networks to read and write data with low overhead from each other's memory.

---

* This research is supported in part by Department of Energy's Grant #DE-FC02-01ER25506; National Science Foundation's Grants #CCR-0204429, #CCR-0311542, #CNS-0403342; grants from Mellanox, Intel, and SUN MicroSystems; and equipment donations from Intel, Mellanox, Ammasso, AMD, Apple, and SUN.

B. Di Martino et al. (Eds.): EuroPVM/MPI 2005, LNCS 3666, pp. 200–208, 2005.

Though most high-speed networks offer user-level communication interfaces that are similar in semantics, they usually differ syntactically from each other. To avoid being tied down to a particular network, most application writers use a communication library abstraction such as MPICH. This approach allows the application writer to exploit highly optimized libraries such as MVAPICH [10], MPICH-GM [8] and MPI/Elan4 [11]. While this approach works well, it might cause delays when a new networking technology gets introduced. The MPI library needs to be implemented on that network before the application can be moved.

The emerging User Direct Access Programming Library (uDAPL) [4] defines a standard and device independent interface for accessing the transport mechanisms of RDMA capable networks. This allows developers to design applications in a network independent fashion. Depending on the design of the underlying uDAPL, the performance impact of using the underlying uDAPL might be negligible.

With the development of uDAPL providers for a variety of networks, it is natural to ask whether a MPI library can be designed with an uDAPL interface. This might potentially have advantages at many levels. It would allow an application using an MPI library designed with an uDAPL transport to move seamlessly from one network to another. In addition, an extensively deployed and tested library may inspire more confidence than a newer implementation. Finally, depending on the performance of the underlying uDAPL provider library, there may be little degradation in performance compared with an application designed with a native interface. Thus, the open challenges are: 1. How to design such an MPI-2 library over the uDAPL interface, and 2. How much performance degradation this approach will lead to compared with an MPI-2 implementation over the native interface.

In this paper, we take on these challenges and design a portable MPI-2 with uDAPL and evaluate the performance trade-offs. We first present some background information on high-performance networks, uDAPL and MPI in section 2. Following that in section 3, challenges and design alternatives are presented. In section 4, the performance in terms of micro-benchmarks and applications is presented. Finally we provide conclusions in section 5.

2   Background

2.1   Overview of Interconnects

In the area of high performance computing, InfiniBand and Myrinet are popular and established technologies. These technologies support Remote Direct Memory Access (RDMA) [12]. The Ammasso 1100 is a relatively new entry. It is a RDMA-enabled Gigabit Ethernet adapter [1].

The InfiniBand Architecture (IBA) [5] defines a System Area Network with a switched, channel-based interconnection fabric. IBA 4X has bandwidth up to 10 Gbps. The Myrinet [9] network technology utilizes programmable network interface cards (NIC) and cut-through crossbar switches with operating system bypass techniques for full-duplex 4Gbps data rates. The Gigabit Ethernet used is a full-duplex 1Gbps Ethernet adapter that also supports RDMA.

The native programming interfaces of the networking technologies are VAPI, GM, and ccil for IBA, Myrinet, and Gigabit Ethernet, respectively. OpenIB is proposed as the next-generation software stack over InfiniBand, which is currently under development.

2.2 uDAPL

User Direct Access Programming Library (uDAPL) is a lightweight, transport-independent, platform-independent user-level library that provides a common API for all RDMA-enabled modern interconnects. The uDAPL library is defined by DAT (Direct Access Transport) Collaborative.

2.3 Message Passing Interface

MPI (Message Passing Interface) [13] is a standard library specification for message passing in parallel applications. MPI-2 [7] is an extension to MPI-1 standard. MPI-2 supports several new types of functionalities, including one-sided communication. One-sided communication requires only one process to specify all communication parameters, and ideally the other process is not involved at all. The two processes need explicit synchronization. We focus on active one-sided communication defined by MPI-2, which includes three functions: *MPI_Put*, *MPI_Get*, and *MPI_Accumulate*.

MVAPICH2 [6] [10] is a high-performance MPI-2 implementation over Infini-Band. It is an implementation of MPICH2's [2] RDMA channel. MVAPICH2 is implemented on top of VAPI. MVAPICH2 together with MVAPICH is currently being used by more than 230 organizations worldwide [10] to extract the benefits of InfiniBand for MPI applications.

3 Design Issues

Our design is adapted from MVAPICH2. As can be seen from Figure 1, MVA-PICH2 has four major components: connection management, communication, memory management, and descriptor management. In the mean time, uDAPL provides many features. Our goal is to design an adaptation layer to allow MVAPICH2 to run smoothly over uDAPL interface for both two-sided and one-sided communication.

In this section we discuss in detail about connection and descriptor management. The design of communication and memory management is inherited from MVAPICH2.

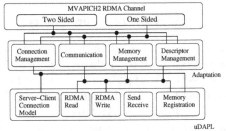

Fig. 1. Design for MVAPICH2 communication interfaces with uDAPL

3.1   Connection Management

MPI communication assumes a fully connected topology. Since MVAPICH2 is based on Reliable Connection (RC) service of InfiniBand, every process needs to establish a connection with every other process in the initialization phase.

Figure 2(a) shows the InfiniBand/VAPI model of establishing a connection between two processes. The detailed discussion is in [5]. Both processes first create *Queue Pairs* (QPs), and then exchange QPs and *Local IDentifiers* (LIDs) through the *Process Management Interface* (PMI) [2]. Then both processes initialize QPs and transit to *Ready-To-Receive* (RTR) state. After a barrier through the PMI, both processes transit to *Ready-To-Send* (RTS) state. And after another barrier through the PMI, a connection is established between these two processes. The uDAPL library provides a server-client model for connection establishment which is totally different from InfiniBand/VAPI's peer-to-peer model, as shown in Figure 2(b). Both processes first create communication *Endpoints* (EPs). Then the server process creates a *Public Service Point* (PSP) which is a listen handle. Each PSP is associated with a system-wide unique *Connection Qualifier*. The server then gives the *Connection Qualifier* to the client through the PMI. After that the server listens on the PSP handle to wait for connection requests. The client issues a connection request (*EP_Connect*) to the PSP. Once the request is accepted by the server, a connection is established between the server and the client.

In order to achieve efficient connection establishment, we need to consider two issues. One is to avoid retransmission. When a client tries to connect to a server, the server should already be listening there, otherwise the connection request will get rejected. The other issue is performance. Since every two processes need to establish a connection, we want this whole process to take place concurrently.

Having these two issues in mind, we propose the following approach, as shown in Figure 3. Every process acts as a server for processes which have higher global rank. At the same time, it also acts as a client for processes with lower global rank. This means a process must listen persistently on the PSP handle while actively issuing connection requests. In order to achieve this, we use a separate

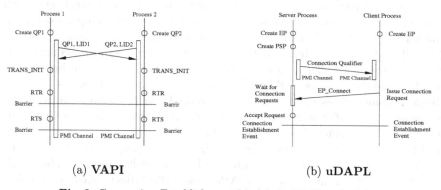

(a) **VAPI**   (b) **uDAPL**

Fig. 2. Connection Establishment Models in VAPI and uDAPL

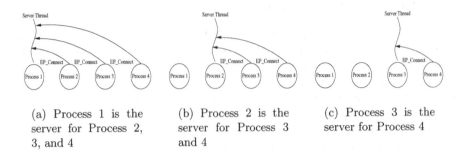

(a) Process 1 is the server for Process 2, 3, and 4

(b) Process 2 is the server for Process 3 and 4

(c) Process 3 is the server for Process 4

Fig. 3. Proposed Thread-based Connection Establishment Scheme. Actions shown in Figures (a), (b), and (c) take place concurrently.

thread for server functionality. Every process first spawns a server thread, then after a synchronization, processes can request connections to their corresponding servers. The server thread exits once it has accepted the correct number of requests so that it will not affect any communication performance later. Using this approach, we can establish connections between every two processes in a reliable and efficient manner.

3.2   Descriptor Management

Both MVAPICH2 and MVAPICH2 with uDAPL use RDMA Write for eager protocol and RDMA Read for rendezvous protocol. When a RDMA operation is posted, information such as local address, segment length, remote address, etc. is encapsulated in a descriptor, and the descriptor is passed as an argument to the underlying VAPI or uDAPL functions. However, VAPI and uDAPL have different requirements for descriptor management.

There is no explicit descriptor management in MVAPICH2. Once a RDMA operation is posted, the descriptor is internally copied by VAPI. But for designing MVAPICH2 with uDAPL, we must carefully manage the descriptors because according to the uDAPL specification, descriptors should not be modified until the corresponding RDMA operation has finished.

For using InfiniBand, buffers used for communication must be registered with InfiniBand Host Channel Adaptor (HCA). Since the registration process is time consuming, MVAPICH2 uses a set of pre-registered buffers for eager protocol. This allows us to use a simple and efficient method for descriptor management. We associate each buffer with a descriptor. Whenever a buffer can be reused - which means the previous RDMA Write associated with this buffer has finished - the corresponding descriptor can be safely reused. For rendezvous protocol, the communication buffer is registered on the fly. A descriptor is dynamically allocated for each buffer. The address of the descriptor is saved as a cookie. An uDAPL cookie is a user-supplied identifier for a *Data Transfer Operation* (DTO), which allows a user to uniquely identify the DTO when it completes. The cookie is passed as an argument to the uDAPL post-RDMA-Read function. When a RDMA Read completes, we can use the cookie to retrieve the corresponding descriptor and free it. This descriptor management approach adds almost no overhead to the overall performance.

3.3   Design Issues in One-Sided Communication

The design issues discussed above also exist in one-sided communication. For connection management, we setup a separate set of EPs for one-sided communication, and connections are established in a similar manner as described in section 3.1. For descriptor management, since buffers are also pre-registered for eager protocol, and registered on the fly for rendezvous protocol, we use the same descriptor management scheme as described in section 3.2.

4   Performance Evaluation

4.1   Experimental Setup

Two clusters are used for the evaluation. Cluster A has 8 nodes. Each node is equipped with dual 3.0 GHz processors and 64-bit 133 MHz PCI-X interfaces. It is connected through MT23108 HCA's to a MT2400 switch. The uDAPL library from IBGD 1.7.0 is used. Each node also has a Myrinet E-card connected to a Myrinet-2000 switch. The uDAPL library 0.94+2 with GM 2.1.9 is used. Ammasso Gigabit Ethernet cards are also present on each node. They are connected to a Foundry switch. The uDAPL library provided by Ammasso Inc. is used.

Cluster B has 32 nodes connected with InfiniBand. Each node has dual 2.6 GHz processors. Other configurations are similar to Cluster A.

4.2   Performance Evaluation over InfiniBand

Micro-benchmark Level Evaluation: Micro-benchmark performance is evaluated on Cluster A. Figure 4 shows the uDAPL-level and MPI-level ping-pong latency results. Small message latency of MVAPICH2/uDAPL/VAPI is around $6.7\mu s$. It is $1\mu s$ higher than the latency of MVAPICH2/VAPI for messages smaller than 256 bytes. This is because MVAPICH2 utilizes *inline data transfer* scheme, where small data is sent in one message with the request, while uDAPL/VAPI doesn't utilize *inline data transfer*. For messages larger than 256 bytes, MVA-PICH2/uDAPL and MVAPICH2 have comparable latency performance. The latency of MVAPICH2/uDAPL/VAPI is also close to that of uDAPL/VAPI for messages smaller than 256 bytes. After 256 bytes, the latency of both MVA-PICH2 and MVAPICH2/uDAPL/VAPI goes up, because MPI-level messages need to be copied from user buffers to pre-registered RDMA buffers, and copy time goes up as message size grows.

From Figure 5 we can see that MVAPICH2/uDAPL/VAPI achieves the same bandwidth capability compared with MVAPICH2. The peak bandwidth is around 867 MillionBytes/second (MB/s).

Figure 6 shows the latency comparison of active one-sided communication. The small message latency of MVAPICH2/uDAPL/VAPI is $8.9\mu s$ for MPI_Put,

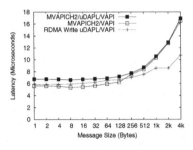

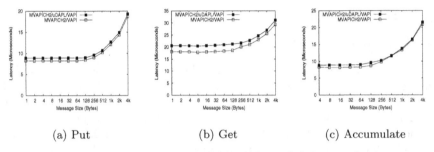

Fig. 4. Small Message Latency Comparison

Fig. 5. Large Message Bandwidth Comparison

(a) Put   (b) Get   (c) Accumulate

Fig. 6. Latency Comparison of Active One-sided Communication

20.5μs for MPI_Get, and 8.9μs for MPI_Accumulate. MPI_Put and MPI_Get latency is about 8% higher than that of MVAPICH2/VAPI, and MPI_Accumulate latency is about 6% higher. This overhead can be associated with the two small synchronization messages needed for one-sided operations which are not inlined.

Application Level Evaluation: In this section, we evaluate the implementation of MVAPICH2/uDAPL/VAPI against MVAPICH2 using some of the NAS Parallel [3] and ASCI Blue [14] benchmarks. To evaluate MVAPICH2/uDAPL and MVAPICH2, the class A size of the applications FT, CG, LU and MG and class C size of IS were used. These are denoted by FT(A), CG(A), LU(A), MG(A) and IS(C), respectively. For the application Sweep3D, the large size 150 was used. This is denoted by S3D(150). All runs were with 64 processes on 32 nodes (Cluster B).

Figure 7 shows the execution time of the applications. For FT(A), CG(A) and S3D(150), MVAPICH2/uDAPL performs slightly worse than MVAPICH2 by approximately 6%, 2.5% and 7%, respectively. FT(A), CG(A) and S3D(150) use small messages in the range of 0-256 bytes. As discussed in section 4.2, uDAPL/VAPI has approximately 1μs higher latency for the small message range of 0-256 bytes compared with the native VAPI. This difference in latency is reflected in the MVAPICH2/uDAPL timing. For the other applications, MG(A), LU(A) and IS(C), MVAPICH2/uDAPL performs comparably with MVAPICH2.

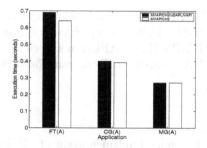
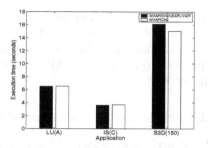

Fig. 7. Execution times of different applications with InfiniBand with 64 processes on 32 nodes using MVAPICH2/uDAPL and MVAPICH2

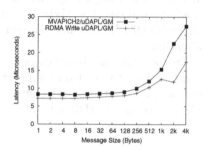
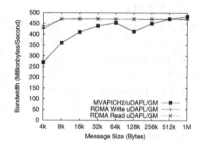

(a) Small Message Latency   (b) Large Message Bandwidth

Fig. 8. Latency and Bandwidth Performance on Myrinet

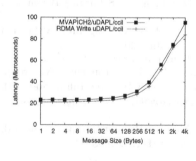
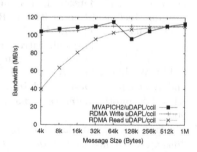

(a) Small Message Latency   (b) Large Message Bandwidth

Fig. 9. Latency and Bandwidth Performance on Gigabit Ethernet

4.3 Performance Evaluation on Myrinet and Gigabit Ethernet

As can be seen from Figure 8, MVAPICH2/uDAPL/GM and uDAPL/GM have comparable latency for messages smaller than 1KB, around 8.3μs. After 1KB, copy overhead makes MVAPICH2/uDAPL/GM latency go higher. With respect to bandwidth, MVAPICH2/uDAPL/GM achieves the same peak bandwidth with uDAPL/GM, which is around 480MB/s.

From Figure 9(a), we can see that the MVAPICH2/uDAPL/ccil latency performance is relatively close to uDAPL/ccil RDMA Write latency. Small message latency is around $21\mu s$. The bandwidth of MVAPICH2/uDAPL/ccil also closely matches with that of the uDAPL/ccil RDMA bandwidth. Peak bandwidth is about 110 MB/s.

5   Conclusions and Future Work

In this paper, we have designed a high-performance implementation of MVA-PICH2 with uDAPL. The performance evaluation has been done on three different interconnects using both micro-benchmarks and applications. For InfiniBand, the implementation of MVAPICH2 with uDAPL performs comparably with that of MVAPICH2 on micro-benchmarks as well as applications. For Myrinet and Gigabit Ethernet (Ammasso), the MVAPICH2 with uDAPL performs comparably with the uDAPL layer in terms of micro-benchmarks.

In the current implementation, MPI collective operations are based on point-to-point communication. We plan on investigating how to efficiently support collective operations using uDAPL. In addition, we plan on studying the impact of moving our current design from the RDMA channel to the ADI3 layer.

References

1. Ammasso, Inc. The Ammasso 1100 High Performance Ethernet Adapter User Guide. http://www.ammasso.com/amso1100_usersguide.pdf, February 2005.
2. Argonne National Laboratory. MPICH - A Portable Implementation of MPI. http://www-unix.mcs.anl.gov/mpi/mpich.
3. D. H. Bailey, E. Barszcz, L. Dagum, and H.D. Simon. NAS Parallel Benchmark Results. Technical Report 94-006, RNR, 1994.
4. DAT Collaborative. uDAPL: User Direct Access Programming Library Version 1.2. http://www.datcollaborative.org/udapl.html, July 2004.
5. Infiniband Trade Association. http://www.infinibandta.org/.
6. J. Liu, W. Jiang, Pete Wyckoff, D. K. Panda, D. Ashton, D. Buntinas, W. Gropp, and B. Toonen. Design and Implementation of MPICH2 over InfiniBand with RDMA Support. In *International Parallel and Distributed Processing Symposium*, 2004.
7. Message Passing Interface Forum. MPI-2: A Message Passing Interface Standard. *High Performance Computing Applications*, 12(1–2):1–299, 1998.
8. MPICH-GM Software. *www.myrinet.com/scs*.
9. N. J. Boden, D. Cohen, R. E. Felderman, A. E. Kulawik, C.L. Seitz, J. Seizovic, and W. Su. Myrinet - a gigabit per second local area network., February 1995.
10. Network-Based Computing Laboratory. MPI over InfiniBand Project. http://nowlab.cis.ohio-state.edu/projects/mpi-iba/index.html.
11. Quadrics Ltd. *www.quadrics.com*.
12. RDMA Consortium. RDMA Protocol Verb Specification. http://www.rdmaconsortium.com/home, April 2003.
13. Marc Snir, S. Otto, S. Huss-Lederman, D. Walker, and J. Dongarra. *MPI–The Complete Reference. Volume 1 - The MPI-1 Core, 2nd edition*. The MIT Press, 1998.
14. The ASCI Blue Benchmarks. http://www.llnl.gov/asci_benchmarks.

Experiences, Strategies and Challenges in Adapting PVM to $VxWorks^{TM}$ Hard Real-Time Operating System, for Safety-Critical Software

Davide Falessi[1], Guido Pennella[2], and Giovanni Cantone[1]

[1] University of Rome "Tor Vergata", DISP, Italy
[2] MBDA Italia SpA, Rome, Italy
falessi@ing.uniroma2.it, guido.pennella@mbda.it, cantone@uniroma2.it

Abstract. The role performed by Open Source Software in safety-critical systems is growing and gaining importance. Due to many, and large variety of, hard real-time constraints and functional requirements that safety-critical applications have to meet, these applications are nowadays composed by logical and physical components, deployed on heterogeneous distributed platforms. This paper is part of a still ongoing project, and is concerned with exploring experimentally the porting of PVM to $VxWorks^{TM}$: the latter has an internal architecture very different from the Unix standard OS(s) (like for example *Linux* or $Solaris^{TM}$), which in turn is the reference OS platform for PVM.

Keywords: Open Source Software (OSS), Safety-Critical Software (SCS), Embedded Hard Real-Time (HRT) Distributed Systems, Parallel Virtual Machine (PVM), Experimental Software Engineering (ESE).

1 Goal and Problem Definition

The role performed by Open Source Software (OSS) is growing and gaining importance in the industrial field. One key factor is the high level of maintainability that OSS offers: in fact, OSS places all the code in the complete control of the software engineer. OSS is hence becoming a key factor in certain areas, due to the need of safety, and the necessity of fixing defects *at home*, quickly and effectively, whatever their software level might be. Due to the large variety of hard real-time (HRT) constraints and functional requirements that industrial safety-critical applications have to meet, these applications are nowadays composed by logical and physical components, deployed on heterogeneous, distributed platforms. The Parallel Virtual Machine (PVM) [1, 2] provides a convenient, open-source, computational architecture for supporting communication in heterogeneous distributed applications through message passing paradigm. The family of standard Unix Operating Systems (OS) constitutes the reference OS platform for PVM. $VxWorks^{TM}$, in its turn, is an OS that is able to provide many advantageous HRT features, and it is widely used by industrial organizations, including the MBDA Italy. This is the Italian site of a multinational

B. Di Martino et al. (Eds.): EuroPVM/MPI 2005, LNCS 3666, pp. 209–216, 2005.

company that works in the domain of electronics real time systems; its name comes from the initials of the founding companies (Matra, BAE, Dynamics, and Alenia); MBDA use $VxWorks^{TM}$ for controlling some actuator devices of the systems that it produces. For those reasons, and in order to have a common, open model of interaction between the actuating platforms (built on $VxWorks^{TM}$) and the Human Machine Interface (HMI) that performs control and command of the system (built on Linux), the Software Research Lab of MBDA took decision to launch experimentation with PVM on $VxWorks^{TM}$. We preferred to start experimentation by using PVM rather than MPI [3] because the former, as required by the application domain, integrates node management and fault-tolerance features, which the latter insufficiently provides (however, in order to give empirical evidence to such our conjecture, we are thinking to conduct a systematic comparison of PVM and MPI for usage in the specific domain). This paper is focused on issues encountered during the first, functional adaptation of PVM to $VxWorks^{TM}$. In the remaining, Section 2 briefly recalls on terms of reference. Section 3 is concerned with the work we made for adapting PVM to $VxWorks^{TM}$, the software process that we enacted, the resulting prototype PVMVX, and its functionality. Section 4 describes in some details the experimental infrastructure and the effort enacted. Section 5 reports on testing PVMVX, and sketches on the analysis of results. One further section presents conclusive remarks and future works.

2   Terms of Reference

Hard Real-Time Systems: It is crucial to distinguish time-critical events from remaining ones, assign each critical event a level of criticality, and evaluate as precisely as possible the handling time of any time-critical event in the worst case [4, 5, 6, 7]. However, the following should be noted: (i) missing a deadline implies a failure in hard real-time systems. Such a failure might have a strong impact on individual or social life when safety-critical systems are concerned; (ii) generally, time duration of software computation is really not predictable with enough precision (approx by a Gaussian), eventually it is not decidable; (iii) increasing speed of computation (i.e.: processor clock) might help, but can be expensive and does not ensure strict determinism. In our reference applications, the overall system is composed by different subsystems, connected via an Ethernet network, where each subsystem has its own real-time constraints (e.g.: the Human/Computer interface subsystem need to react with *human* response time, while *pure* algorithm calculation subsystems need to react with much shorter time-frame). Hence we have heterogeneous requirements that lead to heterogeneous software and hardware platforms.

Embedded Systems: Each embedded system is a logical-physical unit, which works inside a larger system and, in the most common case, is not visible to clients of that system [7]. Embedded systems usually do not support desktop I/O devices but provide environment-specific devices.

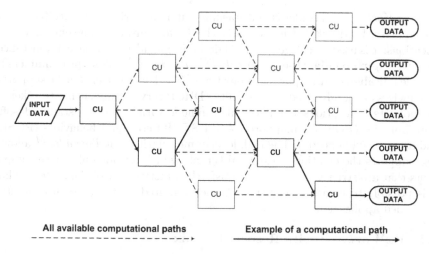

Fig. 1. Computational architecture of a software component for a typical application in the safety-critical domain of reference

Computational Unit: Our reference applications execute on heterogeneous networks of computers, and can be seen as composed by a set of computational units CU(s). This set operates like an orchestra, where each CU collaborates with other CU(s) and performs some specific basic functions (in our case: mathematical computation). The application behavior results from the CU(s) functions available. Figure 1 shows the computational architecture of a typical application in our safety-critical domain, in which: (a) the computational path of the data is selected based on the meaning achieved at run-time, (b) the data is exchanged among CU(s) using a message passing mechanism, (c) the output data of a component could be the input data of other components.

Those CU(s) can be implemented in different languages; they can be modeled by heavy processes or just threads [8, 9]. Each CU runs on any of the suitable computers in the network, and a CU has its own set of data, program code and stack; the scheduler of the OS is responsible for CU management. In order to enhance flexibility, software instructions should refer logical addresses rather than physical ones; the Paging Unit is the hardware unit responsible for mapping the former to the latter. In fact, when a Unix process forks itself, a new CU is born with the forker's logical address; however, variables are not shared because the Paging Unit translates differently the logical addresses of different CU(s).

2.1 Brief Introduction to the $VxWorks^{TM}$ OS

$VxWorks^{TM}$ [10] is a HRT operating system from $WindRiver^{TM}$, and $Tornado^{TM}$ [11] is its integrated development environment, required to build any $VxWorks^{TM}$ application. $VxWorks^{TM}$ has been bred specifically for Embedded HRT applications. $Tornado^{TM}$ is not a native tool; it needs: (i) a software development stations (host); (ii) a direct link to a machine configured as the final computational node, where the application will execute (a.k.a. the target).

The $VxWorks^{TM}$ scheduler is multitasking, interrupt-driven, priority-based and pre-emptive. $VxWorks^{TM}$ is the market leader for hard real-time embedded applications; it is used in a large variety of domains included aerospace: for instance, the rover $Spirit$ on Mars uses $VxWorks^{TM}$. $Tornado^{TM}$ allows programmers to include reusable software parts of many types, like device drivers (cache supports, usb devices etc.), advanced functionality (DHCP server, ping client etc.), kernel components (exceptions handlers, environment variables etc.), standard software libraries (POSIX semaphores, ANSI STDIO etc.) and behaviors (POSIX and other schedulers etc.). The high level of modularity that $Tornado^{TM}$ allows, also supports the creation of essential kernels, which include only those components that are strictly necessary to satisfy the functionality of the system. Thus it is possible to create a very small kernel, required by the a resource-limited embedded application.

2.2 $VxWorks^{TM}$ with Respect to Unix

$VxWorks^{TM}$ names $task$ its basic CU. A task is an active computation entity; its context includes private (not shared) attributes for storing values of program counter, CPU registers and timers, an optional data structure for debugging purposes, standard input and output devices. Let us highlight that $VxWorks^{TM}$ tasks share the same address space. Moreover, $VxWorks^{TM}$ uses physical-memory addresses, while Unix manages physical and logical address. The use of logical memory allows implicit protection of data, while addressing memory directly in physical mode exposes software to data access errors that can lead to fatal failure of the system. Pros of using physical addressing relates to performance: it is much faster than logical access because of the absence of address translation. Other main non-Unix $VxWorks^{TM}$'s characteristics are: (a) $VxWorks^{TM}$ does not provide copy-constructors for tasks; hence, clients cannot create new computational units by coping data from another computational unit (in Unix, the system call $fork()$ provides such a functionality), (b) $VxWorks^{TM}$ computational units are flat; application system programmers cannot structure those units hierarchically, (c) $VxWorks^{TM}$ allows complete control at application level over the existing CU(s) through proprietary API, (d) the context of any $VxWorks^{TM}$ computational unit can be allowed to maintain data for debugging purposes. Thus the $VxWorks^{TM}$'s tasks are extremely different form Unix processes, but have similarities with Unix threads.

3   Adapting PVM to $VxWorks^{TM}$

Let us focus now on showing the problems that we faced, and explaining the solutions that we adopted to allow communication, in heterogeneous computer networks, between Unix and $VxWorks^{TM}$ OS, using PVM as a middleware. In order to meet such a goal, we had one reasonable chance, which is trying to enhance the last available version of PVM (i.e. v. 3.4.4), in the aim of adapting PVM to deal with the last version of $VxWorks^{TM}$ OS (i.e. v. 5.5). Let us highlight that the resulting product, PVMVX, is not to consider as a new middleware but just a PVM extension in terms of portability. In relation to changes that we applied to PVM source code, we defined two different compilation directives to encapsulate all the modifications done: (a) VXWORKS, for including the portions of code that is developed specifically for $VxWorks^{TM}$, (b) CUTFORVXWORKS, for detecting PVM functionality not yet adapted to $VxWorks^{TM}$.

3.1 The Process

We used an evolutionary maintenance process, with each iteration structured in three main phases: 1) Analyzing for detecting the remaining most critical changes to enact in the current iteration; 2) Understanding for Maintaining; 3) Analyzing, designing, enacting, and testing the effects of, the maintenance intervention.

3.2 Analysis

The prospected high level of diversity between $VxWorks^{TM}$ and Unix made the analysis phase very significant. This phase was focused on the following activities: (a) identifying priority changes; (b) understanding the PVM software context for the change, and identifying the affected functionality; (c) configuring the kernel options; (d) locating the identified PVM key functionality in the source code; (e) finding the path, if any, of the system calls required by PVM; (f) setting up the required environment variables. The remaining sub-sections synthesize on changes that we applied.

3.3 Encapsulating PVM Communication Protocol in PVMVX

The key step was to analyze the PVM network communication protocol. To enact that step we used Ethereal (v. 0.10.10) [12], which is an open source sniffer; we preferred such a black box approach, both to have a direct view of what was passing through the net, and in the reasonable expectation of saving with effort to spend. The result was that PVM uses the *Remote Shell* (RSH) [13] for virtual machine initialization and task spawning, while $VxWorks^{TM}$ does not provide RSH server daemons. Consequently, in order to proceed with our work, we had to port RSH to$VxWorks^{TM}$.

3.4 Forking Computation Units

Adapting the Unix primitive $fork()$ to $VxWorks^{TM}$ resulted in a work-task quite strong to enact. As previously mentioned, Unix allows copy-constructors for tasks through the system call $fork()$, which uses *logical* addresses; unfortunately, these features are not supported by $VxWorks^{TM}$. Since $fork()$ is utilized by PVM as the key mechanism to create distributed computational units, we had to develop a similar mechanism by using $VxWorks^{TM}$ features. Concerning this point, let us recall that $VxWorks^{TM}$ does not implement the concept of logical-address. However, we noted that it has so called *private* variables; differently from other $VxWorks^{TM}$ variables, the *private* variables refer different addresses while using common variable names in different CU(s). In order to simulate logical addressing mechanism in $VxWorks^{TM}$, our first approach was hence to use extensively such a feature: defining *private* variables and linking them dynamically to tasks eventually allowed us to simulate $fork()$ mechanism in $VxWorks^{TM}$. Unfortunately, a testing time, we had to verify that such a solution was not sufficient because *private* variables cause significant performance overheads; in fact, the $VxWorks^{TM}$ scheduler loads all the *private* variables every time a task-context change occurs. We estimated in one millisecond the loading time for each *private* variable. Moreover, we realized that PVM utilizes about one-hundred variables of type *global* which need to be privatized. Consequently, in such a preliminary PVMVX, performance overheads resulted in 0.1 seconds for every change of context, which is not acceptable for any HRT system; so we had to reject that approach. The subsequent approach consisted in compacting all unshared variables in one large *private* structure. Concerning implementation, once created the structure of *private* variables, we proceeded to map unshared variables through that new structure. Since the number of necessary *private* variables became one, i.e. the base address of that private structure, the actual lost of performance dropped down up to 1%; however, an overheads of 1 millisecond for every context change seems us still quite a strong charge, and further improvements should be provided by next iterations of the adaptation work.

3.5   Other Implemented Functionality

Others things that we implemented, which are mandatory for PVM usage and $VxWorks^{TM}$ does not provide, are: (a) the User concept at application level, (b) pipe functionality, which is the local communication channel that PVM establishes between the PVM-daemon and the remaining PVM local tasks, (c) parameter passing from the command shell, which PVM uses to set starting parameters of all the PVM CU(s). We implemented all those things by using some $VxWorks^{TM}$ environment variables.

3.6   Not Implemented Functionality

Two PVM functionality are not yet provided by PVMVX: executing *Master* and *Debugger* processes. Our plan is to implement these functionalities in a next iteration of our development.

4   Experimental Laboratory, Software Process, and Effort

The hardware equipment that we used for developing and testing PVMVX includes: (a) one desktop *host* computer, suitable for developing/modifying the PVMVX source code by using $Tornado^{TM}$; (b) one *target* computer, for executing PVMVX on $VxWorks^{TM}$ OS; a COTS Single Board Computer we utilized as target, which is produced by Dy4[14], based on $PowerPC^{TM}$ and $Altivec^{TM}$ technology, and names $SVME/DMV - 181^{TM}$; (c) four desktop computers running PVM, three of them with *Linux RedHat*^{TM} 9.0, and one with *Solaris*^{TM} 9 OS. All those computers were interconnected through a LAN (see Figure 2); the connection of five computers, the ones described at points (*b*) and (*c*) above, constitutes a heterogeneous network that truly replicates the hardware architecture which is common in our safety-critical applications.

The software process that we enacted is Evo (Evolutionary) [15], as ideated by T. Gilb. Evo is utilized by the production lines of the reference company; based on their experience, Evo is a suitable process when requirements are not stable. In fact, when we started this project, requirements were quite vague, we did not know if PVM was adaptable to $VxWorks^{TM}$, and what amount of work we would be enacting. In order to develop this project up to the status described above, we spent an effort of 1.580 man-hours; their split through different phases is: 505 for analysis, 331 for development, 395 for test, and 349 for documentation. PVM understanding for adaptation maintenance involved reading 55 KLOC, while adaptation maintenance involved the development of 3 KLOC.

5   Testing, and Analysis of Results

As already mentioned, we tested PVMVX on the architecture in Figure 2. In order to evaluate performance and some fault-tolerant features, we run three types of tests, as in the followings: (*a*) measuring the computational time required to send and receive up to two-thousands messages of type int from the target system to *Solaris*^{TM} desktop; (*b*) measuring the computational time required to send and receive up to two-thousands messages of type int from the target system to one *Linux* desktop; (*c*) evaluating some fault tolerant features by using a particular distributed application, which involved all the networked systems in Figure 2. The analysis of results shows that: (*i*) mean time to transfer one of the test messages is approximately 0.033 seconds, (*ii*) PVMVX is able to tolerate at least one non master fault by reconfiguring the parallel virtual machine, as typical for PVM behaviors.

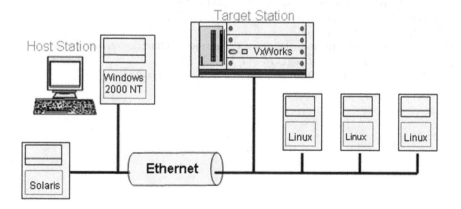

Fig. 2. The PVMVX development and testing architecture

6   Conclusions and Future Works

Based on experience gained in developing a prototypic adapter, this paper presented and discussed the experiences, strategies and challenges in adapting PVM to the hard real-time operating system $VxWorks^{TM}$. Results from this exploratory experimental study seem to confirm that PVM is a promising solution for distributed embedded hard real-time applications. In order to employ PVM in the development lines of safety-critical software at our reference organization, further work is needed, aimed to improve performances of the prototype that we realized. Based on the experience that we gained while conducting the present study, the next step will concern re-engineering again PVM with a more precise focus on $VxWorks^{TM}$: this, in fact, is absolutely not compatible with Unix, which call for further specific PVM modelling.

References

[1] Al Geist, Jack Dongarra, Weicheng Jiang, Robert Manchek, Vaidy Sunderam, *PVM: Parallel Virtual Machine A User's Guide and Tutorial for Networked Parallel Computing*, The MIT Press, 1994.
[2] William Gropp and Ewing Lusk, Goals Guiding Design: PVM and MPI, *Proceedings of the IEEE International Conference on Cluster Computing 2002*.
[3] Introduction to MPI. *www.mpi-forum.org/docs/mpi-11-html/node1.html*.
[4] Jane W. S. Liu, *Real-Time Systems*, Prentice Hall 2000.
[5] Giorgio C. Buttazzo, *Hard Real-Time Computing Systems: Predictable Scheduling Algorithms and Applications*, Springer 1997.
[6] John A. Stankovic, Krithi Ramamritham *Hard Real-Time Systems*, IEEE SCP, 1988.
[7] Qing Li and Caroline Yao, *Real-Time Concepts for Embedded Systems*, CMP, 2003.
[8] Jean Bacon and Tim Harris, *Operating Systems: Concurrent and Distributed Software Design*, Addison Wesley 2003.
[9] Daniel Bovet and Marco Cesati, *Understanding the LINUX Kernel*, O'Reilly 2002.
[10] Wind River, *VxWorks_ programmers_ guide*, 2003.
[11] Wind River, *Tornado_ users_ guide_ windows*, 2003.
[12] http://www.ethereal.com/
[13] http://stuff.mit.edu/afs/athena/astaff/reference/4.3network/rshd/rshd.c
[14] Dy4, *DPK-TechDoc-CD 602716-001* Disk 1, 2004.
[15] T. Gilb, *Evo: The evolutionary Project Managers Handbook*.

MPJ/Ibis: A Flexible and Efficient Message Passing Platform for Java

Markus Bornemann, Rob V. van Nieuwpoort, and Thilo Kielmann

Vrije Universiteit, Amsterdam, The Netherlands
http://www.cs.vu.nl/ibis

Abstract. The MPJ programming interface has been defined by the Java Grande forum to provide MPI-like message passing for Java applications. In this paper, we present MPJ/Ibis, the first implementation of MPJ based on our Ibis programming environment for cluster and grid computing. By exploiting both flexibility and efficiency of Ibis, our MPJ implementation delivers high-performance communication, while being deployable on various platforms, from Myrinet-based clusters to grids. We evaluated MPJ/Ibis on our DAS-2 cluster. Our results show that MPJ/Ibis' performance is competitive to mpiJava on Myrinet and Fast Ethernet, and to C-based MPICH on Fast Ethernet.

1   Introduction

In recent years, Java has gained increasing interest as a platform for high performance and Grid computing [1]. Java's "write once, run anywhere" property has made it attractive, especially for high-performance grid computing where many heterogeneous platforms are used and where application portability becomes an issue with compiled languages like C++ or Fortran.

In previous work on our Ibis programming environment [2], we showed that parallel Java programs can run and communicate efficiently. Ibis supports object-based communication: method invocation on remote objects and object groups, as well as divide-and-conquer parallelism via spawned method invocations [2]. The important class of message-passing applications was not supported so far.

To enable message passing applications, the Java Grande Forum proposed MPJ [3], the MPI language bindings to Java. So far, no implementation of MPJ has been made available. In this paper, we present MPJ/Ibis, our implementation of MPJ on top of the Ibis platform. Being based in Ibis, MPJ/Ibis can be deployed flexibly and efficiently, on machines ranging from clusters with local, high-performance networks like Myrinet or Infiniband, to grid platforms in which several, remote machines communicate across the Internet.

In this paper, we discuss our design choices for implementing the MPJ API. As evaluation, we run both micro benchmarks and applications from the Java-Grande benchmark suite [1]. Micro benchmarks show that on a Myrinet cluster, MPJ/Ibis communicates slower than C-based MPICH, but outperforms MPI-Java, an older Java wrapper for MPI. Using TCP on Fast Ethernet shows that

B. Di Martino et al. (Eds.): EuroPVM/MPI 2005, LNCS 3666, pp. 217–224, 2005.

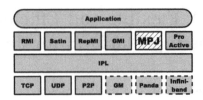

Fig. 1. Design of Ibis. The various modules can be loaded dynamically.

MPJ/Ibis is significantly faster than C-based MPICH. (Unfortunately, MPI-Java does not run at all in this configuration.) With the JavaGrande benchmark applications, MPJ/Ibis is either on-par with MPIJava or even outperforms it. MPJ/Ibis can thus be considered as a message-passing platform for Java that combines competitive performance with portability ranging from high-performance clusters to grids.

2   Related Work

Many attempts were made to bind MPI to Java. MpiJava [4] is based on wrapping native methods like the MPI implementation MPICH with the Java Native Interface (JNI). The API is modeled very closely on the MPI standard provided by the MPI Forum. Due to limitations of the Java language (primitive type arguments cannot be passed as reference), small changes to the original standard had been made. JavaMPI [5] also uses JNI to wrap native methods to Java. It overcomes the argument passing problems using automatically generated C-stub functions and JNI method declarations. The MPIJ [6] implementation is written in pure Java and runs as a part of the Distributed Object Group Metacomputing Architecture (DOGMA) [7]. If available on the running platform, MPIJ uses native marshaling of primitive types instead of Java marshaling.

The first two approaches provide fast message passing, but do not match Java's "write once, run anywhere" property. JavaMPI and mpiJava are not portable enough, since a it requires a native MPI library and the Java binding must be compiled on the target system. MPIJ is written in Java and addresses the conversion of primitive datatypes into byte arrays. However, it does not solve the more general problem of Object serialization, which is a bottleneck.

MPJ [3] proposes MPI language bindings to Java. These bindings merge the earlier proposals mentioned above. In this paper, we present MPJ/Ibis, which is the first available implementation of MPJ. MPJ/Ibis features a pure Java implementation, but can also use high speed networks using some native code. Moreover, MPJ/Ibis uses Ibis' highly efficient object serialization, greatly speeding up the sending of complex data structures.

3   Ibis, Flexible and Efficient Grid Programming

Our MPJ implementation runs on top of Ibis [2]. The structure of Ibis is shown in Figure 1. A central part of the system is the Ibis Portability Layer (IPL) which

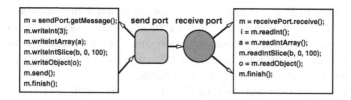

Fig. 2. Send ports and receive ports

consists of a small number of well-defined interfaces. The IPL can have different implementations, that can be selected and loaded into the application *at run time*. The IPL defines both serialization (the conversion of objects to bytes) and communication. Ibis also provides more high-level programming models, see [2]. In this paper, we focus on the MPJ programming model.

A key problem in making Java suitable for grid programming is designing a system that obtains high communication performance while retaining Java's portability. Current Java runtime environments are heavily biased to either portability or performance. The Ibis strategy to achieve both goals simultaneously is to develop reasonably efficient solutions that work "anywhere", supplemented with highly optimized solutions for increased performance in special cases. With Ibis, grid applications can run simultaneously on a variety of different machines, using optimized software where possible (e.g., Myrinet), and using standard software (e.g., TCP) when necessary.

3.1 Send Ports and Receive Ports

The IPL provides communication primitives using send ports and receive ports. A careful design of these ports and primitives allows flexible communication channels, streaming of data, efficient hardware multicast and zero-copy transfers. The layer above the IPL creates send and receive ports, which are connected to form a *unidirectional message channel*, see Figure 2. New (empty) message objects can be requested from send ports, and data items of any type can be inserted. Both primitive types and arbitrary objects can be written. When all data is inserted, the *send* primitive can be invoked on the message.

The IPL offers two ways to receive messages. First, messages can be received with the receive port's blocking *receive* primitive (see Figure 2). It returns a message object, from which the data can be extracted using the provided set of read methods. Second, the receive ports can be configured to generate *upcalls*, thus providing the mechanism for implicit message receipt. An important insight is that zero-copy can be made possible in some important special cases by carefully designing the port interfaces. Ibis allows native implementations to support zero-copy for array types, while only one copy is required for object types.

3.2 Efficient Communication

The TCP/IP Ibis implementation is using one socket per unidirectional channel between a single send and receive port, which is kept open between individual messages. The TCP implementation of Ibis is written in pure Java, allowing to compile an Ibis application on a workstation, and to deploy it directly on a grid. To speedup wide-area communication, Ibis can transparently use multiple TCP streams in parallel for a single port. Finally, Ibis can communicate through firewalls, even without explicitly opened ports.

The Myrinet implementation of the IPL is built on top of the native GM library. Ibis offers highly-efficient object serialization that first serializes objects into a set of arrays of primitive types. For each send operation, the arrays to be sent are handed as a message fragment to GM, which sends the data out without copying. On the receiving side, the typed fields are received into pre-allocated buffers; no other copies need to be made.

4   MPJ/Ibis

MPJ/Ibis is written completely in Java on top of the Ibis Portability Layer. It matches the MPJ specification mentioned in [3]. The architecture of MPJ/Ibis, shown in Figure 3, is divided into three layers. The *Communication Layer* provides the low level communication operations. The *MPJObject* class stores MPJ/Ibis messages and the information needed to identify them, ie. tag and context id. To avoid serialization overhead the *MPJObject* is not sent directly, but is split into a header and a data part. When header and message arrive at the destination, MPJ/Ibis decides either to put the message directly into the receive buffer or into a queue, where the retrieved message waits for further processing.

The *Base Communication Layer* takes care of the basic sending and receiving operations in the MPJ specification. It includes the blocking and nonblocking *send* and *receive* operations and the various *test* and *wait* statements. It is also responsible for group and communicator management. The *Collective Communication Layer* implements the collective operations on top of the *Base Communication Layer*. The algorithms realizing the collectives are shown in table 1.

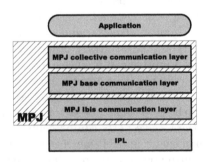

Fig. 3. Design of MPJ/Ibis

4.1   MPJ/Ibis Implementation

MPJ/Ibis tries to avoid expensive operations like buffer copying, serialization and threads where it is possible. On the sender side, MPJ/Ibis analyses the

Table 1. Algorithms used in MPJ to implement the collective operations

Collective Operation	Algorithm
allgather	double ring
allgatherv	ring
alltoall	flat tree
alltoallv	flat tree
barrier	flat tree
broadcast	binomial tree
gather	binomial tree
gatherv	flat tree
reduce	binomial tree
reduceScatter	phase 1:reduce; phase 2: scatterv
scan	flat tree
scatter	phase 1: broadcast; phase 2: filter
scatterv	flat tree

message to find out if there is a need to copy it into a temporary buffer. This is necessary when using displacements, for example. If no copy is required, the message will be written directly to the Ibis send port.

On the receiver side MPJ/Ibis has to decide to which communicator the message is targeted. The receive operation uses a blocking downcall receive to the Ibis receive port, where it waits for a message to arrive. When the message header comes in MPJ/Ibis determines if this message was expected. If it was not (a rare event), the whole message including the header will be packed into a MPJObject and then moved into a queue, copying then is mandatory. Otherwise MPJ/Ibis decides either to receive the message directly into the user's receive buffer or into a temporary buffer from where it will be copied to it's final destination (when displacements are used, for instance). There is no need to use threads for the blocking send and receive operations in MPJ/Ibis, which saves a lot of processor time. In many simple but often occurring cases zero-copying is possible as well. MPJ supports non-blocking communication operations, such as *isend* and *irecv*. These are built on top of the blocking operations using Java threads.

4.2 Open Issues

Since Java provides derived datatypes natively there is no real need to implement derived datatypes in MPJ/Ibis. Nevertheless contiguous derived datatypes are supported by MPJ/Ibis to achieve the functionality of the reduce operations MINLOC and MAXLOC, which need at least a pair of values inside a given one-dimensional array. At the moment MPJ supports one-dimensional arrays. Multidimensional arrays can be sent as an object. In place receive is not possible in this case. MPJ/Ibis supports creating and splitting of new communicators, but intercommunication is not implemented yet. At this moment, MPJ/Ibis does not support virtual topologies.

5   Evaluation

We evaluated MPJ/Ibis on the DAS-2 cluster in Amsterdam, which consists of 72 Dual Pentium-III nodes with 1 GByte RAM, connected by Myrinet and Fast Ethernet. The operating system is Red Hat Enterprise Linux with kernel 2.4.

Table 2. Low-level performance. Latencies in microseconds, throughputs in MByte/s.

network / implementation	Myrinet			Fast Ethernet		
	latency	array throughput	object throughput	latency	array throughput	object throughput
MPICH / C	22	178	N.A.	1269	10.6	N.A.
mpiJava / SUN JVM	84	86	1.2	N.A.	N.A.	N.A.
mpiJava / IBM JVM	41	178	2.7	N.A.	N.A.	N.A.
Ibis IPL / SUN JVM	56	80	4.8	146	11.2	3.0
Ibis IPL / IBM JVM	46	128	12.8	144	11.2	4.4
MPJ / SUN JVM	98	80	4.6	172	11.2	3.0
MPJ / IBM JVM	58	128	12.4	162	11.2	4.4

5.1   Low-Level Benchmarks

Table 2 shows low-level benchmark numbers for the IPL, MPJ/Ibis, MPICH and mpiJava. For the Java measurements, we used two different JVMs, one from Sun and one from IBM, both in version 1.4.2. For C, we used MPICH/GM for Myrinet and MPICH/P4 for Fast Ethernet. MpiJava uses MPICH/GM, we were unable to run it with MPICH/P4. First, we measured the roundtrip latency by sending one byte back and forth. On Myrinet, Java has considerably higher latencies than C. This is partly caused by switching from Java to C using the JNI. On Fast Ethernet MPJ is faster than MPICH/P4 (the latency is more than 7 times lower). In this case, only Java code is used, the JNI is not involved.

Next, we measured the throughput for 64 KByte arrays of doubles. The data is received in preallocated arrays, no new objects are allocated and no garbage collection is done by the JVM. The numbers show that the IBM JVM is much faster than the SUN JVM in this case, because the SUN JVM makes a copy of the array when going through the JNI. This almost halves the throughput. When we compare the mpiJava results on the IBM JVM and Myrinet with MPICH, we see that performance is the same. Ibis and MPJ are somewhat slower, but still achieve 128 MByte/s. On Fast Ethernet, all Java implementations are able to fill the network. MPICH/P4 is marginally slower.

Finally, we use a throughput test that sends binary trees of 1023 nodes, with four integer values payload per node. We show the throughput of the payload. In reality, more data is sent, such as type information and the structure of the tree (pointers to the left and right children). The tree is reconstructed at the receiving side, in newly allocated objects. It is not possible to express this test in C in this way. Ibis and MPJ are much more efficient than mpiJava when sending objects, resulting in a 4.5 times higher throughput, thanks to Ibis' highly efficient serialization implementation. This result is significant, because in Java programs typically send complex graphs of objects.

5.2   Application Performance

Figure 4 shows the speedups achieved with three applications from the Java Grande MPJ benchmarks using the Sun JVM (we found that mpiJava is unstable in combination with the IBM JVM). We also show an additional application, ASP, which is not part of the Java Grande set.

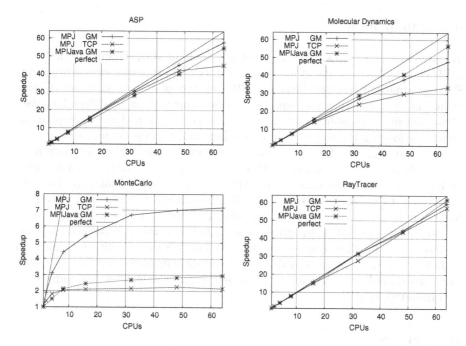

Fig. 4. Speedup of MPJ/Ibis and MPIJava applications

ASP All-pairs Shortest Paths (ASP) computes the shortest path between any two nodes of a given 5952-node graph. In ASP one machine broadcasts an array of data each iteration. Both the MPJ/Ibis and mpiJava obtain excellent speedups, but MPJ/Ibis scales better to larger number of CPUs.

MolDyn is an N-body code. For each iteration, six reduce-to-all summation operations update the atoms. We enlarged the problem size to 19 (27436 particles). Both MPJ/Ibis and mpiJava perform well on this application. Only on 64 machines, mpiJava slightly outperforms MPJ/Ibis.

The **MonteCarlo** application is a financial simulation. Each node generates an array of Vector objects. These arrays of complex objects are sent to CPU 0 by individual messages. We cannot make the problem larger than size B due to memory constraints. With this problem size, neither mpiJava nor MPJ/Ibis scale well. However, MPJ/Ibis clearly is more efficient: it outperforms mpiJava with more than a factor of two, thanks to Ibis' highly efficient serialization mechanism.

Ray Tracer renders a scene of 64 spheres. Each node calculates a checksum over its part of the scene, and a reduce operation is used to combine these checksums into a single value. The machines send the rendered pixels to machine 0 by individual messages. We enlarged the problem to an image of 2000x2000 pixels. MPJ/Ibis and mpiJava perform almost perfectly on this application.

The measurements in this section show that MPJ/Ibis achieves similar performance as mpiJava. In one case (MonteCarlo), MPJ/Ibis outperforms mpiJava by a large margin. The results indicate that the flexibility provided by the MPJ implementation on top of Ibis does not come with a performance penalty.

6 Conclusions

We presented MPJ/Ibis, our implementation of the Java language binding of MPI. Our implementation is based on our Ibis grid programming environment. Putting a message-passing layer like MPJ on top of Ibis provides an efficient environment, allowing message-passing applications in Java. Ibis' flexibility then allows to run these applications on clusters and on grids, without recompilation, merely by loading the respective communication substrate at run time.

We have evaluated MPJ/Ibis using micro benchmarks and applications from the JavaGrande benchmark suite. Our results show that MPJ/Ibis shows competitive or better performance than MPIJava, an older MPI language binding. Comparing to C-based MPICH, MPJ/Ibis is somewhat slower using Myrinet, but outperforms its competitor when using TCP/IP over Fast Ethernet.

To summarize, MPJ/Ibis can be considered as a message-passing platform for Java that combines competitive performance with portability ranging from high-performance clusters to grids. We are currently investigating the use of both MPJ and shared-object communication, paralleling single-sided communication as introduced in MPI-2.

References

1. The JavaGrande Forum: www.javagrande.org (1999)
2. van Nieuwpoort, R.V., Maassen, J., Hofman, R., Kielmann, T., Bal, H.E.: Ibis: an Efficient Java-based Grid Programming Environment. In: Joint ACM Java Grande - ISCOPE 2002 Conference, Seattle, Washington, USA (2002) 18–27
3. Carpenter, B., Getov, V., Judd, G., Skjellum, A., Fox, G.: MPJ: MPI-like Message Passing for Java. Concurrency: Practice and Experience **12** (2000) 1019–1038
4. Baker, M., Carpenter, B., Fox, G., Ko, S.H., Lim, S.: mpiJava: An Object-Oriented Java interface to MPI. In: Intl. Workshop on Java for Parallel and Distributed Computing, IPPS/SPDP, LNCS, Springer Verlag, Heidelberg, Germany (1999)
5. Mintchev, S., Getov, V.: Towards portable message passing in Java: Binding MPI. In: Recent Advances in PVM and MPI. Number 1332 in Lecture Notes in Computer Science (LNCS), Springer-Verlag (1997) 135–142
6. Judd, G., Clement, M., Snell, Q., Getov, V.: Design issues for efficient implementation of mpi in java. In: ACM 1999 Java Grande Conference. (1999) 58–65
7. Judd, G., Clement, M., Snell, Q.: DOGMA: Distributed Object Group Metacomputing Architecture. Concurrency: Practice and Experience **10** (1998) 977–983

The Open Run-Time Environment (OpenRTE): A Transparent Multi-cluster Environment for High-Performance Computing

R.H. Castain[1], T.S. Woodall[1], D.J. Daniel[1]
J.M. Squyres[2], B. Barrett[2], and G.E. Fagg[3]

[1] Los Alamos National Lab
[2] Indiana University
[3] University of Tennessee, Knoxville

Abstract. The Open Run-Time Environment *(OpenRTE)*—a spin-off from the Open MPI project—was developed to support distributed high-performance computing applications operating in a heterogeneous environment. The system transparently provides support for interprocess communication, resource discovery and allocation, and process launch across a variety of platforms. In addition, users can launch their applications remotely from their desktop, disconnect from them, and reconnect at a later time to monitor progress. This paper will describe the capabilities of the OpenRTE system, describe its architecture, and discuss future directions for the project.

1 Introduction

The growing complexity and demand for large-scale, fine-grained simulations to support the needs of the scientific community is driving the development of petascale computing environments. Achieving such a high level of performance will likely require the convergence of three industry trends: the development of increasingly faster individual processors; integration of significant numbers of processors into large-scale clusters; and the aggregation of multiple clusters and computing systems for use by individual applications.

Developing a software environment capable of supporting high-performance computing applications in the resulting distributed system poses a significant challenge. The resulting run-time environment (RTE) must be capable of supporting heterogeneous operations, efficiently scale from one to large numbers of processors, and provide effective strategies for dealing with fault scenarios that are expected of petascale computing systems [7]. Above all, the run-time must be easy to use, providing users with a transparent interface to the petascale environment in a manner that avoids the need to customize applications when moving between specific computing resources.

The Open Run-Time Environment (OpenRTE) has been designed to meet these needs. Originated as part of the Open MPI project [3]—an ongoing collaboration to create a new open-source implementation of the Message Passing

B. Di Martino et al. (Eds.): EuroPVM/MPI 2005, LNCS 3666, pp. 225–232, 2005.

Interface (MPI) standard for parallel programming on large-scale distributed systems [1,8]—the OpenRTE project has recently spun-off into its own effort, though the two projects remain closely coordinated. This paper describes the design objectives that under-pin the OpenRTE and its architecture.

Terminology. The concepts discussed in the remainder of this paper rely on the prior definition of two terms. A *cell* is defined as a collection of computing resources *(nodes)* with a common point-of-contact for obtaining access, and/or a common method for spawning processes on them. A typical cluster, for example, would be considered a single cell, as would a collection of networked computers that allowed a user to execute applications on them via remote procedure calls. Cells are assumed to be persistent—i.e., processors in the cell are maintained in an operational state as much as possible for the use of applications.

In contrast, a *local computer* is defined as a computer that is not part of a cell used to execute the application, although application processes can execute on the local computer if the user so desires. Local computers are not assumed to be persistent, but are subject to unanticipated disconnects. Typically, a local computer consists of a user's notebook or desktop computer.

2   Related Work

A wide range of approaches to the problem of large scale distributed computing environments have been studied, each primarily emphasizing a particular key aspect of the overall problem. LAM/MPI, for example, placed its emphasis on ease of portability and performance [9], while LA-MPI and HARNESS FT-MPI focused on data and system fault tolerance (respectively) [4,5]. Similarly, the Globus program highlighted authentication and authorization to allow operations across administrative zones [6].

The OpenRTE project has drawn from these projects, as well as other similar efforts, to meet objectives designed to broaden the petascale computing user community.

3   Design Objectives

The OpenRTE project embraces four major design objectives: ease of use, resilient operations, scalability, and extensibility.

Ease of Use. Acceptance of a RTE by the general scientific community (i.e., beyond that of computer science) is primarily driven by the system's perceived ease of use and dependability. While both of these quantities are subjective in nature, there are several key features that significantly influence users' perceptions.

One predominant factor in user acceptance is transparency of the RTE—i.e., the ability to write applications that take advantage of a system's capabilities without requiring direct use of system-dependent code. An ideal system should support both the ability to execute an application on a variety of compute resources, and allow an application to scale to increasingly larger sizes by drawing resources from multiple computational systems, without modification. This level of transparency represents a significant challenge to any RTE in both its ability to interface to the resource managers of multiple cells, and the efficient routing of shared data between processes that may no longer be collocated within a highly-interconnected cell (e.g., a cluster operating on a high-speed network fabric).

Several desirable system features also factor into users' perceptions of a RTE's ease of use. These include the ability to:

- Remotely launch an application directly from the user's desktop or notebook computer—i.e., without requiring that the user login to the remote computing resource and launch the application locally on that system. Incorporated into this feature is the ability to disconnect from an application while it continues to execute, and then reconnect to the running application at a later time to monitor progress, potentially adjust parameters "on-the-fly", etc.
- Forward input/output to/from remote processes starting at the initiation of the process, as opposed to only after the process joins the MPI system (i.e., calls MPI_INIT).
- Provide support for non-MPI processes, including the ability to execute system-level commands on multiple computing resources in parallel.
- Easily interface applications to monitoring and debugging tools. Besides directly incorporating support for the more common tools (e.g., TotalView), the RTE should provide interfaces that support integration of arbitrary instrumentation (e.g., those custom developed by a user).

Finally, the RTE should operate quickly (in terms of startup and shutdown) with respect to the number of processes in an application, and should not require multiple user commands to execute. Ideally, the run-time will sense its environment and take whatever action is required to execute the user's application.

Resilient. Second only to transparency in user acceptance is dependability. The RTE must be viewed as solid in two key respects. First, the run-time should not fail, even when confronted with incorrect input or application errors. In such cases, the run-time should provide an informational error message and, where appropriate, cleanly terminate the offending application.

Secondly, the RTE should be capable of continuing execution of an application in the face of node and/or network failures. Current estimates are that petascale computing environments will suffer failure of a node every few hours or days [7]. Since application running times are of the same order of magnitude, an acceptable RTE for petascale systems must be capable of detecting such failures and initiating appropriate recovery or shutdown procedures. User-definable or selectable error management strategies will therefore become a necessity for RTE's in the near future.

Scalable. A RTE for distributed petascale computing systems must be capable of supporting applications spanning the range from one to many thousands of processes, operating across one to many cells. As noted earlier, this should be accomplished in a transparent fashion—i.e., the RTE should automatically scale when adding processes. This will require that users either provide binary-compatible images for each machine architecture in the system, pre-position files and libraries as necessary—or that the run-time be capable of providing such services itself.

Extensible. The design objectives presented thus far have all dealt with the RTE from the user's perspective. However, there are also significant requirements in relation to both developers and the larger computer science community. Specifically, the RTE should be designed to both support the addition of further features and provide a platform for research into alternative approaches for key subsystems.

This latter element is of critical importance but often overlooked. For example, the possible response of the run-time to non-normal termination of a process depends somewhat on both the capabilities of the overall computing environment, the capabilities of the RTE itself, and the nature of the application. The responses can vary greatly, ranging from ignoring the failure altogether to immediate termination of the application or restarting the affected process in another location. Determining the appropriate response for a given situation and application is a significant topic of research and, to some extent, personal preference.

Supporting this objective requires that the RTE allow users and developers to *overload* subsystems—i.e., overlay an existing subsystem with one of their own design, while maintaining the specified interface, in a manner similar to that found in object-oriented programming languages.

4   Architecture

The OpenRTE is comprised of several major subsystems that collectively form an OpenRTE *universe*, as illustrated in Figure 1. A universe represents a single instance of the OpenRTE system, and can support any number of simultaneous applications. Universes can be *persistent* – i.e., can continue to exist on their own after all applications have completed executing – or can be instantiated for a single application lifetime. In either case, a universe belongs to a specific user, and access to its data is restricted to that user unless designated otherwise.

Implementation of the OpenRTE is based upon the Modular Component Architecture (MCA) [3] developed under the Open MPI project. Use of component architectures in high-performance computing environments is a relatively recent phenomenon [2,9,10], but allows the overlay of functional building blocks to dynamically define system behavior at the time of execution. Within this architecture, each of the major subsystems is defined as an MCA *framework* with a well-defined interface. In turn, each framework contains one or more *components*, each representing a different implementation of that particular framework.

Thus, the behavior of any OpenRTE subsystem can be altered by simply defining another component and requesting that it be selected for use, thereby enabling studies of the impact of alternative strategies for subsystems without the burden of writing code to implement the remainder of the system.

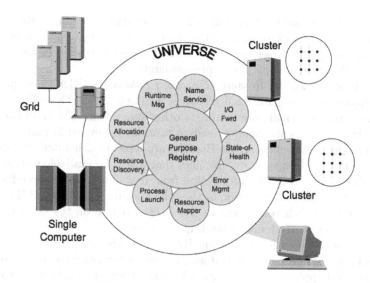

Fig. 1. The OpenRTE architecture

Researchers wishing to study error response strategies, for example, can overlay the standard error manager with their own implementation while taking full advantage of the system-level support from the process launch, state-of-health monitor, and other OpenRTE subsystems.

This design also allows users to customize the behavior of the run-time at the time of application execution. By defining appropriate parameters, the user can direct the OpenRTE system to select specific subsystem components, thus effectively defining system behavior for that session. Alternatively, the user can allow the system to dynamically sense its environment and select the best components for that situation.

OpenRTE's subsystems can be grouped into four primary elements.

General Purpose Registry. At the core of the OpenRTE system is a general purpose registry (GPR) that supports the sharing of data (expressed as key-value pairs) across an OpenRTE universe. Information within the GPR is organized into named *segments*, each typically dedicated to a specific function, that are further subdivided into *containers*, each identified by a set of character string tokens. Collectively, the container tokens, segment names, and data keys provide a searchable index for retrieving data. Users have full access to the system-level information stored on the GPR, and can define their own segments/containers to support their applications.

The GPR also provides a publish/subscribe mechanism for event-driven applications. Users can specify both the data to be returned upon an event, the process(es) and function(s) within the process(es) to receive the specified data, and combinations of actions (e.g., modification, addition, or deletion of data entries) that define the event trigger. Notification messages containing the specified data are sent to the recipients as asynchronous communications via the OpenRTE messaging layer (described below).

Resource Management. Four independent, but mutually supportive, subsystems collectively operate to manage the use of resources by applications within the OpenRTE system. Together, these subsystems provide services for resource discovery, allocation, mapping, and process launch.

True to its name, the *Resource Discovery Subsystem* (RDS) is responsible for identifying the computational resources available to the OpenRTE system and making that information available to other subsystems. The RDS currently contains two components: one for reading hostfiles in several formats covering the common MPI implementations. Hostfiles are typically generated by a specific user and contain information on machines that might be available to that user; and another that obtains its information from a system-level resource file containing an XML-based description of the cells known to the OpenRTE system. Information from each component in the RDS is placed on the GPR for easy retrieval by other subsystems within that universe.

The *Resource Allocation Subsystem*(RAS) examines the environment and the command line to determine what, if any, resources have already been allocated to the specified application. If resources have not been previously allocated, the RAS will attempt to obtain an allocation from an appropriate cell based on information from the RDS. Once an allocation has been determined, the RAS constructs two segments on the GPR – a node segment containing information on the nodes allocated to the application, and a job segment that holds information on each process within the application (e.g., nodename where the application is executing, communication sockets, etc.).

Once resources have been allocated to an application, the application's processes must be mapped onto them. In environments where the cell's resource manager performs this operation, this operation does not require any action by the OpenRTE system. However, in environments that do not provide this service, OpenRTE's *Resource Mapping* (RMAP) subsystem fills this need.

Finally, the *Process Launch Subsystem* (PLS) utilizes the information provided by the prior subsystems to initiate execution of the application's processes. The PLS starts by spawning a *head node process* (HNP) on the target cell's frontend machine. This process first determines if an HNP for this user already exists on the target cell and if this application is allowed to connect to it – if so, then that connection is established. If an existing HNP is not available, then the new HNP identifies the launch environment supported by that cell and instantiates the core universe services for processes that will operate within it. The application processes are then launched accordingly.

Error Management. Error management within the OpenRTE is performed at several levels. Wherever possible, the condition of each process in an application is continuously monitored by the *State-of-Health Monitor* (SOH)[1]. The SOH subsystem utilizes its components to field instrumentation tailored to the local environment. Thus, application processes within a BProc environment are monitored via the standard BProc notification service. Similarly, the SOH might

---

[1] Some environments do not support monitoring. Likewise, applications that do not initialize within the OpenRTE system can only be monitored on a limited basis.

monitor application processes executing on standalone workstations for abnormal termination by detecting when a socket connection unexpectedly closes.

Once an error has been detected, the *Error Manager* (EMGR) subsystem is called to determine the proper response. The EMGR can be called in two ways: locally, when an error is detected within a given process; or globally, when the SOH detects that a process has abnormally terminated. In both cases, the EMGR is responsible for defining the system's response. Although the default system action is to terminate the application, future EMGR components will implement more sophisticated error recovery strategies.

Support Services. In addition to the registry, resource management, and error management functions, the OpenRTE system must provide a set of basic services that support both the application and the other major subsystems. The *name services* (NS) subsystem is responsible for assigning each application, and each process within each application, a unique identifier. The identifier, or *process name*, is used by the system to route inter-process communications, and is provided to the application for use in MPI function calls.

Similarly, the *Run-time Messaging Layer* (RML) provides reliable administrative communication services across the OpenRTE universe. The RML does not typically carry data between processes – this function is left to the MPI messaging layer itself as its high-bandwidth and low-latency requirements are somewhat different than those associated with the RML. In contrast, the RML primarily transports data on process state-of-health, inter-process contact information, and serves as the conduit for GPR communications.

Finally, the *I/O Forwarding* (IOF) subsystem is responsible for transporting standard input, output, and error communications between the remote processes and the user (or, if the user so chooses, a designated end-point such as a file). Connections are established prior to executing the application to ensure the transport of *all* I/O from the beginning of execution, without requiring that the application's process first execute a call to MPI_INIT, thus providing support for non-MPI applications. IOF data is usually carried over the RML's channels.

5 Summary

The OpenRTE is a new open-source software platform specifically designed for the emerging petascale computing environment. The system is designed to allow for easy extension and transparent scalability, and incorporates resiliency features to address the fault issues that are expected to arise in the context of petascale computing. A beta version of the OpenRTE system currently accompanies the latest Open MPI release and is being evaluated and tested at a number of sites.

As an open-source initiative, future development of the OpenRTE will largely depend upon the interests of those that choose to participate in the project. Several extensions are currently underway, with releases planned for later in the year. These include several additions to the system's resource management and fault recovery capabilities, as well as interfacing of the OpenRTE to the Eclipse integrated development environment to allow developers to compile, run, and monitor parallel programming applications from within the Eclipse system.

Interested parties are encouraged to visit the project web site at http://www.open-rte.org for access to the code, as well as information on participation and how to contribute to the effort.

Acknowledgments

This work was supported by a grant from the Lilly Endowment, National Science Foundation grants 0116050, EIA-0202048, EIA-9972889, and ANI-0330620, and Department of Energy Contract DE-FG02-02ER25536. Los Alamos National Laboratory is operated by the University of California for the National Nuclear Security Administration of the United States Department of Energy under contract W-7405-ENG-36. This paper was reviewed and approved as LA-UR-05-2718. Project support was provided through ASCI/PSE and the Los Alamos Computer Science Institute, and the Center for Information Technology Research (CITR) of the University of Tennessee.

References

1. A. Geist et all. MPI-2: Extending the Message-Passing Interface. In *Euro-Par '96 Parallel Processing*, pages 128–135. Springer Verlag, 1996.
2. D. E. Bernholdt et. all. A component architecture for high-performance scientific computing. *to appear in Intl. J. High-Performance Computing Applications.*
3. E. Gabriel et all. Open MPI: Goals, concept, and design of a next generation mpi implementation. In *11th European PVM/MPI Users' Group Meeting*, 2004.
4. R.T. Aulwes et all. Architecture of LA-MPI, a network-fault-tolerant mpi. In *18th Intl Parallel and Distributed Processing Symposiun*, 2004.
5. G. Fagg and J. Dongarra. HARNESS Fault Tolerant MPI Design, Usage and Performance Issues. *Future Generation Computer Systems*, 18(8):1127–1142, 2002.
6. I. Foster and C. Kesselman. Globus: A metacomputing infrastructure toolkit. *Intl J. Supercomputer Applications*, 11(2):115–128, 1997.
7. E.P. Kronstadt. Petascale computing. In *19th IEEE Intl Parallel and Distributed Processing Symposium*, Denver, CO, USA, April 2005.
8. Message Passing Interface Forum. MPI: A Message Passing Interface. In *Proc. of Supercomputing '93*, pages 878–883. IEEE Computer Society Press, November 1993.
9. J.M. Squyres and A. Lumsdaine. A Component Architecture for LAM/MPI. In *10th European PVM/MPI Users' Group Meeting*, 2003.
10. V. Sunderam and D. Kurzyniec. Lightweight self-organizing frameworks for metacomputing. In *11th International Symposium on High Performance Distributed Computing*, Edinburgh, UK, July 2002.

PVM-3.4.4 + IPv6: Full Grid Connectivity

Rafael Martínez-Torres

Departamento de Sistemas Informáticos y Programación,
Universidad Complutense de Madrid, Spain
rafael.martinez@novagnet.com

Abstract. Protocol IPv4 32 bits address space introduces some well known problems when dealing with local *intra-networks*: internal pvm nodes could reach external ones, while the reverse is not always true. Hence, the PVM-GRID is not entirely deployable around the Internet as expected. New version of IP protocol, IPv6 (RFC-2460) provides full connectivity to achieve it. Other properties as multicast contribute to increase performance, and together with mobility and embedded security they yield a new concept: a trusted grid of roaming nodes.

1 Introduction

Some Beowulf clusters are designed so that not all nodes are visible to the external world. This is commonly known as the NAT[2] problem and it is inherently bound to the IPv4 protocol address space of 32 bits. Lacking a wider one, networks administrators are usually forced to setup a *front end, masquerading* node which connects to the world, while the rest of the nodes are kept on an internal network, running private, non-reachable addresses (192.168.0.x). As a transient patch, this allow internal users to run typical client-server services (*http, ftp, ssh...*) in a relatively transparent manner[1]. However, this approach has the disadvantage for PVM users running on an outside computer who want to add the Beowulf computer to their configuration: the PVM code on their computer will be unable to communicate with the Beowulf nodes on the internal network.

Some time ago, the BEOLIN port [4] of PVM was introduced in order to solve this problem by making the parallelism of the Beowulf computer transparent to the user. This is done in a way similar to the MPP ports of PVM for parallel computers such as the IBM SP2, Cray T3D and others.

Today we present a new way to solve this problem based on the new networking protocol: IPv6 [5]. Among new features, address space is virtually increased into "infinite" (128 bits vs. 32 bits). This will eventually enable not only computers, but even cellular phones and any sort of electronic devices to connect the Internet sharing a common address space. Therefore, NAT strategy is to become pointless, and the so called collaborative technologies ("groupware", P2P, GRID...) are expected to be the main beneficiaries of it.

---

[1] Firewall and proxy's policies are not covered here obeying a didactical strategy.

B. Di Martino et al. (Eds.): EuroPVM/MPI 2005, LNCS 3666, pp. 233–240, 2005.

As additional advantage, IPv6 provides the GRID with *mobile computing*, as a complement to *mobile computation* supplied by PVM. Former refers to the fact of *the nodes* of a network moving about, while latter highlights rather the notion that *running programs* need not to be forever tied to single network node. These concepts are widely explained at [10].

This paper is organized as follows: In second section we will show briefly how IPv6 technology can be integrated into PVM's architecture. Those interested on a fully detailed description can download the code from the pointed URL [12]. Next section shows the feasibility of the experiment, and advanced readers will find usefull notes regarding an hypothetic hybrid space IPv4-IPv6. A new emerging GRID paradigm based on addtional IPv6 services is introduced at fourth section, before the final summary of the paper.

2   Porting PVM into IPv6

Porting an application into IPv6 is rather a trivial task [7], provided a layered system such as PVM. Note that two steps are involved (see Fig. 1): the *PVM's runtime-system* (2.1), and the *end user's application* (2.2).

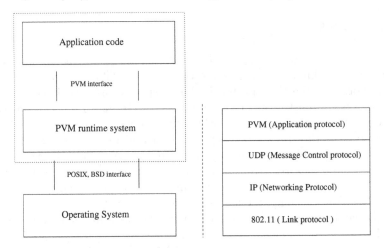

Fig. 1. PVM's Software Layout and Network Layout

2.1   PVM Runtime-System

PVM architecture is explained in detail at [6]. Essentially it relies on the next points:

- A *pvmd daemon* runs on UDP protocol serving as a *message router*[2] and *controller* for application *tasks* on the upper level, while UDP runs in time on the underlying protocol IP.

---

[2] Other further added optimizations, as *pvmd-pvmd' split* and *message direct routing* among tasks, are intentionally excluded here to focus on an more abstract approach. See 4.3 for more implications on performance.

IPv4	IPv6	Protocol Independent (IPv4-IPv6)
struct sockadddr_in	struct sockadddr_in6	struct sockaddr_storage
gethostbyname()	gethostbyname2()	getaddrinfo()
gethostbyaddr()	–	getnameinfo()

Fig. 2. Some basic structures and functions and their new counterparts in IPv6

– Interface between the *task* and the *daemon* is provided by the *library* (namely pvm.h for compilation and libpvm.a for linking issues).

Given such scenario, all we have to do is just to arrange the *runtime system's* code asking the O.S for IPv6 network services. To achieve this, the socket API has been extended (see Fig. 2) to support the new version and most UNIX kernels variants have yet implemented it[3]. Both structure and methods have been introduced to fit the wider address space (128 bits) as well as support for straight and reverse DNS resolution.

In general terms, main changes affecting PVM source code are:

1. Redefining the network address structure, according to Fig. 2, for each host setting up the PVM cluster, namely struct hostd at host.h.
2. Every time we make a socket() call, either to bind the socket as a UDP daemon, or to send a datagram packet trough it, we have to claim explicitly the PF_INET6 protocol for it.
3. The code must be ready to parse the new URL format the nodes exchange among them (see [2001:470:1f01:ffff::8e9]:32770 on the context of Fig. 3)
4. A curious situation: Initially written not to consume network sockets, but UNIX ones, further portability reasons across diverse platforms forced the interface among *tasks* and *daemon* to use a TCP/IP socket on *loop-back interface* (127.0.0.1); its counterpart IPv6 *loop-back* (::1) fulfills its original purpose.

Those interested on a deeper understanding of the code can find at [12] a patch and detailed instructions to make your PVM run on IPv6 .

2.2 End User's Applications

Having PVM's runtime system adapted and ready for IPv6, we focus on particular application's code.

As Fig. 1 suggests, PVM interface was designed to avoid application programmer getting involved with low-level networking issues like connectivity, packet formatting, sending and/or receiving, in order to concentrate in parallelism aspects. As previously pointed, the programmer asks PVM runtime for services through pvm.h interface.

Surprisingly, this interface - their function prototypes, strictly speaking - keeps unchanged, meaning this in practical terms you *have not to modify at all*

---

[3] *Linux, Solaris, FreeBSD, Mac OS, HP are known to.*

your original application source code, but just only to *re-compile and link* it against the new library you get in the previous phase, namely `libpvm.a`, taking care of the new environment variable:

```
$ export PVM_ROOT=/path/to/pvm/ipv6/patch
```

This implies a fast, cheap and easy way to deploy your GRID/PVM applications around a sort of "world wide grid", by means of IPv6 technology.

3   Deployment and Running: Advanced Topics

Fig. 3 shows the fragment of a trace file generated by a PVM-IPv6 cluster of six `i386-unknown-linux` nodes. Tests were run within the framework of research initiatives promoted by Euro6IX project[14].

Nevertheless, some additional services required by the cluster were not fully available on IPv6 yet, thus we were forced to import them via conventional IPv4 (3.2). Following lines are intended to explain how PVM software interacts on an hybrid network space IPv4-IPv6, contributing to solve potential problems beyond PVM software.

```
[pvmd pid4781] 04/23 12:08:47 version 3.4.4
[pvmd pid4781] 04/23 12:08:47 ddpro 2316 tdpro 1318
[pvmd pid4781] 04/23 12:08:47 main() debug mask is 0xff (pkt,msg,tsk,slv,hst,sel,net,mpp)
[pvmd pid4781] 04/23 12:08:47 master_config() null host file
[pvmd pid4781] 04/23 12:08:47 master_config() host table:
[pvmd pid4781] 04/23 12:08:47 ht_dump() ser 1 last 1 cnt 1 master 1 cons 1 local 1 narch 1
[pvmd pid4781] 04/23 12:08:47 hd_dump() ref 1 t 0x0 n "pvmd'" a "" ar "LINUX" dsig 0x4 08841
[pvmd pid4781] 04/23 12:08:47      lo "" so "" dx "" ep "" bx "" wd "" sp 1000
[pvmd pid4781] 04/23 12:08:47      sa [2001:470:1f01:ffff::8e9]:0 mtu 4080 f 0x0 e 0 txq 0
[pvmd pid4781] 04/23 12:08:47      tx 1 rx 1 rtt 1.000000 id "(null)"
[pvmd pid4781] 04/23 12:08:47 hd_dump() ref 1 t 0x40000 n "linux" a "" ar "LINUX" dsig 0x408841
[pvmd pid4781] 04/23 12:08:47      lo "" so "" dx "" ep "$HOME/pvm3/bin/$PVM_ARCH:$PVM_ROOT
[pvmd pid4781] 04/23 12:08:47      sa [2001:470:1f01:ffff::8e9]:0 mtu 4080 f 0x0 e 0 txq 0
[pvmd pid4781] 04/23 12:08:47      tx 1 rx 1 rtt 1.000000 id "(null)"
[t80040000] 04/23 12:08:47 linux ([2001:470:1f01:ffff::8e9]:32770) LINUX 3.4.4
[t80040000] 04/23 12:08:47 ready Sat Apr 23 12:08:47 2005
```

Fig. 3. A trace ("-d255" flag) of PVM-IPv6 execution

3.1   Can an IPv4 Node Join an IPv6 PVM Cluster?

PVM was initially designed to *support heterogeneity at application, machine and network level* [6]. Additionally, as the transition to IPv6 occurs, it is unlikely that IPv4 will be discarded (see Fig. 4). Instead, an hybrid system running both IPv4-IPv6, would be preferable [3].

So the things, the question is how to organize the transition for IPv4 nodes joining the incoming technology PVM-IPv6:

Dual stack approach, i.e., running both IPv4 and IPv6 servers at different ports. Unfortunately, host-table configuration is not shared by both processes unless exported via an IPC mechanism. Indeed, keeping such an heterogeneous table of connectivity complicates its update-maintenance: GRID IPv6-only nodes are not reachable by IPv4-only ones, and viceversa.

Tunneling. A sysadmin can take IPv4's backbone itself as a medium to gain access into IPv6's native world. Given a public IPv4 address and a *tunnel provider* [11], system will be assigned a virtual IPv6 address and integrated via a *tunnel connection* in a transparent manner for PVM applications.

6to4 relay router. A variant on the previous one, this formula enables you to setup a complet IPv6 network from an only IPv4 given address, should you export a whole cluster.

3.2   Additional Network Services

In a distributed GRID computation, PVM does not run alone on network space, but occasionally getting support from auxiliary network protocols, as network file systems, authentication services, name resolution and remote execution.

Some of named services are not fully available for IPv6 yet; others, though implemented, are not recommended in terms of performance, or just not feasible in the context of a local cluster. Following notes are intended as a reference covering such topics:

NFS. A network shared file-system is certainly not a pre-requisite to deploy PVM, and rather discouraged in terms of performance when trying to dispatch the tasks among distant nodes. In a local context, however, it makes easier its scalability as more and more nodes are added to it, having to compile your application once and the rest of the nodes mirroring it.
To our knowledge[4], no distro was available with NFS/IPv6 support, hence every cluster node ran on *dual stack* mode: IPv6 for PVM and IPv4 for NFS. Note that a network interface can hold both types of connectivity.

DNS. Dealing with chunks like 2a05:8004:30a7:231d:1142:2bc4::15 is error-prone and oftenly cause of typos. DNS is ready to manage IPv6 addresses (the so called AAAA registers), no matter what transport protocol used, TCP/IPv6 or TCP/IPv4. Optionally, /etc/host file accepts also entries on such format, should you fail to register all private nodes contributing to the GRID.

RSH. PVM's startup system relies on RSH protocol, which in turn is managed by inetd super-daemon. Somewhat deprecated, this protocol enables remote execution, provided exchange of passwords in plain text. Of course, this is not acceptable in the context of a "world wide grid". Anyway, both RSH and its secure counterpart SSH are known to be ready on IPv6.

4   Towards a Trusted Grid of Roaming Nodes

In addition to the new 128-bits address space, IPv6 covers a range of other complex subjects actually being developed at a considearable rate[9], involving other

---

[4] Date when this report was closed 2005/04/26.

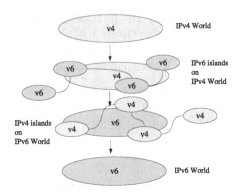

Fig. 4. Transition from IPv4 to IPv6

agents as hardware manufacturers, kernel-module developers, and telecom operators. Let's see how PVM-GRID systems can benefit of it on a medium term:

4.1 Security for the GRID: IPsec

Though SSH (3.2) provides a secure framework for authentication, once pvmd daemons have been booted further PVM messages are exchanged in plain format, at risk of interception or forgery by third parts while in transit.

As it was the case for many other application protocols (*http, ftp*), a classical solution is to implement a SSL-variant of PVM, let's say pvms; indeed, this implies a great effort to reprogram PVM's runtime, including libraries like OpenSSL[13] to support strong cryptography. Fortunately, IPsec provides security services embedded at network level, like *data integrity* and *confidentiality*, requiring no extra effort on PVM side. As a result, we get a *trustable* environment for the GRID, a desirable feature on some reserved contexts.

4.2 Roaming and Mobile IP

One of strongest points in PVM's design is fault detection and possible recovery. When a pvmd determines a peer is no longer reachable (i.e., network link become down), it deletes the node from its host-table, alerting the rest of the grid for that. This strategy, programmed at *application level*, is reinforced at *network layer* with Mobile IP, designed to enable a node moving from one IPv6 subnet to another, preserving the ability to be reached at its permanent address and keeping continuity of its ongoing computations.

This feature increases the number of possibilities for deployment, eventually enabling any moving vehicle (let it be a plane, a ship or a car) to host some nodes contributing to GRID computation. Under certain circumstances, this may be an essential requirement, should any nodes be forced to roam (i.e., those in charge of sampling tasks *ad hoc*).

4.3   Multicast and Performance

In order to increase performance, PVM provides a flag (`PvmRouteDirect`) to exchange a message among tasks through a TCP link, avoiding the overhead caused by pvmds. Of course, this rationale is kept at IPv6 version. A more sophisticated idea is implemented by an interesting function, namely `pvm_mcast()`, designed to send a message to multiple destinations simultaneously, hopefully in less time than several calls to `pvm_send()`. This is done via a two-steps protocol:

- Origin task sends its pvmd a control message containing a list of destination tasks; this, in turn, will alert each pvmd-peer hosting those tasks.
- Next, it sends *one only* data message. As it is processed by pvmd, routing layer *replicates* it, once per each destination pvmd.

Note that this only implements multicasting at *application level*, saving duplicate transmission for tasks eventually hosted on the same node. Indeed, several packets with identical data content are sent, should the destiny tasks allocated at different nodes.

Thanks to IPv6 multicast, the sender must inject the whole content only once into the network, while *network itself* handles duplication and transport to the receivers, which have explicitly joined a so-called "multicast ip group" in order to receive this special content. Such an approach results very efficient in terms of network usage and bandwidth requirements.

5   Summing Up

IPv4 address space limits the possibility of deploying a GRID platform around Internet. Network Address Translation provides a transient solution for client-server applications, but clearly inefficient for GRID paradigm.

Once described PVM's main design lines, we have explained how to support IPv6 technology, designed to provide an "infinite" address space. The final arrangement is available as a source-code patch, and has been successfully tested. Additional notes are given to support future users deploying the new system in the context of an hybrid network space IPv6-IPv4.

Other complementary services provided by IPv6 are embedded security, mobility and multicast. Nowadays subjects of intensive research, they will contribute in medium-term to enrich GRID paradigm, deriving a new formula we have designated as "the trusted grid of roaming nodes".

References

1. LONG Lab. "Over Next Generation Networks", IST-1999-20393.
   http://www.ist-long.com/
2. IETF, *NAT: Network Address Translator* http://www.ietf.org/rfc/rfc1631.txt
3. J. Kennedy, *An introduction to Writing Protocol Independent Code.*
   http://www.gmonline.demon.co.uk/cscene/CS6/CS6-03.html

4. *Readme.Beolin* File describing the Beolin port of PVM, shipped with pvm source distribution. http://www.pvm.org/download/pvm3.4.4.tar.gz

5. IETF, *Neighbor Discovery for IP version 6 (IPv6).*
   http://www.ietf.org/rfc/rfc2460.txt

6. Al Geist, Adam Beguelin, Jack Dongarra, Weicheng Jiang, Robert Manchek, Vaidy Sunderman. *PVM 3 USER'S GUIDE AND REFERENCE MANUAL.* Edited by Oak Ridge National Laboratory .ORNL/TM-121187

7. Eva Castro, Tomas P. de Miguel. *Guidelines for migration of collaborative work applications.* LONG D3.2 v2, July 15 2002.

8. Juan Quemada. *Hacia una Internet de Nueva Generación.* 2003. Universidad Politécnica de Madrid-Telefónica. ISBN: 84-608-0066-0

9. *IPv6 Cluster. Moving to IPv6 in Europe.* Edited by 6LINK with the support of the European Commission and the EC IPv6 cluster. ISBN: 3-00-011727-X

10. Luca Cardelli. *Mobility and Security.* Microsoft Research. http://www.luca.demon.co.uk

11. Hurricane tunnel broker. http://tunnelbroker.net

12. Rafael Martínez Torres. *IPv6 patch for PVM-3.4.4.* http://www.ngn.euro6ix.org/IPv6/pvm

13. OpenSSL http://www.openssl.org

14. Euro6IX project http://www.euro6ix.org

Utilizing PVM in a Multidomain Clusters Environment

Mario Petrone and Roberto Zarrelli

University of Molise, Italy
petrone@unimol.it

Abstract. A cluster is often configured with computational resources where there is only one IP visible front-end machine that hides all its internal machines from the external world. Considering clusters located in different network domains to be used in a computation, it is difficult to exploit all the internal machines of each cluster. This paper presents a PVM extension that enables us to exploit clusters in a multidomain environment so that each clustered machine can take part in a PVM computation. To improve the system performance, the PVM inter-task standard communications has been replaced by a method based on UDP sockets. Moreover, the existing code written for PVM can easily be ported to use these features.

1 Introduction

Recently, cluster architectures are always more used to solve large computational problems since they are constituted by thousands of computational nodes that provide a high performance computing [1]. In addition, it is possible to increase this computational power by joining clusters among themselves and forming a cluster grid [2]. In this approach, the resources are located across multiple networks that are geographically and administratively separate. This usually entails that some parts of the resources are on a non-routable private network and are not available to external clusters.

Among the different distributed programming technologies, such as Web Services [3], problem solving environments [4], Grid enabled software toolkits [5,6,7] or distributed metacomputing frameworks [8,9,10], PVM [11,12] remains a de facto standard programming paradigm for high performance applications. In the PVM model, a collection of heterogeneous machines are viewed and used as a single virtual parallel machine; in fact, PVM transparently handles all message routing, data conversion, and task scheduling across a network of incompatible computer architectures. However, PVM requires that all the computing nodes making up a virtual machine are IP addressable in the same domain and this appears as a serious limitation in a multidomain clusters environment, where each cluster is often configured with only one IP visible front-end machine that hides from the external world all its internal machines.

In this paper a PVM extension is presented, called MD-PVM (Multidomain PVM), which is able to manage computational resources in a multidomain clusters environment through an inter-task communication method based on UDP sockets, that improves the system performance.

B. Di Martino et al. (Eds.): EuroPVM/MPI 2005, LNCS 3666, pp. 241–249, 2005.

2   Background and Related Work

The Beolin [13] extension of PVM is a solution to exploit cluster resources, in fact, in this architecture, a Linux cluster can be added to the virtual machine and the parallel tasks can consequently be spawned onto the individual nodes of the cluster that are on a private network. This port sees a Linux cluster as a single host image, where the PVM daemon starts up on the front-end node. An environment variable is read by pvmd that tells it what cluster nodes are available for use. Each subsequent spawning request that arrives causes a node to be allocated to a single task, then the pvmd daemon marks the target nodes as busy, so that two tasks cannot share the same node. If the user attempts to spawn more tasks than the available nodes, PVM will return an error. Once a task starts on a node, it makes a TCP socket connection to the pvmd daemon, then PVM messages are transferred back and forth through this connection. There are several disadvantages to this architecture, related with the existence of one only pvmd daemon. In fact, the number of machines that can be grabbed in a Linux cluster is limited by the number of connections that the operating system can support. From the PVM user's point of view, the limits are due to the fact that you cannot know how many computational resources are available and, consequently, you cannot allocate tasks on particular nodes inside the cluster. However, the maximum number of tasks that can be spawned is the same as the number of hosts that compose the cluster; in fact, it is possible to create only one task per host.

ePVM [14,15] is a valid PVM extension that solves the problem of managing computational resources in a multidomain clusters environment; in fact, it can build an *Extended Virtual Machine* (EVM) composed of clusters. In contrast to Beolin, in the EVM each node of a cluster can take part in a PVM computation as host. The communications in the same cluster are made through the pvmd daemons while the communications among clusters are made through the epvmd daemons. Each epvmd is a multi-process daemon composed of a *Manager Task* (MT), a *Dispatcher Task* (DT) and a pool of *Workers* (Ws). Consequently, when a message must reach a remote cluster, it passes through these special tasks. However, before a message arrives at destination, it suffers the overhead due to the interaction with the pvmd daemons. In addition, the overhead is increased when we consider messages addressed to remote clusters because MT, DT and the Ws are PVM tasks that communicate in the standard PVM manner.

MD-PVM is a PVM extension that transfers from ePVM the concept of EVM and improves the communication performance through an architecture that avoids the passage of messages through the pvmd daemons.

3   The MD-PVM Architecture

MD-PVM transfers from ePVM the concept of the EVM; in fact, it permits a set of clusters, connected by an IP network, to be used as one only parallel calculator.

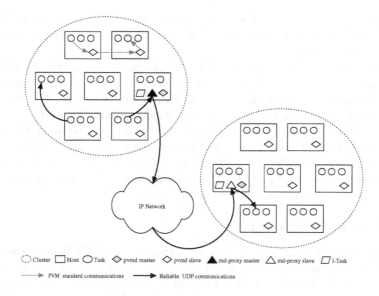

Fig. 1. The MD-PVM architecture

The messages that circulate in MD-PVM can be local or remote to a cluster and using the *Virtual Task Identifier* (VTID), as in ePVM, it is possible to know their characteristics. A VTID is a 32 bit integer that identifies a task in the EVM. It is an extension of the original *Task Identifier* (TID) that contains the *Hostid*, the *Taskid* and the *Cluster Identifier* (CID) fields. The CID is used to identify a cluster in the EVM.

In MD-PVM, each task sends and receives messages through a reliable communication protocol based on UDP sockets. When a task starts, it creates an UDP server on the port K + *Taskid* field (where K is a global constant) then each task knows the ports of every other task in the system. In addition, each task remains in communication both with the pvmd daemon through the TCP socket and both on the UDP socket just created. Consequently, the communication model of MD-PVM is composed of two types of communication mechanisms: PVM standard communications and reliable UDP communications.

The PVM standard communications are only used to manage PVM commands that do not concern exchange of messages among tasks (for example a *pvm_spawn*); in fact, when a task wants to send a message to another task in the same cluster, it sends the message directly with the reliable UDP communication protocol. This protocol permits direct communications among tasks and, in contrast with the PVM *route direct* directive, is more scalable when many tasks communicate among themselves. Consequently, the work of the pvmd daemons, in comparison with PVM, is lighter because they are only responsible for managing the virtual machine configuration of a cluster.

When a task wants to communicate with another task in a remote cluster, it sends a PVM message directly to a proxy module called *md-proxy*, through the reliable UDP communication protocol. This module is responsible for sending messages among clusters and acts in conjunction with the *libmd-pvm* library. This library contains the functions that permits transparent communications based on UDP sockets.

Finally, if a task wants to send a PVM command to a remote cluster, the command is sent through the *md-proxy* and then, when it arrives at the destination cluster, it is sent to the *Interface Task* (I-Task) module. This task is a special PVM task that interacts with the virtual machine of a cluster.

The modules that compose MD-PVM are: the *md-proxy* daemon, the *libmd-pvm* library and the I-task. Thanks to these modules, the reliable UDP protocol implemented and some changes to the original *libpvm3* PVM library (described in detail in the next sections), MD-PVM is able to manage efficiently computational resources in a multidomain clusters environment.

3.1 The Md-Proxy Daemon

The *md-proxy* is a multi-thread daemon, independent of PVM, composed of a receiver thread and a pool of sender threads. The receiver thread intercepts messages that come from: (1) PVM tasks that want to communicate by external clusters; (2) external *md-proxies* located on other clusters. The sender threads are dynamically activated during the sending of a message and their job is to send messages to remote clusters or local tasks in the same cluster.

When an application starts, a master *md-proxy* daemon is created on the front-end node of the cluster from which the application is started, while the slave *md-proxies* are started dynamically on the front-end node of each grabbed cluster. In addition, all *md-proxies* perform the same actions, but only the master can manage the global configuration of the EVM.

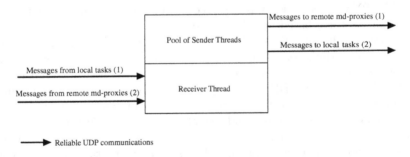

Fig. 2. The md-proxy architecture

3.2 The Libmd-Pvm Library

The library performs a wrap between the "old" PVM functions and the "new" one provided by MD-PVM so that a program written for PVM can easily be ported in this new environment. The PVM code is changed into MD-PVM at the macro compiler level that operates a textual substitution through the macro definitions contained inside the file *md-pvm.h*. Then, the only changes to apply to a PVM application to be set up in MD-PVM, is to replace the *pvm3.h* include

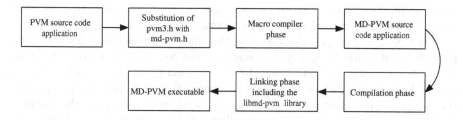

Fig. 3. Porting of a PVM application to MD-PVM

directive with the new *md-pvm.h*. Finally, at the compiler linking phase the *libmd-pvm* library must be included. Moreover, to manage the clusters added to the EVM, the additional functions exported by this library are the same as ePVM (for example *pvm_mycid*, *pvm_addcluster*, etc.).

3.3   The I-Task

The I-Task is a special PVM task that is located on the front-end node of a cluster. It executes remote PVM commands coming from external clusters (for example a *pvm_spawn*). The commands that the I-Task execute do not concern the communication primitives among tasks; in fact, they are directly managed by the *md-proxies*. After a PVM command is executed through the interaction between the I-Task and the pvmd master daemon, the results will be sent to the destination cluster that requested the remote operation.

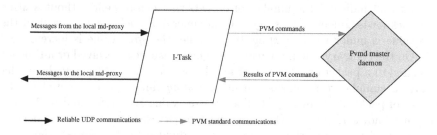

Fig. 4. Interactions between the I-Task and the pvmd daemon

3.4   The Reliable UDP Protocol

The reliable UDP protocol implemented is a connectionless variant based on the selective repeat algorithm with a congestion control method and has a dynamic packet retransmission system with cumulative acknowledgement features. It permits direct communications among tasks and has been developed as an independent library that we have called Communication Network library (CNET library).

The functions exported by the CNET library are: *cnet_init* that is used to initialize a reliable UDP server and to enable a task to send / receive messages; *cnet_close* that is used to stop a reliable UDP server created with the *cnet_init* function; *cnet_send_msg* that is used to send a message to a receiver; *cnet_recv_msg* that is used to receive a message from a sender.

The CNET library permits a receiver to manage multiple messages that come from different senders. In a typical scenario, the sequence of actions that take place when messages are sent using the CNET library functions is as follows: (1) the senders execute the *cnet_send_msg* function; (2) the CNET library fragments the messages into small packets and tries to send these packets to the receiver. If the retry timeout associated to a sending message expires, the *cnet_send_msg* function returns an error code otherwise it returns a success code; (3) the receiver receives the packets from the network through the *cnet_recv_msg* function and reassembles the original messages in dedicated memory buffers. When all the packets related to a message are correctly received, the function returns the message to the caller. In addition, to remove messages that are not completely received because, for example, a sender goes down or there is a network failure between a sender and the receiver, to each dedicated buffer is associated a timeout. So, if one or more of these timeouts expire, the related buffers are freed from the memory.

The reliability is assured by the selective repeat algorithm that associates sequence numbers and timeouts to each packet.

The congestion control algorithm implemented is controlled by two variables: *sfrag* and *maxfrag*. The *sfrag* variable indicates how many fragments have been sent but not yet acknowledged while the *maxfrag* variable indicates the maximum number of packets that can be sent without waiting for an acknowledgment. If a sender wants to send a message, *maxfrag* is initialized to WIN_SIZE (it is a constant that indicates the number of packets that can be sent without waiting for an acknowledgment) and *sfrag* to 0. In general, a packet can be sent, if the *sfrag* value is minor than *maxfrag*. If it is true, the *sfrag* value is incremented by 1. On the contrary, when an acknowledgment packet is received or a timeout associated to a packet expires, the *sfrag* value is decremented by 1. If the sender receives a cumulative acknowledgment, the *sfrag* value is decremented by the number of packets acknowledged. The *maxfrag* value is incremented by 1 every time an acknowledgement packet is received, otherwise if the timeout associated to a sent packet expires, the *maxfrag* value is divided by 2. However, the range within which the *maxfrag* value can fluctuate is 1 to MAX_WIN_SIZE (it is a constant that delimits the maximum number of packets that can be sent without waiting for an acknowledgment).

The acknowledgment system implemented is based on two types of acknowledgement packets: the Single Acknowledgement Packet (SAP) used to acknowledge single packets and the Cumulative Acknowledgement Packet (CAP) used to acknowledge multiple in-order packets. When a receiver receives a packet, it waits up to WAIT_TMOUT milliseconds for the arrival of other in-order packets related to the same message. So, the following events can occur: (a) if the timeout does not expire and all the packets are received in the in-order mode, the receiver sends a CAP packet that acknowledges all these packets; (b) if the timeout expires and no other packets are received, the receiver sends a packet that acknowledges the last in-order packets received; (c) if a non in-order packet related to the same message is received, the receiver sends a packet containing the acknowledgement of the last in-order packets received; (d) if a packet related to a different message is received, the receiver sends a packet containing the acknowledgement of the last in-order packets received related to the previous message. In the events (b)(c)(d), the acknowledgement packets generated can be of CAP or SAP type.

3.5   Changes to the Original Libpvm3 Library

The main updates to the *libpvm3* library are the two new functions: *pvm_getRawMessage* and *pvm_setRawMessage*. The first returns a PVM message, identified by a buffer identifier, as an array of byte. The second executes the opposite work, in fact, it converts a previously saved PVM message through the *pvm_getRawMessage* function in the form of standard PVM message. In addition, the *pvm_setRawMessage* function stores the converted message in the message queue of the PVM task so that the standard PVM functions of message manipulation can work correctly. In MD-PVM these two functions allow the sending of a PVM message through the UDP sockets, bypassing the standard PVM communication protocol.

4   Performance Results

We performed a comparison experiment on two clusters, each composed of 10 nodes, equipped with Pentium 4 2.0 Ghz and interconnected by a 100 Mbit/s Ethernet switch. The machines had 512 MB RAM each and were running Fedora Core Linux 3.

In Test 1 of Figure 5 the round trip delay between two tasks located on two hosts of the same cluster was measured. This test shows the improvement of the MD-PVM architecture through the reliable UDP protocol. In fact, MD-PVM and PVM configured with the *route direct* directive exhibited almost the same behaviour, with differences fluctuating from 5% to 10%. PVM configured with the standard communication system resulted slower, owing to the passage of messages through the pvmd daemons.

Test 2 of Figure 5 shows the round trip delay in a remote task-task communication, where two tasks were located in the front-end machines of the two clusters. MD-PVM has better performances than ePVM, thanks to the direct path communications between tasks and *md-proxies* through UDP socket communications. Moreover, in this test, in comparison with ePVM, we measured an improvement of about 15% for small messages and an improvement of about 50% for messages up to 1MB.

Test 3 of Figure 5 has been conducted taking into consideration an application that was communication and CPU intensive that solves differential heat equations. In this test we spawned 20 tasks, where each task was located on a computational node of the two clusters. In this test, MD-PVM resulted, in the computation time, better than ePVM. Moreover, with the increasing of the

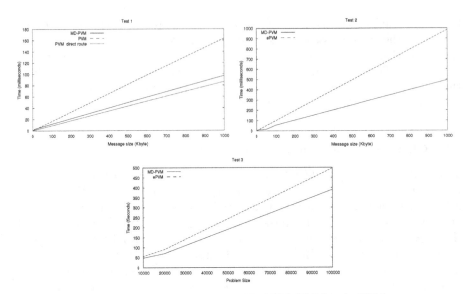

Fig. 5. Comparison tests on MD-PVM, PVM and ePVM

problem size, MD-PVM in comparison with ePVM resulted from 15% to 20% faster.

5   Conclusions

This paper presents MD-PVM, a PVM extension, that is able to manage computational resources in a multidomain clusters environment. MD-PVM, through a communication model based on UDP sockets, avoids the passage of messages through the pvmd daemons and improves the system communication performance. In comparison with the ePVM architecture, MD-PVM results from 15% to 50% faster in the test cases executed. In addition, the existing code written for PVM can easily be ported to use the MD-PVM features.

Acknowledgments

Special thanks to Eugenio Pasquariello, Angelo Iannaccio and Carmelina Del Busso for their useful suggestions in implementing MD-PVM.

References

1. Top 500 supercomputer site. http://www.top500.org.
2. W. Gentzsch, "Grid Computing, A Vendor's Vision", *Procs of the 2nd IEEE/ACM Intl. Symposium on Cluster Computing and the Grid*, Berlin, Germany, May 2002.
3. W3C Consortium, Web Services activity. http://www.w3c.org/2002/ws.

4. S. Agrawal, J. Dongarra et al., "NetSolve: Past, Present, and Future - a Look at a Grid Enabled Server", *Grid Computing: Making the Global Infrastructure a Reality*, Edited by F. Berman, G. Fox, and A. Hey, Wiley Publisher, 2003.

5. I. Foster, N. Karonis, "A grid-enabled MPI: Message passing in heterogeneous distributed computing systems", *In Supercomputing 98*, Orlando, FL, 1998.

6. N. Karonis, B. Toonen and I. Foster, "MPICH-G2: A grid-enabled implementation of the Message Passing Interface", *Journal of Parallel and Distributed Computing*, May 2003.

7. R. Keller, B. Krammer et al., "MPI development tools and applications for the grid", *In Workshop on Grid Applications and Programming Tools*, Seattle, WA, June 2003.

8. D. Kurzyniec, T. Wrzosek et al., "Towards Self-Organizing Distributed Computing Frameworks: The H2O Approach", *Parallel Processing Letters*, 2003.

9. P. Hwang, D. Kurzyniec and V. Sunderam, "Heterogeneous parallel computing across multidomain clusters", *Procs of the 11th EuroPVM/MPI Conference*, Budapest, Hungary, September 2004.

10. D. Kurzyniec, P. Hwang and V. Sunderam, "Failure Resilient Heterogeneous Parallel Computing Across Multidomain Clusters", *International Journal of High Performance Computing Applications*, Special Issue: Best Papers of EuroPVM/MPI 2004, 2005.

11. Al Geist, A. Beguelin et al., "PVM: Parallel Virtual Machine. A Users' Guide and Tutorial for Networked Parallel Computing", *The MIT Press*, 1994.

12. A. Beguelin, J. Dongarra et al., "Recent Enhancements to PVM", *Intl. Journal for Supercomputer Applications*, 1995.

13. P. L. Springer, "PVM Support for Clusters", *IEEE 3rd Intl. Conf. on Cluster Computing*, Newport Beach, California, USA, October 2001.

14. F. Frattolillo, "A PVM Extension to Exploit Cluster Grids", *Procs of the 11th EuroPVM/MPI Conference*, Budapest, Hungary, September 2004.

15. F. Frattolillo, "Exploiting PVM to Run Large-Scale Applications on Cluster Grids", *International Journal of High Performance Computing Applications*, Special Issue: Best Papers of EuroPVM/MPI 2004, 2005.

Enhancements to PVM's BEOLIN Architecture

Paul L. Springer

California Institute of Technology, Jet Propulsion Laboratory,
4800 Oak Grove Drive, Pasadena CA 91109, USA

Abstract. Version 3.4.3 of PVM had previously been enhanced by the addition of a new architecture, BEOLIN, which allowed a PVM user to abstract a Beowulf class computer with a private network to appear as a single system, visible to the outside world, which could spawn tasks on different internal nodes. This new enhancement to PVM handles the case where each node on the Beowulf system may be composed of multiple processors. In this case, the software will, at the user's request, spawn multiple jobs to each node, to be run on the individual processors.

1 Introduction

The BEOLIN architecture support was added to PVM in version 3.4.3. The motivation for this addition was the limitation PVM had before that time in dealing with Beowulf clusters. Prior to that version, a PVM user of a heterogeneous system that included a Beowulf cluster was unable to treat that cluster as a single system image. If the Beowulf nodes were on an internal network, they were invisible to the rest of the PVM system. The only way the nodes could be used was to treat them as individual computers, provided they were visible to the entire PVM system. The disadvantage to this approach was that the user's application was forced to issue PVM *addhost* and *spawn* commands to each individual node. Furthermore, each node incurs the overhead of running the PVM daemon.

The special architecture support that had been added on earlier versions of PVM for individual parallel machines, such as IBM's SP2, Intel's Paragon, and others, provided the inspiration to add the BEOLIN architecture support, to handle Beowulf clusters running Linux. By means of the BEOLIN support, PVM can spawn tasks to a single machine target for the Beowulf cluster. That target knows about, but hides the details of the machine, such as the numbers of nodes available, and the individual IP address of each node. The target handles the details of spawning the tasks onto individual Beowulf nodes for execution.

Since that time hardware enhancements have been made to Beowulf systems, and one such enhancement is the availability of multiple processors per node, where those processors share the node's IP address. The initial BEOLIN release could only handle a single processor per node. This paper describes a new enhancement that supports multiple processors per node.

No paper was presented describing the original BEOLIN release, and so information on that release will be included in this paper to provide the necessary background.

B. Di Martino et al. (Eds.): EuroPVM/MPI 2005, LNCS 3666, pp. 250–257, 2005.

2   Design

Of the Massively Parallel Processor (MPP) architectures previously supported, BEOLIN is most similar to the SP2MPI port, which supported the SP2, using MPI as the underlying message passing protocol. BEOLIN, however, uses sockets instead of MPI.

The following description of the initiation of tasks onto a BEOLIN machine assumes the cluster has a front-end: a node that is visible to the rest of the PVM system, and can itself communicate to the individual nodes that comprise the BEOLIN computer. PVM can begin running on the cluster in different ways, for example by a user logging on to the front-end and running the pvm command on that front-end, or by running a pvm application on an external computer that issues a *pvm_addhosts* command, targeting the cluster. (See figure 1.)

When PVM is initiated on the cluster, the first thing that happens is that the PVM daemon (pvmd) begins running on the front-end. The daemon reads an environment variable to discover the names of the nodes in the cluster, and forms a node pool for later use. The variable in question specifies the initial order of node allocations, as well as the number of tasks that can be executed on a given node. Subsequent spawning requests cause the daemon to allocate nodes from the pool and initiate the requested tasks onto them. As tasks complete, their corresponding nodes are freed back into the pool for future use.

At the time the pvmd daemon on the front-end initiates a task on a cluster node, it forks a copy of itself, with both child and parent running on the front-end. The child's standard output and error are connected to the just-started task. Each child process so generated is visible to pvmd, and the pvm monitor will show the child as a task with the suffix ".host" appended to its name. No copy of the daemon runs on the targeted node.

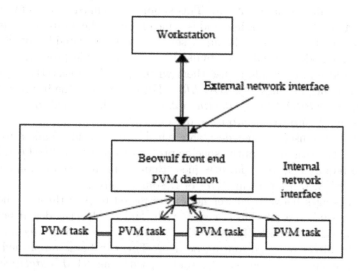

Fig. 1. Beowulf Block Diagram

When the remote task begins execution, it examines a shared file space (/tmps) for a file name beginning with "pvmd" and ending with the user id. That file contains the necessary addressing information for the task to make a connection with the pvmd daemon. PVM messages and commands are relayed through the resulting connection.

3   Installing and Using the BEOLIN Port of PVM

As of version 3.4.3, BEOLIN is a defined architecture for PVM. However, to force the build process to use the BEOLIN files, the environment variable *PVM_ARCH* must be set to the value *BEOLIN*–don't rely on the *pvmgetarch* command for this. If this is not done properly, the build will probably produce a plain LINUX version instead. Once the architecture variable is properly set, build PVM as described in Chapter 9 of [1].

Before the resulting BEOLIN build can be run, a shared file space with the name /tmps must be set up in such a way that any file in this subdirectory is accessible by the front-end as well as the cluster nodes on which tasks may be run by PVM.

In the previous version of BEOLIN, there was a requirement that the nodes had to be able to connect to the front-end using the address returned by the *gethostname()* Linux call on the front-end. In the current version that requirement has been relaxed by the use of an existing command line parameter for PVM. When starting either pvm or pvmd, using the command line parameter -n < *hostname* > tells the BEOLIN code that pvmd should write the IP address of < *hostname* > in the /tmps file it creates, informing the cluster nodes what address they should use in making their connection to the daemon.

The BEOLIN daemon code looks for the environment variable *PROC_LIST* when it starts up on the front-end. This should be defined in the environment used by pvmd when it is running, and is typically set in the .cshrc (or equivalent) file. The value of the variable should be set to a colon separated list of the names of the cluster nodes available for pvmd to use. If multiple processors exist on the node, and it is desirable to use them when spawning tasks, the node name can appear multiple times in *PROC_LIST*. For example, a line in the .cshrc file that read *setenv PROC_LIST n0:n1:n1:n2:n3* would allow pvmd to spawn 1 task onto node n0, 2 onto n1, 1 onto n2 and 1 onto n3.

Note that the user can not designate which processor in a node a task is to be run on. The assignment of the specific processor is left up to the O/S. If more nodes are required by the spawn request than there are slots available in the node pool, PVM will return an "Out of Resources" error.

If all messages between cluster nodes are forced to pass through the pvmd on the front-end, the job will not scale well, and the front-end will become a communication bottleneck. To avoid this situation, it is strongly recommended that PVM's direct message routing be used. This can be accomplished by the application calling the *pvm_setopt()* routine, with the *PVMRouteDirect* parameter. This forces messages from one node to another to go directly, instead of being routed through the daemon.

3.1   Restrictions and Limitations

The BEOLIN version of the PVM monitor or pvmd can be run on the front-end node, but starting either one on one of the other nodes of the cluster has not been well tested, and is advised against.

When the application calls *pvm_addhosts()* from an external computer, to add the cluster, it should only specify the front-end node of the cluster as an argument. The command should not be used to add the other cluster nodes to the virtual machine. Similarly, the *add* command of the PVM monitor should not be used to add individual nodes. Avoid including individual node names in the PVM hostfile. The purpose of this BEOLIN port is to treat the entire cluster as a single machine, to be addressed only by the name of the front-end.

PVM uses its TID (task ID) word to uniquely identify each task running on the virtual machine. To distinguish different tasks running on the same node, the BEOLIN code uses the three bit partition field in the TID. This limits the maximum number of child tasks able to run on a single cluster node to eight.

Even if all the cluster nodes have a direct connection to the external network, this BEOLIN port can still be used to treat the cluster as a single PVM machine. In this particular case, any node can be arbitrarily designated as the front-end.

4   Internals

The bulk of the BEOLIN code is in the pvmdmimd.c file in the BEOLIN source code subdirectory in the PVM package. This section will give an overview of the BEOLIN program code.

4.1   Initialization

When the pvmd daemon starts up on the cluster front-end, the main() routine in pvmd.c calls *mpp_init()*, passing it the argc and argv parameters, in order to initialize the BEOLIN part of PVM. The *mpp_init()* routine first parses the environment variable *PROC_LIST*, and calls *gethostbyname()* for each entry in the list, storing the resulting addressing information internally. The number of times each node is referenced in the list is tracked, and the reference number is stored in the *partNum[]* array.

Several arrays are created in the BEOLIN initialization process, with entries in the arrays corresponding to entries in *PROC_LIST*. The *nodepart* array holds the partition number for the entry, the *nodeaddr* array contains the IP address, the *nodeconn* array is initialized to 0, and the *nodelist* array carries the name as it appears in *PROC_LIST*.

4.2   Spawning

When the cluster's pvmd receives a command to start the spawned tasks on its nodes (by means of a call to *pvm_spawn()*, for example), control is passed to the BEOLIN routine *mpp_load()*. The first thing *mpp_load()* does is to check to see if there are enough node slots in its free node pool, by calling *mpp_new()*. If there are enough slots, *mpp_new()* allocates and returns the set of slots that will be used to spawn the tasks; otherwise it will report an error. As part of the allocation process, *mpp_new()* generates an identifier called ptype that is unique to the set of nodes allocated for this set of tasks, and puts this set of nodes into its busynodes list, marking each node with the ptype value.

Once the node slots are allocated, *mpp_load()* iterates a set of actions for each of the new tasks, whereby it sets up a task structure for that task, and calls *forkexec()* in the main body of PVM to actually start the task running by means of an rsh command, on the appropriate node. Before *forkexec()* runs the rsh command, it first forks to produce the previously described "host" task. The "host" task then executes the rsh command. (See figure 2.)

When first started on the target node, each spawned task must connect with its corresponding "host" task on the front-end. This action is triggered by the first PVM call the spawned task makes, and the action takes place inside the routine *pvmbeatask()*, contained in the lpvm.c module which is part of the PVM code linked to the application. To open a socket connection with pvmd, *pvm-beatask()* calls the version of the *mksocs()* routine in the lpvm.c module. The connection is made by searching for a file name with the pattern pvmd.*userid*, in the /tmps file space shared by the front-end and the nodes. The file is read and the information within it is used to make the connection.

Once the socket connection is made, the spawned task sends a *TM_CONNECT* message to the pvmd. The pvmd's routine *tm_connect()* sends an acknowledgment back to the task, which then causes the task to respond with a *TM_CONN2* message back to the pvmd. The pvmd code enters the *tm_conn2()* routine, which in turn calls the BEOLIN *mpp_conn()* code to determine which of the task spawn requests this current connection process matches. When the match is made, the corresponding entry in the BEOLIN *nodeconn* array is set true.

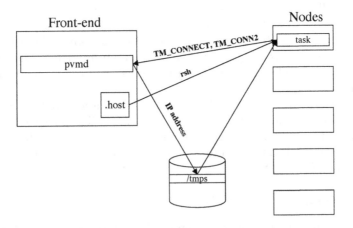

Fig. 2. Task Spawning

4.3 Communication

When the application wants to send a message from one node to another, the routine *pvm_send()* is called. Once the message is constructed, *mroute()* is called to determine the routing and send the message. It first determines whether the direct routing option has been set, allowing the message to bypass the pvmd. If so, and if this is the first time the destination has been requested by this task, it opens a socket to the destination task, and makes an entry (ttpcb) in the task process control block list. The entry includes the socket information, to be used the next time a message is sent to the same destination.

4.4 Task Termination

Task termination is triggered by the application's call to *pvm_exit()*, done for each spawned task that is part of the application. When the pvmd receives the exit request, it calls *task_free()*, which in turn calls BEOLIN's *mpp_free()* routine, passing each task as the argument. The *mpp_free()* routine first finds a match for this task, based on node number and partition number, and then shuts down the "host" task corresponding to the spawned task that was passed, sets a flag to indicate that this task is done executing, and returns. The pvmd code later calls BEOLIN's *mpp_output()* routine, as it does on a regular basis. As *mpp_output()* cycles through each task that has been spawned and is still alive, it checks to see if the "host" task corresponding to this task is still alive. If not, it then takes responsibility for shutting down the spawned task, and returns its nodes to the free pool.

5 Performance

In allowing the use of multiple processors on a node, this latest enchancement to the BEOLIN architecture focuses on increasing the computing power available to a PVM application. It is appropriate, then, to evaluate this version by choosing a benchmark that is computationally intensive, rather than one that relies on communication speed. One benchmark that meets this criteria is based on the POV-Ray program[2]. POV-Ray is a multi-platform program that creates an image of a scene using a ray tracing technique. It uses as its input a POV format text file that describes the objects in the scene as well as the camera and lighting and other effects.

POV-Ray was not originally written to run under PVM, but there is a patch available to do this. The patch was downloaded from [3], and applied to version 3.5 of POV-Ray. As patched, the program runs in a master-slave configuration, with the work parceled out to the slaves whenever a slave has completed its previous work. The number of slaves is configurable at startup.

For the purposes of benchmarking, a standard benchmark.pov file is available as input to the program, as well as a standard initialization file[4]. A large number of both sequential and parallel benchmark numbers have been published. The benchmark was run on a Beowulf cluster composed of 800 MHz dual Pentium

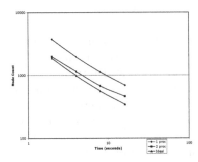

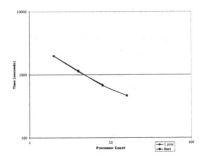

Fig. 3. BEOLIN Performance   **Fig. 4.** Comparative Performance

III nodes. PVM was configured to run on 2, 4, 8 and 16 nodes, using both 1 and 2 processors per node. The results can be seen in the accompanying figures.

Figure 3 shows the improvement in computing power enabled by the new BEOLIN version. The upper trace shows the benchmark times with the old limitation of running just 1 processor per node. In the middle trace, both processors on the node are put to use, and the run times are faster, close to the ideal of being twice as fast (as shown in the bottom trace). The ideal might be missed not because of the BEOLIN implementation, but because the problem itself doesn't scale well. The true explanation can best be determined by looking at figure 4. The two processors per node trace is shown as before. The trace labeled "best" indicates how well the problem scales with the when using the same number of *total processors* (one per node) as the two processors per node case. It can be seen that run times are identical whether the processors used are on the same node or not. This indicates that the slight lack of scaling shown in figure 3 is not the result of using two processors on the same node.

As shown in figure 4, performance falls off slightly when 32 processors are used, in the two processors per node case. The computer used did not have 32 nodes, so the equivalent time for the 1 processor per node case could not be measured. But the benchmark produces a number of statistics, including load balancing information, and analysis of that showed that the run time was slower in this case because work was not evenly distributed, and near the end of the run only a few nodes had work to do.

6   Future Work

The largest number of nodes used so far with this port of PVM has been 31. The software needs to be run on much larger machines in order to characterize its performance at the high end. It is unclear what bottlenecks and limitations exist, and how they will manifest themselves, when one pvmd daemon is controlling a large number of tasks. Changes in the pvmd code to accommodate very large systems may be necessary. One particular area to examine is the way that a host task is forked off on the front-end for each spawned task. This will not scale as the number of tasks becomes very large. To support larger number of tasks, it will be necessary to change the way this is done, or perhaps to spawn limited numbers of additional pvmds on the other nodes.

7   Conclusion

The BEOLIN architecture described here offers new capabilities to the PVM applications programmer. A Linux PC cluster can now be added to the virtual machine, and parallel tasks spawned onto the cluster, even when the individual nodes are on a private network. Clusters with multiple processors on a single node can have tasks spawned to each processor on the node.[1]

References

1. Geist, Al, et al: PVM: Parallel Virtual Machine. The MIT Press, Cambridge, Massachusetts (1996)
2. "POV-RAY - The Persistence of Vision Raytracer." 25 Mar. 2005. Persistence of Vision Raytracer Pty. Ltd. 3 Jul. 2005. $< http://www.povray.org >$
3. "PVM patch for POV-RAY." 2003. SourceForge. 3 Jul. 2005. $< http://pvmpov.sourceforge.net >$
4. "Haveland-Robinson Associates - Home Page." 5 Mar. 2004. Haveland-Robinson Associates. 3 Jul. 2005. $< http://haveland.com/povbench >$

[1] This research was carried out at the Jet Propulsion Laboratory, California Institute of Technology, under a contract with the National Aeronautics and Space Administration. The funding for this research was provided for by the Defense Advanced Research Projects Agency under task order number NM0715612, under the NASA prime contract number NAS7-03001.

Migol: A Fault-Tolerant Service Framework
for MPI Applications in the Grid

André Luckow and Bettina Schnor

Institute of Computer Science,
University Potsdam, Germany
{schnor, drelu}@cs.uni-potsdam.de

Abstract. In a distributed, inherently dynamic Grid environment the
reliability of individual resources cannot be guaranteed. The more
resources and components are involved the more error-prone is the
system. Therefore, it is important to enhance the dependability of the
system with fault-tolerance mechanisms. In this paper, we present Migol,
a fault-tolerant, self-healing Grid service infrastructure for MPI
applications.

The benefit of the Grid is that in case of a failure an application
may be migrated and restarted from a checkpoint file on another site.
This approach requires a service infrastructure which handles the necessary activities transparently for an application. But any migration framework cannot support fault-tolerant applications, if it is not fault-tolerant
itself.

Keywords: Grid computing, fault-tolerance, migration, MPI, Globus.

1 Introduction

The Grid is dynamic by nature, with nodes shutting down respectively coming
up again. The same holds for connections. For long running compute-intensive
applications fault-tolerance is a major concern [2,17]. A benefit of the Grid is that
in case of a failure an application may be migrated and restarted on another site
from a checkpoint file. But a migration framework cannot support fault-tolerant
applications, if it is not fault-tolerant itself.

Here, we present a fault-tolerant infrastructure, i.e. without any single point
of failure, conforming to the Open Grid Service Architecture (OGSA) [13] for
supporting the migration of parallel MPI applications. OGSA builds upon open
Web service technologies to uniformly expose Grid resources as Grid services.
The Open Grid Services Infrastructure (OGSI) [30] is the technical specification
of extensions to the Web services technology fulfilling the requirements described
by OGSA. An implementation of the OGSI is provided by the Globus Toolkit 3
(GT3) [15]. Migol services are built on top of GT3.

B. Di Martino et al. (Eds.): EuroPVM/MPI 2005, LNCS 3666, pp. 258–267, 2005.
© Springer-Verlag Berlin Heidelberg 2005

2   Related Work

Migration and migration strategies were well studied in the context of cycle-stealing clusters in the mid 90s (see for example [25,3]).

Approaches for fault tolerance in MPI applications are discussed in [17]. MPI-FT [11], CoCheck [29], Starfish [1] and MPICH-V [5] provide fault tolerance for MPI applications based on different fault diagnostics. checkpointing, and message logging. These systems operate on a lower level within a MPI application. Therefore, they are for the most part restricted to homogenous cluster environments and are not capable of handling multi-institutional Grids.

Recently, migration frameworks were also developed for Grid environments. The Condor/PGRADE system [20] consists of a checkpointing mechanism for PVM applications and uses Condor-G [10] for scheduling. Fault-tolerance of the service infrastructure is not supported.

In [16] a fault-tolerant mechanism to support divide-and-conquer applications in the Grid is presented. The solution exploits special characteristics of these applications and cannot be generalized.

GridWay [31] and GrADS [24] provide migration capabilities to MPI applications based on the GT and application-level checkpoints. These systems are primarily performance-oriented. Fault-tolerance and OGSI-compliance of the components is not within the focus of these projects.

3   Migol Architecture

The main features of Migol are scheduling, monitoring, and, if necessary, migrating applications in the Grid. The Grid is assumed to be a loosely coupled organization: each site has its own administration and resource management system (for example Condor [22], PBS [18], ...).

The Migol framework builds upon the standard open source solution provided by the Globus Toolkit 3.2.1. The key concepts for achieving failure-tolerance are:

- Clear separation of services with no state, transient and persistent state.
- Replication of persistent services.
- Support for recovery of transient Grid service instances.
- Reuse of services provided by GT3, e. g. security and persistence services.

Fundamental part of Migol is the Grid Object Description Language (GODsL), which defines a generic and extensible information model for describing different aspects of the Grid, e. g. hardware, applications, services or data [21]. For example, when an application is started, a *Grid Service Object* is created which stores all relevant information of the application: resource requirements, location of binaries and checkpoint files, global unique identifier, etc. Every resource, service or application is associated with a Grid object.

At any time, the currently used Grid objects represent the state of the Grid. A special information service was developed, the Application Information Service (AIS), to store Grid objects. The AIS contains substantial information about

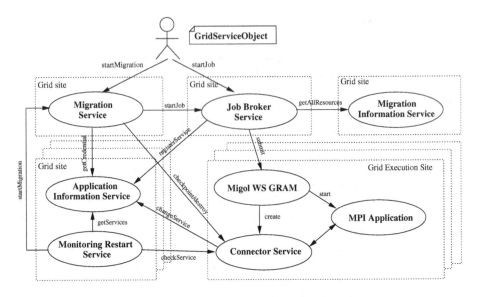

Fig. 1. Migol Service Architecture

running services and available resources. Applications can register and deregister services, files, machines etc. through Grid objects. To avoid a single point of failure the AIS is replicated using a ring-based replication protocol, which ensures the consistency of the stored data (see [23] for more details).

Obviously, a monitoring service is necessary to monitor the status of applications, i.e., *active, pending, staging, inactive, terminated.* In case the monitor service discovers an application to be inactive, it initiates a restart respectively a migration. Hence, we call it Monitoring and Restart Service (MRS). The MRS monitors all applications which have registered themselves at the AIS. The MRS is also a critical component. But replication of the MRS is complicated: This would mean that the monitoring traffic is also replicated and even worse, in case of a failure the different monitoring services would initiate migrations concurrently. Hence, the MRS has to be replicated, but only one instance, e.g. the one with the lowest ID, may be active. The remaining instances operate in stand-by mode.

All other Migol services which are necessary for scheduling and migration are created on demand. Figure 1 shows the Migol services and their interactions. The services will be explained in more detail in the next sections.

3.1   Migration Information Service

The Migration Information Service (MIS) collects static and dynamic data about the Grid. For that purpose the MIS aggregates data of different services, e.g. the Network Weather Service (NWS) [27]. The NWS provides latency and bandwidth information, which are not available through the Globus information services.

There can be multiple deployments of the MIS in a Grid to improve availability. The state of the MIS can be completely re-collected after a failure.

3.2 Connector Service: Grid Services and MPI Applications

At the moment a gap between high performance applications and Grid services exists. There are several approaches for exposing a MPI application as OGSI-compliant Grid service [12,26].

We decided to use a general approach using a Grid service to monitor applications submitted via the Globus Resource Allocation Manager (GRAM) [9]. GRAM provides a standardized interface to different local resource management systems (e.g. PBS, LSF, Condor). Extensions to the WS GRAM service were necessary, as the supplied mechanisms for management and monitoring of applications were not sufficient. For example, there is no possibility to communicate with an application, e.g. to trigger the writing of a checkpoint.

Further, we observed that job states reported by GRAM are partly incorrect. Thus, it is impossible to determine whether a job exited normally or failed and therefore must be recovered. To fill this gap the Connector Service (CS) was developed. The CS is a standard OGSI Grid service, which can be used to virtualizes any MPI application. This approach enables the seamless integration of MPI applications within the service infrastructure and provides a standard Grid service interface for starting, monitoring and controlling.

The communication between Grid service and application is implemented through sockets and transparently encapsulated in a vendor-independent, user-level MPI library which is similar to the approach in CUMULVS and other systems (e.g. [31,24,19]). The user must include the Migol header and link against the library. Library functions are provided via a special API. The application must make an initialization and finalization call to the library.

The Migol library depends on application-level checkpointing. The application is responsible for writing checkpoints. Our first prototype depends on locally stored checkpoints, which have to be accessible to enable auto-recovery. Future versions will support replicated checkpoints.

For every GRAM job one CS instance has to be created. This is done by the Migol WS GRAM service, which transparently extends the WS GRAM of the GT3 and can be plugged in at deployment time.

3.3 Job Broker Service

The Job Broker Service (JBS) integrates different computing resources, making them available to the user as a single point of entry (submit & forget). The JBS performs resource discovery through the AIS. Resources are matched according to the requirements specified by a Grid Service Object. The scheduler module ranks the matched resources based upon the following heuristic:

- The computation time is estimated upon a speed factor (currently the normalized cpu speed).

- The waiting time for a job is evaluated based on the queue lengths of the local resource management systems.
- The proximity between the source and selected hosts has to be considered to avoid the selection of a difficult-to-reach or disruption-prone site. This is done by using a configurable bandwidth-threshold.

The dynamic attributes needed for scheduling are collected from the MIS. The resulting prioritized list is used to dispatch the jobs.

Because of the highly dynamic and complex Grid environment and the lack of local control (site autonomy) this metric can only be considered a rough approximation. Especially the waiting time for a job is, due to different resource requirements of the individual jobs, hardly predictable. In future advanced reservation may fill this gap. At the moment we use a very simple but practical approach to minimize the waiting time. The job is submitted to the top three sites through the Migol WS GRAM service. When one of these jobs gets active all other jobs are canceled. This approach avoids orphan jobs and ensures that the overhead caused by allocation of more than necessary resources will be minimized.

The Community Scheduler Framework (CSF) [28] also provides a framework for meta-scheduling. In practice the CSF proved to be not robust and flexible enough. The available round-robin scheduler plug-in does not suit our needs. A dynamic scheduling mechanism based on GODsL with submission to multiple sites is not feasible with the CSF plug-in mechanism. Furthermore, important features like advanced reservations are only operable in conjunction with LSF.

3.4   Migration Service: Fault-Tolerance of MPI Applications

The Migration Service (MS) is responsible for the automatic relocation of already running MPI applications. A migration can be initiated on user request, on behalf of a self-adaptive application or for recovery.

Migol defines a `Monitorable` port type with a `checkService()` operation, which has to be implemented by the application. The Connector Service and the Migol MPI library provide a default implementation of this port type. A Grid service call to the `checkService()` operation of the CS will initiate a socket ping to all nodes of a MPI application. If all nodes are reachable the CS will return a successful response.

The MRS periodically checks all registered services using the `Monitorable` interface. In case of a failure, the condition is reported to the AIS. After a threshold of failures is reached the recovery of the job is initiated through the Migration Service.

The state of the application is saved in a checkpoint. We use an application-level approach for checkpointing, i. e. applications have to be specially written to accommodate checkpointing and restart. A system-level or user-level checkpointing mechanism cannot be provided due to the great heterogeneity of the Grid. The application developer must implement the `checkpoint()` operation of the `Monitorable` port type.

After checkpointing the application can be securely stopped. The service is then restarted using the JBS. To accommodate an automatic migration the MS uses the AIS to obtain the credentials of the service owner.

3.5   Fault-Tolerance of Migol Services

The Migol services are designed with special regard to fault-tolerance. The MRS as well as the AIS are critical to the Grid. Therefore, these services are replicated. For both services, Migol uses a a ring-based replication strategy to guarantee data consistency (see [23] for more details).

The MIS and the factory services for the JBS and MS are stateless. The availability of these services can therefore be enhanced by deploying these services on multiple, independent sites without any replication protocol for data consistency. While the MIS caches different information about the Grid, this information can be completely rebuild after failure.

The JBS and MS instances created by the factory services are transient and therefore need special treatment. GT3 provides a hosting environment with improved fault-tolerance via a persistence API. JBS and MS instances use this API to record critical internal state information. In the event of a container crash or an instance failure, the container can recreate existing instances. The JBS and MS instances query the persistent attributes at startup in order to determine whether they are a new or restarted instance. If indeed a persistent state is discovered, this state information is utilized to recover the service.

CS and Migol WS GRAM instances represent the MPI application. In case of a failure a migration of these service instances is triggered by the MRS.

3.6   Security

Grid resources are potentially shared among many different virtual organizations, which leads to special security requirements. The Grid Security Infrastructure (GSI) addresses issues like single sign on and provides a certificate based security infrastructure [14]. Migol services extensively use GT3 security services such as authentication, authorization, message protection and credential delegation. All important messages are encrypted. Delegation enables services to act on behalf of a user. The MS and JBS can conduct critical operations, e. g. a job submission, using the delegated credential of a user.

Further, credentials are stored encrypted in the AIS to allow the recovery of failed services. The main reason for incorporating these features into the AIS is that existing solutions are not reliable and flexible enough. For example the MyProxy server [4] cannot be replicated and therefore is a single point of failure. Credentials can only be retrieved using encrypted communication.

4   Measurements

First measurements were conducted to analyze the performance of the implementation. The testbed consisted of two machines with AMD Athlon XP 2000+

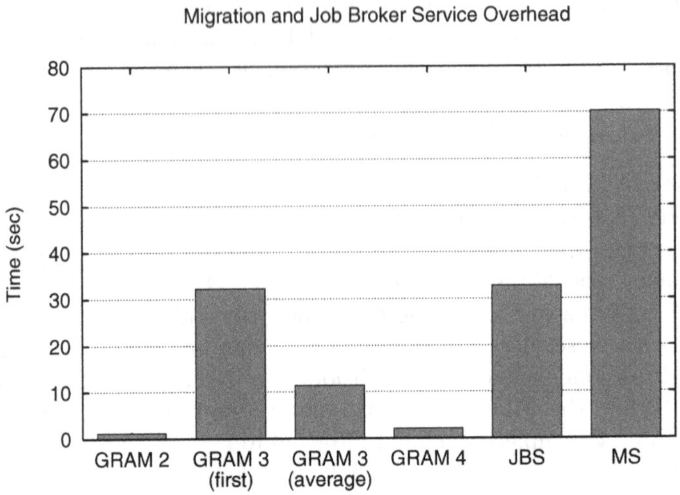

Migration and Job Broker Service Overhead

Fig. 2. Migration and Job Broker Service Overhead

and 256 MB memory connected by Fast Ethernet. The GT3 container was run using Sun Java 1.4.2 on Mandrake Linux 10.0.

The MS and JBS were deployed on machine 1, the AIS on machine 2. On both machines the Migol executing environment, consisting of the WS GRAM with fork jobmanager and the CS, was installed. Our test application is a simple 9-point stencil simulation called cellular automat. We compared the submission times to the JBS and MS against the different Globus GRAM implementations. Measurements have been carried out using the corresponding command line tools.

Figure 2 presents the results of these experiments. The performance of the GT3 GRAM has a clear disadvantage against its GT2 and GT4 counterpart. Since the Migol framework depends on the GT3 GRAM we will focus our examination on this service. We noticed a difference between an initial submission to GT3 GRAM and subsequent job submissions. This behavior is a result of the GT3 GRAM architecture: For each user a new container, the user hosting environment, is started.

We observed an overhead of about 22 s of the JBS compared to GT3 GRAM. This overhead occurs due to the necessary interactions with other Migol services. The scheduler obtains e. g. information from the MIS. Further, the submission of the job to the GT3 GRAM service is very time-consuming (~11 s). Also enabled security has a major impact on performance [6].

The performance of the MS was determined by measuring the time for stopping and restarting the service through the JBS. As we were primarily interested in the overhead imposed by the Grid service environment, the checkpointing time and the transfer time of the checkpoint was not measured. Since the migration service uses the JBS for job submission, the JBS overhead also applies to the MS. In addition, several Grid service calls are necessary, e. g. to authorize users (delegation of credentials), to trigger checkpointing, and to update the corresponding Grid object at the AIS.

In order to assess the future direction of our work we conducted a first evaluation of the GT4 GRAM implementation. Due to a major refactoring the GT4 GRAM provides a much better performance, which is comparable to the performance of the GT2 GRAM.

5   Conclusions and Future Work

We presented a fault-tolerant service infrastructure for MPI applications. The key concept of our approach is a replicated information service (AIS) and a replicated monitoring service (MRS). It is the task of the MRS to monitor the state of the applications and to initiate a restart or migration using the JBS and MS in case of a failure.

Migol's security model is based on GSI of the GT3, which provides a set of security services to implement authentification, authorization, credential delegation and secure conversation.

While the measurements show a substantial overhead for the interaction of the services in case of a migration, we think that this is acceptable compared to the benefits a fault-tolerant self-healing infrastructure offers.

Further studies regarding the performance and scalability especially of the AIS will be necessary. Future work is also the investigation of new scheduling strategies within the Migol Job Broker Service (JBS). Currently, we use an heuristic to determine the best destination. Advanced reservation mechanisms could significantly enhance the capabilities of the JBS.

Further, we are going to evaluate more transparent and reliable solutions for creating and storing checkpoints, e. g. user-level checkpoints and replication of checkpoints as provided by the Replica Location Service [7].

With ongoing standardization in the GGF and OASIS we will evaluate proposed standards on their usability in context of Migol. For example, the Web Service Resource Framework (WSRF), which evolved from OGSI, incorporates Web service standards as WS-Addressing to model stateful Web services. The Globus Toolkit 4 provides a WSRF implementation [8]. Due to the flexible design of Migol the transition to GT4 will be straightforward.

References

1. Adnan Agbaria and Roy Friedman. Starfish: Fault-tolerant dynamic mpi programs on clusters of workstations. In *HPDC '99: Proceedings of the The Eighth IEEE International Symposium on High Performance Distributed Computing*, page 31, Washington, DC, USA, 1999. IEEE Computer Society.
2. Anh Nguyen-Tuong and Andrew S. Grimshaw and Glenn Wasson and Marty Humphrey and John C. Knight. Towards Dependable Grids. Available at: http://www.cs.virginia.edu/~techrep/CS-2004-11.pdf.

3. A. Barak, A. Braverman, I. Gilderman, and O. Laaden. Performance of PVM with the MOSIX Preemptive Process Migration. In *Proceedings of the 7th Israeli Conference on Computer Systems and Software Engineering*, pages 38–45, Herzliya, June 1996.

4. Jim Basney, Marty Humphrey, and Von Welch. The myproxy online credential repository. Available at: http://www.ncsa.uiuc.edu/~jbasney/ myproxy-spe.pdf, 2005.

5. George Bosilca, Aurelien Bouteiller, Franck Cappello, Samir Djilali, Gilles Fedak, Cecile Germain, Thomas Herault, Pierre Lemarinier, Oleg Lodygensky, Frederic Magniette, Vincent Neri, and Anton Selikhov. Mpich-v: toward a scalable fault tolerant mpi for volatile nodes. In *Supercomputing '02: Proceedings of the 2002 ACM/IEEE conference on Supercomputing*, pages 1–18, Los Alamitos, CA, USA, 2002. IEEE Computer Society Press.

6. D. Chen et al. OGSA Globus Toolkit 3 evaluation activity at CERN. *Nucl. Instrum. Meth.*, A534:80–84, 2004.

7. Ann L. Chervenak, Naveen Palavalli, Shishir Bharathi, Carl Kesselman, and Robert Schwartzkopf. Performance and scalability of a replica location service. Available at: http://www.globus.org/alliance/publications/papers/ chervenakhpdc13.pdf, 2004.

8. Karl Czajkowski, Donald F Ferguson, Ian Foster, Jeffrey Frey, Steve Graham, Igor Sedukhin, David Snelling, Steve Tuecke, and William Vambenepe. The WS-Resource Framework. Available at: http://www.oasis-open.org/committees/ download.php/6796/ws-wsrf.pdf, 2005.

9. Karl Czajkowski, Ian T. Foster, Nicholas T. Karonis, Carl Kesselman, Stuart Martin, Warren Smith, and Steven Tuecke. A resource management architecture for metacomputing systems. In *IPPS/SPDP '98: Proceedings of the Workshop on Job Scheduling Strategies for Parallel Processing*, pages 62–82, London, UK, 1998. Springer-Verlag.

10. T. Tannenbaum D. Thain and M. Livny. Condor and the grid. In Fran Berman and A.J.G. Hey, editors, *Grid Computing: Making the Global Infrastructure a Reality*. John Wiley, 2003.

11. Graham E. Fagg and Jack Dongarra. Ft-mpi: Fault tolerant mpi, supporting dynamic applications in a dynamic world. In *Proceedings of the 7th European PVM/MPI Users' Group Meeting on Recent Advances in Parallel Virtual Machine and Message Passing Interface*, pages 346–353, London, UK, 2000. Springer-Verlag.

12. E. Floros and Yannis Cotronis. Exposing mpi applications as grid services. In Marco Danelutto, Marco Vanneschi, and Domenico Laforenza, editors, *Euro-Par*, volume 3149 of *Lecture Notes in Computer Science*, pages 436–443. Springer, 2004.

13. Ian Foster, Carl Kesselman, Jeffrey M. Nick, and Steven Tuecke. The Physiology of the Grid – An Open Grid Services Architecture for Distributed Systems Integration. Available at: http://www-unix.globus.org/toolkit/3.0/ogsa/ docs/physiology.pdf, 2002.

14. Ian T. Foster, Carl Kesselman, Gene Tsudik, and Steven Tuecke. A security architecture for computational grids. In *ACM Conference on Computer and Communications Security*, pages 83–92, 1998.

15. Globus Homepage. Available at: http://www.globus.org, 2005.

16. Jason Maassen Thilo Kielmann Gosia Wrzesinska, Rob V. van Niewpoort and Henri E. Bal. Fault-tolerant scheduling of fine-grained tasks in grid environments. In *Proceedings of the 19th IEEE International Parallel and Distributed Processing Symposium(IPDPS'05)*, Denver, Colorado, USA, April 2005.

17. William Gropp and Ewing Lusk. Fault tolerance in mpi programs. *High Performance Computing and Applications*, 2002.
18. R. Henderson and D. Tweten. Portable Batch System: External reference specification. Technical report, NASA Ames Research Center, 1996.
19. James Arthur Kohl and Philip M. Papadopoulos. Cumulvs version 1.0. Available at http://www.netlib.org/cumulvs/, 1996.
20. Jozsef Kovacs and Peter Kacsuk. A migration framework for executing parallel programs in the grid. In *Proceedings of the 2nd European Across Grids Conference*, Nicosia, Cyprus, January 2004.
21. Gerd Lanfermann, Bettina Schnor, and Ed Seidel. Grid object description: Characterizing grids. In *Eighth IFIP/IEEE International Symposium on Integrated Network Management (IM 2003)*, Colorado Springs, Colorado, USA, March 2003.
22. Michael Litzkow, Miron Livny, and Matthew Mutka. Condor - a hunter of idle workstations. In *Proceedings of the 8th International Conference of Distributed Computing Systems*, June 1988.
23. Michael Mihahn and Bettina Schnor. Fault-tolerant grid peer services. Technical report, University Potsdam, 2004.
24. Rubén S. Montero, Eduardo Huedo, and Ignacio Martín Llorente. Grid resource selection for opportunistic job migration. In Harald Kosch, László Böszörményi, and Hermann Hellwagner, editors, *Euro-Par*, volume 2790 of *Lecture Notes in Computer Science*, pages 366–373. Springer, 2003.
25. S. Petri and H. Langendörfer. Load Balancing and Fault Tolerance in Workstation Clusters – Migrating Groups of Communicating Processes. *Operating Systems Review*, 29(4):25–36, October 1995.
26. Diego Puppin, Nicola Tonellotto, and Domenico Laforenza. Using web services to run distributed numerical applications. In Dieter Kranzlmüller, Péter Kacsuk, and Jack Dongarra, editors, *PVM/MPI*, volume 3241 of *Lecture Notes in Computer Science*, pages 207–214. Springer, 2004.
27. Neil Spring Rich Wolski and Jim Hayes. The network weather service: A distributed resource performance forecasting service for metacomputing. *Journal of Future Generation Computing Systems*, 15(5-6):757–768, 1999.
28. C. Smith. Open source metascheduling for virtual organizations with the community scheduler framework (csf). Technical report, Platform Computing Inc., 2003.
29. Georg Stellner. CoCheck: Checkpointing and Process Migration for MPI. In *Proceedings of the 10th International Parallel Processing Symposium (IPPS '96)*, Honolulu, Hawaii, 1996.
30. Steven Tuecke, Ian Foster, and Carl Kesselman. Open Grid Service Infrastructure. Available at: http://www-unix.globus.org/toolkit/draft-ggf-ogsi-gridservice-33_2003-06-27.pdf, 2003.
31. Sathish S. Vadhiyar and Jack J. Dongarra. A performance oriented migration framework for the grid. In *Proceedings of the 3rd IEEE/ACM International Symposium on Cluster Computing and the Grid*, page 130. IEEE Computer Society, 2003.

Applicability of Generic Naming Services and Fault-Tolerant Metacomputing with FT-MPI

David Dewolfs, Dawid Kurzyniec, Vaidy Sunderam, Jan Broeckhove,
Tom Dhaene, and Graham Fagg

Depts. of Math and Computer Science, University of Antwerp,
Belgium and Emory University, Atlanta, GA, USA
{David.Dewolfs, Jan.Broeckhove, Tom.Dhaene}@ua.ac.be
{dawidk, vss}@mathcs.emory.edu

Abstract. There is a growing interest in deploying MPI over multiple, heterogenous and geographically distributed resources for performing very large scale computations. However, increasing the amount of geographical distribution and resources creates problems with interoperability and fault-tolerance. FT-MPI presents an interesting solution for adding fault-tolerance to MPI, but suffers from interoperability limitations and potential single points of failure when crossing multiple administrative domains. We propose to overcome these limitations by adding "pluggability" for one potential single point of failure - the name service used by FT-MPI - and combining FT-MPI with the H2O metacomputing framework.

Keywords: FT-MPI, H2O, metacomputing, fault-tolerance, heterogeneity.

1 Introduction

Cluster systems have rapidly grown into one of the most popular approaches to supercomputing. Traditionally, these clusters are built within strongly controlled environments, using homogenous resources within a single administrative domain (AD). However, there is a growing interest in clustering resources that feature high levels of geographical distribution across multiple ADs to perform large scale collaborative computations.

MPI is arguably the most popular approach to programming parallel applications on cluster systems. Projects like MPICH-G2 [9] give response to a rising demand to adapt MPI to higher levels of heterogeneity and geographical distribution than those available in traditional cluster systems. These initiatives have proven to be quite successful at the user level [8,9,10]. However, they create a number of issues at the administrative level as resources in different administrative domains must all use common policies and be properly synchronized [5]. Also, these systems lack fault-tolerance features. A number of solutions have been developed to address this issue [2,3,4,6,13]. One such solution is FT-MPI [7]. FT-MPI roughly divides fault recovery into two major phases : MPI level

B. Di Martino et al. (Eds.): EuroPVM/MPI 2005, LNCS 3666, pp. 268–275, 2005.

recovery and application level recovery. After a failure, FT-MPI makes sure that the MPI environment is correctly restored to a functioning state (automatic MPI-level recovery). From there on, it is up to the application itself to restore it's own state instead of relying on automated but potentially unscalable solutions like global distributed checkpointing. This makes FT-MPI an interesting solution for highly geographically distributed, heterogenous resources.

However, FT-MPI is currently confined to single ADs. Also, bottlenecks and potential single points of failure (SPoFs) become an issue when deploying it on multiple ADs, connected by less reliable (WAN) networks than those available within a single AD (LAN, specialized infrastructure). One of the points where such problems might occur is the FT-MPI name service (NS). We address these issues, not by directly providing "yet another NS", but rather by making the NS "pluggable", allowing deployers of FT-MPI to use an existing and generic, potentially "off-the-shelf" NS of their own choice (e.g. ActiveDirectory, OpenLDAP, HDNS[14], ...). Such products often provide built-in fault-tolerance and performance enhancing features, as well as other capabilities that might be of interest to the operator of an FT-MPI system. The goal is to enable an operator to choose an NS that corresponds best to his specific needs and available resources, instead of being bound to a single, custom-made NS as is currently the case. The design includes a hierarchical message delivery mechanism which allows for setting up FT-MPI with only a single connection to the NS per AD. This greatly reduces the network bottlenecks that might occur when massive amounts of machines spread over multiple ADs all have to connect to the NS concurrently. We use the features of the H2O [11] metacomputing framework to address the interoperability issues. H2O gives us the capability to set up, run and use an FT-MPI environment spanning multiple ADs without the need for system accounts - as long as they are accessible via the H2O framework. H2O allows operators to do this in a centralized, transparent and generic manner, hiding the heterogeneity of the underlying systems.

2   Design Overview

2.1   Basic FT-MPI Architecture

FT-MPI offers the user a virtual machine (VM) built around the interactions of the MPI library with three different types of daemons. Roughly speaking : the *MPI library* takes care of the message-passing and other MPI-related issues. The *startup daemons* start up processes on the individual nodes on which they run, and keep a handle for each of those for the duration of the job. The *notifier daemons* are non-critical processes that marshal and manage failure notification.

For the work described in this paper, we focus on the third : the *naming daemon*. Each VM runs exactly one such daemon. The naming daemon provides a NS, which functions as a repository for contact information of nodes in the VM as well as a general data repository. It also plays a crucial role in the recovery mechanism. It is used during VM buildup, job startup and job recovery for setting and determining the currently active nodes within a job or VM. Currently, the naming daemon constitutes a potential SPoF, as it is highly state-retaining and critical in the general functioning of the VM. It is also a possible choke-point when communicating between different ADs, as these are most probably connected by slower, higher latency networks (WAN/internet) than available within a single AD (LAN). Finally, the naming daemon does not support features like replication and load balancing. These features would be desirable to improve scalability at very large VM sizes.

2.2   Extensions to the FT-MPI Architecture

Extended Design. As stated in the introduction, we have decided not to address these issues with the NS by providing a completely new one. Rather, we enable an operator of an FT-MPI VM to use an NS of his own choice. There is a wide range of "off the shelf" NS systems available, ranging from commercially available solutions by major vendors like Sun or Microsoft to open source alternatives like OpenLDAP or research project like HDNS[14]. These provide a wide range of fault-tolerance and performance features like load distribution, replication, checkpointing etc.

Our intent is to enable the operator to choose an NS that is most appropriate w.r.t. his performance and fault tolerance requirements. The ultimate goal is to move all state-retaining functionality away from FT-MPI and into this pluggable NS, thus using any fault-tolerance features in the "plugged" NS to remove the potential SPoF. In order to accomplish this goal, we expanded on the basic FT-MPI design as follows :

- instead of directly contacting the NS, components of FT-MPI now contact a proxy server which resides on the gateway between the single AD and the "outside world"; this proxy server acts as a *"front-end"* to the real NS, translating the internal FT-MPI protocol calls to a format that is understood by the real, *"back-end"* name service; the front-end does not retain internal state - thus, failures can be handled through simple measures like a trivial replication scheme or a restart
- all nodes on a single AD retain an open connection to the NS front-end for that AD, and each NS front-end retains a single connection with the NS back-end (hierarchical message forwarding)
- the NS front-end is implemented as a H2O "pluglet" making it fully remotely deployable by operators on any machine that runs an H2O kernel

The front-end services all of the nodes within the single AD - posing as the real NS - while retaining only a single connection with the back-end. This hierarchical message forwarding approach significantly cuts down the number of connections that would have to cross multiple ADs. On top of this, it potentially enables us to resolve certain calls to the NS locally at the proxy through caching, without having to pass on every call to the back-end. Thus, this approach drastically reduces the chance of bottlenecks on in-transit calls between multiple ADs, and offers room for improvements to scalability. This also solves issues with AD-specific configuration schemes like network address translation (NAT) for the NS (these issues were already addressed in [1] for computational nodes). With the front-end situated on the gateways, it can take in messages from the nodes at the internal interface using virtual IP addresses, and transfer them over the "outside" interface using real IP addresses, or potentially even through a completely different communication fabric (JXTA, ...).

More importantly, this enables the NS to be FT-MPI "agnostic". All FT-MPI related logic is encapsulated within the front-end. Communications between proxy and back-end are done through a generic interface (more on this under "approach"), making the NS back-end fully pluggable. Using proxies also enables us to integrate this design change without a recompile of the FT-MPI sources. Basically, it suffices to start up the proxy and an appropriate back-end instead of the original NS, and then run the rest of FT-MPI without further modifications. Last, but not least, using H2O to implement this design enables us to take away any extra overhead in setting up this scheme from the individual AD administrators. Given that the necessary resources run a H2O kernel, operators of an FT-MPI VM can now deploy this complete setup remotely and in a transparent manner. For remote setup, H2O provides the necessary features to manage the whole process from code staging up to setup and running of the whole mechanism.

Approach. For the implementation of this design we opted for the use of Java. This decision was motivated by a) it being used for H2O, of which we wanted to inherit the useful features described above, b) its own inherent qualities for implementing server-side solutions across heterogenous resources and c) its powerful API. Specifically, we make use of JNDI, the Java Naming and Directory Interface. JNDI provides uniform access to a diverse range of NS solutions, ranging from LDAP (the Lightweight Directory Access Protocol) to DNS (the Domain Name Service). Any provider can make a NS "JNDI-enabled" by implementing a Service Provider Interface (SPI). All of this is fully transparent to the user. Thus, using JNDI allows us to make access to the backend generic towards different NS implementations. The only backend-specific code which needs to be written concerns naming conventions. These widely differ between different NS implementations. Impact of this issue is minimized by separating this functionality into a very basic pluggable name resolver. Providing an implementation of this for the specific back-end is enough to make the whole scheme work transparently for the rest of the process. A basic resolver for LDAP has been implemented, and others are currently in the pipeline.

Extending Upon JNDI. Though it can solve the problems regarding SPoF, decoupling the NS from FT-MPI introduces new issues concerning concurrency. FT-MPI assumes a centralized NS. This NS is basically single-threaded and queues all incoming requests on receive. Thus, it never poses a problem w.r.t. multiple requests accessing the same data resource. However, the new design we propose has the front-ends running and accessing the back-end in parallel. This introduces the issue of atomicity.

To enable for genericity, JNDI assumes a base level, lowest common denominator approach to accessing a NS. However, the calls in the protocol which FT-MPI uses to access its NS are quite high-level in nature, requiring compound operations like atomic increment, atomic compare and set etc. to be resolved in a single call. FT-MPI is highly dependent upon atomicity of these calls. This results in a necessity to resolve certain protocol calls to the NS through multiple primitives in JNDI, requiring separate lookups and subsequent binds. This makes potential concurrent access to shared data resources, and potential resulting problems (e.g. race conditions), an important issue.

We have tackled these issues by building an unreliable named locking scheme on top of JNDI. This locking scheme allows us to acquire named locks, thus giving individual front-ends exclusive access to a given shared resource when needed. Locks are given a limited lease time, and must be retained by the current lock owner. If a lock is not properly retained, it grows stale and can be forcefully taken by another contender for the same resource. By using an abstract interface for the implementation, NS providers are able to supply native locking to the user if it is supported in their product, simply by offering their own extension of the interface. For NS systems that do not support native locking, we offer our own implementation of the interface which provides locks using basic primitives of JNDI (natively implemented locking, of course, potentially yields better performance).

This is accomplished as follows. JNDI guarantees atomicity of bind. Therefore, binding a new named object into the NS will either completely succeed, or completely fail, without leaving the NS in an inconsistent state. On top of this, a regular bind will not succeed if the object to be bound is already present in the NS (using the same name) - you'd have to use a *rebind* operation to do that. We can use these features to bind a "lock" object with a given name into the NS. If the bind succeeds, the lock has been acquired and further operations can proceed. If the bind does not succeed, this means that a lock with that name was already bound by another agent. In this case, a repetitive retry scheme with graceful back-off is used to wait until the lock becomes either available (the owner un-binds it), or stale (the owner does not renew the lease in time). If the lock has grown stale, it is forcefully transferred on the next attempt to acquire it.

However, the locking scheme we use is necessarily unreliable, due to the nature of the intervening medium between the front-ends and the back-end (the network). This could potentially lead to inconsistencies in the state of the back-end when a failure occurs or a lock grows stale, somewhere in the middle of the compound update process.

To overcome this issue, we have adopted an approach to the design of our update algorithms which enables for consistent state up to and until the last bind-operation for a given protocol call. Failure in any phase before the final bind can simply be solved by applying a garbage-collection operation to the back-end. Garbage collection can potentially be piggybacked on other operations for performance gain. On top of this, to make certain that no state-changing operations from a previous lock owner come through after a stale lock was forcefully transferred, the state-changing bind is only done when the remaining lease time on the remote lock is higher than the socket time out. In this way, we can be certain that the "final" bind has either succeeded or failed before we lose the lock, as the socket will have timed out and the bind will have returned an error before expiration of the lease. This does require resetting the socket time-out to more reasonable levels than the default, which is trivial in Java.

3 Evaluation

To evaluate communication overhead generated by the new design in comparison to the old, we created a proof-of-concept implementation of the front-end. We performed a comparison experiment on two nodes : one in Atlanta (Georgia), USA, the other situated in Antwerp, Belgium. The node in Atlanta is a 2.4 GHz Pentium 4 with 1GB memory running Mandrake Linux 10.0. The node in Antwerp is a 1.90GHz Pentium 4 with 256MB memory running Suse Linux 7. Both nodes are connected to the internet through a broadband internet provider, and communicate through plain, unencrypted TCP sockets. This setup was used in order to simulate the conditions which the design is aimed at : geographically distributed, heterogenous resources. For the back-end in the new design, we used a standard installation OpenLDAP version 2.1.25 with BDB (Berkeley Data Base) for storage. The node in Atlanta ran the original FT-MPI NS or OpenLDAP depending on the case being tested. The node in Antwerp ran a basic client program in both cases, plus the front-end in the case of the new design.

We ran two experiments : in the first, we first inserted and then read entries with progressively growing payloads (10-900 B, using 100 B steps from 100 to 900 B) and measured wall-clock time for both insertion and read. In the second, we inserted and read batches with a progressively growing amount of equal-sized

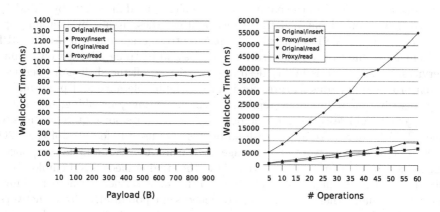

Fig. 1. Evolution of wall-clock time as the payload per transaction (graph) and the amount of operations increase

entries into the NS and again, measured wall-clock time for both cases. These experiments evaluate scalability in terms of transaction size and transaction frequency. They were primarily geared towards testing feasibility, scalability and stability of the new design. Ultimately, we want the new NS to behave as scalable and stable as the original. When it comes to raw performance however, it is to be expected that the original NS will outperform the new design in its current state. This is because a) the original NS was implemented using RAM-based data storage only and b) the original NS uses purely static memory structures as well. Of course, this is not the case with the OpenLDAP backend. On top of this, OpenLDAP will show a heavy bias towards retrieval operations as it optimizes these over modification/insertion.

The experiment successfully ran to completion leaving the backend in a fully consistent state, proving the feasibility of the new design. The results of our experiment are shown in figure 1. As expected, the numbers show a performance advantage for the original NS on insert vs. our new design. Read operations though are almost on an equal level for both cases. The figure also clearly shows that in both cases, the evolution of the measured wall clock time remains linear. This confirms that, despite the need to replace the monolithic features of the original NS by several aggregated operations in the backend (especially on insert), the system remains scalable and delays remain predictable. On top of this, performance remains more than acceptable for regular jobs, as the NS overhead only comes into account at job startup and during fault recovery. Regular job performance will not be hindered by the performance loss on the NS. Even when taking into account the factor 9 performance hit on insert, operation timings remain below 1 second.

4   Conclusions

In this paper, we have discussed issues concerning the deployment of FT-MPI for large scale computations on highly geographically distributed, heterogenous resources. We have shown that "vanilla" FT-MPI poses some limitations in this area due to the nature of the naming service which is used internally by FT-MPI.

We have worked out a design which address these issues by enabling operators of an FT-MPI setup to a) transparently set up and run an FT-MPI system across multiple administrative domains and b) "plug in" their own name services. This feature is highly desirable as existing off-the-shelf name services often provide numerous features for improved fault tolerance and performance (e.g. distribution, redundancy, replication, checkpointing, journalling, automated management and restart etc.). The proposed design does not require changes to the FT-MPI source. To accomplish all of this , we use the features of the H2O metacomputing framework and leverage the features of JAVA and its component for name service management, JNDI. Also, the staging mechanism employed in the design reduces the nr. of connections to the name service to one per administrative domain, thereby reducing bottlenecks on potentially slower network connections between multiple administrative domains and allowing for local optimization through caching.

References

1. D. Kurzyniec and V. Sunderam. Combining FT-MPI with H20: Fault-tolerant MPI across administrative boundaries. In *Proceedings of th HCW 2005-14th Heterogeneous Computing Workshop,* (accepted), 2005
2. A. Agbaria, R. Friedman. Starfish: Fault-tolerant dynamic MPI programs on clusters of workstations. In *Eighth IEEE International Symposium on High Performance Distributed Computing,* 1999, pp. 31
3. A. Bouteiller, F. Cappello, T. Herault, G. Krawezik, P. Lemarinier and F. Magniette. MPICH-V2: a fault tolerant MPI for volatile nodes based on pessimistic sender based message logging. In *ACM/IEEE SC2003 Conference,* 2003, pp. 25
4. Y. Chen, K. Li, J.S. Plank. CLIP : A checkpointing tool for message-passing parallel programs. 1997. Available at http://citeseer,ist.psu.edu/chen97clip.html
5. J. Chin and P.V. Coveney. Towards tractable toolkits for the Grid : a plea for lightweight, usable middleware. Available at http://www.realitygrid.org/lgpaper21.pdf
6. E. Elnozahy and W. Zwaenepoel. Manetho : Transparent rollback-recovery with low overhead, limited rollback and fast output. In *IEEE Transactions on Computers, Special Issue on Fault-Tolerant Computing,* 41(5), May 1992, pp.526-531
7. G. Fagg, E. Gabriel, Z. Chen, T. Angskun, G. Bosilca, J. Pjesivac-Grbovic and J. Dongarra. Process fault-tolerance : Sematics, design and applications for high-performance computing. In *International Journal for High Performance Applications and Supercomputing.* 2004.
8. T. Imamura, Y. Tsujita, H. Koide and H. Takemiya. An architecture of Stampi : MPI library on a cluster of parallel computers. In *7th European PVM/MPI Users' Group Meeting,* 2000, pp. 4-18
9. N. Karonis, B. Toonen and I Foster. MPICH-G2 : A grid-enabled implementation of the Message Passing Interface. In *Journal of Parallel and Distributed Computing (JPDC),* 63(5), May 2003, pp. 551-563
10. R. Keller, B. Krammer, M.S. Mueller, M.M. Resch and E. Gabriel. MPI development tools and applications for the grid. In *Workshop on Grid Applications and Programming Tools,* 2003
11. D. Kurzyniec, T. Wrzosek, D. Drzewiecki and V. Sunderam. Towards self-organising distributed computing frameworks : The H2O approach. In *Parallel Processing Letters,* 13(2), 2003, pp. 273-290
12. S. Louca, N. Neophytou, A. Lachanas and P. Eviripidou. MPI-FT: Portable fault-tolerance scheme for MPI. In *Parallel Processing Letters,* 10(4), 2000, pp. 371-382.
13. G. Stellner. CoCheck: Checkpointing and process migration for MPI. In *10th International Parallel Processing Symposium,* 1996, pp. 526-531
14. T. Tyrakowski, V. S. Sunderam, M. Migliardi. Distributed Name Service in Harness. In *Proceedings of the international Conference on Computational Sciences - Part 1(LNCS Vol. 2073),* 2001, pp. 345-354

A Peer-to-Peer Framework for Robust Execution of Message Passing Parallel Programs on Grids

Stéphane Genaud and Choopan Rattanapoka

ICPS-LSIIT - UMR CNRS-ULP 7005
Université Louis Pasteur, Strasbourg
{genaud, rattanapoka}@icps.u-strasbg.fr

Abstract. This paper presents P2P-MPI, a middleware aimed at computational grids. From the programmer point of view, P2P-MPI provides a message-passing programming model which enables the development of MPI applications for grids. Its originality lies in its adaptation to unstable environments. First, the peer-to-peer design of P2P-MPI allows for a dynamic discovery of collaborating resources. Second, it gives the user the possibility to adjust the robustness of an execution thanks to an internal process replication mechanism. Finally, we measure the middleware performances on two NAS benchmarks.

Keywords: Grid, Middleware, Peer-to-peer, MPI, Java.

1 Introduction

Grid computing offers the perspective of solving massive computational problems using a large number of computers. It involves sharing heterogeneous resources located in different places, belonging to different administrative domains over a network. When speaking of computational grids, we must distinguish between grids involving stable resources (e.g. a supercomputer) and grids built upon versatile resources, that is computers whose configuration or state changes frequently. The latter are often referred to as *desktop grids* and may in general involve any unused connected computer whose owner agrees to share its CPU. Thus, provided some magic middleware glue, a desktop grid may be seen as a large-scale computer cluster allowing to run parallel application traditionally executed on parallel computers. However, the question of how we may program such an heterogeneous cluster remains unclear. Most of the numerous difficulties that people are trying to overcome today fall in two categories.

- **Middleware.** The middleware management of tens or hundreds grid nodes is a tedious task that should be alleviated by mechanisms integrated to the middleware itself. These can be fault diagnostics, auto-repair mechanisms, remote update, resource scheduling, data management, etc.
- **Programming model.** Many projects propose a client/server (or RPC) programming style for grid applications (e.g. JNGI [17], DIET [4] or XtremWeb [8]). However, the *message passing* and *data parallel* programming model are the two models traditionally used by parallel programmers.

B. Di Martino et al. (Eds.): EuroPVM/MPI 2005, LNCS 3666, pp. 276–284, 2005.

MPI [13] is the *de-facto* standard for message passing programs. Most MPI implementations are designed for the development of highly efficient programs, preferably on dedicated, homogeneous and stable hardware such as supercomputers. Some projects have developed improved algorithms for communications in grids (MPICH-G2 [10], PACX-MPI [9], MagPIe [11] for instance) but still, assume hardware stability. This assumption allows for a simple execution model where the number of processes is static from the beginning to the end of the application run[1]. This design means no overhead in process management but makes fault handling difficult: one process failure causes the whole application to fail. This constraint makes traditional MPI applications unadapted to run on grids. Moreover, MPI applications are OS-dependent binaries which complicates execution in highly heterogeneous environments.

If we put these constraints altogether, we believe a middleware should provide the following features: a) self-configuration (system maintenance autonomy, discovery), b) data management, c) robustness of hosted processes (fault detection and replication), and d) abstract computing capability. The rest of the paper shows how P2P-MPI fulfills these requirements. We first describe (section 2) the P2P-MPI middleware through its modules, so as to understand the protocols defined to gather collaborating nodes in order to form a platform suitable for a job request. In section 3 we explain the fault-detection and replication mechanisms in the context of message-passing programs. We finally discuss P2P-MPI behavior in section 4, at the light of experiments carried out on two NAS benchmarks.

2   The P2P-MPI Middleware

2.1   Modules Organization

Fig. 1 depicts how P2P-MPI modules are organized in a running environment. P2P-MPI proper parts are grayed on the figure. On top of diagram, a message-passing parallel program uses the MPI API (a subset of the MPJ specification

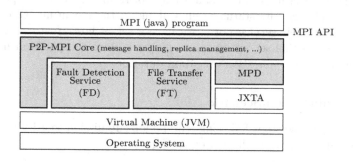

Fig. 1. P2P-MPI structure

---

[1] Except dynamic spawning of process defined in MPI-2.

[5]). The core behind the API implements appropriate message handling, and relies on three other modules. The *Message Passing Daemon* (MPD) is responsible for self-configuration, as its role is either to search for participating nodes or to act as a gate-keeper of the local resource. The *File Transfer Service* (FT) module handles the data management by transferring executable code, input and output files between nodes. The *Fault Detection Service* (FD) module is necessary for robustness as the application needs to be notified when nodes become unreachable during execution.

In addition, we also rely on external pieces of software. The abstract computing capability is provided by a Java Virtual Machine and the MPD module uses JXTA [1] for self-configuration.

2.2   Discovery for An Execution Platform

In the case of desktop grids, the task of maintaining an up-to-date directory of participating nodes is a so tedious task that it must be automated. We believe one of the best options for this task is *discovery*, which has proved to work well in the many *peer-to-peer* systems developed over the last years for file sharing. P2P-MPI uses the discovery service of JXTA. The discovery is depicted in JXTA as an advertisement publication mechanism. A peer looking for a particular resource posts some public advertisement (to a set of decentralized peers called rendez-vous) and then waits for answers. The peers which discover the advertisement directly contact the requester peer.

In P2P-MPI, we use the discovery service to find the required number of participating nodes at each application execution request. Peers in P2P-MPI are the MPD processes. When a user starts up the middleware it launches a MPD process which publishes its *pipe* advertisement. This pipe can be seen as an open communication channel that will be used to transmit boot-strap information.

When a user requests n processors for its application, the local MPD begins to search for some published pipe advertisements from other MPDs. Once at least n peers have reported their availability, it connects to the remote MPDs via the pipe to ask for their FT and FD services ports. The remote MPD acts as a gate-keeper in this situation and it may not return these service ports if the resource had changed its status to unavailable in the meantime. Once enough hosts have sent their service ports, we have a set of hosts ready to execute a program. We call this set an *execution platform* since the platform lifetime is not longer than the application execution duration.

2.3   Job Submission Scenario

We now describe the steps following a user's job submission to a P2P-MPI grid. The steps listed below are illustrated on Figure 2.

(1) The user must first join the grid. By invoking mpiboot, it spawns the MPD process which makes the local node join the P2P-MPI group if it exists, or creates it otherwise.

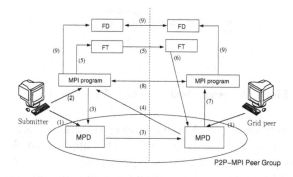

Fig. 2. A submission where the submitter finds one collaborating peer

(2) The job is then submitted by invoking a run command which starts the process rank 0 of the MPI application on local host.

(3) Discovery: the local MPD issues a search request to find other MPDs pipe advertisements. When enough advertisements have been found, the local MPD sends into each discovered pipe, the socket where the MPI program can be contacted.

(4) Hand-shake: the remote peer sends its FT and FD ports directly to the submitter's MPI process.

(5) File transfer: program and data are downloaded from the submitter host via the FT service.

(6) Execution Notification: once transfer is complete the FT service on remote host notifies its MPD to execute the downloaded program.

(7) Remote executable launch: MPD executes the downloaded program to join the execution platform.

(8) Execution preamble: all processes in the execution platform exchange their IP addresses to construct their local communication table.

(9) Fault detection: MPI processes register in their local FD service and starts. Then FD will exchange their heart-beat message and will notify MPI processes if they become aware of a node failure.

3 Replication for Robustness

3.1 Replication

Though absolutely transparent for the programmer, P2P-MPI implements a replication mechanism to increase the robustness of an execution platform. When specifying a desired number of processors, the user can request that the system run for each process[2] an arbitrary number of copies called *replicas*. In practice, it is shorter to request the same number of replicas per process, and we call this

---

[2] Except for rank 0 process. We assume a failure on the submitter host is critical since the user would lose the control on the application.

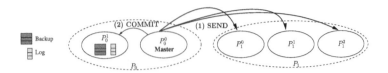

Fig. 3. A message sent from logical process P_0 to P_1

constant the *replication degree*. In the following we name a "usual" MPI process a *logical process*, noted P_i when it has rank i in the application. A logical process P_i is thus implemented by one or several replicas, noted $P_i^0, \ldots, P_i^n$. The replicas are run in parallel on different hosts since the goal is to allow the continuation of the execution even if one host fails.

Of course, replicas behavior must be coordinated to insure that the communication scheme is kept coherent with the semantics of the original MPI program. Ad hoc protocols have been proposed, and our solution follows the *active replication* [6] strategy in which all replicas of the destination group receive the sent message except that we impose coordination on the sender side to limit the number of sent messages.

In each logical process, one replica is elected as master of the group for sending. Fig. 3 illustrates a send instruction from P_0 to P_1 where replica P_0^0 is assigned the master's role. When a replica reaches a send instruction, two cases arise depending on the replica's status:

- if it is the master, it sends the message to all processes in the destination logical process. Once the message is sent, it notifies the other replicas in its logical process to indicate that the message has been correctly transmitted.
- if the replica is not the master, it first looks up a journal containing the identifiers of messages sent so far (*log* on Fig. 3) to know if the message has already been sent by the master. If it has already been sent, the replica just goes on with subsequent instructions. If not, the message to be sent is stored into a *backup* table and the execution continues. (Execution only stops in a waiting state on a receive instruction.) When a replica receives a commit, it writes the message identifier in its log and if the message has been stored, removes it from its backup table.

3.2   Fault Detection and Recovery

To become effective the replication mechanism needs to be notified of processes failures. The problem of failure detection has received much attention in the literature and we have adopted the *gossip-style* protocol described by [14] for its scalability. In this model, failure detectors are distributed and reside at each host on the network. Each detector maintains a table with one entry per detector known to it. This entry includes a counter called heartbeat counter. During execution, each detector randomly picks a distant detector and sends it its table after incrementing its heartbeat counter. The receiving failure detector will merge its local table with the received table and adopts the maximum heartbeat counter for each entry. If the heartbeat counter for some entry has not increased after a certain time-out, the corresponding host is suspected to be down.

When the local instance of the MPI program is notified of a node failure by its FD service, it marks the node as faulty and no more messages will be sent to it. If the faulty node hosts a master process then a new master is elected in the logical process. Once elected, it sends all messages left in its backup table.

4 Experiments

4.1 Experimental Context

Experiment Setup. Though we claim P2P-MPI is designed for heterogeneous environments, a precise assessment of its behavior in terms of performance is difficult because we would have to define representative configurations for which we can reproduce the experiments. Before that, we measure the gap between P2P-MPI and some reference MPI implementations in an homogeneous environment so as to identify potential weaknesses. The hardware platform used is a student computers room (24 Intel P4 3GHz, 512MB RAM, 100 Mbps Ethernet, Linux kernel 2.6.10). We compare P2P-MPI using java J2SE-5.0, JXTA 2.3.3 to MPICH-1.2.6 (p4 device) and LAM/MPI-7.1.1 (both compiled with gcc/g77-3.4.3). We have chosen two test programs with opposite characteristics from the NAS benchmarks [2] (NPB3.2)[3]. The first one is IS (Integer Sorting) which involves a lot of communications since a sequence of one MPI_Allreduce, MPI_Alltoall and MPI_Alltoallv occurs at each iteration. The second program is EP (Embarrassingly Parallel). It does independent computations with a final collective communication. Thus, this problem is closer to the class of applications usually deployed on computational grids.

Expected Behavior. It is expected that our prototype achieves its goals at the expenses of an overhead incurred by several factors. First the robustness requires extra-communications: regular heart-beats are exchanged, and the number of message copies increase linearly with the replication degree as can be seen on Fig. 3. Secondly, compared to fine-tuned optimizations of communications of MPI implementation (e.g. in MPICH-1.2.6 [16]), P2P-MPI has simpler optimizations (e.g. binomial trees). Last, the use of a virtual machine (java) instead of processor native code leads to slower computations.

4.2 Performances

Benchmarks. Fig. 4 plots results for benchmarks IS (left) and EP (right) with replication degree 1. We have kept the same timers as in the original benchmarks. Values plotted are the average total execution time. For each benchmark, we have chosen two problem sizes (called class A and B) with a varying number of processors. Note that IS requires that the number of processors be a power of two and we could not go beyond 16 PCs.

---

[3] We have translated IS and EP in java for P2P-MPI from C and Fortran respectively.

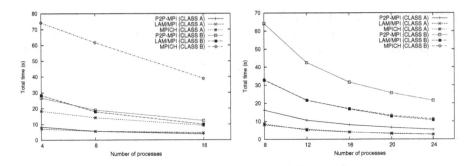

Fig. 4. Comparison of MPI implementations performance for IS (left) and EP (right)

For IS, P2P-MPI shows an almost as good performance as LAM/MPI up to 16 processors. The heart-beat messages seem to have a negligible effect on overall communication times. Surprisingly, MPICH-1.2.6 is significantly slower on this platform despite the sophisticated optimization of collective communications (e.g. uses four different algorithms depending on message size for MPI_Alltoall). It appears that the MPI_Alltoallv instruction is responsible for most of the communication time because it has not been optimized as well as the other collective operations.

The EP benchmark clearly shows that P2P-MPI is slower for computations because it uses Java. In this test, we are always twice as slow as EP programs using Fortran. EP does independent computations with a final set of three MPI_Allreduce communications to exchange results in short messages of constant size. When the number of processors increases, the share of computations assigned to each processor decreases, which makes the P2P-MPI performance curve tends to approach LAM and MPICH ones.

Replication Overhead. Since replication multiplies communications, the EP test shows very little difference with or without replication, and we only report measures for IS. Figure 5 shows the performances of P2P-MPI for IS when each

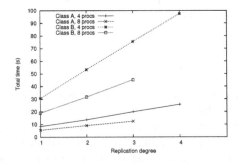

Fig. 5. Performance (seconds) for IS depending on replication degree

logical process has one to four replicas. For example, for curve "Class B, 8 processors", 3 replicas per logical process means 24 processors were involved. We have limited the number of logical processes so that we have at most one replica per processor to avoid load-imbalance or communications bottlenecks. As expected, the figure shows a linear increase of execution time with the replication degree, with a slope depending on the number of processors and messages sizes.

5   Related Work

Since the deployment of message passing applications in unstable environments is challenging, fault-tolerance of MPI has been well studied. Most works are devoted to check-point and restart methods (e.g. [3, 7, 12]) in which the application is restarted from a given recorded state. The replication approach is an alternative which does not require any specific reliable resource to store system states.

The work closest to ours is the P3 project [15], which share common characteristics with P2P-MPI. First, JXTA discovery is also used for self-configuration: *hosts* entities automatically register in a peer group of workers and accept work requests according to the resource owner policy. Secondly, both a master-worker and message passing paradigm are proposed. Unlike P2P-MPI, P3 also uses JXTA for its communications. This allows to communicate without consideration of the underlying network constraints (e.g. firewalls) but incurs performance overhead when the logical route established goes through several peers. In addition, P3 has no integrated fault-tolerance mechanism for message passing programs.

6   Conclusion and Future Work

We have described in this paper the design of a grid middleware offering a message-passing programming model. The middleware integrates fault-detection and replication mechanisms in order to increase robustness of applications execution. Two NAS parallel benchmarks with opposite behavior have been run on a small configuration and compared with performances obtained with LAM/MPI and MPICH. The results show good performance and are encouraging for experiments at large scale on the opening Grid5000 testbed[4] to study the scalability of the system. In-depth study of replication and robustness is also under work. Next developments should also concern strategies for mapping processes onto resources. Though the peer-to-peer model abstracts the network topology, we could use some network metrics (e.g. ping time) to choose among available resources. Also, the mapping of replicas could be based on information about resources capability and reliability.

---

[4] http://www.grid5000.org

References

[1] JXTA. http://www.jxta.org.

[2] D. H. Bailey, E. Barszcz, J. T. Barton, D. S. Browning, R. L. Carter, D. Dagum, R. A. Fatoohi, P. O. Frederickson, T. A. Lasinski, R. S. Schreiber, H. D. Simon, V. Venkatakrishnan, and S. K. Weeratunga. The NAS Parallel Benchmarks. *The Intl. Journal of Supercomputer Applications*, 5(3):63–73, 1991.

[3] A. Bouteiller, F. Cappello, T. Hérault, G. Krawezik, P. Lemarinier, and F. Magniette. MPIch-V2: a fault tolerant MPI for volatile nodes based on the pessimistic sender based message logging. In *SuperComputing 2003*, Phoenix USA, Nov. 2003.

[4] E. Caron, F. Deprez, F. Frédéric Lombard, J.-M. Nicod, M. Quinson, and F. Suter. A scalable approach to network enabled servers. In *8th EuroPar Conference*, volume 2400 of *LNCS*, pages 907–910. Springer-Verlag, Aug. 2002.

[5] B. Carpenter, V. Getov, G. Judd, T. Skjellum, and G. Fox. Mpj: Mpi-like message passing for java. *Concurrency: Practice and Experience*, 12(11), Sept. 2000.

[6] F. Schneider. *Replication Management Using State-Machine Approach. In S. Mullender, Distributed Systems*, chapter 7, pages 169–198. Addison Wesley, 1993.

[7] G. Fagg and J. Dongarra. FT-MPI: Fault tolerant MPI, supporting dynamic applications in a dynamic world. In *EuroPVM/MPI 2000*. Springer, 2000.

[8] G. Fedak, C. Germain, V. Néri, and F. Cappello. XtremWeb : A generic global computing system. In *CCGRID*, pages 582–587. IEEE Computer Society, 2001.

[9] E. Gabriel, M. Resch, T. Beisel, and R. Keller. Distributed Computing in an Heterogeneous Computing Environment. In *EuroPVM/MPI*. Springer, 1998.

[10] N. T. Karonis, B. T. Toonen, and I. Foster. MPICH-G2: A Grid-enabled implementation of the Message Passing Interface. *Journal of Parallel and Distributed Computing, special issue on Computational Grids*, 63(5):551–563, May 2003.

[11] T. Kielmann, R. F. H. Hofman, H. E. Bal, A. Plaat, and R. A. F. Bhoedjang. MagPIe: MPI's collective communication operations for clustered wide area systems. *ACM SIGPLAN Notices*, 34(8):131–140, Aug. 1999.

[12] S. Louca, N. Neophytou, A. Lachanas, and P. Evripidou. MPI-FT: Portable fault tolerenace scheme for MPI. *Parallel Processing Letters*, 10(4):371–382, 2000.

[13] MPI Forum. MPI: A message passing interface standard. Technical report, University of Tennessee, Knoxville, TN, USA, June 1995.

[14] R. V. Renesse, Y. Minsky, and M. Hayden. A gossip-style failure detection service. Technical report, Ithaca, NY, USA, 1998.

[15] K. Shudo, Y. Tanaka, and S. Sekiguchi. P3: P2P-based middleware enabling transfer and aggregation of computational resource. In *5th Intl. Workshop on Global and Peer-to-Peer Computing, in conjunc. with CCGrid05*. IEEE, May 2005.

[16] R. Thakur, R. Rabenseifner, and W. Gropp. Optimization of collective communication operation in mpich. *International Journal of High Performance Computing Applications*, 19(1):49–66, Feb. 2005.

[17] J. Verbeke, N. Nadgir, G. Ruetsch, and I. Sharapov. Framework for peer-to-peer distributed computing in a heterogeneous, decentralized environment. In *GRID 2002*, volume 2536 of *LNCS*, pages 1–12. Springer, Nov. 2002.

MGF: A Grid-Enabled MPI Library with a Delegation Mechanism to Improve Collective Operations*

F. Gregoretti[1], G. Laccetti[2], A. Murli[1,2], G. Oliva[1], and U. Scafuri[1]

[1] Institute of High Performance Computing and Networking ICAR-CNR,
Naples branch, Naples, Italy
[2] University of Naples Federico II, Naples, Italy

Abstract. The success of Grid technologies depends on the ability of libraries and tools to hide the heterogeneous complexity of Grid systems. MPI-based programming libraries can make this environment more accessible to developers with parallel programming skills. In this paper we present MGF, an MPI library which extends the existing MPICH-G2. MGF aims are: to allow parallel MPI applications to be executed on Grids without source code modifications; to give programmers a detailed view of the execution system network topology; to use the most efficient channel available for point-to-point communications and finally, to improve collective operations efficiency introducing a delegation mechanism.

Keywords: MPI, message passing, collective operations, Grid computing, MPICH-G2.

1 Introduction

Computational Grids [1] have increased computational power more than ever before, supplying HW/SW resources at low-cost. The present and future success of Computational Grids depends on the ability of libraries and tools to hide architectural issues from users. MPI-based programming environments can make this technology more accessible for end-users with parallel programming skills because MPI is the widely used de-facto standard for parallel applications development.

Developing Grid-enabled applications is a complex issue. However, by taking advantage of existing parallel/MPI tools and applications developed for low-cost high performance machines such as clusters, it is possible to tackle some of these issues. As many of these tools are already available in Grid environments, the focus of our work was to develop an MPI implementation with inner devices to simplify the extension of MPI applications to the Grids.

Previous work [2] shows that topology-aware communication patterns can improve the efficiency of MPI collective operations in Grid environments. Furthermore the use of communication daemons on the front-end nodes of clusters,

* This work has been partially supported by Italian Ministry of Education,University and Research (MIUR) within the activities of the WP9 workpackage "Grid Enabled Scientific Libraries", part of the MIUR FIRB RBNE01KNFP *Grid.it* project.

B. Di Martino et al. (Eds.): EuroPVM/MPI 2005, LNCS 3666, pp. 285–292, 2005.

like those provided by the PACX-MPI library [3], enables transparent intercluster communications, thus increasing portability of MPI codes from traditional parallel machines to metacomputers and Grids. The use of a delegation mechanism to avoid needless message passing through these daemons is a further improvement for intercluster communications in collective operations.

We have developed a library called MGF based on MPICH-G2, which implements the communications daemons on the PACX-MPI model with the delegation mechanism mentioned above. The use of MGF is transparent to the user in the sense that no modification of the source code is required. Moreover MGF expands the topology description provided by MPICH-G2 by including information about existing private networks.

In section 2 we briefly review the state of the art in Grid-enabled MPI implementations; in section 3 we give a detailed description of MGF architecture; in section 4 we describe experimental results and illustrate the benefits of the delegation mechanism by comparing it with the PACX-MPI daemons implementation; in section 5, finally, we conclude with a discussion of future work.

2   State of the Art

There are several projects for the realization of MPI libraries for Grids [4]: MagPIe, MPICH-G2, MPI_Connect, MetaMPICH, Stampi, PACX-MPI, etc. Many of these implementations allow to couple multiple machines potentially based on heterogeneous architecture for MPI programs execution and to use vendor-supplied MPI libraries over high performance networks for intramachine messaging. The most widespread and complete are MPICH-G2 and PACX-MPI.

- **MPICH-G2** [5] is a grid-enabled implementation of the MPI v1.1 standard which uses grid services provided by the Globus Toolkit for user authentication, resources allocation, file transfer, I/O management, process control and monitoring. MPICH-G2 is based on the MPICH library, which is developed and distributed by the Mathematics and Computer Science Division at Argonne National Laboratory. MPICH-G2 implements topology-aware collective operations that minimize communications over the slowest channels.
- **PACX-MPI** [3] is a complete MPI-1 standard implementation and supports some routines of the MPI-2 standard. PACX-MPI is developed by the Parallel and Distributed Systems working group of The High Performance Computing Center in Stuttgart. PACX-MPI uses daemon processes executing on the front-end nodes of each parallel computer for intermachine communications.

We chose to base our work on MPICH-G2 because we believe that many of its features are very useful in Grid environment. For instance, MPICH-G2 provides the user with an advanced interconnecting topology description with multiple levels of depth, thus giving a detailed view of the underlying executing environment. It uses the Globus Security Infrastructure [6] for authorization and authentication and the Grid Resource Allocation and Management protocol [7] for resources allocation. Further, MPICH-G2 is not cluster-specific and hence enables the use of any type of grid resource (e.g. single hosts) for MPI process execution. Finally it implements multilevel topology-aware collective communications, which have been proven [2] to perform better than the PACX-MPI two-level approach.

However, MPICH-G2 usage becomes complicated for application developers in the presence of clusters where only the front-end node is provided with a public IP address. This is due to the fact that unlike PACX-MPI, MPICH-G2 doesn't provide any routing mechanism among networks. Therefore, MPI processes started on computing nodes belonging to different private networks are unable to contact one another. This prevents the transparent porting of MPI application to Grids where clusters with private networks are used.

3   MGF Library

MGF (MPI Globus Forwarder) is an MPI library based on MPICH-G2 that enables the transparent coupling of multiple Grid resources for the execution of MPI programs. In particular, communication is made possible between clusters with private networks. The principal aims of the library are to allow parallel MPI applications to be executed on Grids without modification of the source code; to give a programmer a detailed view of the underlying network topology of system during execution; to use the most efficient channel available for any point-to-point communication and finally, to implement efficient collective operations.

3.1   Communication Channels

In this context we define a *communication channel* as the network path that a message needs to follow from a source MPI process to a destination. MGF distinguishes between two communication channels classes:

- **direct channels** - implemented by exclusively using network devices
- **indirect channels** - implemented using intermediary processes

MPI processes executing on hosts of the same network use direct channels to communicate (i.e. nodes of the same cluster using its interconnection or front-ends of different clusters using Internet). MPI processes executing on hosts belongings to different private networks use indirect channels. Inside indirect channels one or more processes take care of routing messages between networks. MGF uses MPICH-G2 to handle communication on direct channels and manages indirect communications with the help of the Forwarders.

3.2   Forwarders

Forwarders are service processes executed on the front-end nodes of clusters with private networks. Users enable the execution of the Forwarders by defining an

environment variable named MGF_PRIVATE_SLAN in the rsl [8] used for the
MPICH-G2 job start-up.

MGF introduces a new "world communicator" called MGF_COMM_WORLD,
which includes all MPI processes except Forwarders. When compiling an MPI
application with MGF, the new communicator is automatically substituted to
MPI_COMM_WORLD in the preprocessing phase. Therefore MPI routines that
access communicators (like MPI_Comm_Size and MPI_Comm_Rank) invoked
with the MPI_COMM_WORLD argument, will return information about
MGF_COMM_WORLD hiding the presence of Forwarders to the program.

3.3   Physical Topology

Executing an MPI program on a complex system like a computational Grid
involves different types, levels and topologies of interconnection. It is therefore
useful if the programmer has access to a representation of the underlying physical
topology.

MPICH-G2 describes a topology with a four level array where each level
represents a communication channel: TCP over WAN (level 0), TCP over LAN,
TCP over machine networks and vendor MPI library over high performance
networks (level 3). MPICH-G2 assigns a non-negative integer named "color" to
every process at each level; processes with same colors can communicate over
the corresponding channel [2].

Since MPICH-G2 doesn't support private networks, it is assumed that all
processes must be able to communicate over WAN and hence all have the same
color at level 0. However, this is not true for processes executing on cluster
internal nodes. To overcome this limitation, MGF provides a new data structure
for WAN topology description. This structure is an array of N integers, where N
is the number of computing processes. The i-th array's component describes the
i-th process WAN access: if its value is 0, it means that the process is running on
a host with direct WAN connection; if its value is -1, it means that the process
is running on a cluster internal node. The array can be accessed by the users as
a MGF_COMM_WORLD communicator attribute.

3.4   Point-to-Point Communications

In a point-to-point communication, MGF detects the availability of direct channels by looking up MPICH-G2's topology table and MGF's WAN topology array.
If two processes have the same color at levels 2 or 3 of the MPICH-G2 topology table it means that they belongs to the same cluster and hence they can
communicate over a direct channel. Otherwise the MGF WAN table is accessed:
if both processes have value 0 they can use a direct channel as they both have
direct access to Internet, otherwise an indirect channel is needed.

If the processes can communicate over a direct channel, MGF invokes the
corresponding MPICH-G2 routines (which automatically select the most efficient
channel available between vMPI or TCP).

If the processes are not executing on hosts in the same network, MGF uses
Forwarders to build indirect channels. Forwarders use the native MPI library

(when available) to handle intracluster communications and TCP for communication over LAN and WAN. When the sender/receiver is an internal node of a private network cluster, the message and the destination rank are sent to its cluster Forwarder which takes care of message delivery.

If one of the two processes has direct access to WAN one Forwarder will be used. If both sender and receiver do not have direct access to WAN, that is, they both have value -1 in the MGF WAN table, two Forwarders are needed. The forwarding process is embedded in the point-to-point routines and is completely transparent to the user.

3.5   Delegation Strategy for Collective Communications

MGF inherits the MPICH-G2 topology-aware multi-level implementation of collective operations and introduces two improvements: it enables the execution of collective operations when not possible with MPICH-G2, and it implements a delegation mechanism: collective operations functionalities are delegated to the Forwarders to avoid needless message passing, thus improving performance.

Consider MPICH-G2's broadcast implementation. It follows three steps [2]:

1. The root process sends the message to the master process of each LAN,
2. the LAN master processes then send the message to all machines master processes in the local network.
3. Finally, each machine master process starts a local broadcast operation.

MPICH-G2 broadcast fails when the root process is running on a cluster node, or the communicator encloses any private node of a cluster without containing its front-end.

MGF implementation overcomes these limitations by using Forwarders. When a root process is an internal node of a private network cluster it cannot send messages over WAN, hence the MPICH-G2 broadcast algorithm fails on the first step. However, when using MGF the root process sends its message to the cluster Forwarder together with the ranks of the master processes at level 0. The Forwarder receives the message and performs an unblocking send to each LAN master process. Hence there is no repetition of message passing through the Forwarder for each communication from the root to a every LAN master process.

When a LAN master process is running on a private node, the second step of the broadcast algorithm is completed by its Forwarder, which takes care of sending the message to the other machine master processes in the LAN. This means that there is no repetition of the message passing through the Forwarder for each communication from LAN masters to machines masters processes.

When the broadcast communicator encloses some private nodes of a cluster without containing the front-end, MPICH-G2 fails because the machine master process which is a private node, cannot receive the message from its own LAN master. MGF point-to-point routines ensure that the broadcast message is always delivered.

4   Experimental Results

To evaluate the benefits introduced by using the delegation strategy we compared the efficiency of the MGF broadcast routine to the PACX-MPI implementation which doesn't provide delegation. We used the **broadcast round** and the **OL**$_i$ benchmark methods [9] to perform this comparison. The former measures the time to complete some large number M of broadcast rounds, each consisting of one broadcast by each possible root. The latter measures the operation latency OL$_i$ to each destination i of the broadcast and uses the largest measurement as an estimate of the operation latency OL. Each OL$_i$ is computed by executing M broadcasts. After each broadcast, process i sends an acknowledgement to the root. The root does not start the next broadcast until it receives the acknowledgement. As the overhead of the acknowledgement is the latency of a single point-to-point message from process i to the root, it can be easily measured and subtracted.

As a testbed we used four Beowulf class clusters: Vega (20 nodes Pentium 4 with Fast Ethernet), Beocomp (16 nodes Pentium II with Fast Ethernet), Altair (16 nodes Pentium PRO with two Fast Ethernet networks) and ClusterPC (8 nodes Pentium 4 cluster with Gigabit Ethernet). Only front-end nodes were provided with a public IP address.

All clusters run Red Hat Linux 7.3, with kernel 2.4.20 and GNU libc 2.2.4. The local version of MPI is MPICH 1.2.5.2 with the ch_p4 device. The Globus Toolkit 2.4.3 is used as Grid middleware and the version of MPICH-G2 is 1.2.5.2 while the version of PACX-MPI is 5.0.0rc1. Vega is located at the Naples Branch of the Institute of High Performance Computing and Networking (ICAR-Na CNR), while Altair, Beocomp and ClusterPC are located on the same LAN at the Department of Mathematics and Applications "R. Caccioppoli" of the University of Naples Federico II.

We ran several batteries of tests with various M values. The rsl files used for the tests didn't contain any LAN topology information for MPICH-G2, therefore all clusters were seen as belonging to different LANs. We used this configuration to measure the effects of the delegation strategy without considering the benefits of the MPICH-G2 multi-level topology awareness.

In Figures 1 and 2 we show results from just one experiment for each benchmark method; results in other test cases are similar.

Fig. 1 shows the results obtained by executing the OL$_i$ benchmark with one computing process running on a private node for each of the four clusters. In each broadcast all communications occurred on the WAN/TCP channel. The aim of this test was to measure the effects of the delegation strategy over the WAN/TCP channel.

In Fig. 2 we show results obtained by executing the broadcast round method with four computing process running on private nodes on three clusters (Vega, Beocomp and Altair). In each broadcast both intercluster and intracluster communications occurred. This test aimed to show the overall improvement of the delegation mechanism on the broadcast algorithm.

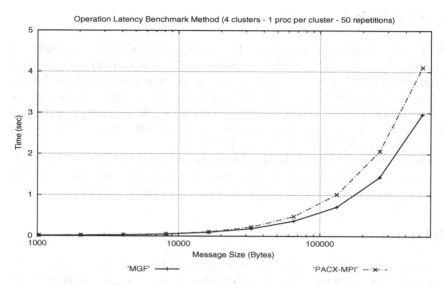

Fig. 1. PACX-MPI/MGF OL_i benchmark comparison, data points are the mean of ten measurements

The results demonstrate the advantages of the delegation strategy. Note in particular that the the benefits are more pronounced as the message size and the number of clusters involved increase.

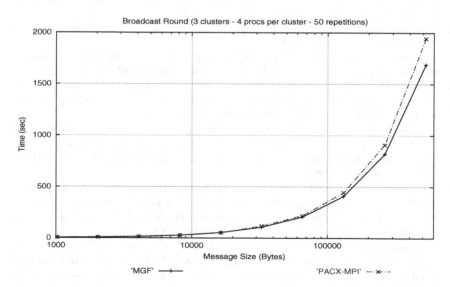

Fig. 2. PACX-MPI/MGF broadcast round test comparison, data points are the mean of ten measurements

5   Conclusions

We have developed a library called MGF based on MPICH-G2 that allows the transparent use of coupled Grid resources to run MPI applications in presence of private network clusters. It implements communications daemons on the PACX-MPI model but introduces a novel delegation strategy to improve collective operation performance. We have shown the benefits of the delegation strategy in the execution of collective operations for the broadcast routine by comparing the performance of MGF with that of PACX-MPI.

Collective operations could be further improved by introducing some kind of network evaluation and by varying the point-to-point fixed communication scheme over a wide area with one based on network performances.

References

1. Foster, I., Kesselman, C.: Computational grids. In Foster, I., Kesselman, C., eds.: The Grid: Blueprint for a New Computing Infrastructure. Morgan Kaufmann (1998) 15–51
2. Karonis, N.T., de Supinski, B.R., Foster, I., Gropp, W., Lusk, E., Bresnahan, J.: Exploiting hierarchy in parallel computer networks to optimize collective operation performance. In: Proceedings of the 14th International Symposium on Parallel and Distributed Processing, IEEE Computer Society (2000) 377
3. Gabriel, E., Resch, M., Beisel, T., Keller, R.: Distributed computing in a heterogeneous computing environment. In: Proceedings of the 5th European PVM/MPI Users' Group Meeting on Recent Advances in Parallel Virtual Machine and Message Passing Interface, Springer-Verlag (1998) 180–187
4. Laforenza, D.: Grid programming: some indications where we are headed. Parallel Comput. **28** (2002) 1733–1752
5. Karonis, N.T., Toonen, B., Foster, I.: Mpich-g2: a grid-enabled implementation of the message passing interface. J. Parallel Distrib. Comput. **63** (2003) 551–563
6. Foster, I., Kesselman, C., Tsudik, G., Tuecke, S.: A security architecture for computational grids. In: Proceedings of the 5th ACM conference on Computer and communications security, ACM Press (1998) 83–92
7. Czajkowski, K., Foster, I., Kesselman, C.: Resource co-allocation in computational grids. In: HPDC '99: Proceedings of the The Eighth IEEE International Symposium on High Performance Distributed Computing, Washington, DC, USA, IEEE Computer Society (1999) 37
8. Czajkowski, K., Foster, I.T., Karonis, N.T., Kesselman, C., Martin, S., Smith, W., Tuecke, S.: A resource management architecture for metacomputing systems. In: IPPS/SPDP '98: Proceedings of the Workshop on Job Scheduling Strategies for Parallel Processing, London, UK, Springer-Verlag (1998) 62–82
9. de Supinski, B.R., Karonis, N.T.: Accurately measuring mpi broadcasts in a computational grid. In: HPDC '99: Proceedings of the The Eighth IEEE International Symposium on High Performance Distributed Computing, Washington, DC, USA, IEEE Computer Society (1999) 4

Automatic Performance Analysis of Message Passing Applications Using the KappaPI 2 Tool[*]

Josep Jorba, Tomas Margalef, and Emilio Luque

Computer Architecture & Operating Systems Departement,
Universidad Autónoma de Barcelona,
08193 Bellaterra, Spain
{josep.jorba, tomas.margalef, emilio.luque}@uab.es

Abstract. Message passing libraries offer the programmer a set of primitives that are not available in sequential programming. Developing applications using these primitives as well as application performance tuning are complex tasks for non-expert users. Therefore, automatic performance analysis tools that help the user with performance analysis and tuning phases are necessary. KappaPI 2 is a performance analysis tool designed openly to incorporate parallel performance knowledge about performance bottlenecks easily. The tool is able to detect and analyze performance bottlenecks and then make suggestions to the user to improve the application behavior.

1 Introduction

Applications running in parallel/distributed environments must achieve certain performance indexes to fulfill the objectives of high performance computing systems. So performance analysis becomes a very significant task in carrying out applications. However, there are still very few really useful performance analysis tools and the most popular approach to carrying out performance analysis is to use visualization tools [1, 2, 3]. These tools provide a set of views of the application execution. The user must carry out the performance analysis process manually. This is a difficult and time consuming task that requires a high degree of expertise to overcome the performance bottlenecks and reach the performance expectations.

To tackle all these problems and help the user in the analysis phase there needs to be more user-friendly tools. In this context, automatic performance analysis tools, like Expert [4], Scalea [5] and KappaPI [6], have been developed.

These tools take a trace file from the application execution, and try to detect performance bottlenecks using performance patterns. This automatic post-mortem approach has the advantage of being able to consider all detailed information gathered while executing the application and, moreover, the analysis phase does not introduce any overheads while the application is being executed; only the application tracing

---

[*] This work has been supported by the MCyT under contract number TIN 2004-03388 and partially funded by the Generalitat de Catalunya – Grup de recerca consolidat 2001-SGR-00218.

B. Di Martino et al. (Eds.): EuroPVM/MPI 2005, LNCS 3666, pp. 293–300, 2005.

introduces some overheads which can be minimized by various techniques [7]. However, current tools have some limitations that must be solved to become useful tools:

- The performance bottleneck specification is hard-coded in the tool or has several constraints. These performance tools cannot be easily extended to detect (or analyze) a larger set of bottlenecks.
- Detecting bottlenecks is quite significant but the precise point where the bottleneck is detected is not identified.
- The information provided to the user does not indicate the actions that should be carried out to overcome the bottlenecks.

In this context our goal is to design and implement a new tool that solves all these difficulties by automatically analyzing trace files from message passing applications carried out using PVM or MPI. The paper is organized as follows: Section 2 presents the basic concepts and the general structure of Kappa-Pi 2. Section 3 introduces the specification of the performance bottleneck knowledge. In Section 4 the internal mechanisms for bottleneck detection and analysis are shown. Some experimental results are shown in Section 5. Section 6 presents some related work. Finally, section 7 summarizes and concludes our work.

2 KappaPI 2

Our goal is to design and implement an automatic performance analysis tool that has the following features:

- Performance knowledge specification: Independent specification mechanisms to introduce new performance bottlenecks.
- The performance bottleneck detection engine must read the performance knowledge specification.
- Independence of the background message passing system: The tool builds abstract entities that are independent from the particular trace file format or the message passing primitives.
- Relate bottlenecks to the source code of the application: one set of quick parsers to search for dependences in the source code must be included to determine why the bottleneck appears.

In KappaPI 2 (Fig. 1), the first step is to execute the application under the control of a tracing tool (for PVM or MPI environments) that captures all the events related to the message passing primitives that occur while carrying out the application. Our tool uses the trace and performance bottleneck knowledge base as inputs to detect the performance bottleneck patterns defined from a structural point of view. Then, it sorts the performance bottlenecks it has found according to some indexes. It carries out a bottleneck cause analysis, based on the application source code analysis, and finally provides a set of recommendations to the user, indicating how to modify the source code to overcome the detected bottlenecks.

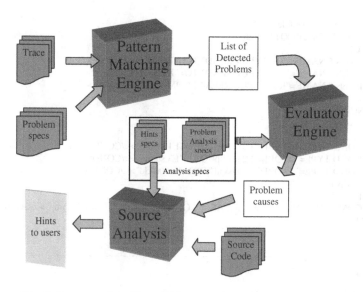

Fig. 1. Structure of the KappaPI 2 automatic performance analysis tool

3 Bottleneck Specification

Performance analysis is an evolving field which must take into account the programming environments, operating systems, and network and hardware facilities currently used. It is not a closed knowledge set, but rather it has to be adapted to the new situations that appear. Therefore, it is important that automatic performance analysis tools are designed openly so that the knowledge related to new performance bottlenecks can be introduced into the tool knowledge base. An open specification is required to add new bottlenecks. In this context it is necessary to develop a specification language that allows us to define new performance bottlenecks. ASL (APART Specification Language) is a good example of this approach [8]. In our case the bottlenecks are specified in a structural way, defining the events involved, the time and location constraints, and some computations to evaluate the "importance" of the bottleneck.

3.1 Bottleneck Specification Language

The specification language used in KappaPI 2 is a simplified XML translation of ASL, using the compound events extensions [4] to describe performance bottlenecks. The specification is based on the event structure of the performance bottleneck, with an initial event (Root event, the first event detected in the root task), followed by some event instances (INSTANTIATION section), and some constraints, such as time considerations or related tasks.

The next example shows a part of the specification code corresponding to the Blocked Sender performance bottleneck (see Figure 2, for a graphical representation). In this bottleneck one task is waiting in a receive operation because its sender task is blocked in another previous communication.

```
<PATTERN Name="Blocked Sender">
  <ROOTTYPE>RECV</ROOTTYPE>
  <INSTANCES>
        <EVENT NAME="S1" TYPE="SEND" TO="ROOT"></EVENT>
        <EVENT NAME="S2" TYPE="SEND" TO="R2"></EVENT>
        <EVENT NAME="R2" TYPE="RECV" FROM="S2"></EVENT>
        ...
  </INSTANCES>
  <CONSTRAINT>
        ...
        <COND TYPE=">" OP1="E2.stamp" OP2="E1.stamp"></COND>
        <COND TYPE=">" OP1="E2.stamp" OP2="E3.stamp"></COND>
        <COND TYPE="=" OP1="E3.taskId" OP2="E2.taskId"></COND>
  </CONSTRAINT>
  <EXPORT>
        <COMPUTE NAME="idle_time" AS="-" OP1="E2.stamp" OP2="E1.stamp"></COMPUTE>
  </EXPORT>
```

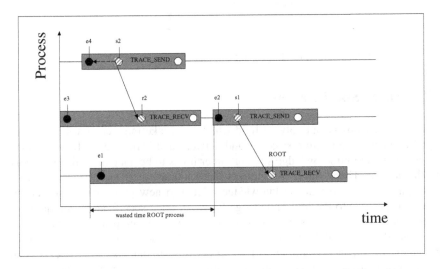

Fig. 2. Graphical structure of a Blocked Sender performance bottleneck

Three tasks participate in this bottleneck and four events are involved. The third task has an idle time due to a blocked receive from a non-initiated sending operation in the second task. This send is not reached because this task is blocked in a receive operation from the first task.

3.2 Performance Bottleneck Catalogue

Our performance Bottleneck catalogue contains a set of bottlenecks defined in the related literature [8][9][10]. The knowledge base includes communication bottlenecks in message passing systems related to point to point communications, and collective/group communications and synchronization.

Point to point communication bottlenecks include: Late Sender (LS), Late Receiver (LR), Blocked Sender (BS), Multiple Output (MO), and Wrong Order (WO). These bottlenecks appear when: one operation is posted later than it is needed (LS and LR); one task delays a second task because it is blocked by a third task (BS); one task sends messages to several tasks (MO); the messages are sent in a different order than expected (WO).

Collective communication bottlenecks include 1-to-N, N-to-1, and N-to-N communications with different collective primitives such as scatter, gather, reduce and All primitives. All these bottlenecks are commonly caused by delays in posting a process involved in the collective operation, or the inherent delays involved in entering (or leaving) tasks between the first task and the last one.

In synchronization, Block at Barrier (BB), is a typical bottleneck when a set of tasks are waiting at the barrier, until the last one reaches it.

4 Detection and Analysis

The mechanism to detect performance bottlenecks is based on searching for patterns that describe the bottlenecks and then classifying them according to importance (number of instances found, global idle time balanced by processors/tasks affected).

KappaPI 2 starts by reading the specification of performance bottlenecks included in the knowledge base then it creates a search tree that is used to detect bottlenecks in trace files. Once the search tree has been created, KappaPi 2 is ready to start the processing trace. The trace streams are processed by matching the events in the search tree and detecting the corresponding bottlenecks.

4.1 Generating Search Trees

Each bottleneck specification includes a set of events which defines its structure in the trace. Each bottleneck has a root event and a set of instances of other related events. A search tree is built by parsing the list of performance bottlenecks. Each bottleneck is a path in the tree from the root node to one particular leaf node.

However, there are some points that must be taken into account:

- It is possible that several bottlenecks share an initial set of common events.
- One bottleneck can be an extension of another one. In this case, the bottleneck has a final node that indicates that the bottleneck has been detected and the indexes can be evaluated, but events related to the extended second bottleneck can appear in the trace, and in this case, it is necessary to continue the process to detect the second bottleneck. If finally the second bottleneck is detected, the indexes must be calculated and the first one should not be shown.
- Some bottlenecks involve an indeterminate number of events of the same type. To solve this problem, it is necessary to create nodes in the tree that can record information about a varying number of events of the same type.

4.2 Detection Engine

Once the search tree has been created it is possible to start reading the trace files from the application execution process. Examples of tracing tools are TapePVM tracer [7] and MPITracer [11] which is based on the DyninstAPI [12] interface.

One key aspect is that the tracing tools must be able to record message passing calls, including the line number and source code file where the events occurred.

When the application trace has been read, an application model is built, taking into account the tasks and execution hosts involved.

The detection engine starts reading the events ordered according to their global time-stamp. Each event is matched in the search tree moving to the next node in the tree. When the leaf node is reached a bottleneck is detected, and the indexes are evaluated and included in the table of detected bottlenecks. The detected bottleneck has all the information needed in order to relate it to the execution history of the application, and the source code lines where the message passing calls can be found.

4.3 Cause Analysis

Once the bottleneck is detected, it is necessary to carry out an analysis to determine its main cause. This cause analysis is carried out using knowledge about the use cases of the bottleneck. Each bottleneck can have various cases, and it is necessary to examine the source code related with the calls to reach a conclusion. This analysis uses a template file of use cases as input, with the suggested recommendation about the code. In each case, the tool must evaluate some simple conditions, such as the data dependence between two variables, or for example if a parameter of a call is a constant, a variable, etc... These analyses are carried out using a set of small parsers to search for the source code.

5   Experimentation

We carried out some tests of our tool to validate the bottleneck detection mechanisms. The tool has been tested on some synthetic applications, some standard benchmarks, and one real application. In table 1 we summarize the test results.

Table 1. Summary of application test suite

Bench	Bottlenecks found
Apart LS (mpi)	1 LS
Apart LR (mpi)	0
Apart BS (mpi)	1 BS
pvmLR (pvm)	1 LR
PingPong (pvm)	5 LS
NAS IS (pvm)	27 LS, 18 BS
IMB-MPI1 p2p (mpi)	3LS, 1BB
Xfire (pvm)	17 LS, 6 BS

The test suite is a set of MPI and PVM applications. We have used some synthetic benchmarks to detect a particular performance bottleneck, or repetitions of the same bottleneck or some simple combinations. These benchmarks (LS, LR and BS) are based on the ATS (APART Test Suite) [13] which is a flexible construction kit for synthetic benchmarks in MPI. Other tests are classic simple benchmarks, like pingpong, or IS (Integer Sort) from NAS benchmarks. The recent Intel benchmark (IMB) suite includes benchmarks considering p2p communication primitive calls. Finally Xfire [14] is a real PVM Application, used for forest fire propagation simulations.

The first series of experiments with APART synthetic benchmarks were used to validate the search for particular bottlenecks. In the LR test no bottleneck was found because the bottlenecks depend on the particular MPI implementation.

In Intel benchmark (IMB) for p2p communications, some tests only use a small set of processors, the others wait at a Barrier.

In bigger applications more bottlenecks are found. Usually, the user does not correctly understand the complexity of the interrelations between tasks, processors, network and communication patterns and this fact causes many bottlenecks.

6 Related Work

Some existing tools related to KappaPI 2 include several automatic performance tools, such as the first version of KappaPI [9], Expert [4] and Scalea [5].

In the first version of KappaPI, detecting performance bottlenecks was focused on idle intervals in the global computation affecting the biggest number of tasks. Processor efficiency was used to measure the execution quality, and idle processor intervals represented performance bottlenecks. The tool examined the intervals to find the causes of the inefficiency. The tool had a closed "hard-coded" set of bottlenecks, and no mechanisms for new bottleneck specification was included. The same limitation also affects root cause analysis.

The Expert [4, 10] tool allows us to specify performance properties using a script language based on an internal API. This API allows us to examine the trace and to look for relations between events. Expert summarizes the indexes of each bottleneck defined in its list of bottlenecks. Expert tries to answer the question: where is the application spending time? It summarizes the performance bottlenecks found and accumulates their times to compare their impact on the total execution time.

Scalea [5] is used with the Aksum tool [15] for multi-experiment performance analysis. This tool uses an interface called JavaPSL to specify the performance properties by using syntax and semantic rules of the Java programming language. The user can specify new properties and formats without changing the implementation of the search tool that uses JavaPSL API.

7 Conclusions

We have discussed a new automatic performance analysis tool oriented towards the end user to avoid the high degree of expertise needed to improve message passing applications. Our KappaPI 2 tool is open to introducing new models for performance bottlenecks. It is able to make suggestions about the source code to improve the application execution time and therefore avoid performance bottlenecks. The experiments carried out, show that the tool detects specified bottlenecks which can be related to the source code.

References

1. W.E. Nagel, A. Arnold, M. Weber, H.C. Hoppe, K. Solchenbach. VAMPIR: Visualization and Analysis of MPI Resources, In Supercomputer 63, vol XII, number 1, Jan. 1996.
2. T. Cortes, V. Pillet, J. Labarta, S. Girona. Paraver: A tool to visualize and analyze parallel code. In WoTUG-18, pages 17-31, Manchester, April 1995.
3. L. De Rose, Y. Zhang, D.A. Reed, SvPablo: A Multilanguage Performance Analysis system. LNCS, 1469 pp352-99, 1998.
4. F. Wolf, B. Mohr, J. Dongarra, S Moore, Efficient Pattern Search in Large Traces Through Successive Refinement, In Euro-Par 2004, LNCS 3149, 2004.
5. HL.Truong, T. Fahringer, G. Madsen, AD. Malony, HMoritsch, S. Shende, On using SCALEA for Performance Analysis of Distributed and Parall el Programs, Supercomputing 2001 Conference (SC2001), Denver, Colorado, USA. November 10-16,2001
6. A. Espinosa, T. Margalef, E. Luque, Automatic Performance Evaluation of Parallel Programs. In IEEE Proceedings of the 6th Euromicro Workshop on Parallel and Distributed Processing. Jan. 1998.
7. E. Maillet, TAPEPVM an efficient performance monitor for PVM applications - user guide. Technical report, LMCIMAG,University of Grenoble, 1995.
8. T. Fahringer., M. Gerndt, G. Riley, J. Larsson, Specification of Performance bottlenecks in MPI Programs with ASL. Proceedings of ICPP, pp. 51-58. 2000.
9. A. Espinosa, T. Margalef, E. Luque. Automatic Performance Analysis of PVM applications. EuroPVM/MPI 2000, LNCS 1908, pp. 47-55. 2000.
10. F. Wolf, B. Mohr, Automatic Performance Analysis of MPI Applications Based on Event Traces, In EuroPar 2000, LNCS, 1900, pp123-132, 2000.
11. V.J. Ivars, Monitor de Aplicaciones MPICH Basado en Dyninst (in spanish), Master Thesis, Universidad Autónoma de Barcelona, 2004.
12. Hollingsworth, J.K., Buck, B. DyninstAPI Programmer's Guide. Release 3.0. University of Maryland, January 2002.
13. M. Gerndt, B. Mohr, JL. Träff, Evaluating OpenMP Performance Analysis Tools with the APART Test Suite, In Euro-Par 2004, LNCS 3149, 2004.
14. J. Jorba, T.Margalef, E.Luque, J.Andre, D.Viegas, Application of Parallel Computing to the Simulation of Forest Fire Propagation. Proceedings of International Conference in Forest Fire Propagation, Vol 1, pp 891-900, Portugal, Nov. 1998.
15. C. Seragiotto, M. Geisller, et al: On Using Aksum for Semi-Automatically Searching of Performance Problems in Parallel and Distributed Programs. Procs. Of 11th Euromicro Conference on Parallel Distributed and Network based Processing (PDP) 2003.

Benchmarking One-Sided Communication
with SKaMPI 5

Werner Augustin[1], Marc-Oliver Straub[2], and Thomas Worsch[3]

[1] IZBS, Universität Karlsruhe, Germany
augustin@ira.uka.de
[2] IAKS, Universität Karlsruhe, Germany
straub@sb-software.de
[3] IAKS, Universität Karlsruhe, Germany
worsch@ira.uka.de

Abstract. SKaMPI is now an established benchmark for MPI implementations. Two important goals of the development of version 5 of SKaMPI were the extension of the benchmark to cover more functionality of MPI, and a redesign of the benchmark allowing it to be extended more easily. In the present paper we give an overview of the extension of SKaMPI 5 for the evaluation of one-sided communication and present a few selected results of benchmark runs, giving an impression of the breadth and depth of SKaMPI 5. A look at the source code, which is available under the GPL, reveals that it was easy to extend SKaMPI 5 with benchmarks for one-sided communication.

Keywords: SKaMPI, MPI benchmark, extensibility, one-sided communication.

1 Introduction

SKaMPI measures the performance of an MPI implementation on a specific hardware and is known for a comprehensive set of benchmarks and advanced methods for measurements, especially for collective operations [12]. By providing not simply one number, but detailed data about the performance of each MPI operation, a software developer can judge the consequences of design decisions regarding the performance of the system to be built.

The recent version 5 of SKaMPI offers two major new features: a significantly improved extension mechanism and support for all aspects of one-sided communication in MPI-2.

In Section 2 we briefly describe some features of SKaMPI 5. Section 3 discusses related work. Section 4 is concerned with simple standard pingpong and related measurements. In Section 5 we give an overview of the more complex measurements SKaMPI 5 provides for one-sided communication and present some example results. The main part of the results of this paper is based on the diploma thesis of the second author [9]. Section 6 concludes the paper.

B. Di Martino et al. (Eds.): EuroPVM/MPI 2005, LNCS 3666, pp. 301–308, 2005.

2   SKaMPI 5

SKaMPI 5 is a major redesign and reimplementation of its predecessors. The main objectives were a much increased flexibility and extensibility. To demonstrate the ease of implementing a measurement function, let's have a look at the *complete* code in C needed for something like MPI_Bcast:

```
void init_Bcast(int count, MPI_Datatype dt, int root) {
  set_send_buffer_usage(get_extent(count, dt));
  set_recv_buffer_usage(0);
  set_reported_message_size(get_extent(count, dt));
  init_synchronization();
}
double measure_Bcast(int count, MPI_Datatype dt, int root) {
  double start_time, end_time;
  start_time = start_synchronization();
  MPI_Bcast(get_send_buffer(), count,dt,root,get_measurement_comm());
  end_time = stop_synchronization();
  return end_time - start_time;
}
```

Apart from a small initialization function which does some administrative stuff like telling SKaMPI about buffer usage there is a small measurement function which has to return the time associated with a particular measurement. start_synchronization() and stop_synchronization() define a synchronous time slot which is reserved for the measured operation [12]. All times are obtained by calls to MPI_Wtime() since this is portable.

Now let's have look at measurement specifications in the configuration file which make use of the above defined measure_Bcast() function:

```
begin measurement "MPI_Bcast-procs-length"
    for procs = 2 to get_comm_size(MPI_COMM_WORLD) do
        for count = 1 to ... step *sqrt(2) do
            measure comm(procs) Bcast(count, MPI_INT, 0)
        od
    od
end measurement
```

The configuration file is now actually a program which is interpreted at runtime and therefore much more flexible than before. Loops can be nested allowing multi-dimensional measurements. Besides the standard data-types int, double and string MPI specific data-types like MPI_Comm, MPI_Datatype or MPI_Info are provided. A "..." limit allows to measure with message sizes as large as fit in the specified buffer. Additional functions can be added easily by the user, allowing to construct derived data types, topologies etc.

This extensibility makes SKaMPI much more valuable for actual front line research, where one doesn't have established measurement functions in advance and doing many experiments is inevitable.

3   Benchmarking One-Sided Communication in MPI

SKaMPI 5 offers more than 60 different functions for investigating different performance aspects of one-sided communication routines in an MPI-2 library. Several groups of functions can be identified.

- Functions for measuring the amount of time needed for calls to synchronization functions like `MPI_Win_fence`. This includes more complicated cases where for example `MPI_Win_wait` is delayed by a specified amount of time after another process has called `MPI_Win_complete`.
- Functions for measuring the amount of time needed for calls to communication functions like `MPI_Put`. Latency and bandwidth of simple pingpong communications implemented with one-sided communication are also easily determined.
- Functions for measuring the amount of time needed for more complex communication patterns, e.g. the shift and exchange patterns mentioned above, one-sided implementations of collective operations, exchange of "ghost cells" etc.

We have chosen a few interesting aspects with an emphasis on those *not* covered by other benchmarks, e.g. those mentioned next.

3.1   Related Other MPI Benchmarks

There are some benchmarks which offer latency and bandwidth measurements. The following ones are available on the WWW and they are all similar in the sense, that they produce data with running times for message exchange operations. Benchmarks like PPerfMark [7] take a different approach.

The MBL library [11] measures the following communication patterns: ping, pingpong, shift and exchange. For the first two one can choose between `MPI_Get`, `MPI_Put` and `MPI_Accumulate` for the communication. In the shift pattern each process i sends to $i + 1$ mod P. Thus a total of P messages is exchanged. The exchange pattern realizes a kind of `MPI_Alltoall` with a total of $P(P-1)$ messages. MBL always uses `MPI_Win_fence`, dedicated and passive synchronization not considered.

NetPIPE [8] is a benchmark measuring latency and bandwidth for a variety of communication mechanisms, including MPI2's one-sided communication. One can choose between uni- or bi-directional communication and between `MPI_Put` and `MPI_Get`. The end of the communication is usually ensured using `MPI_Win_fence`. When using `MPI_Put` one can alternatively request, that the target process watches the last memory location to be written for a specific byte to arrive. This violates the requirements of the MPI standard.

More comprehensive than the above was the Pallas MPI benchmark (PMB) which has been replaced by the Intel MPI benchmark (IMB) [4]. For the benchmarking of one-sided communication a variable number of M messages are sent. One can choose whether `MPI_Win_fence` is called after each message or only after the last one. In the latter case it is ensured that for each message a different part of the target buffer is accessed. Furthermore the user may select MPI_Get or MPI_Put and uni- or bi-directional communication. There are also a few further possibilities, but measurements using dedicated or passive synchronization are not possible.

3.2 Related Papers

Gabriel et al. [3] report latency and bandwidth numbers for several MPI implementations. They also show the influence of using MPI_Alloc_mem. The authors sound somewhat reluctant when it comes to recommending one-sided communication. On the other hand Matthey et al. [6] show that there are situations where significant speedups can be obtained by replacing two-sided by one-sided communication.

Träff et al. [10] differs from the above mentioned papers in that it emphasizes that pingpong measurements are definitely not the only way to assess the quality of an implementation of one-sided communication, but more complex communication patters should be considered, too. SKaMPI 5 offers this possibility.

The paper by Luecke et al. [5] compares implementations of collective operations using SHMEM and MPI-2 one-sided communication. The broadcast algorithm in this paper violates the MPI-2 standard: data retrieved with MPI_Get are accessed locally before the communication epoch is closed.

The same problem is present in the paper by Desplat [2]. It describes the differences between SHMEM and MPI-2 one-sided communication (blocking versus non-blocking) and how to make the transition from the first to the second.

In the following sections we will report on some results from benchmark runs of SKaMPI 5. Of course it is possible to do the standard latency and bandwidth measurements. But in order to stress the much greater breadth and depth of what is possible with SKaMPI 5 we will spend some space on results which can (to the best of our knowledge) not be obtained with the other benchmarks mentioned above. And what really has to be emphasized here is the fact, that

SKaMPI 5 allows to provide and use such measurements very easily.

4 Latency and Bandwidth

Though it is probably not the most important aspect of one-sided communication [10], we start with a simple pingpong measurement in Section 4.1. But SKaMPI 5 allows to do much more; because of the very strict page limit for the paper only one example can be presented in Section 4.2.

4.1 Pingpong Using MPI_Put

The code for the standard bandwidth benchmark looks like this:

```
Process 0                                  Process 1
MPI_Win_fence(...);                        MPI_Win_fence(...);
t1=start_synchronization();                t1=start_synchronization();
MPI_Put(...,1,...);
MPI_Win_fence(...);                        MPI_Win_fence(...);
                                           MPI_Put(...,0,...);
MPI_Win_fence(...);                        MPI_Win_fence(...);
time=stop_synchronization()-t1;            time=stop_synchronization()-t1;
```

Putting the calls to MPI_Put in above code in if statements checking the rank of the process already gives the complete code one has to write in order to have SKaMPI 5 provide the measurement [1]. The rest is provided by the SKaMPI 5 framework.

Figure 1 shows some results on a NEC SX6 and an HP XC6000. Surprisingly on the latter for large messages staying inside a node is *slower* than going outside. (All times reported in this paper have been obtained with the vendor supplied MPI libraries.)

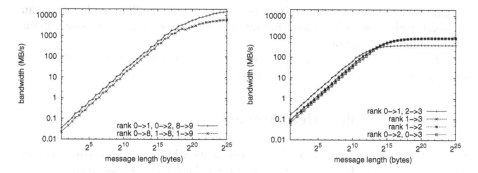

Fig. 1. Bandwidth achieved with pingpong of messages using MPI_Put on a NEC SX6 (8 processors per node, left) an HP XC6000 (2 processors per node, right)

4.2 Call Duration of MPI_Put

The code for determining the duration of a call to MPI_Put is basically a

```
t1 = MPI_Wtime();    MPI_Put(...,1,...);    time = MPI_Wtime() - t1;
```

on process 0. Obviously this method can't be more precise than the resolution of MPI_Wtime, but we don't know of any other portable, more accurate method of time measurement. Results for an NEC SX6 are shown in Figure 2. In the first case processes with ranks 0 and 1 (on the same node) communicate, in the second case process 0 and 8 (on different nodes). In each figure there are four lines: MPI_Put is compared with MPI_Isend for the cases where message buffers were allocated using MPI_Alloc_mem and where they were allocated using malloc. The large variations of times spent in calls to MPI_Put for different

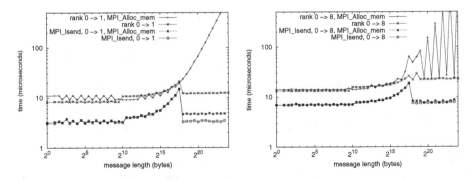

Fig. 2. Time needed for calling `MPI_Put` on a NEC SX6 compared to `MPI_Isend` for intra-node (left) and extra-node (right) communication. Data are shown for the case of message buffers allocated using `MPI_Alloc_mem` and the case when that was not done.

message lengths (with `MPI_Alloc_mem`) can be explained as follows: If the message length is exactly a multiple of 4, `MPI_Put` blocks and immediately does the communication; if the message length is different, `MPI_Put` only makes some arrangements for the transmission to be carried out later. (This explanation can be confirmed by additional measurements, which were enabled by some minor changes in the configuration file).

This characteristic may not be disclosed by a benchmark which only uses message lengths which are a power of two.

5   More Complex Measurements

In the last section we deal with measurements which are more complicated. SKaMPI 5 includes functions for investigating the following questions:

- Does the MPI implementation delay the actual data transfer triggered by e.g. an `MPI_Put` and when? Does it try to combine several small messages into a larger one and transmit that?
- How do simple self-made "one-sided collective operations" perform compared to the built-ins like `MPI_Bcast`?
- What happens in the case of high network load? How much bandwidth per processor is still available?
- What happens if several processes want to access the same window and request an `MPI_LOCK_EXCLUSIVE`?
- How is the performance of one-sided (versus two-sided) communication in simple applications, where dynamic load balancing is to be achieved by splitting the local work and giving part of it to other processors upon request?
- How fast can the exchange of border ("ghost" or "halo") cells in one- or higher-dimensional grid decompositions be realized using one-sided (versus two-sided) communication?

As one example we now have a look at some interesting results from measurements of a self-made "one-sided Alltoall".

Sometimes it is clear, that there will be a communication epoch for some one-sided communication operations. If in addition there is the need for an alltoall exchange of data whose results will only be needed after the end of the epoch, it may be feasible not to use the built-in `MPI_Alltoall` but realize it using one-sided communication, too. For example, a simple variant of one-sided alltoall might be implemented like this:

```
void onesided_alltoall(void* buffer, int count, MPI_Datatype datatype,
                       int displacement, MPI_Win win) {
  [... declarations and initializations ...]
  displacement = displacement + myrank * count;

  for(i = 1; i <= size; i++) {
    rank = (myrank + i) % size;
    rank_buffer = buffer + rank * (count * datatype_size);
    MPI_Put(rank_buffer, count, datatype, rank, displacement, count,
            datatype, win);
  }
}
```

Figure 3 shows the running times for this operation on a NEC SX6 for different sizes of the communicator and for four different lengths of messages exchanged between any two processes. For comparison the time needed by `MPI_Alltoall` for 1 kb messages is also presented. As one can see the self-made implementation is faster if the communicator is larger than one node (with 8 processors). When increasing the data size from 1 kByte by a factor of 256, the running time increases by a factor of 10 on the NEC, but by a factor of 160 on a HP XC6000.

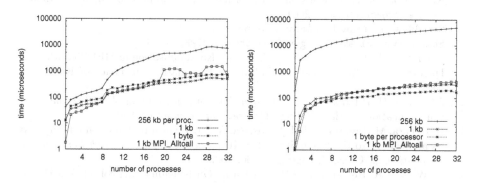

Fig. 3. Measurements of a "one-sided alltoall" and `MPI_Alltoall` on a NEC SX6 (left) and on a HP XC6000 (right)

6   Conclusion

Recently the number of MPI implementations providing the full MPI-2 standard has grown considerably, i.e. a growing number of machines allow the use of one-sided communication operations. While this gives the application developer more freedom to implement his program as he thinks it would be appropriate (and "look fast") it also makes it considerably harder to aim for peak performance because it makes the performance model of the machine much more complicated. Therefore the results of a tool like SKaMPI 5 are essential for learning what's going on. In addition to the many already provided measurement methods it is very easy to write some new ones, which are specifically tailored and focused to the individual problems an application developer might face.

References

1. W. Augustin, M. Haller, M.-O. Straub, and T. Worsch. SKaMPI — towards version 5. In E. Krause, W. Jäger, and M. Resch, editors, *High Perf. Comp. in Science and Engineering '04*, pages 371–382. Springer-Verlag, 2005.
2. J.-C. Desplat. Porting SHMEM codes to MPI-2. Technical Report EPCC-TR01-01, EPCC, Univ. of Edinburgh, 2001.
3. E. Gabriel, G. Fagg, and J. Dongarra. Evaluating the performance of MPI-2 dynamic communicators and one-sided communication. In J. Dongarra, D. Laforenza, and S. Orlando, editors, *Proc. EuroPVM/MPI*, LNCS 2840. Springer-Verlag, 2003.
4. Intel. MPI benchmarks 2.3 distribution. Available at: `http://www.intel.com/software/products/cluster/downloads/IMB_2.3.tar.gz`, 2004.
5. G. R. Luecke, S. Spanoyannis, and M. Kraeva. The performance and scalability of SHMEM and MPI-2 one-sided routines on a SGI Origin 2000 and a Cray T3E-600. *Concurrency and Computation: Practice and Experience*, 16:1037–1060, 2004.
6. T. Matthey and J. P. Hansen. Evaluation of MPI's one-sided communication mechanism for short-range molecular dynamics on the Origin2000. In *Proc. PARA 2000*, LNCS 1947, pages 356–365, 2001.
7. K. Mohror and K. L. Karavenic. Performance tool support for MPI-2 on Linux. In *Proc. Supercomputing*, 2004.
8. Scalable Computing Laboratory. Netpipe. Available at: `http://www.scl.ameslab.gov/Projects/NetPIPE/`, 2004.
9. M.-O. Straub. Leistungsmessung einseitiger Kommunikation in MPI-Bibliotheken. Diploma thesis (in German), Fakultät für Informatik, University of Karlsruhe, 2004.
10. J.L. Träff, H. Ritzdorf, and R. Hempel. The implementation of MPI–2 one-sided communication for the NEC SX. In *Proc. Supercomputing*, 2000.
11. H. Uehara, M. Tamura, and M. Yokokawa. An MPI benchmark program library and its application to the Earth Simulator. In *Proc. Int. Symp. on High Performance Computing*, LNCS 2327, pages 219–230, 2002.
12. T. Worsch, R. Reussner, and W. Augustin. On benchmarking collective MPI operations. In D. Kranzlmüller, P. Kacsuk, J. Dongarra, and J. Volkert, editors, *Proc. EuroPVM/MPI*, LNCS 2474, pages 271–279, 2002.

A Scalable Approach to MPI Application Performance Analysis

Shirley Moore[1], Felix Wolf[1], Jack Dongarra[1],
Sameer Shende[2], Allen Malony[2], and Bernd Mohr[3]

[1] Innovative Computing Laboratory, University of Tennessee,
Knoxville, TN 37996-3450 USA
{shirley, fwolf, dongarra}@cs.utk.edu
[2] Computer Science Department, University of Oregon,
Eugene, OR 97403-1202 USA
{malony, sameer}@cs.uoregon.edu
[3] Forschungszentrum Jülich, ZAM, 52425 Jülich, Germany
b.mohr@fz-juelich.de

Abstract. A scalable approach to performance analysis of MPI applications is presented that includes automated source code instrumentation, low overhead generation of profile and trace data, and database management of performance data. In addition, tools are described that analyze large-scale parallel profile and trace data. Analysis of trace data is done using an automated pattern-matching approach. Examples of using the tools on large-scale MPI applications are presented.

1 Introduction

Parallel computing is playing an increasingly critical role in advanced scientific research as simulation and computation are becoming widely used to augment and/or replace physical experiments. However, the gap between peak and achieved performance for scientific applications running on large parallel systems has grown considerably in recent years. The most common parallel programming paradigm for these applications is to use Fortran or C with MPI message passing to implement parallel algorithms. The complex architectures of large parallel systems present difficult challenges for performance optimization of such applications. Tools are needed that collect and present relevant information on application performance in a scalable manner so as to enable developers to easily identify and determine the causes of performance bottlenecks.

Performance data encompasses both profile data and trace data. Profiling involves the collect of statistical summaries of various performance metrics broken down by program entities such as routines and nested loops. Performance metrics include time as well as hardware counter metrics such as operation counts and cache and memory event counts. Tracing involves collection of a timestamped sequence of events such as entering and exiting program regions and sending and receiving messages. Profiling can identify regions of a program that are consuming the most resources, while detailed

B. Di Martino et al. (Eds.): EuroPVM/MPI 2005, LNCS 3666, pp. 309–316, 2005.

tracing can help identify the causes of performance problems. On large parallel systems, both profiling and tracing present scalability challenges.

Collecting either profile or trace data requires the application program to be instrumented. Instrumentation can be inserted at various stages of the program build process, ranging from source code insertion to compile time to link time to run time options. Although many tools provide an application programmer interface (API), manual insertion of instrumentation library calls into application source code is too tedious to be practical for large-scale applications. Thus our tools support a range of automated instrumentation options. Once an instrumented version of the program has been built, only a few environment variables need to be set to control runtime collection of profile and/or trace data.

Collecting profile data for several different metrics on a per-process and per-routine basis, possibly for several runs with different numbers of processors and/or different test cases and/or on different platforms results in a data management problem as well as a presentation and analysis problem. Similar profile data may be collected by different tools but be incompatible because of different data formats. Our solution to the data management problems is a performance data management framework that sits on top of a relational database and provides a common profile data model as well as interfaces to various profile data collection and analysis tools. For presentation of profile data we have developed graphical tools that display the data in 2-dimensional and 3-dimensional graphs. Our tools also support multi-experiment analysis of performance data collected from different runs.

Event tracing is a powerful method for analyzing the performance behavior of parallel applications. Because event traces record the temporal and spatial relationships between individual runtime events, they allow application developers to analyze dependencies of performance phenomena across concurrent control flow. While event tracing enables the identification of performance problems on a high level of abstraction, it suffers from scalability problems associated with trace file size. Our approach to improving the scalability of event tracing uses call-path profiling to determine which routines are relevant to the analysis to be performed and then traces only those routines.

Graphical tools such as Vampir, Intel Trace Analyzer, and Jumpshot are available to view trace files collected for parallel executions. These tools typically show a time-line view of state changes and message passing events. However, analyzing these views for performance bottlenecks can be like searching for a needle in a haystack. Our approach searches the trace file using pattern-matching to automatically identify instances of inefficient behavior. The performance bottlenecks that are found and related to specific program call-paths and node/process/thread locations can then be focused on using one of the previously mentioned trace file viewing tools.

The remainder of this paper is organized as follows. Section 2 describes our automated approach to insertion of performance instrumentation and collection of performance data. Section 3 describes our performance data management framework. Section 4 describes our scalable approaches to analyzing profile data, including techniques for multi-experiment analysis. Section 5 describes our scalable automated approach to trace file analysis. Section 6 contains conclusions and directions for future research.

2   Automated Performance Instrumentation

TAU (Tuning and Analysis Utilities) is a portable profiling and tracing toolkit for parallel threaded and or message-passing programs written in Fortran, C, C++, or Java, or a combination of Fortran and C [3]. TAU can be configured to do either profiling or tracing or to do both simultaneously. Instrumentation can be added at various stages, from compile-time to link-time to run-time, with each stage imposing different constraints and opportunities for extracting program information. Moving from source code to binary instrumentation techniques shifts the focus from a language specific to a more platform specific approach.

Source code can be instrumented by manually inserting calls to the TAU instrumentation API, or by using the Program Database Toolkit (PDT) and/or the Opari OpenMP rewriting tool to insert instrumentation automatically. PDT is a code analysis framework for developing source-based tools. It includes commercial grade front end parsers for Fortran 77/90, C, and C++, as well as a portable intermediate language analyzer, database format, and access API. The TAU project has used PDT to implement a source-to-source instrumentor (tau_instrumentor) that supports automatic instrumentation of C, C++, and Fortran 77/90 programs.

The TAU MPI wrapper library uses the MPI profiling interface to generate profile and/or trace data for MPI operations. TAU MPI tracing produces individual node-context-thread event traces that can be merged to produce SLOG, SDDF, Paraver, VTF3, or EPILOG trace formats.

TAU has filtering and feedback mechanisms for reducing instrumentation overhead. The user can specify routines that should not be instrumented in a selective instrumentation file. The tau_reduce tool automates this specification using feedback from previously generated profiling data by allowing the user to specify a set of selection rules that are applied to the data.

3   Performance Data Management Framework

TAU includes a performance data management framework, called PerfDMF, that is capable of storing parallel profiles for multiple performance experiments. The performance database architecture consists of three components: performance data input, database storage, database query, and analysis. The performance profiles resident in the database are organized in a hierarchy of *applications*, *experiments*, and *trials*. Application performance studies are seen as constituting a set of experiments, each representing a set of associated performance measurements. A trial is a measurement instance of an experiment. Raw TAU profiles are read by a profile translator and stored in the database. The performance database is an object-relational DBMS specified to provide a standard SQL interface for performance information query. MySQL, PostgreSQL, or Oracle can be used for the database implementation. A performance database toolkit developed with Java provides commonly used query and analysis utilities for interfacing performance analysis tools. ParaProf (described in the next section) is one of the tools capable of using this high-level interface for performance database access. Other performance analysis tools that have been interfaced with PerfDMF include mpiP, Dynaprof, HPM, gprof, and KOJAK.

4 Scalable Display and Analysis of Profile Data

ParaProf is a graphical parallel profile analyzer that is part of the TAU toolkit. Figure 1 shows the ParaProf framework architecture. Analysis of performance data requires representations from a very fine granularity, perhaps of a single event on a single node, to displays of the performance characteristics of the entire application. ParaProf's current set of displays range from purely textual based to fully graphical. Many of the display types are hyper-linked enabled, allowing selections to be reflected across currently open windows.

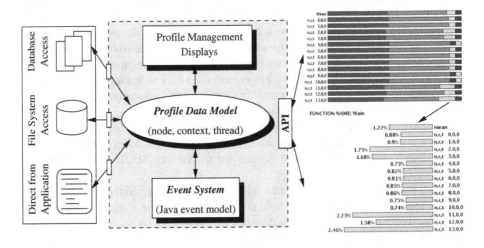

Fig. 1. ParaProf Architecture

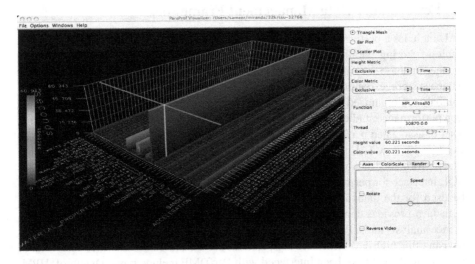

Fig. 2. Scalable Miranda Profile Display

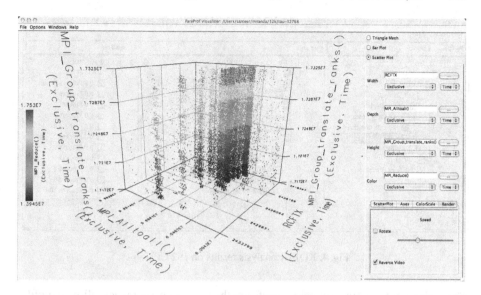

Fig. 3. 3-D Scatter Plot of Performance Metrics for 32K Miranda Processes

Recently, the TAU project has focused on how to measure and analyze larg-scale application performance data. A significant amount of performance data can be generated for large processor runs. We have been experimenting with three-dimensional displays of large-scale performance data. For instance, Figure 2 shows the entire parallel profile measurement for a 32K processor run. The performance events (i.e., functions) are along the x-axis, the threads are along the y-axis, and the performance metric (in this case, the exclusive execution time) is along the z-axis. This full performance view enables the user to quickly identify major performance contributors. Figure 3 is of the same dataset, but in this case each thread is shown as a sphere at a coordinate point determined by the relative exclusive execution time of three significant events. The visualization gives a way to see clustering relationships.

5   Automated Analysis of Trace Data

KOJAK is an automatic performance evaluation system for parallel applications that relieves the user from the burden of searching large amounts of trace data manually by automatically looking for inefficient communication patterns that force processes into undesired wait states. KOJAK can be used for MPI, OpenMP, and hybrid applications written in C/C++ or Fortran. It includes tools for instrumentation, event-trace generation, and post-processing of event traces plus a generic browser to display the analysis results. The instrumentation tools complement those supplied by TAU.

After program termination, the trace file is analyzed offline using EXPERT [5], which identifies execution patterns indicating low performance and quantifies them according to their severity. These patterns target problems resulting from inefficient communication and synchronization as well as from low CPU and memory performance.

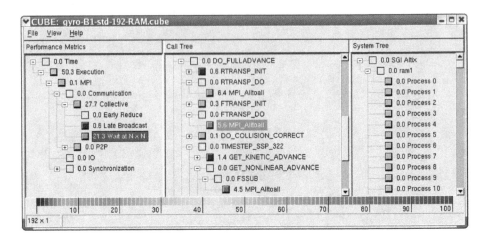

Fig. 4. KOJAK analysis results on 192 CPUs

The analysis process automatically transforms the traces into a compact call-path profile that includes the time spent in different patterns.

Finally, the analysis results can be viewed in the CUBE performance browser [4], which is depicted in Figure 4. CUBE shows the distribution of performance problems across the call tree and the parallel system using tree browsers that can be collapsed and expanded to meet the desired level of granularity. TAU and KOJAK interoperate in that TAU profiles can be read by CUBE, and CUBE profiles can be read by ParaProf and exported to PerfDMF.

We recently used KOJAK to investigate scalability problems observed in running the GYRO MPI application [1] on the SGI Altix platform using a specific input data set. We used TAU in combination with PDT to automatically insert appropriate EPI-LOG API calls into the GYRO source code to record entries and exits of user functions. Unfortunately, non-discriminate instrumentation of user functions can easily lead to significant trace-file enlargement and perturbation: A TAU call path profile taken of a fully-instrumented run with 32 processes allowed us to estimate the trace file size above 100 GB.

As a first result, we present an automated strategy to keep trace-file size within manageable bounds. It was notable that shortly-completed function calls without involving any communication accounted for more than 98 % of the total number of function-call events. Since the intended analysis focuses on communication behavior only, we automatically generated a so-called *TAU include list* from the call path profile using a script. The include list allowed us to instrument only user functions directly or indirectly performing MPI operations. Based on this include list we took trace files of with 32, 64, 128, and 192 processes. Trace file sizes varied between 94 and 562 MB and did not present any obstacles to our analysis.

GYRO's communication behavior is dominated by collective operations - in particular n-to-n operations, where every process sends to and receives from every other process. Due to their inherent synchronization, these operations often create wait states when processes have to wait for other processes to begin the operation. KOJAK defines

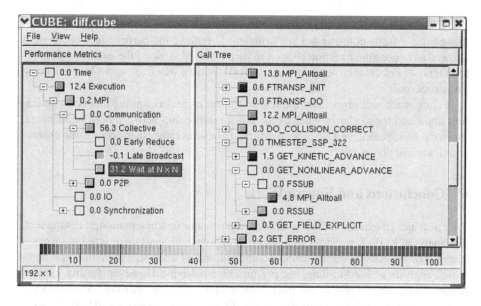

Fig. 5. Differences between the 128- and the 192-processor run

a pattern called *Wait at N × N* that identifies this situation and calculates the time spent in the operation before the last participant has reached it.

Figure 4 shows the KOJAK result display for the 192-processor trace. All numbers shown represent percentages of the execution time accumulated across all processes representing the total CPU-allocation time consumed. The left tree contains the hierarchy of patterns used in the analysis. The numeric labels indicate that 21.3 % was spent waiting as a result of Wait at N × N as opposed to 27.7 % spent in actual collective communication. The middle tree shows the distribution of this pattern across the call tree. Thus, about 1/2 of the collective communication time is spent in wait states, a diagnosis that is hard to achieve without using our technique.

To better understand the evolution of these phenomena as processor counts increase, we compared the analysis output of the 128-processor run against the 192-processor run using KOJAK's performance algebra utilities [4]. The difference-operator utility computes the difference between the analysis results belonging to two different trace files. The difference can be viewed using the KOJAK GUI just in the same way the original results can be viewed. Figure 5 shows the 128-processor results subtracted from the 192-processor results. According to our measurements with KOJAK, the total accumulated execution time grows by 72.3 % when scaling from 128 to 192 processes, indicating a significant decrease in parallel efficiency. Figure 5 (left tree) shows the composition of the difference classified by performance behavior with all numbers being percentages of the total time difference. About 2/3 can be attributed to waiting in all-to-all operations, while about 1/3 can be attributed to the actual communication in collective operations. The increase in computation is negligible.

After writing short script for EARL [2], a high-level read interface to KOJAK event traces, we found that the performance problems observed when running with

192 CPUs happen in relatively small communicators not exceeding a size of 16. Although this information was not yet sufficient to remove the performance problem, it allowed us to pose the question about it more clearly by showing the evolution of hard-to-diagnose performance behavior (i.e., wait states) in a way that cannot be done using traditional tools.

Future work will investigate the reason for the increased waiting and communication times and try to clarify whether there is a relationship between both phenomena. Since the code scales well on other platforms, platform rather than application characteristics might play a role.

6 Conclusions and Future Work

TAU provides an extensible framework for performance instrumentation, measurement, and analysis. KOJAK provides an automated approach to analysis of large-scale event traces. The benefits of our research include automated trace-size reduction and automated analysis of hard-to-diagnose performance behavior. However, further work is needed on integrating profile and trace data analysis and on supporting additional tools such as multivariate statistical analysis tools. Further work is also needed to process the trace files in a distributed parallel manner in order to scale to terascale and petascale systems of the future.

Acknowledgements

This research is supported at the University of Tennessee by the U.S. Department of Energy, Office of Science contract DE-FC02-01ER25490, and at the University of Oregon by the U.S. Department of Energy, Office of Science contract DE-FG02-05ER25680.

References

1. J. Candy and R. Waltz. An Eulerian gyrokinetic Maxwell solver. *J. Comput. Phys.*, 186:545, 2003.
2. N. Bhatia F. Wolf. EARL - API Documentation. Technical Report ICL-UT-04-03, University of Tennessee, Innovative Computing Laboratory, October 2004.
3. S. S. Shende. *The Role of Instrumentation and Mapping in Performan ce Measurement*. PhD thesis, University of Oregon, August 2001.
4. F. Song, F. Wolf, N. Bhatia, J. Dongarra, and S. Moore. An Algebra for Cross-Experiment Performance Analys is. In *Proc. of the International Conference on Parallel Processing (ICPP)*, Montreal, Canada, August 2004.
5. F. Wolf, B. Mohr, J. Dongarra, and S. Moore. Efficient Pattern Search in Large Traces through Successive Refinement. In *Proc. of the European Conference on Parallel Computing (Euro-Par)*, Pisa, Italy, August - September 2004.

High-Level Application Specific Performance Analysis Using the G-PM Tool*

Roland Wismüller[1], Marian Bubak[2,3], and Włodzimierz Funika[2]

[1] University of Siegen, Hölderlinstr. 3, D-57068 Siegen, Germany
[2] Institute of Computer Science, AGH, al. Mickiewicza 30, 30-059 Kraków, Poland
[3] Academic Computer Centre – CYFRONET, Nawojki 11, 30-950 Kraków, Poland
roland.wismueller@uni-siegen.de
{bubak, funika}@agh.edu.pl

Abstract. The paper presents an approach to overcome a traditional problem of parallel performance analysis tools: performance data often is too low level and cannot easily be mapped to the application, e.g. its execution phases. The G-PM tool offers the user an easy but flexible means to define his own high-level, application specific metrics based on existing metrics and application events. In a case study based on a real world medical application from the CrossGrid project, we demonstrate this concept as well as its usefulness in practice.

1 Introduction

Today, most of the applications that require high computing performance are based on parallel programming using the message passing paradigm, as it is supported by MPI. For this class of applications, tools that allow to measure and improve their performance characteristics are vital for the application's success. Generally, performance analysis tools can be based on three different techniques: tracing, profiling, and on-line analysis. The latter can be viewed as a compromise between tracing and profiling, since – as with profiling – the tool computes performance *metrics* (e.g. time spent in the MPI_Send routine) during the program's execution, instead of storing all communication *events*. On the other hand, – as with tracing – the information is still resolved in time, instead of just summarizing the whole execution. Different from both other approaches, on-line analysis tools present the performance metrics *while* the application is executing.

Although a number of sophisticated performance tools exists[1], it is still difficult for programmers to optimize their applications based on the provided information. This has two major reasons: First, the information is often too low-level, since it is usually related to communication or even hardware events. Second, linking the displayed performance data to the source code and the programmer's mental model of the application is rather difficult. The latter includes the structuring of the application's execution in well defined phases, e.g., iterations of a numerical solver, different phases within one iteration, etc. While tracing has some advantage here, since the traced events can contain a link to the

---

* Partially funded by the European Commission (project IST-2001-32243, CrossGrid) and KBN (grant 4 T11C 032 23).

[1] See e.g. [5] for a summary of tools in the context of Grid computing.

B. Di Martino et al. (Eds.): EuroPVM/MPI 2005, LNCS 3666, pp. 317–324, 2005.

source code, it suffers the problem of extreme amounts of data, which need to be stored and analyzed.

An ambitious approach to solve these problems is automatic performance analysis, which points the programmer to the exact cause and location of a bottleneck. Research in this field is done in e.g., the APART[2], KappaPi [2], Paradyn [8], PERIDOT [6], and SCALEA [10] projects. A more pragmatic solution is the provision of higher-level metrics. E.g., the EXPERT tool [15] computes reasons for performance loss and represents them in a three-dimensional hierarchy, which the user can navigate. Ideally, these higher-level metrics should not be hard-coded in the tool, but definable by the user. The G-PM tool presented in this paper exploits this idea. It features the Performance Metrics Specification Language PMSL to allow users to create high-level, application specific metrics, which also can relate performance data to program phases. The main contribution of this paper is a case study, which shows the usefulness of PMSL for the analysis of a real world application from the medical domain.

Configurable metrics have already been used in related tools: Paradyn uses the language MDL [7] to define all the on-line metrics it allows to measure. In a similar spirit, but using a trace based approach, EXPERT supports configurable metrics via the EARL language [14]. In both cases, however, metrics are defined at a low implementation-oriented level, unsuitable for supporting user-defined metrics. A high-level specification of performance *properties* used for automatic bottleneck detection is supported by ASL [3] and JavaPSL [4]. Although performance properties are different from metrics, some features of PMSL have their roots in ASL. Like G-PM, also Paraver[3] and Pablo[4] support user defined data analysis. In contrast to G-PM, these tools are based on off-line processing of trace data and offer weaker support for application specific metrics.

In the next section, we outline the main concepts of user defined metrics in G-PM. Section 3 then presents the mentioned case study on G-PM and PMSL. For details on the syntax, semantics, and implementation of PMSL we refer the reader to [12,11,1].

2   User Defined Metrics in G-PM

G-PM is an on-line performance analysis tool for parallel MPI applications on the Grid. It can measure a wide range of predefined metrics, mostly related to MPI usage (such as communication volume and delay). Metrics can be measured for the whole application, but can also be narrowed down along the three dimensions "location in the system", "location in the code" and "location in time". E.g., metrics can be measured for all processes, a single process, or a set of processes running on given hosts, for the whole program or just selected functions, and as an aggregate over the whole execution or as a time series. The necessary instrumentation for these measurements is performed automatically; the application just needs to be relinked with additional libraries (see [12] for details). As a pure on-line tool, G-PM performs a measurement only after the user explicitly requested it by specifying the metrics, the measurement parameters, and

---

[2] http://www.fz-juelich.de/apart/
[3] http://www.cepba.upc.es/paraver/
[4] http://www-pablo.cs.uiuc.edu/Project/Pablo/PabloDataAnalysis.htm

the required visualization. The measurement results can be displayed in various ways, e.g., bar graphs (c.f., Fig 1), multi-curve plots (c.f., Fig. 2), or matrix diagrams.

Besides predefined metrics, the G-PM tool also offers user-defined metrics in order to face the problems stated in Sect. 1. The ingredients of such a metrics are (1) *existing metrics*, (2) *application specific events* (which optionally can provide application specific performance data), and (3) *numerical operations*, including set operations for data aggregation in space and time. User defined metrics thus can interrelate and/or aggregate any already existing metrics. The important feature, however, is that they can interrelate performance metrics with application events. The key to this feature is a special operator (AT), which takes the value of a metrics when a specified event occurs. This allows to determine performance indicators at specific points in the program execution, and in turn to measure performance for program phases. It also allows to compute completely new metrics, which are derived from application events.

Naturally, the necessary instrumentation for application specific events can not be done fully automatically. In G-PM, the programmer has to insert special function calls (*probes*) at those places in the code where an event should be raised. The probe functions are generated automatically by the G-PM environment and are linked to the application. Each probe receives a *virtual time* and, optionally, application specific performance data as its parameters. The virtual time is an arbitrary, but monotonically increasing integer value, which can be used to determine associations between corresponding events in different processes, but also between different kinds of events, like the beginning and the end of a program phase.

Although inserting probes requires to recompile the application, it does not contradict G-PM's philosophy of on-line measurements, since a probe just marks an important place in the code; it does not yet define any metrics or measurements. Usually, a large variety of different metrics can be defined *at run-time*, based on the very same instrumentation. Via the PMSL language, these metrics are specified either interactively or by reading them from a file containing a previously assembled metrics library for a particular application. Unlike MDL and EARL, PMSL is purely functional and declarative, i.e., users do not need to care about complex implementation details, like combining data that originates from different hosts at different times. This makes PMSL rather intuitive to understand, thus, in the next section we will just present examples instead of a full description. A second attractive consequence of a functional language is the fact that it can easily be translated to a data-flow graph representation. This in turn allows an automatic, distributed evaluation of user defined metrics, as outlined in [11] and [13].

The use case in the next section shows that G-PM's combination of user-defined metrics and application specific events enables the creation of high-level, application oriented metrics, which are meaningful for application domain specialists and even application users, as well as a phase-based performance analysis of parallel programs.

3   A Use Case

Within the international Grid project CrossGrid[5], we analyzed the performance of a parallel application for pretreatment planning in vascular interventional and surgical

---

[5] http://www.crossgrid.org

procedures[6] [9]. It uses an iterative solver based on the Lattice Boltzmann method to simulate the blood flow in a patient's arteries (and envisioned bypasses). The application is based on a one-dimensional decomposition of the simulation volume. In each iteration of the solver, neighboring processes exchange their intermediate results using MPI_Sendrecv. Thus, the communication pattern is a bidirectional ring, which leads to a (loose) synchronization of the iterations in all processes.

The application is clearly structured into phases: First, each iteration itself is a relevant phase, second, the iterations are again structured into phases. Most importantly there is a compute phase which calculates one time step of the blood flow and an output phase which stores the flow data for subsequent visualization. In order to take this structure into account for the performance analysis, we inserted three probes into the solver's source code: one at the beginning of an iteration, one at the transition from compute to output phase, and a last one at the end of the iteration. All probes receive the iteration count as virtual time.

After this preparation it was possible to create several metrics useful for a domain specialist. A rather simple metrics computes the number of iterations executed so far:

```
Loop_Executions(Process p, TimeInterval t) : Unit("iterations")
{
    PROBE iteration_end(Process, VirtualTime);
    VirtualTime vt; Value[] val;
    val[vt] = 1 AT iteration_end(p,vt);
    return SUM(val[vt] WHERE val[vt].time IN t);
}
```

It simply produces a constant "1" at the end of each iteration and computes their sum over the measurement interval. This allows to display the application's progress in terms of the number of executed iterations. Fig. 1 shows the result of an experiment where we started the simulation on two different sites of the CrossGrid testbed. We can immediately see that on the Slovak site (labeled SK), the simulation runs more than twice as fast than on the Polish one (PL). This is mostly because of different hardware performance.

In this experiment, we also assessed the influence of load imbalance on the simulation performance. Originally, the simulation volume was distributed evenly among the processes. This may result in load imbalance, because the amount of computation depends on the actual geometry of the arteries. Since the iterations of the simulation synchronize via MPI_Sendrecv communication, the following metrics could be used:

```
Waiting_Time_in_Iteration(Process p, VirtualTime vt) : Unit("s")
{
    PROBE iteration_begin(Process, VirtualTime);
    PROBE iteration_end(Process, VirtualTime);
    return Sendrecv_delay(p, [START,NOW]) AT iteration_end(p, vt)
        - Sendrecv_delay(p, [START,NOW]) AT iteration_begin(p, vt);
}
Load_imbalance(Process[] procs, TimeInterval t) : Unit("s")
{
    Process p; Value[] imbal; VirtualTime vt;
    imbal[vt] = MAX(Waiting_Time_in_Iteration(p, vt) WHERE p IN procs)
            - MIN(Waiting_Time_in_Iteration(p, vt) WHERE p IN procs);
    return SUM(imbal[vt] WHERE imbal[vt].time IN t);
}
```

---

[6] http://www.crossgrid.org/products/applications/medical.html

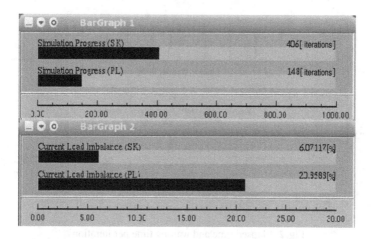

Fig. 1. Simulation progress and load imbalance on two different sites

The first metrics computes the waiting time in one iteration of a given process by subtracting the current total waiting time at the end of an iteration from the one at the beginning. The second metrics determines (for each iteration) the maximum and minimum waiting time among all processes, takes the difference, and then aggregates the result over the measurement interval. For the experiment in Fig. 1 we used a measurement mode that divides the result by the length of the measurement interval, thus, the result is the performance loss in percent caused by load imbalance. We realized that the original version had a performance loss of over 20% (bar labeled with PL). For comparison, we used an optimized data distribution in Slovakia (SK), which lead to a much better load balance with only 6% performance loss.

Although creating own metrics requires some skills, even end users from the application domain can use G-PM to check the performance of their simulations, since metrics, measurements and displays can also be loaded from a pre-assembled configuration file. The results are easy to understand, thanks to the high-level metrics.

In addition, user-defined metrics also allow to inspect the performance behavior in more detail. In Fig. 2, we used the `Waiting_Time_in_Iteration` metrics and a similarly defined one to measure the elapsed time and the waiting time for each iteration of the solver. In this graph, where each displayed value comprises exactly one iteration, we can see three effects: First, we see the load imbalance again (different waiting time in the four processes). Second, we realize that regularly, some iterations take more time to complete than usual. It turned out that this is due to some extra statistics computed in every 20th iteration. Finally, we can also see increased times in random iterations, which coincide with increased waiting time in all processes but one. This is caused by OS background activities on the compute nodes, which delay one of the simulation processes. As a result of the synchronous operation, all other processes have to wait for the delayed one. With a larger number of processes this effect can easily result in a notable performance degradation. Thus, we will try to loosen the synchronization between the simulation processes.

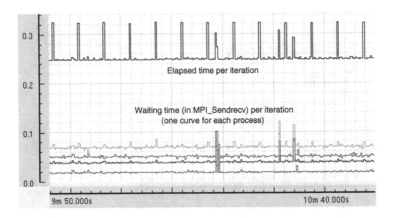

Fig. 2. Elapsed time and waiting time per iteration

In Fig. 3, we finally investigated the behavior of the compute and output phases in the solver's iterations via user-defined metrics similar to the ones shown before. We can see that the output phase is active only in the last iterations of a simulation, but then largely dominates the time and communication requirements. The output phases actually account for about 20% of the solver's total execution time. It turned out that this performance problem has two reasons: First, all the output is written by process 0 (this also causes the high amount of communication), second, the output is ASCII formatted, which makes it rather compute intensive.

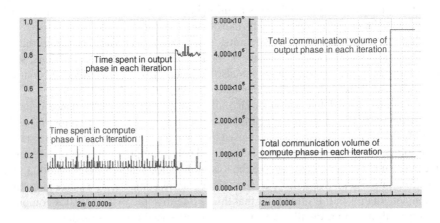

Fig. 3. Behavior of compute and output phase in each iteration

By enabling an already available binary output format and using an optimized data decomposition, we were finally able to speed up the solver by 14%. Further improvements will be possible by writing the output in parallel using, e.g., MPI-I/O.

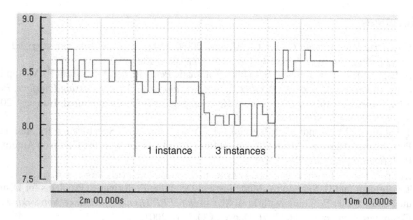

Fig. 4. Overhead of the measurements in Fig. 2

To complete our case study, we finally assessed the overhead induced in the application by the phase-based measurements of G-PM. For this purpose, we first measured the processing speed of the undisturbed application (in iterations per second) using G-PM. Then, after three minutes, we defined the measurements from Fig. 2. The result is shown in Fig. 4. In order to increase the measurement accuracy, we later defined another two instances of the same measurements. Now the performance decreased by about 6% (from 8.5 to 8 iterations/s), which shows that the overhead of one instance of the measurements is just 2%. While this is quite acceptable, it is nevertheless noticeable. The main reason for the overhead is the fact that the current version of G-PM evaluates user-defined metrics centrally, which requires each probe event to be sent to G-PM. We have recently implemented a distributed evaluation [13], which will considerably reduce this overhead in future versions of G-PM.

4   Conclusion and Future Work

The G-PM tool provides user defined metrics, which are based on existing metrics, but can also take into account application specific events. Case studies have shown that this concept allows to overcome two traditional problems of performance analysis tools. Firstly, the application programmer can define high-level metrics, which have a strong meaning in the application domain and thus are easy to understand by domain specialists. Second, they support the mapping between performance data and the application, by enabling measurements for specific, relevant phases of an application.

The current version of G-PM for MPI programs on the Grid is available under GNU Public License, together with extensive documentation. An adaptation of G-PM to support distributed Java/RMI applications is currently being prepared. In a recently finished prototype, an efficient, distributed evaluation of user-defined metrics has been implemented. Out next steps will be the further development of the distributed evaluation scheme, as well as the improvement of the metrics specification language PMSL.

References

1. *CrossGrid User Manual Guide: G-PM*, Nov. 2004. http://www.eu-crossgrid.
   org/user_manuals/CG2.4.1-v0.1-CYF-G-PMUserManual.pdf.
2. A. Espinosa, T. Margalef, and E. Luque. Automatic Performance Analysis of PVM Applications. In *Recent Advances in Parallel Virtual Machine and Message Passing Interface, Proc.
   7th European PVM/MPI Users Group Meeting*, pages 47–55, Balatonfüred, Hungary, 2000.
   Springer Verlag. LNCS 1908.
3. T. Fahringer, M. Gerndt, G. Riley, and J. L. Träff. Knowledge Specification for Automatic
   Performance Analysis. APART Technical Report, ESPRIT IV Working Group on Automatic
   Performance Analysis, Nov. 1999.
   http://www.fz-juelich.de/apart-1/reports/wp2-asl.ps.gz.
4. T. Fahringer and C. Seragiotto. Modeling and Detecting Performance Problems for Distributed and Parallel Programs with JavaPSL. In *9th IEEE High-Performance Networking and
   Computing Conference, SC'2001*, Denver, CO, Nov. 2001.
5. M. Gerndt et al. *Performance Tools for the Grid: State of the Art and Future*. Shaker Verlag,
   Aachen, Jan. 2004. http://www.lpds.sztaki.hu/~zsnemeth/apart/
   repository/gridtools.pdf.
6. M. Gerndt, A. Schmidt, M. Schulz, and R. Wismüller. Automatic Performance Analysis on
   Hitachi SR8000. In S. Wagner et al., editors, *High Performance Computing in Science and
   Engineering*, pages 443–452, Munich, Germany, 2003. Springer Verlag.
7. J. R. Hollingsworth, B. P. Miller, M. J. R. Gonçalves, Q. Xu, O. Naim, and L. Zheng. MDL:
   A Language and Compiler for Dynamic Program Instrumentation. In *Proc. International
   Conference on Parallel Architectures and Compilation Techniques*, San Francisco, CA, USA,
   Nov. 1997. ftp://grilled.cs.wisc.edu/technical_papers/mdl.ps.gz.
8. B. P. Miller et al. The Paradyn Parallel Performance Measurement Tools. *IEEE Computer*,
   28(11):37–46, Nov. 1995.
   http://www.cs.wisc.edu/paradyn/papers/overview.ps.gz.
9. P. Sloot, A. Tirado-Ramos, A. Hoekstra, and M. Bubak. An Interactive Grid Environment for
   Non-Invasive Vascular Reconstruction. In *2nd Intl. Workshop on Biomedical Computations
   on the Grid (BioGrid'04)*, Chicago, Illinois, USA, Apr. 2004. IEEE.
10. H.-L. Truong and T. Fahringer. SCALEA: A Performance Analysis Tool for Distributed
    and Parallel Programs. In *Euro-Par 2002 Parallel Processing, 8th International Euro-Par
    Conference*, pages 75–85, Paderborn, Germany, Aug. 2002. Springer-Verlag. LNCS 2400.
11. R. Wismüller, M. Bubak, W. Funika, T. Arodz, and M. Kurdziel. Support for User-Defined
    Metrics in the On-line Performance Analysis Tool G-PM. In *Grid Computing – Second Eu-
    ropean AcrossGrids Conference AxGrids2004*, pages 159–168, Nicosia, Cyprus, Jan. 2004.
    Springer-Verlag. LNCS 3165.
12. R. Wismüller, M. Bubak, W. Funika, and B. Balis. A Performance Analysis Tool for Inter-
    active Applications on the Grid. *Intl. Journal of High Performance Computing Applications*,
    18(3):305–316, Fall 2004.
13. R. Wismüller, H. Mehammed, M. Gerndt, and A. Bode. Performance Monitoring and Analy-
    sis for the Grid. In B. D. Martino et al., editors, *Engineering the Grid*. American Scientific
    Publishers, 2005. In print.
14. F. Wolf and B. Mohr. EARL - A Programmable and Extensible Toolkit for Analyzing Event
    Traces of Message Passing Programs. In A. Hoekstra and B. Hertzberger, editors, *Proc. of
    the 7th International Conference on High- Performance Computing and Networking (HPCN
    99)*, pages 503–512, Amsterdam, The Netherlands, 1999.
15. F. Wolf and B. Mohr. Automatic Performance Analysis of MPI Applications Based on Event
    Traces. In *Euro-Par 2000 Parallel Processing, 6th International Euro-Par Conference*, pages
    123–132, Munich, Germany, Aug. 2000. Springer Verlag. LNCS 1900.

ClusterGrind: Valgrinding LAM/MPI Applications

Brett Carson and Ian A. Mason

School of Mathematics,
Statistics and Computer Science, University of New England

Abstract. Debugging distributed applications using message passing libraries can be extremely difficult. We have implemented a set of tools collectively called `ClusterGrind` which interface to a GNU licensed debugger, `valgrind`, to ease the debugging process. By generating useful, customisable reports, we believe the time spent debugging large distributed Linux applications can be reduced significantly. Profiling the running programs is also possible to find coding inefficiencies, to aid in improving the overall application performance.

Keywords: `valgrind`, debugging, profiling, message passing, reporting.

1 Introduction

Debugging parallel applications is notoriously difficult, especially in distributed environments like the cluster. The same problems exist with distributed applications that are rampant in any other piece of software, but there is also an added level of complexity and difficulty. Errors can be difficult to reproduce, and the sources and locations of such errors troublesome to find. On occasion, said problems can be solved by reducing the number processors or cluster nodes, and reducing the size of the problem. However, this approach does not always reproduce the specific error(s) and can lead to time and effort being wasted. It is therefore desirable to maintain the conditions as close as possible to those causing the abnormal or unwanted behaviour, and debugging the application appropriately. Debugging may not be the only application monitoring required, performance profiling is also a serious issue that often requires attention.

This paper outlines a set of tools collectively called `ClusterGrind` [1] that can be used to enable `valgrind` [2,3] for debugging LAM/MPI [4,5] as well as PVM [5] applications on a cluster, and generate meaningful reports. Some past experiences and difficulties are discussed in section 2, along with a brief look at available commercial packages for distributed debugging. Section 3 provides a brief overview of `valgrind`, including its various functions, examples of its use, and the output reports that it can generate. This is followed by a walk-through of `ClusterGrind` in section 4, and examples of its use with LAM/MPI on the cluster. The examples focus on setting up the cluster environment for operation, collecting `valgrind` output, and generating meaningful reports based upon this output.

B. Di Martino et al. (Eds.): EuroPVM/MPI 2005, LNCS 3666, pp. 325–332, 2005.

2   Debugging Message Passing Applications

Past experience in debugging applications [6,7] using message passing libraries including LAM/MPI as well as PVM has resulted in many hours spent bug-hunting for what are the majority of the time, very small errors in the code. Often one process out of hundreds distributed across a cluster will die silently, and logging facilities may not provide any clues as to how or why the process died. This can result in the use of debugging print statements littered throughout code, and then further sifting through the output generated in an attempt to find the code location where a crash has occurred. Such a tedious approach does not reveal the reason for the crash, or how it actually happened. *Single program multiple data*, SPMD, applications can be even more troublesome, as the different nature of the data on different processes can cause unpredictable outcomes.

Currently, the two leading debuggers/profilers for cluster computing are Totalview [8] and Distributed Debugging Tool, DDT [9]. Both packages provide debugging and profiling options for multiple programming languages including C and C++, MPI programs, and multi-threaded programs. Unfortunately both of these packages are commercially licensed, and limitations apply to the number processors/processes that can be operated under each environment depending on the chosen licensing scheme. It is this limitation that has led our work in adding functionality to a GNU licensed debugger and profiler, valgrind. Whilst valgrind does not natively support cluster applications, it does contain functionality that can be adapted to the distributed environment.

3   Valgrind - An x86 Profiling and Debugging Tool

Valgrind is a programmable framework for creating program supervision tools such as bug detectors and profilers [2]. Although originally designed for the Linux/x86 platform, recent development has been aimed toward architecture abstraction. Reasonably stable ports to FreeBSD have been produced, and support for 64-Bit AMD processors and PowerPC/Linux will appear in future versions.

3.1   Basic Features of Valgrind

Based upon a virtual CPU approach, valgrind is designed to instrument binary code and emulate various operations such as tracking the malloc/new and free/delete memory operations. This method does tend to slow down the execution of programs being debugged/profiled, but the benefits of error detection may reduce total debugging time. These operations are performed by the valgrind core, but to produce meaningful information about the execution of a process special tools need to be used, some of which are described in section 3.2.

Highly detailed verbose analysis is provided by valgrind and its various tools. The output of the analysis can be managed using valgrind's logging functionality, by using either log files for each process, or sending the messages across the network to a listener on a remote host. These methods for logging can be used to reduce the cluttered output that can be produced when many processes are being traced by `valgrind`.

Whilst `valgrind` is first and foremost a serial debugger, support is provided for tracing and profiling different thread implementations and child processes created by the `clone` and `fork` system calls. Tracing multiple processes in such a way is what allows us to use `valgrind` on the cluster for different message passing implementations.

These are some of the features that make `valgrind` an attractive free package for debugging distributed applications on clusters. By utilizing the available tools, cluster applications using message passing libraries can be debugged and profiled without too much difficulty on the user's behalf.

3.2 Valgrind Tools

The key to producing results with `valgrind` is the use of tools. Each tool interacts with the `valgrind` core to gain meaningful information about the program being profiled. The interaction will usually consist of the tool registering events that it is interested in, and the core subsequently providing information to the tool when the events occur. The events could be the application allocating memory via the `calloc` or `malloc` system calls, signals being delivered, or various thread operations. Some of the tools useful for debugging and profiling distributed software with the `valgrind` core are:

- **Memcheck** - a memory debugger, tracks allocations and deallocations, reads and writes to the heap, and can perform leak analysis.
- **Cachegrind** - a detailed cache profiler, emulates the various CPU caches to detect cache misses in code.
- **Massif** - a heap profiler, it monitors the program's heap at various stages of execution, and in doing so is able to provide graphs of memory usage over time.

3.3 Valgrind Examples

Programs that require debugging or profiling under `valgrind` are passed to it as an argument, along with a number of options including the tool to be used. There may also be some specific tool options, for example, one may turn the leak checking option on when using the `memcheck` tool. Here is a simple example using the `memcheck` tool with leak checking enabled and verbose output on the program `memerror`:

```
valgrind -v --tool=memcheck --leak-check=yes memerror 5 10
```

During execution, `valgrind` will report any errors generated by the application, as well as any errors that occur within the shared libraries that the program links to. This can result in a lot of errors being reported, but it is possible to suppress certain errors. For the developer this is often a necessity, as they may not have the ability to fix the errors in the shared libraries. `Valgrind` is capable of generating very detailed reports depending on the tool being used. The output generated by the command given above would produce output including some useful error messages:

```
==4490== 10 errors in context 8 of 9:
==4490== Invalid write of size 4
==4490==    at 0x80495C5: main (memerror.c:59)
==4490== Address 0x1B92F094 is 4 bytes after a block of size 80 alloc'd
==4490==    at 0x1B9053FD: calloc (vg_replace_malloc.c:176)
==4490==    by 0x804956F: main (memerror.c:52)
```

The offending piece of code causing the error can be seen in Fig 1. Here we can see that the write error is happening because the assignment is going beyond the end of the array, which can be easily fixed by changing the stopping condition to j<dim.

```
for(i=0;i<dim;i++){
  v = (double*)calloc(sizeof(double), dim);        //creates leaks
  if(v == NULL){
    fprintf(stderr,"calloc failed: v\n");
    exit(1);
  }
  for(j=0;j<=dim;j++)
    v[j] = 0;                                        //write errors
}
```

Fig. 1. Troublesome code of the memerror program

Included with the valgrind command was the option --leak-check=yes. This option will keep track of mallocs and frees, and report on any memory leaks found during the execution of the program, marking memory as definitely lost, possibly lost, or still reachable (there is still a pointer to the allocated memory). This information is included as part of the leak check summary:

```
==4490== LEAK SUMMARY:
==4490==    definitely lost: 720 bytes in 9 blocks.
==4490==    possibly lost:   4096 bytes in 1 blocks.
==4490==    still reachable: 29248 bytes in 8 blocks.
```

The sources of the listed leaks are also supplied:

```
==4502== 720 bytes in 9 blocks are definitely lost in loss record 7 of 10
==4502==    at 0x1B9053FD: calloc (vg_replace_malloc.c:176)
==4502==    by 0x804956F: main (memerror.c:52)
```

The v array is reallocated at the beginning of each iteration but never freed, so pointers to the allocated blocks will no longer exist.

4   ClusterGrind

Whilst valgrind is a very useful debugging and profiling package, using it on the cluster can be somewhat cumbersome. Tedious setup of scripts and configuration may be required, which is dependent on the message passing software used. Such setup needs to be performed when any of the options to valgrind change, such as the tool being used, general options, or tool-specific options. Collecting and reporting on the output can also be difficult, as there are many sources for it distributed across a number of hosts. These inconveniences have led to the development of ClusterGrind, a set of tools designed for dealing with the issues described. There are three distinct parts of ClusterGrind:

1. Setup and configuration of the message passing environment for valgrind. This tends to vary depending on the message passing software used, so there are a number of different implementations of the setup tool. The task of this tool is to take a user's valgrind options, and any configuration for their message passing software, and combine the two in such a way that any processes using the particular message passing software will be running under valgrind. An implementation of this utility for LAM is discussed in greater detail in section 4.1.
2. Collection of valgrind output. This involves a simple listener running on a single host, which will collect and store output sent to it across the network. The setup tool forces any process to send valgrind output to the listener. The listener sorts the output it receives and stores it in a way to make reporting much simpler.
3. Reporting on the valgrind output. This tool enables the user to look for specific errors, or generate summaries of all errors found. It also allows the user to locate a particular error of interest, without having to search through the output of every process.

The goal of ClusterGrind is not to necessarily create a free clone of the earlier mentioned commercial debuggers, but to provide a method for using the valgrind framework with LAM/MPI programs. This allows specific functionality to be provided by tools that can be implemented when required.

4.1   Setup Tools

For valgrind to operate on LAM/MPI processes, the LAM environment needs to be specially configured. Ideally, this configuration should be transparent to the user so as to be able to minimise any possible changes an application programmer has to make to the code in order for it to undergo debugging or profiling. It is necessary to run valgrind on the new processes at time of creation, which normally means the daemon has to have some method for invoking valgrind.

As mentioned in section 3.1, valgrind can trace children created by the fork() and clone() system calls. In LAM processes are created in this way by daemons, whether they be on the local node or a remote node. The idea is then, that the daemons be run under valgrind as no option exists for the processes to be started under valgrind.

Setting Up the LAM/MPI Environment. LAM allows MPI tasks to be spawned under a debugger. This functionality is enabled when processes are invoked with the mpirun command. The debugger will be the process spawned by mpirun, and it must at some stage start a MPI program [10]. Unfortunately, this does not cover any processes started with the MPI_Comm_spawn or MPI_Comm_spawn_multiple functions. The solution to this issue is to run valgrind on the LAM daemon, lamd, with the option enabled to trace its children so that both dynamically spawned processes and processes started via mpirun will start under valgrind.

Lamboot. The LAM universe or environment is created by running the lamboot command. Given a file containing a list of hosts and possible options, lamboot will add the hosts to the universe, allowing any processes to be created on those hosts. This works by lamboot executing a LAM program called hboot on each host it encounters in the hostfile. The hboot program is used to start lamd with options set in a configuration file called lam-conf.lamd. A default configuration is located in /etc/lam-conf.lamd, but an instance of the file in the user's home directory will be given precedence. It is this configuration file which is manipulated to enable all lamds in the universe to be executed under valgrind. The lam-conf.lamd file used with valgrind looks like:

```
/tmp/vg_wrapper/valgrind-wrapper.pl lamd $inet_topo \
$debug $session_prefix $session_suffix
```

This single line tells hboot to run a valgrind wrapper script on lamd and it's arguments, which are the variables identified with the $ prefix. These variables are known to hboot, and are substituted when the valgrind wrapper is executed. It is possible to run valgrind directly, however this is not feasible as valgrind's command line options would conflict with the hboot options.

After executing lamboot, each of the daemons in the LAM universe will be running under valgrind. The --trace-children=yes option for valgrind is automatically set so that any applications started by either mpirun, MPI_Comm_spawn and MPI_Comm_spawn_multiple will also be running under valgrind.

Generating Valgrind Scripts. The valgrind wrapper discussed in the previous section is dynamically generated when the LAM universe is setup. From the user-provided options, the script is generated along with the required options:

> --trace-children=yes this enables all processes started by the daemons to be traced by valgrind.
> --log-socket this option will write the valgrind output to a socket, given as host:port.

The wrapper script is then copied to all hosts in the universe, and this script will be pointed to in the lam-conf.lamd file, which is copied to the user's home directory on all nodes.

4.2   Collecting and Reporting Debugging Information

Depending upon the options passed to `valgrind`, a great amount of detailed debugging output can be generated. Under normal use, this information will be printed to the terminal where `valgrind` was launched from. However, in cases where many processes are involved, and where the processes are distributed across a network as when `MPI` is involved, different measures must be taken to collect and organise the debugging information in a meaningful way.

Recall from section 4.1 that the `--log-socket` option is included in the `valgrind` wrapper script. This option redirects debugging output to be written across the network, to an address where a *listener* is running. The listener is a simple server which waits for running `valgrind` processes to send it large chunks of debugging text. The listener sorts the text by host origin, and creates an output file for each process.

After collecting and sorting the debugging output from the various processes distributed across the cluster, reports can be generated to facilitate easier navigation through the files of process information. By examining each process output file one at a time, *eXtensible Markup Language*, XML, documents are produced. This allows custom-formatting of the reports to be generated. Currently, the most user-friendly form of report based on the generated XML is a HTML document. Each node of the cluster is represented in a tree, with each tree element being able to be expanded to display the list of processes that were run under `valgrind` on that node, as seen in Figure 2. When the report is generated, the user is able to specify certain errors of interest which will be highlighted like the first error in the example shown. Beside each process is a number in parentheses, this relates to the number of errors of interest found for that particular process.

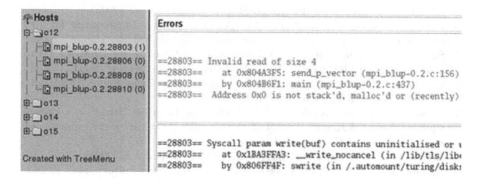

Fig. 2. The `ClusterGrind` HTML report - cluster nodes are represented as elements in a tree menu

The `valgrind` output for each process can be viewed by choosing the desired process from the list. Output shown will be filtered to show error contexts and the counts of each error, and the various summaries which are provided by the specific tool being used. To provide specific information on process interaction, a tool for monitoring MPI function calls is required, which is currently in early stages of developement.

5 Conclusions and Future Directions

Early development versions of ClusterGrind have proven very useful in minimising the amount of setup required by the user to perform debugging and profiling using valgrind on LAM/MPI applications. The collection and reporting functions work well, with some customisation available for reports, to highlight specific errors occurring in processes. ClusterGrind is available at:

http://mcs.une.edu.au/~bcarson/ClusterGrind/,

and all feedback is certainly welcomed.

Possible future improvements include integration into one or more of the graphical user interfaces available to valgrind. Support for different MPI implementations will also be considered if there is demand for it. Supporting different cluster configurations is desirable, as at this stage testing has only been performed on a homogeneous Beowulf cluster.

References

1. Brett Carson, B., Mason, I.A.: ClusterGrind.
   http://mcs.une.edu.au/~bcarson/ClusterGrind/ (2005)
2. Nethercote, N., Seward, J.: Valgrind: A Program Supervision Framework. Electronic Notes in Theoretical Computer Science **89** (2003)
3. Seward, J.: The Valgrind Homepage. http://valgrind.org/ (2005)
4. Burns, G., Daoud, R., Vaigl, J.: LAM: An Open Cluster Environment for MPI. Proceedings of Supercomputing Symposium (1994) 379–386
5. Squyres, J.M., Lumsdaine, A.: A Component Architecture for LAM/MPI. Proceedings of the 10th European PVM/MPI Users' Group Meeting (2003) 379–387
6. Carson, B., Murison, R., Mason I.A.: Computational Gains Using RPVM on a Beowulf Cluster. R News **3** (2003) 21-26,
7. Carson, B., Murison, R., Mason I.A.: Estimating Breeding Values on a Beowulf Cluster. Technical Report, School of Mathematics, Statistics and Computer Science, University of New England (2005)
8. Etnus LLC: The Totalview Debugger. http://www.etnus.com (2005)
9. Allinea Software: Distributed Debugging Tool. http://www.absoft.com/ (2005)
10. Open Systems Lab: LAM/MPI User's Guide (2004)

MPISH2: Unix Integration for MPI Programs

Narayan Desai, Ewing Lusk, and Rick Bradshaw

Mathematics and Computer Science Division,
Argonne National Laboratory, Argonne, IL 60439

Abstract. While MPI is the most common mechanism for expressing parallelism, MPI programs remain poorly integrated in Unix environments. We introduce MPISH2, an MPI process manager analogous to serial Unix shells. It provides better integration capabilities for MPI programs by providing a uniform execution mechanism for parallel and serial programs, exposing return codes and standard I/O stream information.

1 Introduction

The shell is the most familiar interface to Unix systems. In general, it is the first contact that users have with Unix systems. Its ubiquity makes it the dominant mechanism through which command execution occurs.

Unix shells provide a rich environment for task automation, exposing command exit codes, providing control flow constructs, and organizing disparate programs into complex command pipelines. Users are familiar with the decomposition of complex tasks into the invocation of single-function utilities using these mechanisms.

While MPI is not as ubiquitous as Unix shells, it is the dominant mechanism used to express parallelism in scalable applications. Many high-performance implementations of MPI exist; indeed, MPI is so pervasive that a good MPI implementation is frequently cited as one of the requirements for new large-scale computational science machines.

Unfortunately, process management systems that can start MPI programs have not provided or exposed sufficient information for their composition with their serial analogues or even with each other. To address this issue, we have implemented MPISH2, a MPI process manager that provides a user interface and composition capabilities nearly identical to the Bourne shell.

2 Related Work

Unix shells have long been studied. Starting with the original shell included with early Unix systems [14], shells have been augmented into relatively full-featured programming languages, including data types [9]. Because of the familiarity of the shell interface to Unix users, many attempts have been made to present a shell-like interface for program execution on parallel systems.

B. Di Martino et al. (Eds.): EuroPVM/MPI 2005, LNCS 3666, pp. 333–342, 2005.

- PDSH [12] is a program that uses `rsh` or `ssh` to execute tasks in parallel on many systems.
- The C3 tools [8] provide a similar execution mechanism that also runs tasks through `rsh`.
- Gridshell [15] provides a shell-like interface that enables access to Grid resources, including the queueing of jobs. It doesn't natively support the execution of parallel process; rather, it is subject to the limitations of the underlying resource management system used to implement this functionality.

These tools do an admirable job of starting processes scalably; however, they do not expose any of the Unix process information needed to embed parallel commands in more complex execution units. Discrete exit statuses are not returned for each process executed. Most important, none of these tools supports MPI process startup.

Historically, MPI startup mechanisms have scaled poorly and performed badly overall [5]. Two systems have addressed these issues over the past several years.

- MPD [4] uses a group of daemons, arranged in a ring topology, to scalably start MPICH2 processes.
- YOD [2] provides similar capabilities in the Cplant software stack.

Both of these process management systems provide highly scalable process startup services needed to start MPI processes, but neither system provides adequate information for use in shell-style programming. MPD provides access to all exit statuses and to standard I/O multiplexed into single streams. YOD provides similar access to standard I/O but fails to provide any access to return codes.

The work we present here has been motivated largely by the gains in system software scalability afforded by the use of MPI in system tools [6,7]. This approach also has proven quite positive in terms of overall performance gains. More surprising, tools implemented by using MPI-based scalable components have proved far easier to troubleshoot and debug than their ad hoc analogues. The need to execute large numbers of small scalable tools brings execution issues clearly into focus.

3   Design

When we were considering how to integrate MPI programs into a Unix environment, our highest priority was to retain standard Unix shell semantics. The overall goal was to support the execution of parallel programs using an interface indistinguishable from that used for serial programs. With such a uniform execution interface, parallel reimplementations of serial utilities could be automatically used by existing scripts.

Our initial implementation of MPISH [7] had all of the startup functions required to natively execute MPICH2 programs, however, the specification of execution locations was unwieldy. For that reason, we redesigned in the interface and input language for MPISH2.

The Bourne shell [1] was chosen as the language basis for our shell. Two major aspects of the Bourne shell are important: process semantics and control flow constructs. Unix process semantics provide access to information about child processes, including access to return codes, the ability to setup child process environment variables, and the ability to arrange commands into command pipelines that run concurrently. However, this data is available only for child processes that have been directly started. Without the existence of a Unix parent/child relationship, this information is not available and cannot be influenced in any way. Hence, the preservation of this relationship was an important design goal.

The control flow constructs available in the Bourne shell are fairly standard, including while, if, for, and case. Since these are the real workhorses of shell scripting, we attempted to keep their semantics as close as possible to the Bourne shell. However, minor enhancements were required in order to support startup of parallel processes.

3.1   MPISH2: A Parallel Shell

The difference between a normal shell and MPISH2 is that MPISH2 is a parallel program, consisting of multiple communicating Unix programs. A shell script, given to MPISH2, is executed by all of the MPISH2 processes. The MPISH2 processes communicate with each other (in a scalable fashion) using MPI. That is, MPISH2 is itself an MPI program. Therefore, MPISH2 must be started by the startup mechanism of the proper MPI implementation. We assume in this paper that mpiexec invokes this mechanism. Thus, a 100-process instance of MPISH2 is started by a command line something like the following.

```
mpiexec -n 100 mpish2
```

In a cluster environment, the specification of which nodes MPISH2 is run on depends on the particular MPI implementation being used. We have used MPICH2 [11], but MPISH2—being an MPI program—can be run by using any MPI implementation. Note, however, that because of the nonstandard nature of MPI startup, programs started by MPISH2 must use MPICH2.

MPISH2 scripts are Bourne-shell scripts (with some extensions described in Section 3.2) that are presented to the standard input of each MPISH2 process. MPISH2 must be parallel in order to properly provide all information about child processes. For example, using a traditional MPI process manager to run two parallel programs in a pipeline would look like the following.

```
mpiexec -np 10 prog1 | mpiexec -np 10 prog2
```

This command runs prog1 and sends the standard output of the first mpiexec to the second invocation of mpiexec. Handling of standard output is not specified by the MPI standard; however, many MPI process managers provide multiplexed standard output from all processes to the standard output of mpiexec. Likewise, mpiexec typically, though not universally, sends standard input of mpiexec to some number of the parallel process instances.

Under `MPISH2`, a similar command is used, together with a process management system for `MPISH2` startup.

```
mpiexec -np 10 mpish2
```

Once `MPISH2` is running, a command pipeline can be executed by using the following script.

```
prog1 | prog2
```

This script is run by every `MPISH2` instance, resulting in 10 instances of both `prog1` and `prog2`, connected rankwise into a pipeline. That is, standard output produced by the rank 0 instance of `prog1` is fed into the standard input of the rank 0 instance of `prog2`, and so forth. Additional utilities are provided, allowing interrank manipulation of I/O streams. These execution semantics provide more flexibility than those afforded by traditional process management systems.

3.2   Enabling Parallelism

Parallel process managers work in much the same way as serial process managers. They are responsible for post-fork/pre-exec process setup and the setup of standard I/O. The main difference between serial and parallel process managers is the need for parallel library bootstrapping. This bootstrapping consists of two main parts: the description of the parallel process topology and the communication setup.

Many mechanisms describe initial process topology at the time of parallel process startup. Typically, the topology specification consists of process count and some set of resources, usually a list of nodes on which the processes should be executed. This corresponds closely to the common arguments to `mpirun`. Alternatively, one can use `mpiexec`, specified by the MPI standard [10], for supplying the same data. Whatever the input format, this information is used for the same purpose: the description of MPI_COMM_WORLD for the new process. Each communicator has a specific size, and each component process has a specified rank in that communicator. This initial topology description is what differentiates one 32-node program from thirty-two 1-node programs.

`MPISH2` describes the initial communicator in terms of the parallel execution context of the client program. The notion of control flow groupings is maintained across the parallel shell. For example, if a parallel program is run on the first line of a script, it will be run on all processors, with an initial communicator identical to MPI_COMM_WORLD of the parent shell. Control flow constructs all affect this execution context for parallel programs. Their behavior can be most easily described in terms of `MPI_Comm_split`:

- *if* performs a two-way split, corresponding to the truth value of the predicate. Programs run in either branch will be grouped into parallel processes with the other processes executed in the same branch. For example, when *if* is executed in an 8-process context, resulting in a 4-node true, 4-node false split, processes run in the true branch will be grouped into 4-node parallel processes with the other processes executed on the true branch. Similarly, the processes executed on the false branch will be grouped into a 4-process parallel process with the others started on the false branch.

- *case* performs an *N*-way split, operating similarly to *if*.
- *while* creates an execution context corresponding to all ranks for which the condition evaluates as true. All programs run in each iteration will be grouped according to this initial evaluation. The condition will be evaluated at the start of each iteration, continuing until all ranks evaluate false.
- *for* has no effect on parallel execution context because it is not conditional.

Note that each control flow statement now implicitly includes a synchronization barrier at its conclusion. This approach has the distinct advantage of retaining the character of serial Bourne shell control flow operations. In fact, for one-node executions of MPISH2, the behavior is precisely that of a serial Bourne shell.

The second important aspect of parallel process startup is communication bootstrapping. For disparate processes to begin acting as a single parallel entity, communication must be established. This is accomplished in different ways with different parallel libraries. MPICH2 uses an interface called PMI, or Process Manager Interface, to provide this information to client programs. PMI takes the form of a distributed database, providing standard *put, get* and *fence* operations. The client program is provided with connection information for its PMI instance and can use that data to connect to other processes.

4 Implementation

The implementation of MPISH2 is based on a modified version of the Minix [13] shell, included with Busybox [3]. Three main modifications have been made.

The first was driven by the fact that MPISH2 is a parallel, not serial, process. A parallel execution context—that is, a grouping of discrete processes in order to form a parallel process—must be tracked on each instance of MPISH2. Initially, it corresponds to MPI_COMM_WORLD; however, as the script executes, the execution context is modified by control flow constructs, as described in Section 3.2. Changes in the parallel execution context are tracked by using an MPI communicator. This communicator is passed to any new PMI instances created, thereby maintaining cohesion between parallel processes executed in the same context.

The second modification was the creation of a PMI implementation to service requests from client processes. In order to support parallel library bootstrapping, a discrete PMI implementation is provided for each program started by the shell. Setup of this instance consists of initial data structure creation and socket setup. During client execution, the client program will connect and submit commands. Many of the commands, like *put*, which stores a value in a distributed database, will be serviced locally; however, some, like *get*, or *fence* may require communication with other parts of the same PMI instance. All communication operations are implemented by using MPI collective and asynchronous operations. *Fence* is implemented by using MPI_Barrier. The implementation of *get* is more complicated. When a PMI instance receives a *get* request, it checks whether the value is already stored locally. If it is, the request is immediately serviced. If not, a message is sent to the PMI instance with the next higher rank. Each process also receives queries for unknown values asynchronously. If the local process has the value, it responds to the querier; otherwise, it forwards the request to the next rank in the PMI instance. Disparate PMI instances in the same MPISH2 instance are differentiated based on a private communicator. This communicator is MPI_Comm_dup'ed at PMI instance initiation time, so each PMI instance has a unique one.

The third, and perhaps most complex, modification was to the control flow construct to provide topology descriptions for client processes. In a typical serial shell, control flow constructs use only return codes and have no side effects. In MPISH2, however, control flow constructs also affect the parallel execution context by calling MPI_Comm_Split after predicate execution. For example, in serial shells, the shell executes the *if* predicate and either the true or false branch depending on a zero or nonzero return code, respectively. MPISH2 executes the same operations, but with the addition of a call to MPI_Comm_Split using zero/nonzero exit status. Other control flow constructs were similarly modified.

None of these modifications proved complicated, and the overall semantics of the MPISH2 remains very close to the semantics of the Bourne shell. At the same time, these modifications provide a wealth of new capabilities to Unix users.

4.1   Utilties

A parallel execution environment isn't really complete without a set of parallel programs useful for writing basic programs. These programs are analogous to test or wc for serial shells. We have implemented a variety of small utilities, suffixed with the .mpi extension, to address this issue. The following is a list of basic parallel commands, with a short description of each.

- rank.mpi displays the process's rank in the current execution context.
- size.mpi displays the size of the current execution context.
- once.mpi exits with a return code of 0 once per physical node present.
- zoom.mpi provides access to scalable numeric reductions for the provided argument.
- pflatten.mpi sends all stdout streams to process 0.
- ptee.mpi forwards stdin from process 0 to all processes. It functions like a parallel version of tee.
- pcoalesce.mpi coalesces stdout from all nodes, producing rank delimited lines on processor 0.
- bcast.mpi broadcasts the data from one process, specified as an argument, to all other processes. This data is produced on stdout.
- stagein.mpi downloads a file from a http server and broadcasts to all nodes, eventually writing it to disk on each.
- stageout.mpi uploads files, tagged with rank, to the fileserver from all clients.

- `rsync.mpi` synchronizes files from process 0 to all other processes. This program can handle all regular and special files.
- `time.mpi` times the execution of a parallel program, producing a single wall time result.

Each of these programs is a simple MPI program. Nothing special is required to write a utility, as MPISH2 can run arbitrary MPI programs.

5 Usage Examples

MPISH2 is useful across the same broad range of problems as are standard shell scripts, with the added ability to run concurrent, parallel programs. It can easily be used for tasks ranging from the most trivial to those that can strongly benefit from access to parallelism and scalable tools.

5.1 Job Script

This example is a job script for a queueing system. This script runs the prologue, epilogue and file staging commands once per physical node (hostname). Of these commands, the prologue and epilogue are serial, while the file staging commands are parallel. Once setup has completed, the user job is run (under the user's UID), and cleanup is performed. Not only can serial and parallel programs be interchanged, but standard shell scripting mechanisms (like the use of su) can also be used with parallel programs.

```
#!/usr/bin/env_mpish2
user="${1}"
userscript="${2}
indir}="${3}"
outdir}="${4}

once.mpi
once=' '${?}''
if[  ''${once}''-eq 0] ; then
    #_run_the_prologue_once_per_node
    /usr/sbin/prologue
    if [! -z "${indir}" ] ; then
        su"${user}" stagein.mpi "${indir}"
    fi
fi

su "${user}" mpish2 "${userscript}"

if [''${once}'' -eq 0]; then
    if[! --z " ${outdir}"]; tnen
        su "${user}" stageout.mpi "${outdir}"
```

```
    fi
    /usr/sbin/epilogue
fi
```

Several active execution contexts are used in this program. Two instances of a context containing each physical node are created by the script. The first is used for job setup (e.g., prologue and file staging), and the second is used for job cleanup. The user's job script is executed in the global execution context.

5.2 Benchmarking Scripts

This example provides a basic illustration of concurrency. Benchmarking scripts are often implemented as a *for* loop that sequentially executes program runs with different sizes, for example, a script such as the following.

```
#!/bin/sh
for i in 2 4 8 16 32; do
 time mprium -np $i program
done
```

Such a script does a reasonable job of running benchmarks; however, numerous processor resources are wasted in the first few iterations of the loop if the full number of nodes is reserved for the full duration of the execution.

This process can be run far more efficiently if test cases are executed concurrently. First, the application is run on all nodes. Second, the nodes are grouped into partitions, each with a different power of two size, up to half the total number of nodes. Each of these partitions runs a different size test case concurrently. The following example is a concurrent benchmarking script. It is assumed that the script is run on largest size being benchmarked, in this case 32 nodes.

```
#!\usr/bin/env mpish2
rank='rank.mpi'

slot="0"
basenum="2"
count="1"

time.mpi -t "size=32" progname
while [ "$slot" -eq "0"]; do
    remainder='expr "$rank" - "$basenum"'
    if[ "$remainder" -lt "$basenum" ] ; then
       slot="$count"
    else
       basenum='expr "$basenum" "*" "2"'
       count='expr $count + 1'
    fi
done
```

```
case $slot
   1)
     time.mpi -t "size=2" progname
     ;;
   2)
     time.mpi -t "size=4" progname
     ;;
   3)
     time.mpi -t "size=8" progname
     ;;
   4)
     time.mpi -t "size=16" progname
     ;;
esac
```

6   Conclusions and Further Work

We have presented MPISH2, a parallel process manager for MPI programs that provides an interface almost indistinguishable from the standard Unix Bourne shell. It enables the use of MPI in Unix environments in a seamless manner not previously possible. The addition of scalable utilities and simple, Bourne shell style control to Unix environments enables a variety of system and user tasks to be implemented in a scalable and elegant fashion.

The current design and implementation have two main limitations. The first is that all control flow constructs now impact parallel execution context, so serial conditional execution must be separated from parallel conditional execution. The second limitation is that control flow constructs have implicit barriers around them. This can reduce the amount of concurrency available to users. Both of these issues bear further examination.

Acknowledgments

This work was supported by the Mathematical, Information, and Computational Sciences Division subprogram of the Office of Advanced Scientific Computing Research, Office of Science, U.S. Department of Energy, under Contract W-31-109-ENG-38.

References

1. S. R. Bourne. An introduction to the unix shell. *Bell System Technical Journal*, 57(2):2797–2822, Jul-Aug 1978.
2. Ron Brightwell and Lee Ann Fisk. Scalable parallel application launch on cplant. In *Proceedings of SC 2001*, 2001.
3. Busybox home page. http://www.busybox.net.

4. R. Butler, N. Desai, A. Lusk, and E. Lusk. The process management component of a scalable system software environment. In *Proceedings of IEEE International Conference on Cluster Computing (CLUSTER03)*, pages 190–198. IEEE Computer Society, 2003.

5. R. Butler, W. Gropp, and E. Lusk. A scalable process-management environment for parallel programs. In Jack Dongarra, Peter Kacsuk, and Norbert Podhorszki, editors, *Recent Advances in Parallel Virutal Machine and Message Passing Interface*, number 1908 in Springer Lecture Notes in Computer Science, pages 168–175, September 2000.

6. Narayan Desai, Rick Bradshaw, Andrew Lusk, and Ewing Lusk. MPI cluster system software. In Dieter Kranzlmuller, Peter Kacsuk, and Jack Dongarra, editors, *Recent Advances in Parallel Virutal Machine and Message Passing Interface*, number 3241 in Springer Lecture Notes in Computer Science, pages 277–286. Springer, 2004. 11th European PVM/MPI Users' Group Meeting.

7. Narayan Desai, Andrew Lusk, Rick Bradshaw, and Ewing Lusk. MPISH: A parallel shell for MPI programs. In *Proceedings of the 1st Workshop on System Management Tools for Large-Scale Parallel Systems (IPDPS '05)*, Denver, Colorado, USA, april 2005.

8. R. Flannery, A. Geist, B. Luethke, and S. L. Scott. Cluster command & control (c3) tools suite. In *Proceedings of the Third Distributed and Parallel Systems Conference*. Kluwer Academic Publishers, 2000.

9. David G. Korn, Charles J. Northrup, and Jeffery Korn. The new Korn shell. *The Linux Journal*, 27, July 1996.

10. Message Passing Interface Forum. Document for a standard message-passing interface. Technical Report CS-93-214 (revised), University of Tennessee, April 1994. Available on **netlib**.

11. MPICH2. http://www.mcs.anl.gov/mpi/mpich2.

12. Pdsh:parallel distributed shell. http://www.llnl.gov/linux/pdsh/pdsh.html.

13. Andrew Tannenbaum. *Operating Systems, Design and Implementation*. Prentice Hall, 1987.

14. K. Thompson. The unix command language. *Structured Programming*, pages 375–384, 1975.

15. E. Walker, T. Minyard, and J. Boisseau. Gridshell: A login shell for orchestrating and coordinating applications in a grid enabled environment. In *Proceedings of the International Conference on Computing, Communications and Control Technologies*, pages 182–187, 2004.

Ensemble-2: Dynamic Composition of MPMD Programs

Yiannis Cotronis and Paul Polydoras

Department of Informatics and Telecommunications,
National and Kapodistrian University of Athens,
Panepistimioupolis, 15784 Athens, Greece
{cotronis, p.polydoras}@di.uoa.gr

Abstract. Ensemble-1 has been proposed for composing applications, consisting of MPI based components and external composition directives. Composed applications may be executed on any MPI distribution. Ensemble-1 followed the "static" principles of MPI-1; there is no dynamic process creation or destruction. In this paper, we propose Ensemble-2 supporting the composition of dynamically created processes following MPI-2 standard. The composition is based on construction of intercommunicators and establishing communication channels using either the parent/child or the client/server process relationship model.

1 Introduction

MPI [4, 5] has been become the de-facto standard for parallel distributed memory applications such as simulations and modeling of physical phenomena, molecular structures, etc. Due to the advanced complexity of these models, the respective programs are developed by different research teams; they focus on specific aspects of their scientific area and are designed, primarily, to run individually. However, in order to produce more accurate results and explore new areas of interest, these programs need to be composed. In order to accomplish a successful result, the programs interaction must be flexible, easily maintainable and should respect research teams' reluctance to share their code.

Ensemble-1 has been proposed [1, 2] for MPI code coupling. Applications in Ensemble-1 consist of MPI based components which may be composed in various configurations following external directives and executed directly on any MPI distribution, such as MPICH [6, 7]. However, in Ensemble-1 there is no dynamic process creation or destruction. In this paper, we propose Ensemble-2 supporting the composition of dynamically created processes, following MPI-2 standard. The composition is based on construction of intercommunicators and establishing communication channels using either the parent/child or the client/server process relationship model.

The structure of the paper is as follows: In section 2, we outline the composition principles of Ensemble. In section 3, we describe the core of Ensemble-2, which supports composition of dynamically spawned processes, communicating within parent/child or client/server intercommunicators. Finally, in section 4 we present our conclusions and future work.

B. Di Martino et al. (Eds.): EuroPVM/MPI 2005, LNCS 3666, pp. 343–350, 2005.

2 Ensemble Principles

Program composition in Ensemble-1 separates programming into external directives and codes to be composed. Directives are passed to each process by command line arguments (CLAs). Upon process spawning, a routine interprets the directives and establishes the composition by controlling two aspects of MPI: (i) the construction of communicators (intra, inter or topology), the processes of which may have been spawned from different executables (constructing MPMD intra or topology communicators) and (ii) the communication between processes. In traditional programming a decision has to be made on which type of communicator is to be used, which is then hard coded. Components in Ensemble-1 do not need any explicit communicator constructors and only involve point-point and collective communication (data movement, reduction and synchronization). Components are neutral to the communicator type (intra, inter, topology); a process spawned from such a component may be placed in a topology or in an inter-communicator according to the composition directives.

Let us outline the basic mechanics of Ensemble-1. We associate with each process a unique integer, called Unique Ensemble Rank (UER). MPI ranks are obtained from UERs: we construct the Ensemble_Comm_World (ECW) communicator by "splitting" all processes in MPI_COMM_WORLD with a common color using local key=UER. The new communicator just reorders processes, with MPI ranks equal to UERs. For constructing other communicators we pass an integer indicating the color for splitting ECW. To find the rank of a process in a constructed communicator we use MPI_Group_translate_ranks and its UER in ECW. We use UERs in CLAs to specify "actual communication parameters" for point-to-point operations and roots.

Each process parses its CLAs by calling SetEnvPars function and constructs the communicators, translates UERs into ranks and generally interprets CLAs to construct MPI bindings, which are stored in a structure, called EnvPars. Formal communication parameters in MPI communication calls (ranks, tags and communicators) are referring to elements of EnvPars. For point-to-point communication, the concept of *port* is introduced. A port is an abstract representation of the triplet: communicator, rank and tag. Ports which refer to a common communicator, they form *multiports*.

EnvPars is, actually, an array of contexts, where each context represents a specific intracommunicator that a particular process belongs to. The process rank is stored in the field MyRank. Each context includes an array of roots, which holds the ranks of root processes for collective calls in this intracommunicator, and an array of multiports. This array includes an array of ports and an AnyTypeComm field. The array of ports contains the remote process rank and the message tag used in point-to-point communication. The AnyTypeComm field represents a possible distinct communicator of the one which multiport refers to, either an intra or an intercommunicator.

In order to reduce software engineering effort and make the code more readable, we employ virtual names for contexts, multiports and roots, which can be used in MPI calls. These macros are expanded into the proper MPI bindings, before they are used. For example, the macro ENSRoot(Ocn,RootA) is expanded into: EnvPars[1]. Roots[0]. Root, EnvPars[1].IntraComm.ActualComm. The expansion assumes that 1 is the position of Ocn in the array of EnvPars, while RootA is the in the first position of the array of roots in the Ocn context. This macro can be use in collective communication function calls, e.g. MPI_Bcast.

3   External and Port Inter-communicators in Ensemble-2

3.1   Categorization of Intercommunicators

Ensemble-2 makes extensive use of intercommunicators. According to our terminology, intercommunicators may be distinguished into three categories based on the method they are established:

1. **Internal intercommunicators:** they are communicators being created by two intracommunicators within the same MPI_COMM_WORLD.
2. **External intercommunicators:** they are constructed between parent and child processes. They do not belong to the same MPI_COMM_WORLD.
3. **Port intercommunicators:** they refer to a special communication domain which is constructed asymmetrically, in a server/client fashion.

3.2   Avoiding Deadlocks Upon Creation of Communicators

In constructing intercommunicators there are three deadlock situations, caused by the possible wrong order of communicator construction, as specified in CLA directives. This problem however, is inherent to MPI programming and is not caused by Ensemble. Even in SPMD programming the designer has to be careful on specifying the order of splits. In MPMD applications, where codes are developed by different teams, we cannot assume a universal "safe" order, as different composition requirements may demand different orders of communicator constructions. We may distinguish three deadlocking situations.

The first deadlocking situation may occur when trying to construct intracommunicators which have common processes. Suppose that intracommunicators Intra1 and Intra2 both contain processes A and B. Process A may try to construct Intra1 and then Intra2, whilst B Intra2 and then Intra1. We avoid the deadlock by reordering split directives by their color.

The second deadlocking situation may occur when trying to construct intercommunicators. Suppose that process A participates in Intra1 and Intra2, while process C in Intra3 and Intra4. There are two intercommunicators: Inter1 between Intra2 and Intra3, and Inter2 between Intra1 and Intra4. A deadlock occurs as A constructs Intra1 and waits for the remote communicator of Inter2 (requiring Intra4); D constructs Intra2 and waits for the remote communicator of Inter1 (requiring Intra3). We avoid the deadlock by constructing intercommunicators after intracommunicators.

The third deadlocking situation is caused when trying to construct intercommunicators which have common intracommunicators. The situation is similar to the first case. Suppose that processes A and B participate in intercommunicators Inter1 and Inter2. A may try to construct Inter1 and then Inter2, whilst B first Inter2 and then Inter1. We avoid the deadlock by forcing an ordering on the construction of intercommunicators based on the respective construction tag value.

The three policies for avoiding deadlocks have been implemented in SetEnvPars routine. Consequently, CLAs specify the creation of communicators and not their order of creation.

3.3 Methodology Analysis by Example

We demonstrate Ensemble-2 methodology using an example, the objective of which is to create an environmental simulation program. The participant programs are: a generic atmospheric model that computes environmental quantities, such as wind velocity, temperature and humidity, and also simulates the rain model, if calculations indicate rain occurrence; an ocean model which is combined with the atmospheric one in computing these quantities; a snow model, used if atmospheric calculations indicate snow, which simulates the snow phenomenon and computes the water quantity being accumulated to the ground, due to snow; a hydrologic model, which receives water amounts from either snow or atmospheric model. In the rest of the paper we will focus on how model coupling procedure is accomplished, examining each coupling configuration individually. The following figure depicts the model interaction scenario:

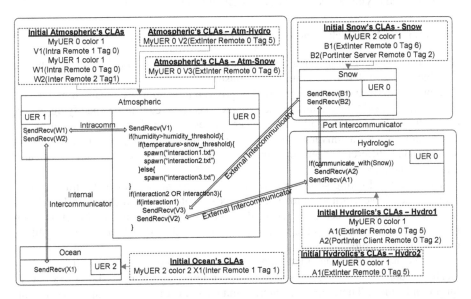

Fig. 1. Each square section indicates a common MPI_COMM_WORLD. Each arrow symbolizes a communication channel. We note that the pseudo code of the two atmospheric processes is not the component's source code, since it should be the same, but the code being actually executed by each process.

In the figure there are five components, one for each model, which establish a number of communication channels, using all three types of intercommunicators, along with an intracommunicator. We depict two processes from the atmospheric program, and one process from each of the others. Atmospheric with UER=1 and the ocean process communicate through the W2-X1 internal intercommunicator channel. Atmospheric processes communicate with each other through the intra channel W1-V1. In the atmospheric process with UER=0 we show the code for spawning processes from hydrologic and snow models. It communicates with the snow process through the external intercommunication channel V3-B1, and with the hydrologic process through the external inter channel V2-A1. Snow and hydrologic processes communicate through the port channel A2-B2.

Since the decision of process spawning from snow and hydrologic components is made by the atmospheric process at runtime, the communication establishment information being delivered to it, through CLAs, should not be predefined. So, we cannot use a single CLA set. In general, in order to solve this problem, we should first be able to numerate all possible child components. Then we describe the communication configuration for each possible child component into separate CLA sets, just as in other process communication cases. Each parent/child interaction is assigned with a unique id. In our example, the atmospheric process with UER=0, may spawn processes from the hydrologic and snow components. The atmospheric-hydrologic interaction is assigned with "*Atm-Hydro*" id and the atmospheric-snow interaction with "*Atm-Snow*" id. When certain criteria are satisfied the atmospheric process spawns a process from either component, using the respective set of CLAs.

However, the child processes should be provided with their own initial CLAs, something that is responsibility of the parent process. So, if we come up with rain, then only one hydrologic process is spawned, using the "*Hydro2*" set of CLAs. Alternatively, if snow phenomenon occurs, then the snow process establishes its communication channels, described in "*Snow*" set of CLAs, whereas the hydrologic process uses the "*Hydro1*" set of CLAs. The difference between "*Hydro1*" and "*Hydro2*" is that in the first case, since a snow process also partixipates, an additional communication channel is described, with that process, through a port intercommunicator.

However, the code presented in figure 1 does not contain any hard coded information about any kind of CLAs or children components. When the atmospheric process 0 spawns child processes, it uses the spawn function, actually a wrapper of MPI_Comm_spawn, but instead of passing the child executable name as the first function's argument, it uses a different configuration file for every possible case. For example, if only the hydrologic process should be spawned, then the "*interaction3.txt*" file is employed. This file contains the "*Hydro*" child component name, the "*Hydro2*" child CLA set and the "*Atm-Hydro*" parent CLA set, which should be used in this interaction scheme. Similarly, the "*interaction2.txt*" contains the same information, except from the "*Hydro1*" child CLA set. Finally, "*interaction1.txt*" defines "*Snow*" as the child component, with "*Snow*" as its CLA set and "*Atm-Snow*" as the parent CLA set. So, the atmospheric parent process can spawn children processes, establishing any desirable connection pattern, coded in a respective configuration file.

3.4 Support of External Intercommunicators

In Ensemble-2, communication among dynamically created processes is handled by the external intercommunicators. In our example, the atmospheric processes communicate with hydrologic and snow processes through two separate external intercommunicators. As far as their construction is concerned, a number of problems arise, particularly due to the asymmetric nature of the whole procedure; parent and child processes do not coexist under a common MPI_COMM_WORLD, the splitting of which would construct new communicators. However, there is a common communicator which comes from the MPI_Comm_spawn, on the parent part, and the

MPI_Get_parent, on the children part. This communicator is an external one and is the construction basis of any new communicator between parent and child processes.

In Ensemble-1 all communicators are constructed in SetEnvPars, called after MPI_Init. In Ensemble-2 parent processes construct new external intercommunicators by calling our External_intercomm_create function, after MPI_Comm_spawn, while child processes construct them by calling SetEnvPars after MPI_Init, as in Ensemble-1. In order to pass communication establishment information, described in CLAs, to this function, we employ a temporary structure which stores this piece of information, just as in the case of internal intercommunicators. This structure is also used for eliminating deadlock situations, examined in section 3.2, by being ordered based on the construction tag of each external intercommunicator.

For constructing of external intercommunicators, a complex mechanism is adopted: In order for parent processes to calculate the ranks of the remote child processes, they use their ECW, translate UERs to actual process ranks and then inform them of their own ranks. This is done because child processes do not have access to their parent ECW, something that does not stand for the parent side.

The additions to CLAs, in order to support construction of external intercommunicators are: (i) the symbolic intercommunicator name, (ii) a tag for the construction of intercommunicator and (iii) the UERs of local and remote group leaders. However, the CLAs of child processes do not include the UER of remote group leader, since they do not have access to the parent ECW, in order to make the translation from UER to the actual process rank. The communication CLAs are exactly the same as in other communicator types. As shown in figure 1, the atmospheric process with UER=0 communicates with the hydrologic one, through local port V2 with message tag 5. On the other side, snow process uses local port A1 with the same message tag.

3.5 Support of Port Intercommunicators

Another important feature of MPI-2 specification is the communication establishment between pairs of process groups that do not share a communicator, through MPI ports, in a client/server manner. In our example of figure 1, these two groups are the hydrologic and the snow. The server component (in our case, snow) opens the port and waits for the client component (hydrologic) to request a connection. Then a new intercommunicator is created, upon which the two groups interact. This procedure is cloaked under Ensemble methodology, according to which programmers just describe which two parts participate and the rest is done automatically and in a transparent way. Apart from their creation method, port intercommunicators are the same as other intercommunicators and therefore are handled by our methodology similarly.

The construction mechanism of port intercommunicators is much simpler that these of other communicators, but also much more time demanding. The time consuming part of this procedure is not for constructing communicators, but that of determining remote process ranks. In all other communicator constructions, there was always a way for determining remote ranks from UERs. However, in this case, the two process groups, which share a port intercommunicator, do not have access to the remote ECW, in order to perform calculations concerning remote process ranks. The only solution is all processes from one group send messages with data their ranks and tag their UERs to all processes of the other group, through the new port intercommunicator. Then each process receives the rank-UER pairs and does the matching.

The additions to CLAs, in order to support port intercommunicators construction are (i) the symbolic intercommunicator name, (ii) a client/server indication and (iii) a unique service name that refers to each port intercommunicator. Due to the asymmetric nature of the construction of these communicators, the server group of processes opens a connection port, through MPI_Open_port, publishes this port with a specific service name and waits for any connection. The client group of processes looks up for this service name and tries to connect to server port. The communication CLAs are exactly the same as in other communicator types. As shown in figure 1, the hydrologic process communicates with the snow one, through local port A2 with message tag 2. On the other side, snow process uses local port B2 with the same message tag.

Similar to the construction of the other types of communicators, there is possibility for deadlocks. The solution we employed is a temporary structure that stores all construction data of the port intercommunicators, ordered by the service name.

4 Conclusions and Future Work

In this paper, we propose Ensemble-2 which supports composition of dynamically created processes, following MPI-2 standard. The Ensemble mechanism, described above, has been developed and tested under Linux environment using LAM-MPI library [9], but can be ported to any programming environment that supports MPI-2. Ensemble alleviates a number of complexities and offers new possibilities for MPMD programming:

1. MPMD intracommunicators may be constructed. Intercommunicators are designed for establishing point-point communication between process groups spawned by different executables. Ensemble demonstrates that SPMD intracommunicators are not an essential aspect of MPI, but just a matter of programming convenience. In Ensemble it is simple to construct MPMD intracommunicators taking advantage of the use of collective communications, if necessary.
2. Processes spawned from any number of executables may be grouped in any configuration of communicators (intra, inter or topology) transparently, as directed by external directives. We have proved that all deadlocking situations are avoided.
3. Application performance depends on specific characteristics of machines. The choice of the communication context(s) may be important. In Ensemble this choice could be specified in the external directives in accordance with the mapping of processes, specified outside MPI programs. If an application is executed on an inhomogeneous environment, such as a grid, it is possible to group processes in different communicator types, best suited to the host machine.
4. In Ensemble, the communication data are not hard coded but can be controlled externally by directives. Components may specify only the possibility of communication not the actual communication, opening the possibility for using processes in any communication pattern.
5. We have extended Ensemble to support dynamically created processes. We compose applications based on external and port intercommunicators. Ensemble-2 constitutes a powerful software package that provides the possibility for composing components dynamically, creating complex MPMD applications.

Up to now we have focused our effort on the basic low level mechanisms for supporting dynamic MPMD applications. We plan to develop a high level interface abstracting tedious low level details, based on the High Level Composition Environment of Ensemble-1 which has been recently implemented in Python [10]. A programmer may use the Ensemble-API within standard Python scripts to specify process compositions and machine mappings. All necessary scripts and files incorporating the low level directives (CLAs) for composing MPI programs are generated.

We also plan to introduce into Ensemble-2 a level of abstraction of low-level ports, by using data quantities in meshes, used successfully in MpCCI [6] for MPI-1 applications. Programmers may design dynamic compositions by defining data-meshes and the physical quantities, which may be either required or produced by components. Composition couples required and produced quantities. We have already applied this coupling in the virtualization of MPI components as grid OGSA services [3]. We are currently redesign and extend this virtualization using WSRF services.

Acknowledgement

We like to thank Special Accounts of National and Kapodistrian University of Athens for the financial support.

References

1. Cotronis, J.Y, *Composition of Message-Passing Interface Applications Over MPICH-G2*, The International Journal of High Performance Computing Applications, Vol. 18, No. 3, Fall 2004, pp. 327-339.
2. Cotronis, J.Y., *Application Composition in Ensemble using Intercommunicators and Process Topologies* PVM/MPI'03, LNCS 2840, pp. 482-490, Springer.
3. E. Floros, Y. Cotronis, *Exposing MPI Applications as Grid Services*, Euro-Par 2004, LNCS Vol. 3149 Springer 2004, pp. 436-443.
4. Message Passing Interface Forum *MPI: A Message Passing Interface Standard*. International Journal of Supercomputer Applications, 8(3/4): 165-414, 1994.
5. Message Passing Interface Forum. *MPI-2: Extensions to the Message Passing Interface*, July 1997, www.mpi-forum.org.
6. Ahrem, Regine; Post, Peter; Steckel, Barbara; Wolf, Klaus. MpCCI: A Tool for Coupling CFD with Other Disciplines. In Proceedings of the 5th World Conference in Applied Fluid Dynamics, CFD-Efficiency and the Economic Benefit in Manufacturing (June 17-21, 2001),
7. Gropp, W. and Lusk, E. 2001. *Installation and User's Guide for MPICH, A Portable implementation of MPI*, ANL-01/x. Argonne National Laboratory.
8. Gropp, W. and Lusk, E. March 2005. *Installation and User's Guide for MPICH2*. Mathematics and Computer Division. Argonne National Laboratory.
9. Greg Burns and Raja Daoud and James Vaigl, *LAM: An Open Cluster Environment for MPI*, Proceedings of Supercomputing Symposium, 379-386, 1994.
10. G. Van Rossum, F. L. Drake, *Python Reference Manual Release 2.4.1*, March 2005, www.python.org.

New User-Guided and `ckpt`-Based Checkpointing Libraries for Parallel MPI Applications[*],[**]

Paweł Czarnul and Marcin Frączak

Faculty of Electronics, Telecommunications and Informatics,
Gdansk University of Technology, Poland
pczarnul@eti.pg.gda.pl, marcin.f@wp.pl
http://fox.eti.pg.gda.pl/~pczarnul

Abstract. We present design and implementation details as well as performance results for two new parallel checkpointing libraries developed by us for parallel MPI applications. The first one, a user-guided library requires from the programmer to support packing and unpacking code with an easy-to-use API using MPI constants. It uses MPI-2 collective I/O calls or a dedicated master process for checkpointing. The other version is a technically advanced parallel implementation of checkpointing based on the user-level `ckpt` library. It uses wrappers for MPI calls in the user program which enables to run a shadow MPI application just for communication purposes. Communication between original processes and the shadow MPI code is done via shared memory segments to which communication buffers are mapped. We present checkpoint/restart times for the two approaches and subversions proposed by us compared to an available LAMMPI/BLCR checkpointing solution for MPI applications. The performance of all the versions and I/O optimizations are discussed for a 4-node, 16-processor cluster with NFS and specifically for single SMP nodes with a local file system.

1   Introduction and Goals

Checkpointing of applications can allow an application to be checkpointed periodically to be restarted after a system crash. Secondly, it enables a process migrate to another node to balance load or make some nodes available to the user. For checkpointing of parallel MPI programs, the solution must either handle all pending communication at the time a checkpoint signal is issued or assume a simplified model in which checkpoints are generated at designated points where there is no application data in the buffers and the network. The following checkpointing methods can be distinguished:

1. user guided ([1], [2], [3]) – programmer specifies what data needs to be included or excluded in/from the checkpoint,
2. user level libraries like ckpt ([4]), Condor ([5]) and Libckpt ([6]) – usually require linking a library to a program with slight or no modifications to the code. Do not require root privileges but are often limited in handling system calls, threads etc. [7] announces Hector (alpha version) – checkpointing for MPI with Dynamite 2.0,

---

[*] Task WP.13 of 6 T11 2003 C/06098 "CLUSTERIX - The National Linux Cluster". Calculations carried out at the Academic Computer Center in Gdansk, Poland.
[**] Partially covered by the Polish National Grant KBN No. 4 T11C 005 25.

B. Di Martino et al. (Eds.): EuroPVM/MPI 2005, LNCS 3666, pp. 351–358, 2005.

3. hybrid approaches – [8] presents an interesting work in which the programmer inserts calls to `PotentialCheckpoint` in the code where checkpoint can be invoked. The whole memory space is saved automatically though.
4. modifications or extensions of existing implementations e.g.: LAM MPI/BLCR (LAM – [9] coupled with kernel-level BLCR – [10]) or MPICH-V ([11], MPICH coupled with Condor – [5]). This solution is especially attractive for parallel MPI programs as can offer a truly transparent solution but is tighly coupled with the internals of a particular implementation. May require root privileges as LAM/BLCR.

The contribution of our work are **two advanced implementations of checkpointing libraries**, using coordinated ([8]) algorithms[1]:

PARUG: flexible user-guided version with fast MPI-2 file system calls saving checkpoint collectively or through a dedicated master process, only necessary data needs to be packed which is selected by the knowledgeable programmer in the code. Thus it has the potential of saving only data necessary after restart.

PARCKPT: extensible, fast and transparent **checkpointing version for any MPI implementation using a sequential checkpointing library – a `ckpt`-based ([4]) version was developed with checkpoints saved locally or through a checkpoint server (`ckpt` feature).** Contrary to other transparent solutions like LAMMPI/ BLCR or MPICH-V, PARCKPT can be used with any MPI implementation giving transparent checkpointing, can also be adapted for use with other sequential checkpointing libraries/tools examples of which are Condor ([5]) and Libckpt ([6]).

These benefits come at the cost of a slightly limited **application model – synchronous parallel application** – it is assumed that all processes successively reach the same points in the program code where checkpoint can occur and pending messages are not considered at those points. This shortens time to checkpoint and the model is suitable for a wide range of applications e.g. SPMD ([12]) or synchronous Master-Slave.

2   Proposed User-Guided Checkpointing

In PARUG, collective routine `CX_CheckCheckpoint()` (sample code shown in Figure 1) needs to be inserted by the programmer into potential points where checkpoints can occur denoting **iterations**, which can, if ordered by a signal earlier, save the states of the processes using MPI-2. Alternatively, the state of the application can be saved by one designated process if the programmer does not provide synchronized operations.

The sequence of actions for the checkpointing procedure is as follows (Figure 1):

1. Send a SIGUSR1 signal to any MPI process which sets a global flag.
2. As soon as the process calls function `CX_CheckCheckpoint()` (potentially in a loop of SPMD computations), the flag is read and asynchronous messages are sent to the other processes. An iteration number (Paragraph 1, hidden from the programmer) for checkpointing to occur at is also propagated (received by `MPI_Irecv` calls).

---

[1] Available from authors, to be released at http://fox.eti.pg.gda.pl/~pczarnul.

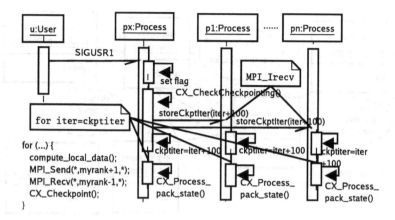

Fig. 1. Proposed User-guided Approach: Inter Process Communication Schema

3. In the following operating modes of PARUG, corresponding actions occur:
   – `CX_SYNCHRONIZED`: when the program reaches the required iteration number all processes save data pointed by the programmer to one checkpoint file. All processes can synchronize on `CX_CheckCheckpoint()` for checkpointing;
   – `CX_LOOSE`: only one selected master process saves data to the checkpoint file at the first call to `CX_CheckCheckpoint()`.

Independently from the above, the library can operate in two modes:

1. **Parallel data write (default) – checkpoint data is written/read by `MPI_File _write_at_all`/ `MPI_File_read_at_all` MPI-2 functions** which can speed-up data write times/access by grouping, collective buffering etc. ([13]).
2. **Data write through a master process** – all checkpoint data from all processes is sent to the process with MPI rank 0 which then writes data to a file using MPI-2 calls.

3   Proposed `ckpt`-Based Parallel MPI Checkpointing Library

In PARCKPT, no code changes are required but all MPI functions are replaced by wrappers (preceded by `RES_`, sample code shown in Figure 2). The wrappers for MPI communication routines denote the aforementioned potential checkpoint points and count **iterations** internally which is used to calculate a global iteration number for checkpointing. Thus, currently, the library can be used with synchronous applications with a uniform number of communication actions per per process per iteration.

In this solution, a static library is linked with the original user application instead of an MPI library. The new library includes functions substituting MPI functions (preceded by `RES_`). Original MPI functions are called by another process, a wrapper. This makes it possible to checkpoint the original processes using `ckpt` since the processes do not call true MPI functions. The wrapper also prepares the MPI world before the start and after the restart of the user application. For each process of the application a separate wrapper process is created (Figure 2).

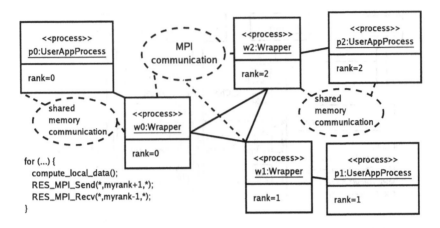

Fig. 2. Proposed `ckpt`-based Checkpointing: Inter Process Communication Schema

Application–wrapper communication uses signals and shared memory i.e. the user application only passes data through shared memory to the wrapper which calls MPI functions. However, copying of user data into/from shared memory regions when passing/fetching it to/from the wrapper for sending/receiving decreases performance. We attach the shared memory 'window' to the memory region that already contains user data – the buffers. This is done using the shmat function with the SHM_REMAP flag set which removes all other memory mappings from that memory region though. Thus the buffer data is saved in a temporary buffer and restored after the attachment. This causes a serious slowdown once, but speeds things up if the same buffer is used repeatedly.

A typical scenario for start/checkpoint/restart looks as follows:

1. Preprocessing the application source code, substituting any calls to MPI functions with their RES_ substitutes and page aligning data buffers.
2. Start of the wrapper which starts the user application process.
3. SIGUSR1 signal to an application process starts checkpointing.
4. During the next MPI action after receiving the signal, the process that received it sends a specific MPI message to every other process of the application. The message defines the checkpoint at some iteration in the future.
5. At the defined iteration processes order their wrappers to leave the MPI world gracefully (call MPI_Finalize()) and exit.
6. The application processes checkpoint (using ckpt, [4]) and exit.
7. Upon restart the wrappers restart the application processes which continue without noticing any checkpoint/restart.

Similarly to PARUG, we distinguish two subversions of the implementation:

1. **standard – all processes simply write data in the local file system.**
2. **a ckpt server (implemented in ckpt) is used to which checkpoints are sent over TCP and then saved locally).**

4   Experimental Results

4.1   Testbed Environment and Parallel MPI Application

All simulations used a 16-processor cluster (four 4-processor nodes, 512MB RAM each) with Pentium III Xeons and Ethernet switches. On one node checkpoints were saved locally (node g55) while the other nodes (g52-g53) saved to g55 via NFS.

We used an SPMD MPI application (LAMMPI 7.0.6, BLCR 0.3.1 for LAM/BLCR) which runs 1000 time steps in which cells of a 2D domain are updated. The domain is divided equally among the processors. Between iterations, processes exchange boundary cell data. We varied the size of the domain from 32MB to 128MB. The implementation corresponds to parallel applications like electromagnetic modeling or medical simulations ([12]). For PARUG we pack the whole domain data. PARCKPT and LAM/BLCR pack communication buffers etc. additionally. In practice, they will save more data than PARUG. We aimed at the assessment of checkpoint/restart costs for all the methods.

4.2   Proposed User-Guided Approach vs Checkpointing with ckpt Library

Figure 3 presents PARUG's execution times with one checkpoint/restart executed after 500 out of 1000 iterations. Within one node (2 and 4 processors on g55) the parallel data write method was faster i.e. MPI-2 calls were more efficient than routing data through one process on this node. However, for larger configurations, writes through a master residing on node g55 were faster than even MPI-2 collective calls. The internode NFS throughput appeared to be lower compared to native MPI send/recvs and fast disk access from node g55 or the rcp internode throughput (measured).

Figure 4 shows execution times for the standard and ckptserv versions of PAR-CKPT, with one checkpoint/restart executed after 500 out of 1000 iterations. The ckpt server version is faster for configurations larger than one node. On one node standard writes to separate files are faster than routing through one local process.

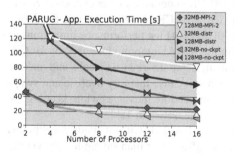

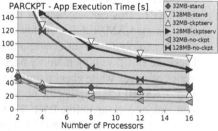

Fig. 3. PARUG: Execution Times of the Testbed Application with One Checkpoint/Restart

Fig. 4. PARCKPT: Execution Times of the Testbed Application with One Checkpoint/Restart

4.3 Comparison of Parallel MPI Checkpointing Methods

Finally, we compared both PARUG and PARCKPT (combinations of best subversions) against each other and LAM/BLCR (Table 1, Figures 5 and 6). Both PARCKPT and LAM/BLCR use sequential checkpointing libraries in a parallel MPI environment.

Table 1. Comparison of Tested Parallel Checkpointing Methods

Method	Subversion	Ckpt+Restart Time	Features
PARUG	MPI-2 version	slow on NFS, fast on SMP node	flexible, not transparent to programmer, fast, packs only necessary data, can restart on different no of processes than checkpointed, limited programming model (extendable with some programming effort)
	designated master	fast on NFS, slow on SMP node	
PAR CKPT	std local writes	slow on NFS, fast on SMP node	theoretically (almost) fully transparent although requires synchronous operations, limited set of MPI functions supported now, fast, checkpoints larger than PARUG and LAM/BLCR, uses LINUX-specific memory mappings for high performance
	ckptserv	fast on NFS, slow on SMP node	
LAMMPI/BLCR		slower than PARUG, faster than PARCKPT for smaller sizes, slower for larger	fully transparent to programmer, easy-to-use, checkpoints smaller than PARCKPT, only slightly larger than PARUG, *for 1-node checkpoints of app processes appeared several seconds earlier than the mpirun checkpoint. The application processes were working until that time. The former yields times very close (but longer) to PARUG (although the mpirun checkpoint is required to restart).

PARUG is the fastest (see LAM/BLCR note * in Table 1) since it packs/unpacks least data and apparently because of fast collective MPI-2 calls within one SMP node. On larger configurations it uses one designated master on node g55. It is followed by:

– On 2 and 4 processors (one node): LAM/BLCR generates smaller checkpoints than PARCKPT. This can account for faster LAM/BLCR for smaller sizes. For larger checkpoints LAM/BLCR was slower. [10] and [14] list performance limitations of BLCR, namely for larger checkpoints in the VMADump module used by BLCR. These are ([10]) writing memory pages using separate write() calls and making copies of pages while checkpointing which can cause memory overuse and swapping. LAM/BLCR empties the network from pending messages while keeping the application working unlike PARCKPT. Also see note * in Table 1.

– On 8-16 processors: PARCKPT is faster than LAM/BLCR – it sends checkpoints from processes via TCP to ckptsrv on node g55 rather than saving locally via the slow NFS as LAM/BLCR does. LAM/BLCR failed to run any MPI application on more than two nodes (cr_init() failed).

We also assessed the overhead of the following components for the testbed application without checkpointing, compared to a standard MPI application without checkpointing: LAMMPI with BLCR – no measurable overhead, ckptserv – no measurable

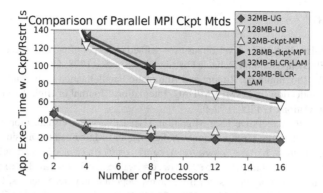

Fig. 5. Comparison of Checkpointing Approaches: Execution Times of the Testbed Application with One Checkpoint/Restart

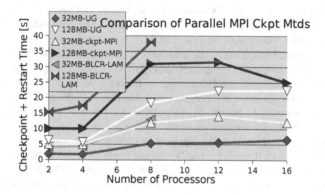

Fig. 6. Comparison of Checkpointing Approaches: Checkpoint/Restart Times

overhead compared to the standard PARCKPT version, PARCKPT – the overhead due to the additional wrappers and shared memory communication and signal synchronization – from 2% on 4 processors to 6% on 16 processors for the domain size of 128MB.

5   Summary and Future Work

We have presented two new checkpointing libraries, their design and showed they offer better performance than LAM/BLCR for large checkpoints in a specific (NFS) environment, at the cost of a constrained application model. For the two solutions, we have investigated two subversions with fast MPI-2 calls/designated master process for PARUG and local writes/ckptserv for PARCKPT. We showed the latter options are faster on a shared NFS on two or more nodes while the former on single SMP nodes. NFS optimizations will be investigated as well. We plan on the incorporation of other checkpointing libraries into the PARCKPT scheme. Currently our PARCKPT supports a limited set of MPI functions which will be extended. We have also developed a parser for user applications which replaces MPI calls with PARCKPT-specific wrappers.

References

1. Silva, L., Silva, J.: System-level versus user-defined checkpointing. In: Proceedings. Seventeenth IEEE Symposium on Reliable Distributed Systems. (1998) 68–74
2. Czarnul, P.: Programming, Tuning and Automatic Parallelization of Irregular Divide-and-Conquer Applications in DAMPVM/DAC. International Journal of High Performance Computing Applications **17** (2003) 77–93
3. CUMULVS: (Collaborative User Migration, User Library for Visualization and Steering) Distributed Computing Group, Computer Science and Mathematics Division, Oak Ridge National Laboratory, http://www.csm.ornl.gov/cs/cumulvs.html.
4. Zandy, V.C.: (ckpt library) http://www.cs.wisc. edu/~zandy/ckpt/.
5. Condor Team, Attention: Professor Miron Livny, Dept of Computer Sciences, 1210 W. Dayton St., Madison, WI 53706-1685, (608) 262-0856 or miron@cs.wisc.edu Condor Team, Computer Sciences Department, University of Wisconsin-Madison, Madison, WI: (The Condor Project, CondorâĂŹs Checkpoint Mechanism)
6. J.S.Plank, M.Beck, G.Kingsley, K.Li: libckpt: Transparent Checkpointing Under UNIX. Conference Proceedings USENIX Winter 1995 Technical Conference (1995)
7. Romanov, S., Malashonok, D.Y., Iskra, K., Gubala, T.: The Dynamite checkpointer 2.0. Faculty of Science, Informatics Institute. (2003) http://www.science.uva.nl/research/scs/Software/ckpt/#hector.
8. Bronevetsky, G., Marques, D., Pingali, K., Stodghill, P.: Automated application-level checkpointing of mpi programs. In: Proceedings of the ninth ACM SIGPLAN symposium on Principles and practice of parallel programming, San Diego, California, USA (2003) 84–94
9. Sankaran, S., Squyres, J., Barrett, B., Lumsdaine, A., Duell, J., Hargrove, P., Roman, E.: The lam/mpi checkpoint/restart framework: System-initiated checkpointing. Los Alamos Computer Science Institute (LACSI) Symposium (2003)
10. Duell, J., Hargrove, P., Roman, E.: The Design and Implementation of Berkeley Lab's Linux Checkpoint/Restart. In: Future Technologies Group white paper. (2003)
11. Franck Cappello, Project Leader at al.: (Mpich-v: Mpi implementation for volatile resources) http://www.lri.fr/~bouteill/MPICH-V.
12. Czarnul, P., Grzeda, K.: Parallel Simulations of Electrophysiological Phenomena in Myocardium on Large 32 and 64-bit Linux Clusters. In: 11th European PVM/MPI Users Group Meeting Budapest, Hungary, September 19 - 22, 2004. Proceedings. (Volume 3241/2004.)
13. Message Passing Interface Forum: MPI-2: Extensions to the Message-Passing Interface. (1997) University of Tennessee, Knoxville, Tennessee.
14. Sankaran, S., Squyres, J., Barrett, B., Lumsdaine, A., Duell, J., Hargrove, P., Roman, E.: The LAM/MPI Checkpoint/Restart Framework: System-Initiated Checkpointing, Los Alamos Computer Science Institute (LACSI) Symposium (2003)

Performance Profiling Overhead Compensation for MPI Programs

Sameer Shende[1], Allen D. Malony[1], Alan Morris[1], and Felix Wolf[2]

[1] Department of Computer and Information Science,
University of Oregon
{sameer, malony, amorris}@cs.uoregon.edu
[2] Innovative Computing Laboratory, University of Tennessee
fwolf@cs.utk.edu

Abstract. Performance profiling of MPI programs generates overhead during execution that introduces error in profile measurements. It is possible to track and remove overhead online, but it is necessary to communicate execution delay between processes to correctly adjust their interdependent timing. We demonstrate the first implementation of a onle measurement overhead compensation system for profiling MPI programs. This is implemented in the TAU performance systems. It requires novel techniques for delay communication in the use of MPI. The ability to reduce measurement error is demonstrated for problematic test cases and real applications.

Keywords: Performance measurement, analysis, parallel computing, profiling, message passing, overhead compensation.

1 Introduction

When a parallel program is profiled, measurement operations generate overhead that affects the performance observed. We call this *performance intrusion* [2]. Performance profiling tools typically report intrusion as a percentage slowdown of total execution time, but the intrusion effects themselves will be distributed throughout the profile results. While performance intrusion can alter program execution and, thus, perceived performance (i.e., *performance perturbation*), performance profiling tools rarely attempt to adjust the performance measurements to compensate for the intrusion.

In earlier work [3], we present a technique to measure overhead on each process of a parallel computation and remove its local effects. But these are not the only effects overhead intrusion can cause. Due to inter-process communication, the delays introduced by intrusion will propagate between processes. In more recent work [4], we specify models for parallel overhead compensation and the algorithms that must be used when profiling message passing parallel programs. These models show why it is necessary to communicate intrusion delays with every message communication. However, this is not so easy to accomplish.

This paper presents our results for the implementation of the parallel profile overhead compensation models in an MPI environment. Such overhead compensation techniques have never been implemented before. Here we outline our approach to piggyback

B. Di Martino et al. (Eds.): EuroPVM/MPI 2005, LNCS 3666, pp. 359–367, 2005.

intrusion delay on messages. We demonstrate the technique with applications and show that measurement error due to overhead can be removed, locally and globally.

Section §2 provides a brief background on the problem. Our solution approach is presented in Section §3. Section §4 outlines our experimental environment and shows the results of our validation tests. Conclusions and future work are given in Section §5.

2   Problem Background

Events are actions that occur during program execution. Typical events include *interval events* that are characterized by a pair of actions marking *entry* and *exit*, and *atomic events* that occur at a single place or time. Tools insert measurement code to track the performance of a parallel program as made visible by the instrumented events. Executing the measurement code introduces overhead. If an event trace is collected, there are techniques to analyze the trace and compensate the measurement overhead [7,8,9,10], including the correction of perturbation effects.

Unfortunately, this type of analysis is not possible with profile-based measurements where all compensation decisions must be made at the time the event occurs, more or less. This raises the problem of how to track delays between processes. Basically, measurement overhead occurring on one process affects events on other processes that are causally related [12]. Consider a process that executes a measured routine a large number of times (and thus incurs a large overhead associated with the entry and exit instrumentation), and then sends a message to another process that blocks waiting for the message. If the receiving process has accrued little measurement overhead before the receive operation, the message receipt will be delayed. Without the receiver knowing that that delay was due to the measurement overhead in the sender, the receiver's profile will end up accounting for the sender's overhead as *its waiting time*, when in reality, it may not have waited at all.

To accurately re-construct events in all the processes, we must account for the time spent executing instrumentation calls in the sender process and subtract this time from the wait time in the receive, if any. To do so, we propose a scheme where local delays are propagated along with inter-process communication events in the form of *piggyback* messages. The delay value represents how much sooner a process would have executed the given communication operation if there was no measurement overhead in the process. The receiving process extracts this piggyback message and adjusts its local delay. What is interesting is that the adjustment cannot be any greater than the waiting time for the *current* receive.

Consider two cases. In the first case, the remote delay is equal to or exceeds the sum of the waiting time and the local delay. In this case, the waiting time in the absence of instrumentation is zero, as the message would be received as soon as the receive call is executed. In the second case, the remote delay is less than the sum of the local delay and the waiting time. Here, depending on whether the remote delay is less than or greater than the local delay, the uninstrumented waiting time may be more or less than the current waiting time. On receiving the piggyback message, the receiver compares the local delay with the remote delay. The adjustment of wait time is the difference in the remote delay and the local delay: $Adjustment = remote(delay) - local(delay)$.

This adjustment is subtracted from the original wait time to give the new waiting time: $W_{new} = W - Adjustment = W - (remote(delay) - local(delay))$.

If we consider the beginning of the wait time and the receipt of the message as two events, then, if there is no delay in the local receiving process, the point that corresponds to the beginning of the wait routine is shifted to the left by the amount equal to the local delay. And the point where the message is received shifts by an amount equal to the remote delay. The distance between the two points is the new waiting period. The adjustment (or the difference between the remote and local delays) may be positive, negative, or zero corresponding to when the remote process experiences more, less or the same delay as compared to the local process. Correspondingly, the wait time may decrease, increase, or remain the same respectively, but it can never be adjusted to be negative. This adjustment of waiting time is propagated along the callstack of the receiving process, so the inclusive time spent in all ancestor routines is adjusted accordingly. This is a necessary calculation in order to properly compute profiling measurements. When we compute the local delay (at both sender and receiver processes), we assess the measurement overhead and then subtract the waiting time adjustments that have been made in the program. This value is then sent along with a message. Thus, delays from one process reach all processes that have causally related events.

The full details of our parallel profile compensation algorithms are described in our earlier paper [4].

3 Implementation

To test our models of parallel overhead compensation, we built a prototype using the TAU performance system [5] and the Message Passing Interface (MPI). Our goal was to produce a widely portable prototype that could be efficiently implemented and easily applied. We chose MPI as the communication substrate due to its wide acceptance in the parallel computing community as the de-facto message communication standard, as well as due to its portable tool support.

3.1 MPI Profiling Support

MPI supports creation of portable profiling and tracing tools using its profiling interface, PMPI. This interface allows a tool to interpose a library between the application and the MPI substrate and intercept one or more MPI calls. MPI provides a name-shifted interface to all its calls. For example, an MPI call such as MPI_Send() is also available as PMPI_Send(). Both are guaranteed by the MPI standard to provide the same functionality. Furthermore, if a tool defines an MPI_Send() call, it takes precedence over the MPI library's MPI_Send() call (this is done by using weak bindings for defining the library's calls). The tool can then define one or more MPI bindings and create measurement timers and start and stop them around the name-shifted version of the corresponding MPI call. Every MPI implementation must implement this profiling interface to conform to the MPI standard. This mechanism allows vendors of parallel systems to optimize the implementation of MPI to their target platforms and at the same time expose the hooks for tracking MPI performance to tool builders without providing them access to their proprietary source code.

3.2   Schemes to Piggyback Delay

To transmit the local delays encountered in a process (due to program instrumentation) to other processes, we examined several alternatives. The first scheme modifies the source code of the underlying MPI implementation by extending the header sent along with a message in the communication substrate (Photon [11] uses this approach). Unfortunately, it is not portable to all MPI implementations and relies on a specially instrumented communication library. The second scheme sends an additional message containing the delay information for every data message. This scheme only requires changes to the portable MPI wrapper interposition library for the tool. While it is portable to all MPI implementations, it has a performance penalty associated with transmitting an additional message, a penalty not incurred by the first scheme. As a result, the overhead caused by the additional message would require further compensation.

The third scheme copies the contents of the original message and creates a new message with our own header that would include the delay information. This scheme has the portability advantage of the second scheme and avoids the second scheme's transmission of an additional message. However, copying contents of a message could prove to be an expensive operation, especially in the context of large messages that are transmitted in point-to- point communication operations.

We implemented a modification of the third scheme, but instead of building a new message and copying buffers in and out of messages (at the sender and the receiver), we create a new datatype. This new datatype is a structure with two members. The first member is a pointer to the original message buffer comprised of n elements of the datatype passed to the MPI call. The second member is a double precision number that contains the local delay value. Once created, the structure is committed as a new user-defined datatype and MPI is instructed to send or receive one element of the new datatype. Internally, MPI may transmit the new message by composing the message from the two members by using vector read and write calls instead of its scalar counterparts. This efficient transmission of the delay value is portable to all MPI implementations, sends only a single message, and avoids expensive copying of data buffers to construct and extract messages.

3.3   TAU Overhead Compensation Prototype

To test the validity of our parallel profile compensation models, we built the portable prototype within the TAU performance system [5]. We previously implemented local overhead compensation, and now included the parallel compensation support. TAU computes parallel profile data during execution for each instrumented event. At runtime, TAU maintains an event callstack for each thread of execution. This callstack has performance information for the currently executing event (e.g., a routine entry) and its ancestors. We compute the delay that a process sees locally by first adding the number of completed calls to half the number of entries along the thread's callstack. We assume that an *enter* profile call takes roughly the same time as an *exit* profile call, which is true is most cases. Once we know the total number of timer calls and the total overhead associated with calling the enter and exit methods (see [3] for details), their product gives the local timer overhead. We keep track of adjusted wait times in a process, as explained earlier and subtract it from the local overhead to compute the local delay. This delay value is then piggybacked with a message.

Mapping MPI Calls. The essence of our parallel overhead compensation scheme is that whenever two processes interact with each other, the receiver is made aware of the sender's delay value, or how much sooner the communication operation would have taken place in the absence of instrumentation. We have discussed above how this scheme operates for synchronous message communication operations using MPI_Send and MPI_Recv. In this section we explore how other MPI calls can be made aware of remote delays.

Asynchronous Operations. When storing or retrieving the piggyback value, we create an auto variable on the stack in our wrapper routines for MPI_Send or MPI_Recv. Synchronization operations involve loads or stores to this variable. The logic to process the piggyback value when it is received is incorporated in the MPI_Recv wrapper routine. Here, we compare the local and remote delays to arrive at how much adjustment needs to be made to the waiting time. Now let us examine the asynchronous MPI_Isend and MPI_Irecv calls. When the user issues the MPI_Isend call, we compute the local delay and create a global variable where this is stored. The location of this global piggyback variable in the heap memory is used when we create our struct for a new datatype for sending the message.

On the receiving side, a similar arrangement of the piggyback value is used. When the message is finally received, MPI automatically copies the contents of the piggyback value into the heap where this value is to be stored. We also create a map that links the address of the MPI request to the address of this piggyback value. The logic that compares the local and remote delays cannot be incorporated in the MPI_Irecv wrapper due to the very nature of the asynchronous operation (the values are not received when the routine executes). Hence, we do not adjust the time spent in MPI_Irecv as we did for MPI_Recv. Instead, an asynchronous message is visible to the program only after executing the MPI_Wait, MPI_Test, or variants of these calls (Waitall, Waitsome, Testall, Testsome) to wait for or test one or more requests. When a request is satisfied, we examine the map and retrieve the value of the piggyback variable where the remote process' overhead is stored. Then, a comparison of local and remote delays and an adjustment of waiting time is made on the receiving side. When more than one message is received by the process, we need to examine all the remote delays to determine how much time the process would have waited in the absence of instrumentation. We discuss this in more detail next with collective operations.

Collective Operations. Consider the class of collective operations supported by MPI. Let us first examine the MPI_Gather call where each process in a given communicator provides a single data item to MPI. The process designated with the rank of root gathers all the data in an array. It is important to communicate the local delays from each process to the root process. To do this, we form a message with the piggyback delay value and call a single MPI_Gather call. At the receiving end, we receive a single contiguous buffer where the application data and the delay values are put together in a single buffer. We extract the piggyback values out of this buffer and construct the application buffer with the rest. Once we get an array of the delay values from each process we compute the minimum delay value from the group of processes. Since the collective operation cannot complete without the message with the minimum delay, it must adjust its waiting time based on this value. So, the collective operation reduces to the case where the receiver gets a message from one process that has the least delay in the communicator. We can now apply the performance overhead compensation model as described in the previous section.

When broadcasting a message from one task to several, MPI_Bcast is modelled based on the two process overhead compensation model (see [4]). We create a new datatype, on the root process, that embeds the original message and the local delay value. This message is sent to all other members of the group. Each receiver compares the remote delay with its local delay and makes adjustments to the waiting time and local overhead, as if it had received a single message from the remote task. We use the model described earlier to do this.

To model MPI_Scatter, which distributes a distinct message to all members of the group, we create a new datatype that includes the overhead from the root process. This is similar to the MPI_Gather operation. After the operation is completed, each receiver examines the remote overhead and treats it as if it had received a single message from the root node, applying our previous scheme for compensating for perturbation.

MPI_Barrier requires all tasks to block until all processes invoke this routine. MPI_Barrier is implemented as a combination of two operations: MPI_Gather and MPI_Bcast, sending the local delay from each task to the root task (arbitrarily selected as the process with the least rank in the communicator). This task examines the local delay and compares it with the task with the least delay, adjusts its wait time and then sends the new local delay to all tasks using the MPI_Bcast operation. This mechanism preserves the efficiencies that the underlying MPI substrate may provide in implementing a collective operation. By mapping one MPI routine to another, we exploit those efficiencies.

4   Experimental Results

We validate our parallel performance intrusion compensation model using a prototype implemented within the TAU performance system. To illustrate the problem, we examine a parallel MPI application that computes the value of π using the Monte-carlo integration algorithm. The program calculates the area under the pi function curve from 0 to 1. The program comprises of a master (or server) task that generates work packets with a set of random numbers. The master task waits for a request from any worker and sends the chunk of randomly generated numbers to it. For each pair of numbers that is given to a particular worker, it finds out if the pair of cartesian co-ordinates represented by the numbers is below or above the pi function curve. Then, collectively, the workers estimate the value of pi iteratively until it is within a given error range. This simple example highlights how instrumentation overheads accumulated at the worker tasks are communicated to the master task. We execute the application in four modes: when there is no TAU instrumentation, with instrumentation without any compensation, with local perturbation compensation, and finally, with parallel perturbation compensation. As shown in table 1, these experiments are shown as distinct columns and we show

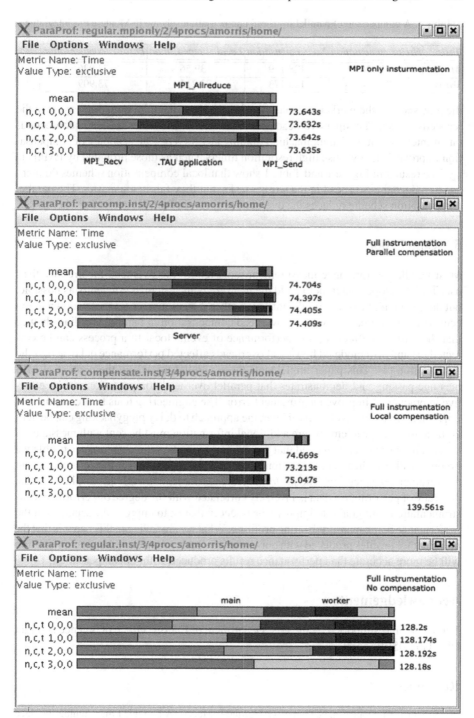

Fig. 1. Parallel overhead compensation in TAU

Table 1. A comparison of parallel overhead compensation scheme in Monte-carlo integrator

Task	No instrumentation	No compensation	Local compensation	Parallel compensation
Master	73.926	128.179	139.56	73.926
Worker	73.834	128.173	73.212	73.909

the time spent in the worker and master tasks. We show the minimum times spent in the respective tasks. The timer overhead associated with a TAU timer was 480 nanoseconds on an Intel®Itanium2 Linux machine running at 1.5 GHz. The accuracy of compensation improves when we use high resolution timers, such as those provided by PAPI[1].

The results in Figure 1 and Table 1 show that local compensation schemes do manage to reduce the overhead in the worker tasks, but they fail in the master. The parallel compensation scheme reduces the overhead properly in both master and worker tasks.

5   Conclusion

Most parallel performance measurement tools ignore the overhead incurred by their use. Tool developers attempt to build the measurement system as efficiently as possible, but do not attempt to quantify the intrusion other than as a percentage slowdown in execution time. Our earlier work on overhead compensation in parallel profiling showed that the intrusion effects on the performance of events local to a process can be corrected [3] and also modeled how local overheads affected performance delay across the computation [4]. This paper implements those parallel models in the context of MPI message passing and demonstrates that parallel overhead compensation can be effective in practice to improve measurement error. The engineering feats to accomplish the implementation are novel. In particular, the approach to delay piggybacking can be generalized to other problems where additional information must be sent with messages.

It is important to understand that we are not saying that the performance profile we produce with overhead compensation represents the actual performance profile of an uninstrumented execution. The *performance uncertainty principle* [2] implies that the accuracy of performance data is inversely correlated with the degree of performance instrumentation. Our goal is to improve the tradeoff, that is, to improve the accuracy of the performance being measured during profiling. What we are saying in this paper is that the performance profiles produced with our models for performance overhead compensation will be more accurate than performance results produced without compensation.

Acknowledgements

This research is supported by the U.S. Department of Energy, Office of Science contract DE-FG02-05ER25680.

References

1. S. Browne, J. Dongarra, N. Garner, G. Ho, and P. Mucci, "A Portable Programming Interface for Performance Evaluation on Modern Processors," *International Journal of High Performance Computing Applications*, **14**(3):189–204, Fall 2000.

2. A. Malony, "Performance Observability," Ph.D. thesis, University of Illinois, Urbana-Champaign, 1991.
3. A. Malony and S. Shende, "Overhead Compensation in Performance Profiling," *Euro-Par Conference*, LNCS 3149, Springer, pp. 119–132, 2004.
4. A. Malony and S. Shende, "Models for On-the-Fly Compensation of Measurement Overhead in Parallel Performance Profiling," (to appear) *Euro-Par Conference*, LNCS, Springer 2005.
5. A. Malony, S. Shende, "Performance Technology for Complex Parallel and Distributed Systems," In G. Kotsis, P. Kacsuk (eds.), *Distributed and Parallel Systems, From Instruction Parallelism to Cluster Computing, Third Workshop on Distributed and Parallel Systems (DAPSYS 2000)*, Kluwer, pp. 37–46, 2000.
6. A. Malony, et al., "Advances in the TAU Performance System," In V. Getov, M. Gerndt, A. Hoisie, A. Malony, B. Miller (eds.), *Performance Analysis and Grid Computing*, Kluwer, Norwell, MA, pp. 129–144, 2003.
7. A. Malony, D. Reed, and H. Wijshoff, "Performance Measurement Intrusion and Perturbation Analysis," *IEEE Transactions on Parallel and Distributed Systems*, 3(4):433–450, July 1992.
8. A. Malony and D. Reed, "Models for Performance Perturbation Analysis," *ACM/ONR Workshop on Parallel and Distributed Debugging*, pp. 1–12, May 1991.
9. A. Malony, "Event Based Performance Perturbation: A Case Study," *Principles and Practices of Parallel Programming (PPoPP)*, pp. 201–212, April 1991.
10. S. Sarukkai and A. Malony, "Perturbation Analysis of High-Level Instrumentation for SPMD Programs," *Principles and Practices of Parallel Programming (PPoPP)*, pp. 44–53, May 1993.
11. J. Vetter, "Dynamic Statistical Profiling of Communication Activity in Distributed Applications," *ACM SIGMETRICS Joint International Conference on Measurement and Modeling of Computer Systems*, ACM, 2002.
12. L. Lamport, "Time, Clocks, and the Ordering of Events in a Distributed System," Communications of the ACM, 21(7), 558–565, July 1978.

Network Bandwidth Measurements and Ratio Analysis with the HPC Challenge Benchmark Suite (HPCC)

Rolf Rabenseifner, Sunil R. Tiyyagura, and Matthias Müller

High-Performance Computing-Center (HLRS), University of Stuttgart,
Allmandring 30, D-70550 Stuttgart, Germany
{rabenseifner, sunil, mueller}@hlrs.de
www.hlrs.de/people/rabenseifner/, .../sunil/, .../mueller/

Abstract. The HPC Challenge benchmark suite (HPCC) was released to analyze the performance of high-performance computing architectures using several kernels to measure different memory and hardware access patterns comprising latency based measurements, memory streaming, inter-process communication and floating point computation. HPCC defines a set of benchmarks augmenting the High Performance Linpack used in the Top500 list. This paper describes the inter-process communication benchmarks of this suite. Based on the effective bandwidth benchmark, a special parallel random and natural ring communication benchmark has been developed for HPCC. Ping-Pong benchmarks on a set of process pairs can be used for further characterization of a system. This paper analyzes first results achieved with HPCC. The focus of this paper is on the balance between computational speed, memory bandwidth, and inter-node communication.

Keywords: HPCC, network bandwidth, effective bandwidth, Linpack, HPL, STREAM, DGEMM, PTRANS, FFTE, latency, benchmarking.

1 Introduction and Related Work

The HPC Challenge benchmark suite (HPCC) [5,6] was designed to provide benchmark kernels that examine different aspects of the execution of real applications. The first aspect is benchmarking the system with different combinations of high and low temporal and spatial locality of the memory access. HPL (High Performance Linpack) [4], DGEMM [2,3] PTRANS (parallel matrix transpose) [8], STREAM [1], FFTE (Fast Fourier Transform) [11], and RandomAccess are dedicated to this task. Other aspects are measuring basic parameters like achievable computational performance (again HPL), the bandwidth of the memory access (STREAM copy or triad), and latency and bandwidth of the inter-process communication based on ping-pong benchmarks and on parallel effective bandwidth benchmarks [7,9].

This paper describes in Section 2 the latency and bandwidth benchmarks used in the HPCC suite. Section 3 analyzes bandwidth and latency measurements

B. Di Martino et al. (Eds.): EuroPVM/MPI 2005, LNCS 3666, pp. 368–378, 2005.

submitted to the HPCC web interface [5]. In Section 4, the ratio between computational performance, memory and inter-process bandwidth is analyzed to compare system architectures (and not only specific systems). In Section 5, the ratio analysis is extended to the whole set of benchmarks to compare the largest systems in the list and also different network types.

2 Latency and Bandwidth Benchmark

The latency and bandwidth benchmark measures two different communication patterns. First, it measures the single-process-pair latency and bandwidth, and second, it measures the parallel all-processes-in-a-ring latency and bandwidth.

For the first pattern, ping-pong communication is used on a pair of processes. Several different pairs of processes are used and the maximal latency and minimal bandwidth over all pairs is reported. While the ping-pong benchmark is executed on one process pair, all other processes are waiting in a blocking receive. To limit the total benchmark time used for this first pattern to 30 sec, only a subset of the set of possible pairs is used. The communication is implemented with MPI standard blocking send and receive.

In the second pattern, all processes are arranged in a ring topology and each process sends and receives a message from its left and its right neighbor in parallel. Two types of rings are reported: a naturally ordered ring (i.e., ordered by the process ranks in MPI_COMM_WORLD), and the geometric mean of the bandwidth of ten different randomly chosen process orderings in the ring. The communication is implemented (a) with MPI standard non-blocking receive and send, and (b) with two calls to MPI_Sendrecv for both directions in the ring. Always the fastest of both measurements are used. For latency or bandwidth measurement, each ring measurement is repeated 8 or 3 times – and for random ring with different patterns – and only the best result is chosen. With this type of parallel communication, the bandwidth per process is defined as total amount of message data divided by the number of processes and the maximal time needed in all processes. The latency is defined as the maximum time needed in all processes divided by the number of calls to MPI_Sendrecv (or MPI_Isend) in each process. This definition is similar to the definition with ping-pong, where the time is measured for the sequence of a *send* and a *recv*, and again *send* and *recv*, and then divided by 2. In the ring benchmark, the same pattern is done by all processes instead of a pair of processes. This benchmark is based on patterns studied in the effective bandwidth communication benchmark [7,9].

For benchmarking latency and bandwidth, 8 byte and 2,000,000 byte long messages are used. The major results reported by this benchmark are:

- maximal ping pong latency,
- average latency of parallel communication in randomly ordered rings,
- minimal ping pong bandwidth,
- bandwidth per process in the naturally ordered ring,
- average bandwidth per process in randomly ordered rings.

Additionally reported values are the latency of the naturally ordered ring, and the remaining values in the set of minimum, maximum, and average of the ping-pong latency and bandwidth.

Especially the ring based benchmarks try to model the communication behavior of multi-dimensional domain-decomposition applications. The natural ring is similar to the message transfer pattern of a regular grid based application, but only in the first dimension (adequate ranking of the processes is assumed). The random ring fits to the other dimensions and to the communication pattern of unstructured grid based applications. Therefore, the following analysis is mainly focused on the random ring bandwidth.

3   Analysis of HPCC Uploads

Fig. 1 is based on base-run uploads to the HPCC web-page. Therefore, the quality of the benchmarking, i.e., choosing the best compiler options and benchmark parameters was done by the independent institutions that submitted results. The authors have added two results from the NEC SX-6+ and some results for fewer number of processes on NEC SX-8 and Cray XT3 [12]. For IBM BlueGene, an additional optimized measurement is also shown in some of the figures. The measurements are sorted by the random ring bandwidth, except that all measurements belonging to some platform or network type are kept together at the position of their best bandwidth.

The diagram consists of three bands: 1) the ping-pong and random ring latencies, 2) the minimal ping-pong, natural ring, and random ring bandwidth-bars together with a background curve showing the accumulated Linpack (HPL) performance, and 3) the ratios *natural ring* to *ping-pong*, *random ring* to *ping-pong*, and additionally *random ring* to *natural ring*.

The systems on the upper part of the figure have a random ring bandwidth less than 300 MB/s, the systems on the lower part are between 400 MB/s and 1.5 GB/s. Concentrating on the lower part, one can see that all systems show a degradation for larger CPU counts. Cray and NEC systems are clusters of SMP nodes. The random ring bandwidth benchmark uses mainly inter-node connections whereas the natural ring bandwidth uses only one inter-node connection in both directions and all other connections are inside of the SMP nodes. Therefore one can see a significant difference between the random ring and the natural ring bandwidth. One exception is the multi-threaded measurement on a NEC SX-6+ (0.5625 GHz); here, all three bandwidth values are nearly equivalent because on each SMP, only one MPI process is running. The ratio natural ring to ping-ping bandwidth varies between 0.4 and 1.0, random ring to ping-pong between 0.1 and 0.45, and random to natural ring between 0.1 and 0.7. With the IBM High Performance Switch (HPS), the reported random ring bandwidth values (0.72-0.75 GB/s) are nearly independent from the number of processes (64 to 256 [1.07 Tflop/s]), while the Cray X1 shows a degradation from 1.03 GB/s with 60 MSPs (0.58 Tflop/s) to 0,43 GB/s with 252 MSPs (2.38 Tflop/s). For some systems, the random ring latency and performance is summarized in Tab. 1.

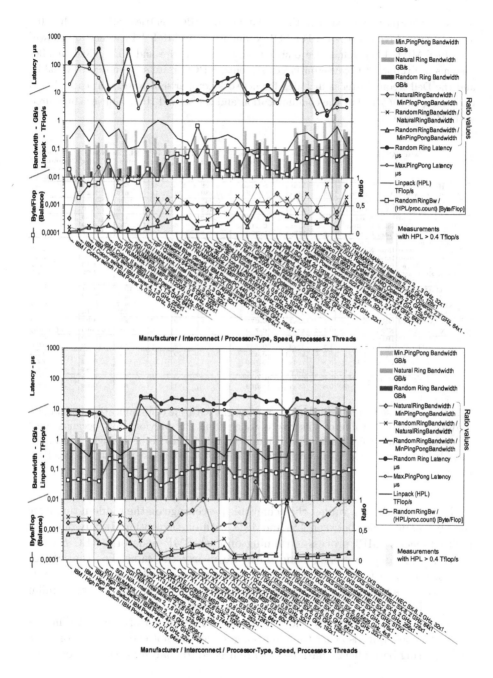

Fig. 1. Base runs of the HPC Challenge bandwidth and latency benchmarks, Status June 27, 2005

Table 1. Comparison of bandwidth and latency on HPCC entries with more than 0.4 Tflop/s with three exceptions: For SGI Numalink, only MPT 1.10 values are shown, the older MPT 1.8-1 values are omitted, and for Sun Fire and NEC SX-6, smaller systems are reported because on larger systems, HPCC results are not yet available, and the Dell Xeon cluster is included for network comparison. Note, that each thread is running on a *CPU*, but the communication and the second HPL value are measured with *MPI processes*.

Switch	CPU	Proc. Speed GHz	Number of MPI processes x threads	Random Ring Bandw. GB/s	Ping-Pong Bandw. GB/s	Rand. Ring Lat. μs	Ping-Pong Lat. μs	HPL Linpack accumu-lated Gflop/s	per process Gflop/s	Balance: Communi./ Comput. byte/kflop
IBM Colony	IBM Power 4	1.3	256x1	0.0046	0.108	374	87	654	2.55	1.8
Quadrics switch	Alpha 21264B	1.0	484x1	0.023	0.280	40	16	618	1.28	17.8
Myrinet 2000	Intel Xeon 3	3.066	256x1	0.032	0.241	22	22	1030	4.02	8.1
Sun Fire Link	Ultra Sparc III	0.9	128x1	0.056	0.468	9	5	75	0.59	94.5
Infiniband	Intel Xeon	2.46	128x1	0.156	0.738	12	12	413	3.23	48.2
SGI Numalink	Intel Itanium 2	1.56	128x1	0.211	1.8	6	3	639	4.99	42.2
SGI Altix 3700 Bx2	Intel Itanium 2	1.6	128x1	0.897	3.8	4	2	521	4.07	220.
Infiniband, 4x, InfinIO 3000	Intel Xeon	2.4	32x1	0.178	0.374	10	7	101	3.17	56.3
Myrinet 2000	Intel Xeon	2.4	32x1	0.066	0.245	19	9	97	3.03	21.7
SCI, 4x4 2d Torus	Intel Xeon	2.4	32x1	0.048	0.121	9	4	100	3.13	15.2
Gigabit Ethernet, PowerConnect 5224	Intel Xeon	2.4	32x1	0.038	0.117	42	37	97	3.02	12.5
NEC SX-6 IXS	NEC SX-6	0.5	192x1	0.398	6.8	30	7	1327	6.91	57.5
NEC SX-6 IXS	NEC SX-6	0.5	128x1	0.429	6.9	27	7	905	7.07	60.7
NEC SX-6 IXS	NEC SX-6	0.5	64x1	0.487	5.2	26	7	457	7.14	68.1
NEC SX-6 IXS	NEC SX-6	0.5	32x1	0.661	6.9	18	7	228	7.14	92.6
NEC SX-6+ IXS	NEC SX-6+	0.5625	32x1	0.672	6.8	19	7	268	8.37	80.3
NEC SX-6+ IXS+	NEC SX-6+	0.5625	4x8	6.759	7.0	8	6	(268)	(66.96)	(100.9)
IBM HPS	IBM Power 4+	1.7	64x4	0.724	1.7	8	6	1074	16.79	43.1
IBM HPS	IBM Power 4+	1.7	32x4	0.747	1.7	8	6	532	16.62	45.0
Cray X1	Cray X1 MSP	0.8	252x1	0.429	4.0	22	10	2385	9.46	45.3
Cray X1	Cray X1 MSP	0.8	124x1	0.709	4.9	20	10	1205	9.72	72.9
Cray X1	Cray X1 MSP	0.8	120x1	0.830	3.7	20	10	1061	8.84	93.9
Cray X1	Cray X1 MSP	0.8	64x1	0.941	4.2	20	9	522	8.15	115.4
Cray X1	Cray X1 MSP	0.8	60x1	1.033	3.9	21	9	578	9.63	107.3

+ This row is based on an additional measurement with the communication benchmark software. The HPL value of this row is taken from the previous row because there isn't a benchmark value available and significant differences between single- and multi-threaded HPL execution are not expected. The last two columns are based on this HPL value.

For the bandwidth values, the achievable percentage on the random ring from the ping-pong varies between 4 % and 48 % with one exception: If only one (but multi-threaded) MPI process is running on each SMP node of a NEC SX-6+, random ring and ping-pong bandwidth are nearly the same. For the latency values, the ratio ping-pong to random varies between 0.23 and 0.99. On only a few systems, the ping-pong latency and the random ring latency are similar (e.g., on Infiniband, IBM HPS, NEC SX-6+ multithreaded).

These examples not only show the communication performance of different network types, but also that the ping-pong values are not enough for a comparison. The ring based benchmark results are needed to analyze these interconnects.

4   Balance of Communication to Computation

For multi-purpose HPC systems, the balance of processor speed, along with memory, communication, and I/O bandwidth is important. In this section, we analyze the ratio of inter-node communication bandwidth to the computational speed. To characterize the communication bandwidth between SMP nodes, we use the random ring bandwidth, because for a large number of SMP nodes, most MPI processes will communicate with MPI processes on other SMP nodes. This means, with 8 or more SMP nodes, the random ring bandwidth reports the available inter-node communication bandwidth per MPI process. To characterize the computational speed, we use the HPL Linpack benchmark value divided by the number of MPI processes, because HPL can achieve nearly peak on cache-based and on vector systems, and with single- and multi-threaded execution. The ratio of the random ring bandwidth to the HPL divided by the MPI process count expresses the communication-computation balance in byte/flop (see in Fig. 1) or byte/kflop (used in Tab. 1). Although the balance is calculated based on MPI processes, its value should be in principle independent of the programming model, i.e., whether each SMP node is used with several single-threaded MPI processes, or some (or one) multi-threaded MPI processes, as long as the number of MPI processes on each SMP node is large enough that they altogether are able to saturate the inter-node network [10].

Table 1 shows that the balance is quite different. Currently, the HPCC table lacks of the information, how many network adapters are used on each SMP nodes, i.e., the balance may be different if a system is measured with exactly the same interconnect and processors but with a smaller or larger amount of network adapters per SMP node. On the reported installations, the balance values start with 1.8 / 8.1 / 17.8 B/kflop on IBM Colony, Myrinet 2000 and Quadrics respectively. SGI Numalink, IBM High Performance Switch, Infiniband, and the largest Cray X1 configuration have a balance between 40 and 50 B/kflop. High balance values are observed on Cray XD1, Sun Fire Link (but only with 0.59 Gflops per MPI process), NEC SX-6 and on Cray X1 and X1E. The best values for large systems are for Cray XT3 and NEC SX-8 (see also Fig. 2).

For NEC SX-6, the two different programming models *single-* and *multi-threaded* execution were used. With the single-threaded execution, 25 % of the random ring connections involve only intra-node communications. Therefore only 0.504 GB/s (75 % from 0.672 GB/s) represent the inter-node communication bandwidth per CPU. The inter-node bandwidth per node (with 8 CPUs) is therefore 4.02 GB/s respectively. The balance of inter-node communication to computation is characterized by the reduced value 60.2 byte/kflop. With multi-threaded execution, all communication is done by the master-threads and is inter-node communication. Therefore, the random ring bandwidth is measured per node. It is significantly better with the multi-threaded application programming scheme (6.759 GB/s) than with single-threaded (4.02 GB/s). Implications on optimal programming models are discussed in [10].

Fig. 2 shows the scaling of the accumulated random ring performance with the computational speed. For this, the HPCC random ring bandwidth was multiplied with the number of MPI processes. The computational speed is benchmarked with HPL. The left diagram shows absolute communication bandwidth, whereas the right diagram plots the ratio of communication to computation speed. Better

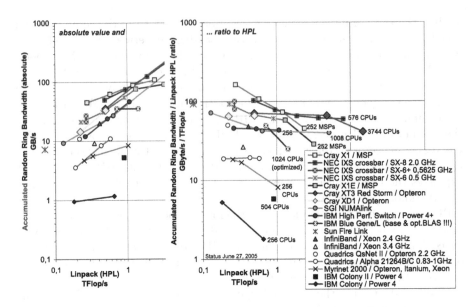

Fig. 2. Accumulated random ring bandwidth versus HPL Linpack performance

scaling with the size of the system is expressed by horizontal or a less decreasing ratio curve. E.g., the Cray X1 and X1E curves show a stronger decrease than the NEC SX-6 or SX-8. Interpolation at 3 TFlop/s gives a ratio of 30 B/kflop on Cray X1E, 40 B/kflop on SGI Altix 700 Bx2, 62 B/kflop on NEC SX-8, and 67 B/kflop on Cray XT3.

5   Ratio-Based Analysis of All Benchmarks

Fig. 3 compares the memory bandwidth with the computational speed analog to Fig. 2. The accumulated memory bandwidth is calculated as the product of the number of MPI processes with the embarrassingly parallel STREAM triad HPCC result. There is a factor of about 100 between the best and the worst random ring ratio values in Fig. 2, but only a factor of 25 with the memory bandwidth rations in Fig. 3 (right diagram). But looking at the systems with the best memory and network scaling the differences in the memory scaling are more significant. E.g., while NEC SX-8 and Cray XT3 have both shown best network bandwidth ratios, here, NEC SX-8 provides 2.4 times more memory bandwidth per Tflop/s than the Cray XT3. The CPU counts also indicate that different numbers of CPUs are needed to achieve similar computational speed.

Fig. 4 and Fig. 5 are comparing the systems based on several HPCC benchmarks. This analysis is similar to the current Kiviat diagram analysis on the HPCC web page [5], but it uses always embarrassingly parallel benchmark results instead of single process results, and it uses only accumulated global system values instead of per process values. If one wants to compare the balance of systems with quite different total system performance, this comparison can be done

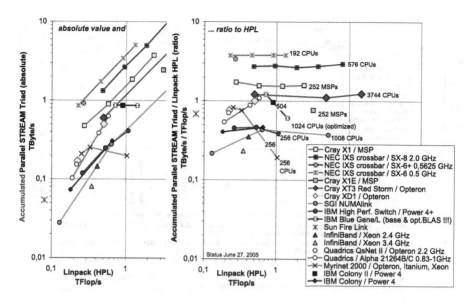

Fig. 3. Accumulated stream triad bandwidth versus HPL Linpack performance

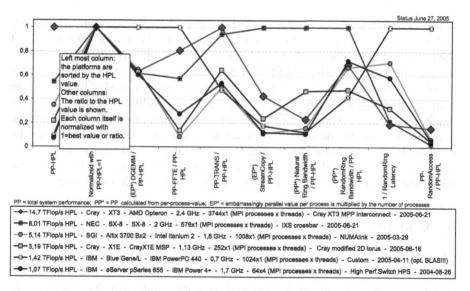

Fig. 4. Comparing the largest clusters in the HPCC list. Each system is normalized with its HPL value.

hardly on the basis of absolute performance numbers. Therefore in Fig. 4 and Fig. 5, all benchmark results (except of latency values) are normalized with the HPL system performance, i.e., divided by the HPL value. Only the left column can be used to compare the absolute performance of the systems. This normaliza-

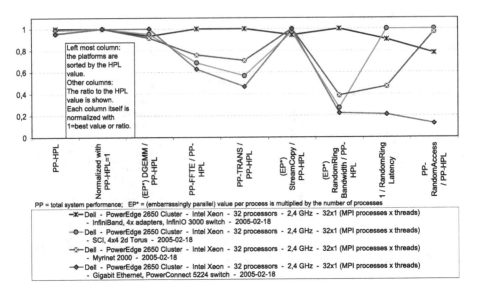

Fig. 5. Comparing different node interconnects

tion is also indicated by normalized HPL value in the second column that is per definition always 1. Each column itself is additionally divided by largest value in the column, i.e., the best value is always 1. The columns are sorted together to show influences: HPL and DGEMM are reporting performance with high temporal and spatial locality. FFT has a low spatial locality, and PTRANS a low temporal locality. FFT and PTRANS are strongly influenced by the memory bandwidth benchmark (EP STREAM copy) and the inter-process bandwidth benchmark (random ring). The two right-most columns are latency based: The reciprocal value of the random ring inter-process latency, and the Random Access benchmark ratio. Fig. 4 compares systems with more than 1 Tflop/s, whereas Fig. 5 analyzes the four different networks on a Dell Intel Xeon cluster.

6 Conclusions

The HPC Challenge benchmark suite and the uploaded results from many HPC systems are a good basis for comparing the balance between computational speed, inter-process communication and memory bandwidth. The figures presented in this paper clearly show the strengths and weaknesses of various systems. One can see that several systems provide a similar balance between computational speed, network bandwidth and memory bandwidth, although hardware architectures vary between MPP concepts (e.g., Cray XT3, IBM BlueGene/L), clusters of vector SMP nodes (NEC SX-8, Cray X1E), constellations (IBM), and ccNUMA architectures (SGI). One can also see that the gap between best and worst balance ratios is more than 25. The number of CPUs needed to achieve similar accumulated performance and network and memory bandwidth is also quite different. Especially the curves in Fig. 2–3 can be used for interpolation and to some extent also for extrapolation. Outside of the scope of the HPCC database is the price-performance ratio. In this paper, most scaling was done on the basis of the HPL system performance. In a procurement, relating the performance data additionally to real costs will give additional hints on pros and cons of the systems.

Acknowledgments

The authors would like to thank all persons and institutions that have uploaded data to the HPCC database, Jack Dongarra and Piotr Luszczek, for the invitation to Rolf Rabenseifner to include his effective bandwidth benchmark into the HPCC suite, Holger Berger for the HPCC results on the NEC SX-6+ and helpful discussions on the HPCC analysis, David Koester for his helpful remarks on the HPCC Kiviat diagrams, and Gerrit Schulz and Michael Speck, student co-workers, who have implemented parts of the software.

References

1. John McCalpin. *STREAM: Sustainable Memory Bandwidth in High Performance Computing.* (http://www.cs.virginia.edu/stream/)

2. Jack J. Dongarra, Jeremy Du Croz, Sven Hammarling, Iain S. Duff: A set of level 3 basic linear algebra subprograms. *ACM Transactions on Mathematical Software (TOMS)*, 16(1):1–17, March 1990.

3. Jack J. Dongarra, Jeremy Du Croz, Sven Hammarling, Iain S. Duff: Algorithm 679; a set of level 3 basic linear algebra subprograms: model implementation and test programs. *ACM Transactions on Mathematical Software (TOMS)*, 16(1):18–28, March 1990.

4. Jack J. Dongarra, Piotr Luszczek, and Antoine Petitet: The LINPACK benchmark: Past, present, and future. *Concurrency nd Computation: Practice and Experience*, 15:1–18, 2003.

5. Jack Dongarra and Piotr Luszczek: *Introduction to the HPCChallenge Benchmark Suite.* Computer Science Department Tech Report 2005, UT-CS-05-544. (http://icl.cs.utk.edu/hpcc/).

6. Panel on HPC Challenge Benchmarks: An Expanded View of High End Computers. SC2004 November 12, 2004
(http://www.netlib.org/utk/people/JackDongarra/SLIDES/hpcc-sc2004-panel.htm).

7. Alice E. Koniges, Rolf Rabenseifner and Karl Solchenbach: *Benchmark Design for Characterization of Balanced High-Performance Architectures.* In IEEE Computer Society Press, proceedings of the 15th International Parallel and Distributed Processing Symposium (IPDPS'01), Workshop on Massively Parallel Processing (WMPP), April 23-27, 2001, San Francisco, USA, Vol. 3. In IEEE Computer Society Press (http://www.computer.org/proceedings/).

8. Parallel Kernels and Benchmarks (PARKBENCH)
(http://www.netlib.org/parkbench/)

9. Rolf Rabenseifner and Alice E. Koniges: *Effective Communication and File-I/O Bandwidth Benchmarks*. In J. Dongarra and Yiannis Cotronis (Eds.), Recent Advances in Parallel Virtual Machine and Message Passing Interface, proceedings of the 8th European PVM/MPI Users' Group Meeting, EuroPVM/MPI 2001, Sep. 23-26. Santorini, Greece, pp 24-35.

10. Rolf Rabenseifner: *Hybrid Parallel Programming on HPC Platforms*. In proceedings of the Fifth European Workshop on OpenMP, EWOMP '03, Aachen, Germany, Sept. 22-26, 2003, pp 185-194

11. Daisuke Takahashi, Yasumasa Kanada: High-Performance Radix-2, 3 and 5 Parallel 1-D Complex FFT Algorithms for Distributed-Memory Parallel Computers. *Journal of Supercomputing*, 15(2):207–228, Feb. 2000.

12. Nathan Wichmann: *Cray and HPCC: Benchmark Developments and Results from Past Year*. Proceedings of CUG 2005, May 16-19, Albuquerque, NM, USA.

A Space and Time Sharing Scheduling Approach for PVM Non-dedicated Clusters*

Mauricio Hanzich[2], Francesc Giné[1], Porfidio Hernández[2], Francesc Solsona[1], and Emilio Luque[2]

[1] Departamento de Informática e Ingeniería Industrial,
Universitat de Lleida, Spain
{sisco, francesc}@eps.udl.es

[2] Departamento de Informática, Universitat Autònoma de Barcelona, Spain
{porfidio.hernandez, emilio.luque}@uab.es, mauricio@aomail.uab.es

Abstract. Wasted resources are a common reality in open laboratories in any University today [2]. Our aim is to take advantage of those resources to do parallel computation without disturbing the local tasks excessively. In order to implement a system that lets us execute parallel applications in a non-dedicated cluster, we propose a new environment, termed CISNE, that integrates Time and Space Sharing scheduling over non-dedicated PVM clusters.

1 Introduction

The studies in [2] indicate that the workstations in a NOW are under-loaded. Our aim is to take advantage of those idle resources to do parallel computation without disturbing the local tasks. Many alternatives have been proposed for dealing with such a non-dedicated environment, including migration as Condor [4], load balancing as Mosix [1], etc. The main drawback of Condor and Mosix is the high cost in doing process migration. Also, they do not consider the relationship between distributed and local applications to enhace the resource utilization. Furthermore, Condor allows parallel jobs to run only when a workstation's owner is not using this machine. Thus, such systems focus on coarse-grained idle periods (on a scale of minutes or hours). However, there are other cycles available that such systems do not harvest.

Due to the above reasons, our proposal is oriented towards the *Job Scheduling* alternative. Parallel job scheduling in a non-dedicated cluster can be performed at two different levels, Space and Time Sharing. Space Sharing (SS) scheduling deals with three different kinds of problems. The first class, faces the problem of selecting the best set of nodes for executing an application (*Node Selection* policies), considering a non-dedicated cluster and its state. The second set of policies deals with the *Job Selection* process (i.e. First Fit, Best Fit, Just First,

---

* This work was supported by the MCyT under contract TIC 2001-2592 and partially supported by the Generalitat de Catalunya -Grup de Recerca Consolidat 2001SGR-00218.

B. Di Martino et al. (Eds.): EuroPVM/MPI 2005, LNCS 3666, pp. 379–387, 2005.

etc) from a waiting queue, while the third set deals with the *Job Ordering* or prioritization process (i.e. FCFS, Shortest Job First - SJF, Smallest Number of Processors First - SNPF, etc). It worth pointing out that the influence of the *Job Ordering* policy over the resulting scheduling is fully determined by the *Job Selection* policy.

Once the parallel job has been assigned to a set of nodes, the Time Sharing (TS) scheduling deals with the problem of distributing the CPU time among the parallel and local tasks. Based on the previous work performed in [5], a coscheduling approach is selected. Coscheduling deals with minimizing synchronization/communication waiting time between remote processes [3]. Thus, coscheduling may be applied to reduce message waiting time and to make good use of the idle CPU cycles by executing distributed applications in a cluster or NOW system.

The solution presented in this paper is an integral environment, called CISNE. CISNE is made up of LoRaS (a Space Sharing system) and CCS [6] (the coscheduling component). CCS implements an implicit coscheduling technique based on identifying the coscheduling necessities of cooperating tasks by gathering and analyzing implicit runtime information, basically communication events and memory requirements. Also, a social contract [7] is provided in the CCS system. This contract is based on insulating a workstation owner from the effects of parallel jobs by allocating local resources increasing the priority of local users. LoRaS provides the ability to join Space Sharing techniques with coscheduling ones. The main objective in doing so is to find the combination that provides the best parallel job performance. In order to enhace the coscheduling profits, LoRaS considers the cluster state, the job requirements and the local load.

The integration of Space and Time sharing systems in a non-dedicated environment is a complex problem and has several aspects to deal with. The lack of studies in this field has driven us to evaluate several scheduling alternatives with the help of CISNE, which also provides a means for selecting and configuring several Space Sharing combinations.

The remainder of this paper is outlined as follows: in section 2, the CISNE system is presented. In section 3, the SS policies implemented by CISNE are depicted. The efficiency measurements of CISNE under our scheduling alternatives are performed in Section 4. Finally, the main conclusions and future work are explained in Section 5.

2   CISNE

CISNE (Cooperative & Integrated Scheduler for Non-dedicated Environments) is a framework which provides a means to merge Space Sharing (SS) and Time Sharing (TS) scheduling towards non-dedicated cluster systems. Figure 1.left depicts the general architecture of the CISNE system. In this figure, the CCS module represents our TS system, carried out by an implicit coscheduling technique. The SS system is supplied by the LoRaS schema using a client-server approach for launching applications and controlling the cluster state. Besides, the interaction between both systems is depicted in the figure.

CCS provides a coscheduling technique for parallel jobs without excessively disturbing the response time of local tasks. Furthermore, CCS controls the assignment of the resources between parallel and local applications by balancing them according to a social contract, in which the minimum resource requirements of the local applications are guaranteed. Given that the CCS system was extensively presented in the past [6], in the present work we center our attention in the newer SS system.

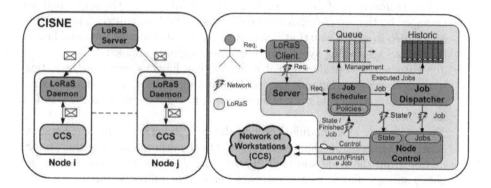

Fig. 1. CISNE general architecture (left) and LoRaS internal structure (right)

The LoRaS (Long Range Scheduler) schema deals with the SS scheduling. It is responsible for mapping both MPI and PVM distributed applications. Nevertheless, we are currently restricted to launching PVM jobs, because a part of CCS is implemented inside the PVM daemon. The information considered by LoRaS for taking its scheduling decisions is the following:

- *Cluster state.* It comprises information for each cluster node, such as load average, memory occupancy, local user activity and the MultiProgramming Level (MPL) of parallel applications running in the cluster.
- *The resource requirements* of the distributed applications waiting in the queue to be launched.
- *CCS characteristics.* It comprises the level of resources given to the parallel applications by the social contract applied by CCS.

The LoRaS components and their main features are (see figure 1.right):

- *Client*: sends a Job Execution Request (JER) to the server module on behalf of a parallel user.
- *Server*: the admittance of new JERs to be executed in the system is performed by the *server* module. This JER is then forwarded to the *Job Scheduler* module.

- *Job Scheduler*: this module is responsible for making job scheduling decisions when a new JER is received in the server. This scheduling decision is carried out by three different policies (explained in depth in the next section), applied in the following order:
  - First of all, the JERs are arranged according to one *job ordering policy*, settled by the system administrator.
  - Next, one job is selected from the queue according to the *job selection policy*.
  - Finally, the best set of nodes to run the chosen job is selected according to the *node selection policy*.
- *Policy (submodule of Job Scheduler)*: this module allows the set of policies managed by the *Job Scheduler* to be established and configured. This module is designed in such a way that it is easy to change its functionality, and hence, the LoRaS scheduling system for implementing our proposals.
- *Job Dispatcher*: considering that every job can have its own characteristics, it is necessary to configure the job before launching it. This module is responsible for doing this work before sending it to be executed in the selected set of nodes.
- *Node Control*: this module has two different functions. On one hand, it launches and controls the job execution. On the other hand, it gathers information from the node state and informs the *Job scheduler* (and hence, the *policy* submodule) so that it can take better scheduling decisions.

It is worth pointing out that LoRaS has been implemented integrally in the user space, including the CCS system which is a time critical system. Thus, the portability of the CISNE system is guaranteed.

3   Implemented SS Policies

The *Job Scheduler* module of LoRaS selects the next parallel job and the best set of nodes according to the Job Ordering (JO), job selection (JS) and the node selection policies, all of which are described below.

The parallel jobs are arranged into the waiting queue according to one of the following JO policies: *FCFS* (First Come First Serve), *SJF* (Shortest Job First), *SNPF* (Smallest Number of Processors First) and *SCDF* (Smallest Cumulative Demand First). The *SNPF* policy orders the queue by *increasing number of requested processors*. On the other hand, the *SCDF* policy orders the queue by the increasing product of the requested processors and time.

One job is selected from the waiting queue according to one of the following JS policies: *Best Fit* , *First Fit* and *Just First*.

It is important to remark that both kinds of JO and JS policies could be combined. The way they are merged determines the resulting policy for selecting a job from the queue. Table 1 shows the resultant policies from the merging job selection policies with the ordering ones.

It should be noted that the resulting policy can be exclusively oriented to JS or JO policies. This is so because the JO policy is irrelevant if the JS process

Table 1. The resultant policies from the merging of JS and JO policies

		Job Ordering			
		FCFS	SJF	SNPF	SCDF
Job Selection	JFirst	FCFS	SJF	SNPF	SCDF
	FFit	FCFS+FFit	SJF	SNPF+FFit	SCDF+FFit
	BFit	BFit	BFit	BFit	BFit

searches the whole queue to select a job, as the *Best Fit* (BFit) policy does. In this case, the resulting policy (JS + JO) is oriented towards the maximization of the resource usage without considering the priority of the queued jobs. On the other hand, a policy like *Just First* (JFirst), selects the first job in the queue, hence the resulting policy is the same as the JO one. With such a policy, the priorities of the jobs are totally preserved at the cost of some possible resource wasting. A compromise case is the *First Fit* (FFit) policy, which searches the queue until it finds a suitable job to execute. In this case, the JO policy has an intermediate influence on the resulting policy. It means that for a pair of fitting jobs, the policy will select the highest prioritary.

On the other hand, we still need to determine the best set of nodes for executing a given job and the current cluster state. This is done according to two different *Node Selection* policies:

- *Uniform.* it is characterized by the following: (a) it merges communication and computation bound applications in the same node and (b) it runs applications in an ordered manner, whenever possible. By *ordering the applications* we mean to launch tasks making up a pair of parallel applications *in the same set of nodes,* balancing the workload across the cluster. An example can be seen in figure 2.a, where J_3 shares its nodes with, and only with J_2.
- *Normal.* unlike the uniform policy, it merges the parallel job independently of its communication/computation characteristics and placement over the cluster. An example can be observed in figure 2.b, where the J_3 shares its nodes with J_1 and J_2.

In order to help CCS to perform better, both *Uniform* and *Normal* policies stop launching applications over those nodes with a load over the established social contract.

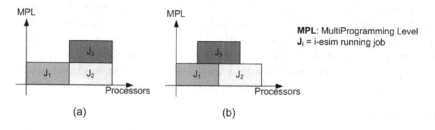

Fig. 2. *Uniform* (a) and *Normal* (b) job distribution

4   Experimentation

In order to carry out the experimentation, we first describe our environment, the exercised workloads and metrics. In the second subsection, we present a set of results that shows how CISNE performs under our defined SS policies.

4.1   Environment

In order to evaluate our assumptions we have to represent a non-dedicated cluster. On one hand, we need to emulate some local user activity and, on the other hand, we need some parallel applications that arrive at some intervals.

The local workload was carried out by running a synthetic benchmark. This allowed the CPU load, memory requirements and network traffic used by the local user to be fixed. In order to assign these values in a realistic way, we monitored the average resources used by real users. According to this monitoring, we defined two local user profiles. The first profile identifies 65% of the users with high needs on inter-activeness (called *XWindows* user: 15% CPU, 35% Mem., 0,5KB/sec LAN), while the other profile distinguishes 35% of the users with web navigation needs (called *Internet* user: 20% CPU, 60% Mem., 3KB/sec. LAN). This benchmark alternates CPU activity with interactivity by means of running several system calls and different data transfers to memory. According to the level of occupation of our monitored laboratories, we loaded 25% of the nodes with local workload in our experiments.

The parallel workload was a list of 90 PVM NAS parallel applications (CG, IS, MG, BT) with a size of 2, 4 or 8 tasks that reached the system following a Poisson distribution [3]. The parallel applications were merged so that the entire workload had a balanced requirement of computation and communication. It is important to note that the MPL reached for the workload depends on the system state at each moment, but in no case surpassed an MPL = 4. This threshold is based on the results shown in [6].

Both workloads were executed in a Linux cluster using 16 P-IV (1,8GHz) nodes with 512MB of memory and a fast ethernet interconnection network.

In order to show the CISNE performance under our proposed policies, we divide the experimentation into two parts. In a first step, we show the behavior of the mixed policies from table 1 by means of the applications *turnaround* time, and fixing the node selection policy to *Normal* with a MPL ≤ 4. Besides, some results showing the stability, and hence predictability, of those policies are presented using the *standard deviation*. In a second step, we compare a FCFS-JFirst-Normal policy, with an MPL = 1, termed *Basic*, with our *Normal* and *Uniform* policies. In this case, the MPL is greater than 1 and again the other policies are set to FCFS-JFirst. In this second experiment we also measure the application turnaround time as a user metric and the workload *makespan* (turnaround of the whole workload) as a system metric.

4.2   Results

In the first part of the experimentation, we show the CISNE performance for the different policies set in table 1. It is important to remark that these experiments are much more extensive [8], but for space reasons we only include a few here. In figure 3.left it is possible to observe how those policies that use BFit as their job selection sub-policy increase the turnaround time of the applications. This is due a resource usage maximization that increase the average MPL, and hence, the average application execution time is incremented affecting the turnaround. Considering that we are using resources that would otherwise be wasted, the usage maximization is not as important as applications performance. Therefore, we can see how policies that take into account the job priorities (those combined with JFirst or FFit), enhance the job turnaround time obtaining almost the same results, which are better for the application performance. The only difference we can find among them is for FCFS policy that forces a strict application arrival ordering and hence increases the application waiting time, resulting in a greater turnaround time.

In figure 3.right we present some results showing the stability, and hence, predictability of each mix. From the figure it is clear that any policy merged with FCFS is more stable, although more restrictive, than the other merges. Considering the policy, the mix with a FFit policy is a little more stable than the others policies in general. On the other hand, a mix with a JFirst policy (i.e. just considering the job ordering), gives us a more unstable merge.

The next step in our evaluation refers to the *Node Selection* policies. In figure 4.left it is possible to observe that the benefits of incrementing the parallel MPL are around 200% comparing our proposals with the *Basic* policy, from the applications turnaround point of view. This is due to the incredible reduction in the waiting time (800%), while the execution time is not incremented by more than 35%. Besides, we could also observe how a uniform distribution of jobs enhances the execution time. This effect is related to an enhancement of the coscheduling system performance that can now approximate a global context switch due to the same load characteristics throughout the cluster.

From the system point of view, figure 4.right shows the workload makespan. From this figure, it is clear that there is a relation between the turnaround and

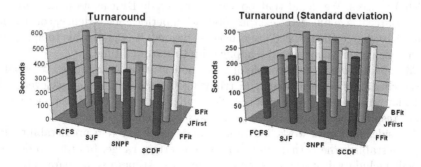

Fig. 3. Parallel application turnaround time (left) and turnaround standard deviation (right), for the evaluated policies

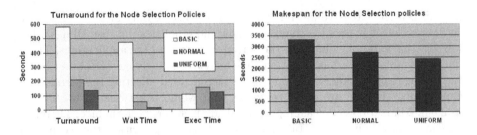

Fig. 4. Application wait, execution and turnaround time (left) and workload makespan (right), for the evaluated node selection policies, using CISNE

the makespan. Again a uniform policy gives us the best results. The reason is the execution time reduction that also impact on the waiting time and hence in the overall turnaround time. Nevertheless, the difference is not as much noticeable as in the case of the turnaround. This is due to the jobs inter-arrival time, which sets a minimum for the makespan value, and could not be less than the arrival time of the last job to the system. Hence, the makespan value tends to be more homogeneous than the individual applications turnaround time.

5   Conclusions and Future Work

This work presents a set of new SS policies oriented towards dynamically-cosche-duled, non-dedicated clusters. Besides, it describes a new integral system, named CISNE, that considers both SS and TS concerns, which is applied on a non-dedicated PVM cluster. Using this framework (CISNE + proposed policies), the paper analyzes how the performance of a dynamic coscheduling system could be affected by the SS policy over a non-dedicated cluster. With this aim, we evaluated several policies oriented towards *Node Selection* (*Uniform, Normal*), *Job Selection* (*BFit, FFit* and *JFirst*) and *Job Ordering* (*FCFS, SJF, SNPF* and *SCDF*) process. We found that policies that include BFit as its policy enhance the resource usage but decrease the application performance. Hence, considering that the resources being used are otherwise wasted, it is preferable to apply any other scheme that enhances the application turnaround time.

Moreover, we present a couple of other policies that are intended to enhance the coscheduling performance by distributing the applications more cleverly. We find that an ordered arrangement of the applications (i.e. *Uniform* policy) gives better results from the parallel user and system point of view.

Considering future work, we want to increase the system predictability, letting us establish the turnaround time into a certain range. In order to do this, we will include a historical system that lets us estimate some parameters for the executing jobs. Besides, we will study the characterization of the parallel applications and how this could be included into the estimating schema. On the other hand, we want to extend the system for considering MPI applications.

References

1. A. Barak, S. Guday and R.G. Wheeler. "The MOSIX Distributed Operating System: Load Balancing for Unix". *Springer-Verlag . LNCS, vol. 672. 1993.*
2. A. Acharya and S. Setia. "Availability and Utility of Idle Memory in Workstation Clusters". *In Proc. of the 1999 ACM SIGMETRICS, Vol. 27, pp 35/46. 1999.*
3. D. G. Feitelson. Packing schemes for gang scheduling. *In Job Scheduling Strategies for Parallel Processing, D. G. Feitelson and L. Rudolph (Eds.), Springer-Verlag. LNCS, vol. 1162, pp. 89-110. 1996.*
4. M. Litzkow, M. Livny and M. Mutka. "Condor - A Hunter of Idle Workstations". *In Proceedings of the 8th International Conference of Distributed Computing Systems, pp. 104–111. 1988.*
5. F. Solsona, F. Giné, P. Hernández, E. Luque. "Implementing Explicit and Implicit Coscheduling in a PVM Environment". *EuroPar 2000, LNCS, vol. 1900, pp 1164-1170. 2000.*
6. M. Hanzich, F. Giné, P. Hernández, F. Solsona and E. Luque. "Coscheduling and Multiprogramming Level in a Non-dedicated Cluster". *EuroPVM'2004, LNCS, vol. 3241, pp. 327-336. 2004.*
7. R.H. Arpaci, A.C. Dusseau, A.M. Vahdat, L.T. Liu, T.E. Anderson and D.A. Patterson. "The Interaction of Parallel and Sequential Workloads on a Network of Workstations". ACM SIGMETRICS'95, *pp.267-277. 1995.*
8. M. Hanzich. "Combining Space and Time Sharing on a Non-Dedicated NOW". *Master's thesis. Universitat Autònoma de Barcelona. 2004.*

Efficient Hardware Multicast Group Management for Multiple MPI Communicators over InfiniBand*

Amith R. Mamidala, Hyun-Wook Jin, and Dhabaleswar K. Panda

Department of Computer Science and Engineering,
The Ohio State University
{mamidala, jinhy, panda}@cse.ohio-state.edu

Abstract. MPI provides a set of primitives that allow processes to dynamically create communicators on the fly. This set of primitives can be exploited by the applications where only a certain group of processes need to participate at any given time. Also, these primitives play an important role in the context of dynamic process management of MPI-2. Special attention has to be paid in creating MPI communicators with InfiniBand's hardware multicast support as it involves the high overhead of interaction between the application and an external multicast management entity. In this paper, we propose different design alternatives of efficiently creating the communicators dynamically. The basic idea behind the schemes proposed is to remove most of the overhead of the hardware multicast group construction from the critical path of the application. Our results indicate that by using Multicast Pool and Lazy approaches of group construction proposed in the paper, we can significantly reduce the overhead by a factor of as much as 4.8 and 3.9, repectively, compared to the Basic approach.

Keywords: MPI, Communicator, Multicast, InfiniBand and Subnet Management.

1   Introduction

Message Passing Interface(MPI) [10] programming model has become the de-facto standard to develop parallel applications. MPI provides a rich collection of *point-to-point* and *collective* communication primitives for the application to take advantage of. These primitives are associated with a well defined *Communicator* object in MPI. Communicators provide a mechanism to construct distinct communication spaces for process groups to operate, isolating them from the rest of the communication flow. Also, they encapsulate several internal communication data structures during the program execution.

* This research is supported in part by Department of Energy's Grant #DE-FC02-01ER25506; National Science Foundation's grants #CCR-0204429, #CCR-0311542 and #CNS-0403342; grants from Intel and Mellanox; and equipment donations from Intel, Mellanox, AMD and Apple.

B. Di Martino et al. (Eds.): EuroPVM/MPI 2005, LNCS 3666, pp. 388–398, 2005.

InfiniBand Architecture (IBA) [6] which is emerging as the next generation interconnect for I/O and interprocessor communication, has several features
which directly impact the performance of the application. One of the notable features of InfiniBand is its support for hardware multicast. By using this feature,
a message posted to a hardware multicast group is delivered to all the processes
attached to this group in an efficient and scalable manner. In our earlier research,
we have shown that significant performance can be achieved by leveraging this
primitive to implement collective operations like MPI_Bcast, MPI_Barrier and
MPI_Allreduce [7] [8]. One primary assumption taken in the above approaches
is that all the processes communicate within a single communicator context.
Thus, it was suffice to construct a single hardware multicast group statically at
the initialization phase serving all the processes.

However, majority of the applications use more than one communicator object during their execution. This is because all the processes may not need to
communicate with each other. Also, the creation of a new communicator is im-
perative in the context of dynamic process management of MPI-2, where new
processes can be spawned from an already existing group of processes. To utilize
the hardware multicast of InfiniBand, these communicators have to be mapped
to hardware multicast groups and this mapping needs to be done on the fly.
More importantly, the multicast groups have to be dynamically set up.

In IBA, construction of hardware multicast groups involves a series of management actions. Some of these involve the interaction of the MPI processes
with an external IBA multicast management and the rest pertain to the fabric configuration by the multicast management entity. Only after the successful completion of these management actions the multicast group can be used.
Depending on the size of the hardware multicast group and the IBA fabric,
all these tasks can take considerable amount of time. From the MPI application perspective, the overhead of these operations should be as minimal as
possible.

In this paper, we present several ways of constructing hardware multicast
groups dynamically. We propose several design alternatives to efficiently map the
communicators to these newly created hardware multicast groups. Our designs
of using Multicast Pool and Lazy approaches of group construction outperform
the Basic approach by a factor of as much as 4.8 and 3.9, respectively, on a 32-
node cluster. We have implemented our proposed designs and integrated them
into MVAPICH [2], a popular implementation of MPI over InfiniBand which is
being used by more than 230 organizations world-wide. The rest of the paper is
organized as follows. In Section 2 we provide the background, Section 3 provides
the motivation for our work, Section 4 presents the various design alternatives
followed by performance evaluation, related work and conclusion.

2   InfiniBand Hardware Multicast Groups

The InfiniBand Architecture (IBA) [6] defines a switched network fabric for
interconnecting processing nodes and I/O nodes. It provides a communication

and management infrastructure for inter-processor communication and I/O. Especially, it provides support for hardware multicast. A hardware multicast group in IBA is realized as a set of ports connected together using a logical spanning tree. Each hardware multicast group has a unique Multicast Group IDentifier (MGID). The routing of multicast packets posted on a multicast group is handled using routing tables present in all the participating switches of the IBA fabric. The nodes join and leave a multicast group through a management action involving Subnet Management and Subnet Administration classes of IBA management. In the remaining part of the paper we use the term multicast management entity to describe the body which implements the functionality of these classes.

The multicast management entity is responsible for handling all the operations specific to multicast group construction from the end nodes. These operations are the following: **Multicast Group Create** which is issued by an end node to create a multicast group. This is an explicit operation in IBA to provide a single control of group characteristics like Message Transfer Unit, etc. and allow members to join subversively. **Multicast Group Join** which is issued by the end node to join the multicast group and **Multicast Group Leave** which is issued for leaving the group. All these requests are transported using MAnagement Datagrams called MADs. The multicast management entity on receiving the Join/Leave requests, constructs the multicast spanning tree and updates the participating switches in the IBA fabric with the new routing information.

3   Motivation: Mapping Between Multicast Groups and MPI Communicators

Communicators play an important role during MPI communication. Communicator objects encapsulate information about all the processes that communicate with each other. This is required for the underlying MPI implementation which interacts with the network device in the forwarding of the messages.

One important information which is required in a communicator to support hardware multicast is that of Multicast Group IDentifier (MGID). Consider a scenario where one process wants to send a message to all the other processes in the communicator. This process issues a MPI_Bcast call with the communicator object as one of its parameter. The underlying MPI layer then posts the message to the multicast group identified by MGID and the actual forwarding is automatically taken care of by the IBA layer.

Figure 1 illustrates the relationship between the communicators and the hardware multicast groups. Let us consider an MPI application consisting of five processes, (P0-P4) as an example. These processes are launched on a subnet consisting of four end nodes (N0-N3) connected by a switch. Processes with global ranks three, four and five (i.e., P2, P3 and P4) are present in one communicator. The local ranks of these processes in the communicator are indicated in the

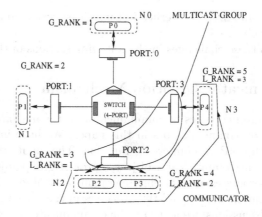

Fig. 1. Mapping between IBA Multicast Groups and MPI Communicators

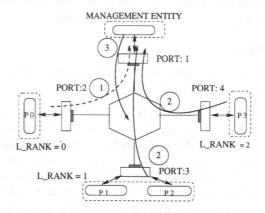

Fig. 2. Multicast Group Setup Operations

figure. For these processes to use hardware multicast, the communicator has to be mapped to the hardware multicast group consisting of port numbers 2 and 3.

In the remaining sections of the paper, we explain how this mapping is done during communicator creation. An important factor to consider is that issuing the Create/Join requests mentioned in the earlier section does not imply that the hardware multicast group is ready for use. This is because the multicast management entity has to first process these requests and construct a spanning tree containing the participating ports. Second, the routing tables in the fabric have to be updated to reflect the logical tree topology. The IBA specification does not define any specific mechanism of informing the processes of the completion of these tasks. Moreover, on large scale clusters, setting up multicast routing information can take considerable time if the size of the multicast group is comparable to the cluster size. This leads to the following questions:

1. How can the MPI application know when the multicast group is ready for use?

2. Can we minimize the overhead of multicast group construction from the MPI perspective?

We address these challenges in the following sections of the paper.

4   Communicator Creation Mechanism

Though there are two types of communicators *intra* and *inter* defined in MPI, we focus on *intra* communicators in this paper. We have implemented all our designs using the MPI_Comm_create function. The inputs to this function are an already existing communicator object, a process group object comprising of a new subset of processes and the final communicator object. MPI_Comm_create is a collective call invoked by all the processes in the existing communicator. In the following discussions, we focus on the communicator creation in the context of mapping these to the hardware multicast groups. All the other steps like the assignment of a unique context and the local ranks have already been done by the time we start constructing the multicast group.

4.1   Basic Design

The following steps are involved in the basic communicator construction. All of these are illustrated in Figure 2.

Multicast create and join: In this step, the process whose local rank is zero issues a create request to the multicast management entity specifying the Multicast Group IDentifier (MGID)(step 1 in Figure 2). The remaining processes then issue join requests to the multicast management entity using the same MGID (step 2 in Figure 2). All these requests carry the port identifiers so that the management entity knows which all ports would like to join a multicast group. The multicast management entity after receiving and validating the requests computes a logical spanning tree containing the ports specified in the requests. It then updates all the routing tables of the participating switches in the fabric (step 3 in Figure 2). At this point of time, the set up of hardware multicast group is complete.

However, the participating processes have no knowledge of this information. One approach to accomplish this would be to let the multicast management entity notify the MPI application after updating the routing tables. Another approach would be to let the MPI application discover about the completion independently. We have taken the latter approach in all our designs as it does not depend on any particular implementation of the multicast management entity. We refer to this approach as *multicast testing*.

Multicast testing: In this approach, the following algorithm is implemented by all the processes after they finish issuing the requests. Process with rank zero who is the root, posts a multicast *ping* message to the new hardware multicast group and waits for Acks from all the other processes. If the routing has been done, the message is received by all the processes and these processes soon post the Acks to the root. On the other hand, if routing is not complete then the message may not arrive at some of the processes. These processes block waiting for the *ping* message. Meanwhile, the root retransmits the *ping* message after a certain time-out interval. This process repeats until everyone has received the *ping* message.

4.2   Lazy Approach

Although the Basic design is good for its simplicity, it is blocking in nature. The application has to wait for the multicast management entity to process the requests and update the routing tables. Until then, all the processes block in the *multicast testing*. Depending on the size of the cluster and the multicast group this can take a considerable amount of time. Instead of doing the *multicast testing* in an eager fashion within the communicator creation call, we do this in a lazy manner by calling this routine every time a collective call is made. We do this until the *multicast testing* phase is over. We accomplish this by making the *multicast testing* as a non-blocking routine.

Asynchronous return: The new *multicast testing* is implemented in the following manner. The root process posts the *ping* message and checks for the arrival of the Acks from the rest of the processes. It does not block for the Acks to arrive. In the subsequent collective calls to this routine, it repeatedly checks for the progress of the Acks. It reposts the *ping* message only if the timeout is exceeded. The root keeps an estimate of the time elapsed by recording the time-stamps in the communicator object. The remaining processes behave in a similar fashion. They check for the *ping* messages in a non-blocking fashion and post the Acks soon after discovering the *ping* message.

Point-to-Point fall back: One important issue requiring detailed attention is the progress of the collective communication call before the communicator is ready for hardware multicast. In our approach, all the collective communication traffic is transmitted via point-to-point messaging until the root discovers that the routing has been done.

This approach overcomes the drawbacks of the Basic design. Due to the asynchronous nature of the *multicast testing* routine, overlap of computation as well as communication is easily achievable.

4.3   Multicast Group Pool Based Design

Though the Lazy approach can effectively hide the overhead of hardware multicast group construction in the MPI application, it still has some drawbacks. The benefits of hardware multicast in an application is reduced if the set-up time of the multicast groups is high and the collective communication follows the setting up of these communicators. Using our earlier design, the communication traffic falls back to point-to-point if the multicast groups are not set up. But, this does not improve the performance of the application.

Multicast Group Pool: We overcome the drawback mentioned above using a *complementary* approach of setting up communicators explained as follows. The basic idea in this design is to have a certain pre-defined pool of multicast groups already constructed. These groups contain all the processes to begin with. In the communicator construction routine, instead of participating nodes joining the multicast group, the non-participating nodes leave a multicast group chosen from the pool. There are several advantages of using this approach. First of all, since the multicast groups are already set-up the routing tables in the fabric are in place. So, when the application calls communicator creation function we can use the multicast group directly and we avoid the overhead of the *multicast testing* phase. This approach considerably improves the utility of the hardware multicast groups in an application. Secondly, the multicast pool can be maintained easily as most of the overhead is due to the multicast management entity and can be done in the background. We now explain the steps involved in this design.

When a call to the communicator creation is made, first a multicast group is chosen from the available list of multicast groups already constructed. If this pool is empty we fall back to the Lazy approach explained in the previous section. Once an available multicast group is obtained, the non-participating processes issue leave requests to the management entity. The list of non-participating process can be easily obtained by subtracting the set of the processes involved in the communicator from the global set involving all the processes. This global set is the MPI_GROUP_WORLD process group in MPI. Once a multicast group is consumed from the pool, it is immediately replenished by making all the processes issue requests for group construction. We also need to check for *multicast testing* before including the group in the pool. However, this check is done in the background by the application. The initial pool can be either constructed by the management entity or by the MPI application in the initialization phase. We have taken the latter approach in our implementation.

5   Performance Evaluation

Each node in our experimental testbed has dual Intel Xeon 2.66 GHz processors, 512 KB L2 cache, and PCI-X 64-bit 133 MHz bus. They are equipped with MT23108 InfiniBand HCAs with PCI-X interfaces. An InfiniScale MTS14400 switch is used to connect all the nodes. OpenSM, version 1.7.0, is the multicast management entity used in our tests.

5.1   Basic Hardware Group Setup Latencies

OpenSM has two parameters which affect the performance of multicast group creation. These are: 1) timeout which is the time for transaction timeouts in milliseconds and 2) maxMADs which is the number of MADs that can be outstanding on the wire at any given point of time. We measure *multicast testing* to tune these parameters as this reflects the time taken by OpenSM to configure routing tables. Figure 3 shows these results. From these we have chosen 10 ms for timeout and the number of outstanding MADs is set to maximum for OpenSM to deliver best performance.

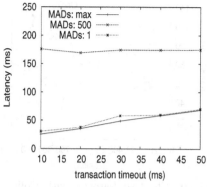
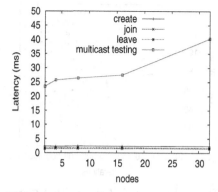

Fig. 3. Tuning of Multicast Testing

Fig. 4. Overhead of Basic Multicast Group Operations

Figure 4 indicates the results of the basic multicast group operations like create, join and leave. We also present the *multicast testing* time for varying number of nodes. As the figure indicates, *multicast testing* overhead is very high compared to the latencies of issuing create, join or leave requests. This is because as explained in the previous sections, after the requests are issued the management entity has to compute the spanning tree and update routing information of the switches in the fabric.

5.2 Effective Latency of Suggested Schemes

To compare the different schemes suggested in the paper we have measured the effective latency which is the latency of MPI_Bcast operation together with the communicator creation time. We have chosen the size of the message to be 1024 bytes in all our tests. The benchmark is constructed by calling communicator creation followed by the communication calls as many as the number of iterations specified. This is done for communicator sizes of 16 and 32 respectively.

In Figure 5 we measure the effective latencies for varying number of iterations for all the three schemes: Basic, Lazy and Pool. We have also taken the traditional point-to-point collectives as the reference. We refer to this as the Original design in the figures. As shown in the figure, the Pool based design outperforms all the rest. This is because *multicast testing* phase can be fully overlapped with the communicator creation operations and also the multicast group is immediately available. For the Lazy approach, we see the benefits of hardware multicast with the increasing number of iterations. This is because of the increasing percentage of communication using hardware multicast rather than point-to-point. The basic design performs poorly compared to all the designs. This is due to the high overhead associated with the *multicast testing* which is not overlapped with communication. Figure 6 shows the same trend for communicator size of 32. Note that the latencies of Pool and Lazy are almost the same for 16 and 32 for higher number of iterations. This is due to the scalability of hardware multicast.

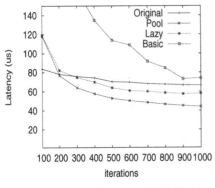

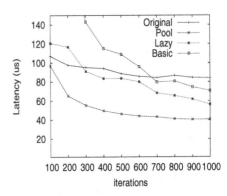

Fig. 5. Effective Latency with Collectives 16 processes

Fig. 6. Effective Latency with Collectives 32 processes

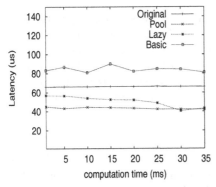

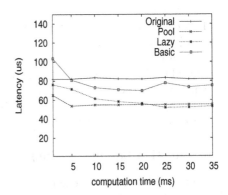

Fig. 7. Effective Latency with Computation and Collectives 16 processes

Fig. 8. Effective Latency with Computation and Collectives 32 processes

To understand the overlap with computation we have introduced some computation between the communicator creation and the communication in the benchmark used for the above experiments. Figures 7 and 8 show the trend with increasing computation for sizes 16 and 32 respectively. The Lazy approach due to its asynchronous nature can overlap communicator creation with computation where as the Basic cannot. The Pool based design on the other hand can immediately take the benefits of hardware multicast. However, the initial latencies for size 32 are higher than for size 16 due to the increased overhead of creating larger hardware multicast group. As Figure8 indicates, the Pool based design and the Lazy approaches improve the effective latency by a factor of 4.9 and 3.8, respectively.

6   Related Work

Various aspects of subnet management like subnet discovery, routing and setting up of forwarding tables have been studied using simulation techniques by the authors in [3] [4] [9]. Paper [5] deals with implementing MPI collective operations using IP multicast over Fast Ethernet. In [11], the authors propose different designs for constructing IP multicast groups. Also, collectives have been implemented using hardware multicast and NACK-based schemes in [1]. Our work differs from these as we provide dynamic schemes of hardware multicast group construction in the context of InfiniBand and we overlap these with the application progress.

7   Conclusions and Future Work

In this paper, we propose efficient schemes of dynamically constructing communicators with hardware multicast support in InfiniBand. The basic idea behind the schemes is to overlap the group construction with the progress of the application. The Multicast Pool and the Lazy approaches proposed in this paper move most of the overhead of multicast group creation out of the critical path of the application execution. We have evaluated these schemes together with the Basic scheme and found that the Multicast Pool performs the best of all the three followed by the Lazy scheme. Multicast Pool and Lazy schemes improve the Effective Latency by a factor of 4.9 and 3.8 respectively. In our future work, we would like to evaluate the impact of these schemes on a range of MPI applications with and without dynamic process creation.

Acknowledgements. We would like to thank Eitan Zahavi, Dror Goldenberg and Eitan Rabin from Mellanox for providing helpful comments.

References

1. Multicast collectives. http://vmi.ncsa.uiuc.edu.
2. MVAPICH: MPI over InfiniBand Project.
   http://nowlab.cis.ohio-state.edu/projects/mpi-iba/.
3. A. Bermudez, R. Casado, F. J. Quiles, T. M. Pinkston, and J. Duato. Evaluation of a Subnet Management Mechanism for InfiniBand Networks. In *Proceedings of ICPP*, 2003.
4. A. Bermudez, R. Casado, F. J. Quiles, T. M. Pinkston, and J. Duato. On the InfiniBand Subnet Discovery Process. In *Proceedings of Cluster Computing*, 2003.
5. H. A. Chen, Y. O. Carrasco, and A. W. Apon. MPI Collective Operations over IP Multicast. In *Workshop PC-NOW 2000*, 2000.
6. InfiniBand Trade Association. InfiniBand Architecture Specification, Release 1.2. http://www.infinibandta.org, October 2004.
7. J. Liu, A. R. Mamidala, and D. K. Panda. Fast and Scalable MPI-Level Broadcast using InfiniBand's Hardware Multicast Support. In *Proceedings of IPDPS*, 2004.
8. A. R. Mamidala, J. Liu, and D. K. Panda. Efficient Barrier and Allreduce Infini-Band Clusters using Hardware Multicast and Adaptive Algorithms . In *Proceedings of Cluster Computing*, 2004.
9. J. C. Sancho, A. Robles, and J. Duato. Effective Strategy to Compute Forwarding Tables for InfiniBand Networks. In *Proceedings of ICPP*, 2001.

10. M. Snir, S. Otto, S. Huss-Lederman, D. Walker, and J. Dongarra. *MPI–The Complete Reference. Volume 1 - The MPI-1 Core, 2nd edition.* The MIT Press, 1998.
11. X. Yuan, S. Daniels, A. Faraj, and A. Karwande. Group Management Schemes for Implementing MPI Collective Communication over IP-Multicast. In *The 6th International Conference on Computer Science and Informatics, Durham, NC*, March 8-14 2002.

Assessing MPI Performance on QsNet^{II}

Pablo E. García[1], Juan Fernández[1],
Fabrizio Petrini[2], and José M. García[1]

[1] Departamento de Ingeniería y Tecnología de Computadores,
Universidad de Murcia, 30071 Murcia, Spain
{pablo.garcia, juanf, jmgarcia}@ditec.um.es
[2] CCS-3 Modeling, Algorithms & Informatics,
Los Alamos National Laboratory, Los Alamos, NM 87545, USA
fabrizio@lanl.gov

Abstract. To evaluate the communication capabilities of clusters, we must take into account not only the interconnection network but also the system software. In this paper, we evaluate the communication capabilities of a cluster based on dual-Opteron SMP nodes interconnected with QsNet^{II}. In particular, we study the raw network performance, the ability of MPI to overlap computation and communication, and the appropriateness of the local operating systems to support parallel processing. Experimental results show a stable system with a really efficient communication subsystem which is able to deliver 875 MB/s unidirectional bandwidth, 1.6 μsec unidirectional latency, and up to 99.5% CPU availability while communication is in progress.

1   Introduction

Clusters have become the most successful player in the high-performance computing arena in the last decade. At the time of this writing, many of the fastest systems in the Top500 list [14] are clusters. These systems are typically assembled from commodity off-the-shelf (COTS) components. In particular, there is a growing interest in those systems assembled with SMP nodes based on 64-bit processors –mainly Itanium2 and Opteron– interconnected with high-performance networks, such as Infiniband [11], Myrinet [12] or Quadrics [13].

Performance of large-scale clusters is determined by the parallel efficiency of the entire system as a whole rather than by the peak performance of individual nodes. In order to achieve a high degree of parallel efficiency, there must be a proper balance over the entire system: processor, memory subsystem, interconnect, and system software. If we focus on the communication capabilities, we must pay attention not only to the interconnection network but also to the communication system software. In this case, while Myrinet, Infiniband and Quadrics are the preferred choices for the cluster interconnect, MPI [10] is the *de facto* standard communication library for message-passing, and Linux is the most popular choice for operating the cluster nodes.

The traditional approach to evaluate the communication capabilities of a cluster primarily relies on bandwidth and latency tests. Even though the bandwidth and latency figures are significant, they are not enough to characterize

B. Di Martino et al. (Eds.): EuroPVM/MPI 2005, LNCS 3666, pp. 399–406, 2005.
© Springer-Verlag Berlin Heidelberg 2005

the system behavior when running scientific and engineering applications in a cluster. There are other aspects that have a great impact in the communication performance as well. On the one hand, the ability of the MPI layer to overlap communication and computation may limit the CPU availability for user applications. An efficient MPI implementation should leverage modern network interface cards to offload protocol processing. On the other hand, the commodity OSes running in every node may interfere with user-level processes. The nodes of a cluster should be properly tuned in order to minimize the computational noise introduced by unnecessary dæmons, services, and tools. In this paper, we focus on analyzing all these aspects on a cluster based on dual-Opteron SMP nodes interconnected with QsNet^{II} from Quadrics.

The rest of the paper is organized as follows. The next section presents the main features of QsNet^{II}. In particular, both the Elan4 network interface card and the Elite4 switch are described. In Section 3, different performance aspects of a QsNet^{II}-based cluster assembled with Opteron nodes are analyzed. Finally, we conclude with some remarks and an outline of future work.

2 QsNet^{II}

QsNet from Quadrics have become one of the preferred choices to interconnect large-scale clusters since its appearance in 1996 [14]. This success is due to the fact that QsNet provides low-latency and high-bandwidth interprocessor communication through a standard interface for systems based on commodity processing nodes. The latest version, QsNet^{II}, was released in 2004. Some large-scale systems based on QsNet^{II}, such as *Thunder* [14], have already been deployed.

QsNet^{II} consists of two ASICs: Elan4 and Elite4. Elan4 is the core for the QsNet^{II} network interface card (NIC). The Elan4 NICs connect commodity processing nodes to the QsNet^{II} network through a standard interface, PCI-X. In turn, Elite4 is capable of driving eight bidirectional links at 1.3 GB/sec each way. The Elite4 switches form a multistage network to interconnect the Elan4 NICs attached to the processing nodes. These are the most salient aspects of QsNet^{II}:

- 64-bit architecture. Both the Elan4 and the Elite4 components have an internal 64-bit architecture and fully support a 64-bit virtual address space.
- PCI-X interface. Elan4 NICs implement a 64-bit, 133 MHz, PCI-X interface.
- DMA engine. User processes can perform read/write from/on remote memory locations by just issuing Remote DMA (RDMA) commands to the Elan4.
- Event engine. Events are used for synchronization purposes and available to the Elan4 processors and the main CPU. Events are triggered upon completion of communication operations (e.g. RDMA transactions).
- Virtual operation. QsNet^{II} extends the conventional virtual memory mechanism so that user processes can transfer data directly between their virtual address spaces.
- Programmability. In addition to the internal command processor, the Elan4 NICs provide a 64-bit RISC programmable thread processor.

- Support for collectives. QsNet^{II} provides hardware-supported multicast operations over subnets of nodes to cut down the synchronization time.
- Reliability. QsNet^{II} implements a packet-level link protocol in hardware which is able to detect faults, route packets around faulty areas, and even retransmit lost packets. Moreover, packets are CRC-protected to detect data corruption.

2.1   Elan4

The Elan4 NICs are in charge of injecting and receiving packets into and from the network. In addition, every Elan4 NIC incorporates a programmable thread processor to offload higher-level protocol processing to the NIC. The Elan4 NIC functional units are interconnected using several separate 64-bit data buses to increase concurrency and reduce latency operation. The main Elan4's functional units are the *command processor*, the *thread processor*, the *DMA engine*, the *event engine*, and the *Short Transaction Engine* (STEN). The *command processor* processes commands from either the main CPU or other Elan4's functional units. Under this model, there is a command port mapped directly into the user process's address space. Each command port is no more than a command queue where user processes can directly issue one or more commands to the Elan4 without OS intervention. In this way, the command processor executes commands from different user processes on their behalf. Also, the command processor controls the thread processor, the DMA engine, the event engine and the STEN. The *thread processor* is a 200 MHz, 64-bit RISC programmable thread processor enriched with special instructions to support lightweight threads. This thread processor can be programmed in C and is used to aid the implementation of communication libraries without explicit intervention from the main CPU. The *Short Transaction Engine* (STEN) is closely integrated with the command processor. This specialized functional unit is optimized to handle short messages. For further details about Elan4 see [5].

2.2   Elite4

The Elite4 is an eight-port crossbar, with two virtual channels per link, that can deliver 1.3 GB/sec each way. QsNet^{II} connects the Elite4 switches in a quaternary fat-tree topology. The Elite4 switches use source routing to implement an *up/down* routing algorithm which takes advantage of this topology. The routing tags can identify either a single output link or a group of links for multicast transfers. The routing algorithm is adaptive in the *up* phase and deterministic in the *down* phase. The implementation of this routing algorithm is highly efficient and introduces a delay of approximately 20 *ns* per switch.

At the link level, packets are divided into smaller 32-bit flits to use wormhole flow control. Every packet transmission creates a virtual circuit between the source and the destination node. The virtual circuit is closed after the destination node acknowledges packet reception.

3   Performance Evaluation

In this section, we present the performance results obtained in our initial evaluation of a cluster based on dual-Opteron SMP nodes interconnected with QsNet^{II}. In particular, we have conducted experiments to measure the basic network performance , the ability of QsNet^{II}'s MPI to overlap computation and communication, and the level of intrusiveness of the local OS in user-level computation. Table 1 summarizes the experimental setup.

Table 1. Experimental Setup

Characteristic		Description
Nodes	Processor	2xAMD Opteron 244 1.8 GHz
	Chipset	AMD-8131 HyperTransport
	I/O Bus	64-bit PCI-X (66, 100, and 133 MHz)
	BIOS	AMIBIOS 08.00.10
	Memory	1x1GB DDR400
Interconnect	NIC	QM500b A02 PCI-X Elan4
	Switch	QS8A B01 8-port Elite4
Software	Kernel	2.4.21-178.x86_64 / 2.4.21-4.16qsnet
	OS	SuSE Linux 9.0 (x86-64)
	Libraries	qsnet2libs-1.6.9-0 / qsnetmpi-1.24-37
	Compiler	gcc-3.3.1-23
	Launcher	SLURM 0.3.8-1

3.1   Network Performance

To expose the network performance of QsNet^{II} as seen by parallel applications, we wrote our microbenchmarks at the MPI level. We perform two different experiments to characterize the network in terms of latency and bandwidth. Unidirectional bandwidth and latency are computed using a simple microbenchmark where two processes residing in two different nodes exchange messages. In this case, both processes invoke alternatively MPI_Send and MPI_Receive operations in a loop for different message sizes. In turn, bidirectional bandwidth and latency are obtained using a similar experiment where both processes send and receive messages simultaneously using MPI_Isend and MPI_Irecv operations. Figure 1(a) shows the MPI unidirectional and bidirectional bandwidth. The peak unidirectional bandwidth, obtained as half of the measured bidirectional traffic, is 875 MB/s, whereas in the bidirectional case is 857 MB/s. Figure 1(b) shows the MPI unidirectional and bidirectional latency. The minimum achievable latency is 1.58μsec for unidirectional traffic and 3.41μsec for bidirectional traffic. This remarkable low latencies are made possible by the QsNet^{II} software infrastructure which provides zero-copy, user-level network access. Finally note that these results indicate that Quadrics has improved the internal design of QsNet^{II} over the previous version of QsNet which showed a significant gap between unidirectional and bidirectional figures [7]. This improvement is due to the fact that Elan4 incorporates several separate 64-bit data buses [5].

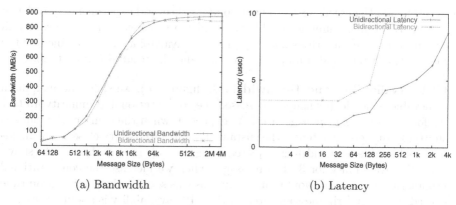

(a) Bandwidth   (b) Latency

Fig. 1. Network performance

3.2   Overlapping Computation and Communication

Network interface cards for modern cluster interconnects, such as Myrinet [1] or Quadrics [7], provide programmable processors and substantial memory. This trend opens a wide range of design possibilities for communication protocols since this added capability allows the host processor to delegate certain tasks to the NIC [9]. This offloading of protocol processing has two significant benefits. First, moving communication protocol processing to the NIC increases the availability of the host processor for use by application programs, that is, to overlap computation and communication. Second, NIC-based collectives show dramatically reduced latency and increased consistency over host-based versions when used in large-scale clusters [6].

In this section, we measure the ability of QsNet^{II}'s MPI to overlap computation and communication. This capability is influenced not only by the characteristics of the underlying network, but also by the quality of the MPI implementation. In order to characterize the overlapping of computation and communication, we measure the processor availability –how much the processor is available to application programs– while communication is in progress. To do so, we use a test, namely *post-compute-wait* test [4], which combines computation and MPI communication.

Post-Compute-Wait Test. This test consists of a *worker process* and a partner *support process* running on two separate nodes. The *worker process* posts a non-blocking send and a receive directed to the partner process, performs some parametric amount of computation, and waits for the pending send and receive calls to complete. The *support process* posts a matching send and receive. The *worker process* code is instrumented to time the non-blocking call phase, the compute phase, and the wait phase. The compute phase is a do-nothing loop which keeps the host processor busy, without timers or system calls, for a predefined amount of time. This loop performs neither memory accesses nor I/O in order to avoid operations which might introduce non-determinism in the experiments. Using this test, we have conducted several experiments in order to obtain (i) the maximum achievable bandwidth given specific message sizes and computational granularities and (ii) the CPU availability figure defined as the ratio between the total compute phase time and the total execution time.

CPU Availability and Bandwidth. In figure 2(a), we show the maximum achievable bandwidth when we increased the computational granularity up to 1 ms for four different message sizes. As expected, when the computation time is shorter than message latency, the sustained bandwidth is very close to the maximum achievable bandwidth as depicted in figure 1(a). In turn, figure 2(b) shows the CPU availability for 32 KB messages when we increase the computational granularity up to 1 ms. Note that, in this case, as soon as the computation time exceeds the roundtrip message latency, the CPU availability is about 95.9% and grows up to 99.5%.

Finally, it is worth noting that this test provides an additional check of whether the MPI library complies with the progress rule of the MPI standard. This rule determines that the non-blocking send and receive calls must complete independently of a process making MPI library calls. As we have shown, QsNet^{II}'s MPI perfectly complies with this rule.

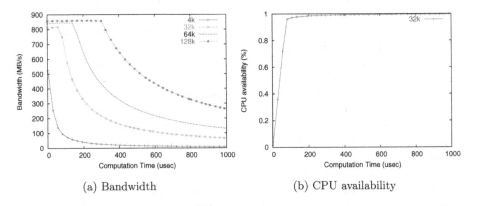

(a) Bandwidth   (b) CPU availability

Fig. 2. Post-compute-wait test

3.3   Computational Noise

Performance of many parallel applications is limited by the ability of the entire system to globally synchronize all nodes. However, local operating systems running on the cluster nodes lack global awareness about parallel applications. Local OS kernels and system dæmons are randomly scheduled across cluster nodes. This unpredictable behavior has a significant impact on tightly-coupled applications in which activities on the compute nodes are highly synchronized. Moreover, this performance bottleneck get worse as cluster size increases [8].

Several techniques have been proposed in the literature to minimize the impact of system activities on the overall performance of a parallel application.

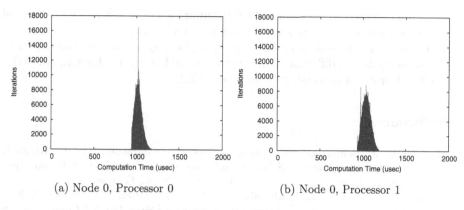

(a) Node 0, Processor 0   (b) Node 0, Processor 1

Fig. 3. Computational noise test

On the one hand, several authors have proposed different co-scheduling schemes to synchronize system activities across cluster nodes [2,3]. On the other hand, performance may be improved by just measuring the computational noise due to periodic system activities which may harm performance in order to remove them or ameliorate their impact [8]. The use of the above mentioned coscheduling techniques is not commonly used, since it requires kernel-level modifications. Therefore, in this section, we follow the second approach. We use a simple microbenchmark which quantifies computational noise due to system activities. In this microbenchmark, each node performs 1 million iterations of a synthetic computation which performs neither memory accesses nor I/O. Each synthetic computation has been calibrated to take about 1 ms in the absence of noise, that is, the run time for each iteration should always be the same in a noiseless machine.

In figures 3(a) and 3(b), we show the results for the first and the second processor on the master node, respectively[1]. From these results we can derive two interesting conclusions. First, the nodes of our cluster are noiseless even in the worse case which corresponds to the master node. Second, the distribution for the second processor has the very same shape but it is slightly displaced to the right on both nodes. This indicates that some kind of system activity is taking place only in the second processor. In this case, the only candidate is the kernel which might be descheduling the microbenchmark process in the second processor more often.

4   Conclusions and Future Work

In this paper, we have presented our initial analysis of the communication capabilities for a cluster based on dual-Opteron SMP nodes interconnected with QsNet^{II}. Our experimental results show that this platform has an extraordinary potential for high-performance cluster computing. The communication subsystem provides can rely on an extremely efficient network, a high degree of

---

[1] Results for slave nodes are similar and we omit them in the sake of brevity.

overlapping between computation and communication, and a really low level of computational noise due to system software interference.

Future work include scalability analysis for larger configurations, performance analysis for different traffic patterns, and study of different scenarios to offload protocol processing to the Elan4 NIC.

References

1. Nanette J. Boden, Danny Cohen, Robert E. Felderman, Alan E. Kulawik, Charles L. Seitz, Jakov N. Seizovic, and Wen-King Su. Myrinet: A Gigabit-per-Second Local Area Network. *IEEE Micro*, 15(1):29–36, February 1995.
2. Juan Fernández, Eitan Frachtenberg, and Fabrizio Petrini. BCS-MPI: A New Approach in the System Software Design for Large-Scale Parallel Computers. In *Proceedings of IEEE/ACM Conference on SuperComputing*, Phoenix, AZ (USA), November 2003.
3. Terry Jones, William Tuel, and Brian Maskell. Improving the Scalability of Parallel Jobs by adding Parallel Awareness to the Operating System. In *Proceedings of IEEE/ACM Conference on SuperComputing*, Phoenix, AZ (USA), November 2003.
4. William Lawry, Christopher Wilson, and Arthur B. Maccabe. COMB: A Portable Benchmark Suite for Assessing MPI Overlap. In *Proceedings of IEEE International Conference on Cluster Computing*, Chicago, IL (USA), September 2002.
5. Quadrics Supercomputers World Ltd. *Elan4 Reference Manual*.
6. Adam Moody, Juan Fernández, Fabrizio Petrini, and Dhabaleswar K. Panda. Scalable NIC-Based Reduction on Large-Scale Clusters. In *Proceedings of IEEE/ACM Conference on SuperComputing*, Phoenix, AZ (USA), November 2003.
7. Fabrizio Petrini, Wu chun Feng, Adolfy Hoisie, Salvador Coll, and Eitan Frachtenberg. The Quadrics Network: High-Performance Clustering Technology. *IEEE Micro*, 22(1):46–57, January/February 2002.
8. Fabrizio Petrini, Darren J. Kerbyson, and Scott Pakin. The Case of the Missing Supercomputer Performance: Achieving Optimal Performance on the 8192 Processors of ASCI Q. In *Proceedings of ACM/IEEE Conference on SuperComputing*, Phoenix, AZ (USA), November 2003.
9. Piyush Shivam, Pete Wyckoff, and Dhabaleswar K. Panda. EMP: Zero-copy OS-bypass NIC-driven Gigabit Ethernet Message Passing. In *Proceedings of IEEE/ACM Conference on SuperComputing*, Denver, CO (USA), November 2001.
10. Marc Snir, Steve Otto, Steven Huss-Lederman, David Walker, and Jack Dongarra. *MPI: The Complete Reference*. MIT Press, 1996.
11. www.infinibandta.org. Infiniband Trade Association.
12. www.myri.com. Myricom, Inc.
13. www.quadrics.com. Quadrics Supercomputers World Ltd.
14. www.top500.org. Top500 Supercomputing Sites.

Optimised Gather Collectives on QsNet^{II}

Duncan Roweth and David Addison

Quadrics Ltd, Bristol, United Kingdom

Abstract. In this paper we describe the implementation of the gather and allgather collectives on QsNet^{II}. Results from a cluster of 980 4-CPU nodes show good latencies, bandwidths and scaling, with a 3920 process, 8-byte, gather completing in 88 microsecs.

Keywords: Gather, AllGather, QsNet, Parallel, Collective.

1 Introduction

Many scientific applications exhibit the need for communication patterns that involve global data movement and global control[1]. Barrier synchronization, broadcast, gather, scatter, reduce and global exchange are typical examples of collective communication patterns.

QsNet^{II} is the latest generation of Quadrics interconnect (see [2] and [3], it consists of two ASICs: Elan4 and Elite4. The Elan4 communication processor forms the interface between a processing node and a high performance multistage network. It has a 64 bit internal architecture and supports 64 bit virtual addresses. The Elan4 generates and accepts packets to and from the network. In addition, it provides local processing power to implement the high-level message passing protocols required in parallel processing. The network is constructed from Elite4 switch components that are capable of switching eight bi-directional communications links. Each link carries data in both directions simultaneously at 1.3 Gbytes/sec. The link bandwidth is shared between two virtual channels. The network supports point-to-point transfer between arbitrary nodes and broadcast across selected ranges of nodes.

In this paper we describe how the QsNet^{II} components are used to implementing gather operations, information on the implementation of other collectives can be found in[4].

2 QsNet

The main features of QsNet^{II} are low latency, high bandwidth, scalability, reliable transmission and a commodity host adapter interface. Many scientific applications are very sensitive to the MPI[5] communication latency. The Elan adapter minimizes this latency by providing specialized units to quickly pipeline small messages into the network, perform protocol processing and notify the completion of the communication primitive. Its pipelined DMA engine uses split transaction reads to maximize host adapter bandwidth. QsNet^{II} is designed to scale to thousands of nodes, both in terms of hardware capability and system

B. Di Martino et al. (Eds.): EuroPVM/MPI 2005, LNCS 3666, pp. 407–414, 2005.

software design. QsNet^{II} is used in Thunder at Lawrence Livermore National Lab (the fifth most powerful computer in the world at the time of writing[6]) and mpp2 at Pacific Northwest National Lab, both 1024-node systems with IA64 CPUs. QsNet^{II} implements a reliable transmission protocol in hardware, and is able to detect faults, route packets around faulty switches and re-transmit packets in the presence of data errors. The Elan4 network adapter contains the following functional units: a pipelined DMA engine, a 64-bit microprocessor, an MMU, 32KB of 4-way set associative cache, a command processor that defines a virtual command queue interface, a short message-processing unit called STEN (Small Transaction Engine), a 1.3 Gbyte/sec each way network interface connection and a PCI-X, 64-bit, 133 MHz host interface.

User processes can perform remote read/write memory operations by issuing DMA commands to the Elan4. The DMA engine services a queue of outstanding DMA requests and ensures that they are completed reliably. The DMA engine can handle arbitrary source and destination buffer alignment as well as endian conversions. In addition, there are facilities to issue broadcasts and queued DMAs. The DMA engine processes 2 DMAs concurrently to overlap the start-up/finish latency and maintain full PCI-X read bandwidth.

The completion of a data transfer can be signalled by setting of an event in both source and destination processes. Events can be a simple word in memory (allowing a process to poll them), but the event engine allows a number of more sophisticated operations to be performed. A main processor interrupt can be generated (allowing a process to poll an event for some period of time and then sleep in the device driver) or a copy can occur. Event copies are of particular interest, as they can be used to initiate further DMAs. For example, the completion of one DMA can be used to copy the descriptor for another to a command queue. Its completion can trigger a third and so on. The main processor need only prepare the DMA descriptors and issue the first. It can then perform other work until finally waiting on completion of the last DMA. QsNet^{II} supports counted events; a wait event of 3 for example will fire when 3 set events have occurred. This mechanism allows us to construct arbitrary trees of processes where an event in the parent fires when each child has completed an operation.

The Elan adapters connect each node to a multi-stage switch network constructed from Elite switches. Each switch is an 8 by 8 bi-directional full crossbar with 2 virtual channels per link. Networks are constructed using in a fat tree topology. This network maintains full bi-section bandwidth.

The Elite4 switch supports broadcast operations in hardware. A DMA packet input on one link can be sent on to a range of output links. Broadcast packets are routed up the broadcast tree (any one of the many trees in the QsNet^{II} network) to a point high enough that all destination nodes can be reached, then down to all of the nodes in the range. Acknowledgements are combined back up this tree and a single success or failure token is returned to the source. This mechanism allows data to be sent to all nodes in much the same time as it can be sent to any one.

The Quadrics software stack includes both MPI and Shmem[7] interfaces. These libraries are implemented with libelan, which provides inter-process communication primitives and libelan4, the device specific command issue library. The interface between MPI and libelan is device independent, allowing the same MPI library to be used for both Elan4 adapters and the older Elan3 adapters installed in many AlphaServer SC and Linux clusters. Dynamic libraries are used throughout, allowing the same user binary to run on different systems of either generation.

MPI collectives operate on arbitrary subsets of processes called communicators. The COMM_WORLD communicator contains all processes; it is created by the call to MPI_Init(). Further communicators containing a subset of the processes can be created and destroyed dynamically. Quadrics libraries map MPI communicators onto a group structure. Each group structure will have its own data Elan events, buffers etc.

MPI collectives do not require each process to supply symmetric addresses (a symmetric variable is at the same address in each process); a gather for example, can collect data from different addresses in each process to a buffer whose address is only known at the root. Shmem collectives however, require the use of symmetric variables. The Elan library collectives handle both cases, with a flags field being used to indicate whether source and destination addresses are symmetric.

3   Gather Algorithms

In the simplest gather algorithms each process sends a fixed size element to the root using a message passing routine such as MPI_Send and the root process receives them using MPI_Recv. The time taken to complete the call scales linearly with the message passing latency and the number of processes.

The simple gather algorithm should work well for large element sizes where bandwidth into the root node controls performance, but its latency is poor. An alternative is to used a tree based gather algorithm, in which each process sends its elements and those of children up to its parent. In the MPICH 1.25 release a recursive doubling algorithm is used to gather elements on a binary tree[5]. In libelan we use trees of variable branching ratio.

The result of a gather is available only at the root process, in an allgather all processes receive the result. This can be achieved using a gather followed by a broadcast or using a ring algorithm[9]. In the ring algorithm each process sends its own elements to its neighbour. It then sends on the elements it has received from its neighbour, and so on in n steps until each process has received the elements from all n processes. We would expect the ring algorithm to perform relatively poorly for small element sizes, but for large element sizes it should achieve close to optimal bandwidth.

4   Gather Implementation

In the gather function the root process collects a fixed size block of data from each process.

```
void elan_gather(ELAN_GROUP *g, void *sbuf, void *dbuf,
                 size_t size, int root, ELAN_FLAGS flags);
```

If the destination address is symmetric this can be implemented in a straight-forward manner using a barrier followed by a call to elan_doput(). The barrier is required to ensure that the root process has entered the collective. If the destination buffer is not known to be symmetric then point-to-point message passing primitives are used instead.

The performance of this algorithm is shown by the plots labelled "Simple" in figures 1 and 2. Latency increases linearly with the number of processes, but is low (less than one microsecond per element) because of the Elan adapter's ability to process multiple such operations concurrently. The host interface limits gather bandwidth to between 820 and 910 Mbytes/sec depending upon the node type. Full bandwidth is sustained for large numbers of processes; it is not reduced by end point contention.

Tree based gather algorithms have been implemented to improve latency, we use a balanced tree of branching ratio 4 by default. Leaf nodes send data to their parents. Intermediate nodes add their own data and forward to their parents, and so on to the root. On completion of the gather the root process performs a broadcast set event releasing the other processes from the collective. In the first tree algorithm (labelled "Main" in figures 1 and 2), the intermediate processes poll in the main CPU until data arrives from their children and then send their data and that of their children on to their parent. This results in a significant reduction in latency (112 microsecs for a 3920 process 8-byte gather on 980 4-CPU nodes rather than 6598). Scaling of this algorithm is good, but the need to involve the main CPU on intermediate nodes can lead to poor performance

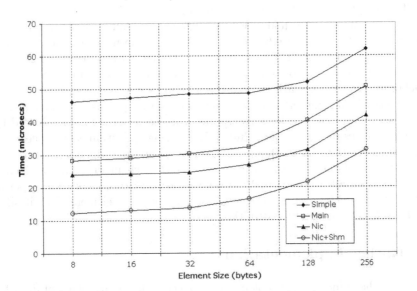

Fig. 1. Gather times as a function of element size for 64 processes

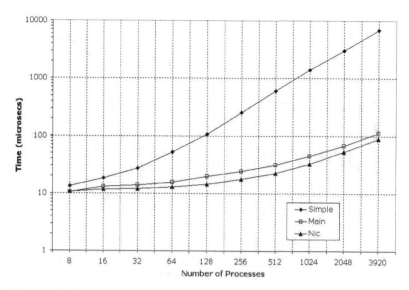

Fig. 2. Gather times as a function of number of processes, 8 byte elements

on large systems as all processes must be scheduled in order to progress the collective[10].

We have implemented a second tree based gather algorithm in which the Elan Event processor is used to chain the puts, reducing main processor involvement and PCI traffic. Each child sends its elements to adapter memory in its parent and then sets an event to signal completion of the transfer. The parent copies in its data and sets the event again. When the event count reaches the number of children plus one, the event fires and the adapter sends the data on up the tree without involving the main CPU. This algorithm reduces the gather latency further; with the time for a 3920 process 8-byte gather on 980 falling to 88 microsecs.

QsNet is generally deployed in multi-CPU nodes (2-16 CPUs is typical) with one MPI or Shmem process running on each CPU. Where a gather operation involves multiple processes per node we can first gather the local elements to per-node a shared memory buffer. The gather is then performed on a tree formed from the first process on each node. This approach reduces the number of network transfers. It works well for small element sizes.

Note that MPI allows non-trivial mapping of subgroups to nodes. When gathering over nodes it is necessary to unpick the mapping of processes to nodes and copy data out from a library buffer to the user's buffer. This limits the element size at which the shared memory optimisation is efficient.

Our tree algorithms are primarily to reduce gather latency, but they also deliver good bandwidth, by using buffers the NIC they avoid the need to copy data across the PCI bus on intermediate nodes. However, the adapter memory requirements scale as element size group size / branching ratio. This restricts the maximum element size as adapter SDRAM is a limited resource. The library

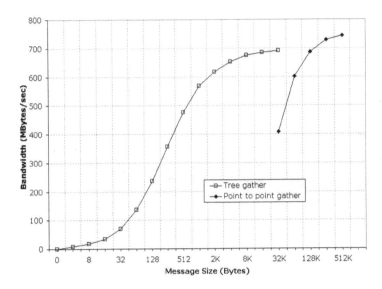

Fig. 3. Gather bandwidths for 64 HP RX2600 nodes

switches to the put based algorithm as the element size increases, as illustrated in figure 3.

5   AllGather Implementation

The QsNet hardware broadcast allows the root process in a gather to broadcast the result to all nodes with a simple put. In figure 4 we compare the results of

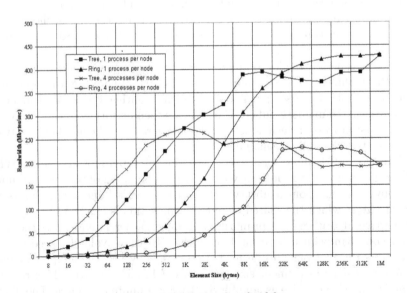

Fig. 4. AllGather bandwidths

this implementation of Allgather with those obtained using the ring algorithm. The single process per node results show that the tree gather and broadcast algorithm achieves significantly higher bandwidth at small element sizes. As the element size increases the gap narrows. Both algorithms achieve 400-450 Mbytes a second for large element sizes as we would expect. However, most MPI and Shmem applications run one process per CPU. Data received from the network must be copied to each local process. On commodity nodes allgather bandwidth is dominated by the node's memory copy performance.

In figure 4 we see that the aggregate allgather bandwidth with 4 processes per 4-CPU node peaks at 275 Mbytes per second for elements of size 1K, but drops to 200Mbytes/sec for large element sizes. Optimal QsNet hardware broadcast bandwidth is achieved when an application is run on a set of nodes that occupy physically contiguous range of network ports. Network broadcast bandwidth drops when the application is run on non-contiguous nodes. We use the gather and broadcast allgather algorithm for small element sizes, switching to the ring algorithm for large sizes if the application where hardware broadcast is less efficient.

6   Conclusions and Further Work

We have shown that QsNet^{II} systems can perform an 8-byte gather over 256 process on 64 4-CPU nodes in 24 microsecs and 3920 process on 980 nodes in 88 microsecs. We have also shown that both gather and allgather saturate link bandwidth for large element sizes. Allgather performance on SMP nodes is limited by memory copy performance, not network bandwidth.

We are working on optimisations for multi-rail systems in which each node has 4 or 16 CPUs and 2,4 or 8 adapters. We are also looking into using fnomial trees, which should reduce the time before data starts arriving at the root node.

7   Benchmark Programs

The results reported in this paper were all obtained using the gping benchmark program supplied with the QsNet^{II} release. This program measures the time taken to complete 1000 repetitions of the gather collective, reporting the average. Source code is included in the open source release, available from the Quadrics website. Similar results are obtained using other MPI benchmark suites, with for example gping, PMB and SkaMPI all reporting 18-19 microsecs for a 256 process, 8 byte AllGather.

The authors would like to thank Lawrence Livermore National Lab and Pacific Northwest National Lab for access to the systems used in preparing this paper. We would also like to acknowledge Adam Moody of Lawrence Livermore National Lab, Fabrizio Petrini of Los Alamos National Lab and Jon Beecroft of Quadrics for their help and encouragement.

References

[1] Fabrizio Petrini, Salvador Coll, Eitan Frachtenberg and Adolfy Hoisie. Hardware - and Software Based Collective Communication on the Quadrics Network. In Proceedings of the 2001 IEEE International Symposium on Network Computing and Applications (NCA 2001) Cambridge, Mass, October 8-10, 2001.

[2] David Addison, Jon Beecroft, David Hewson, Moray McLaren and Fabrizio Petrini. Quadrics QsNet^{II}: A network for Supercomputing Applications. In Hot Chips 15, Stanford University, CA, August 2003

[3] Jon Beecroft, David Addison, David Hewson, Moray McLaren, Fabrizio Petrini and Duncan Roweth. Quadrics QsNet^{II}: Pushing the Limit of the Design of High-Performance Networks for Supercomputers. In IEEE Micro. To appear, 2005.

[4] Duncan Roweth, Ashley Pittman and Jon Beecroft. Optimised Collectives on QsNet^{II} www.quadrics.com/documentation.

[5] Marc Snir, Steve Otto, Steven Huss-Lederman, David Walker and Jack Dongarra. MPI: The Complete Reference. The MIT Press, 1998, Volume 1, The MPI Core. The MIT Press, Cambridge, Massachusetts, 2nd edition, September 1998 ISBN0-262-69215-5.

[6] Hans W. Meuer, Erich Strohmaier, Jack J. Dongarra and Horst D. Simon. Top500 Supercomputer Sites June 2003 Available from www.top500.org

[7] Cray Man Page Collection: Shared Memory Access (SHMEM) S?2383?23, available from the Cray http://website www.cray.com/craydoc.

[8] Elan Programming Manual. Available from www.quadrics.com/documentation

[9] Gregory D Benson, Cho-Wai Chu, Qing Huang and Sadik G Caglar. A Comparison of MPICH Allgather Algorithms on Switched Networks in Lecture Notes in Computer Science Springer-Verlag GmbH, October 2003, pp. 335-343.

[10] Fabrizio Petrini, Darren Kerbyson and Scott Pakin. The Case of the Missing Supercomputer Performance: Achieving Optimal Performance on the 8,192 Processors of ASCI Q. In IEEE/ACM SC2003, Phoenix, AZ, November 2003.

An Evaluation of Implementation Options for MPI One-Sided Communication

William Gropp and Rajeev Thakur

Mathematics and Computer Science Division,
Argonne National Laboratory,
Argonne, IL 60439, USA
{gropp, thakur}@mcs.anl.gov

Abstract. MPI defines one-sided communication operations—put, get, and accumulate—together with three different synchronization mechanisms that define the semantics associated with the initiation and completion of these operations. In this paper, we analyze the requirements imposed by the MPI Standard on any implementation of one-sided communication. We discuss options for implementing the synchronization mechanisms and analyze the cost associated with each. An MPI implementer can use this information to select the implementation method that is best suited (has the lowest cost) for a particular machine environment. We also report on experiments we ran on a Linux cluster and a Sun SMP to determine the gap between the performance that could be achievable and what is actually achieved with MPI.

1 Introduction

Over the past decade, one-sided communication has emerged as a promising paradigm for high-performance communication on low-latency networks. The advantage of one-sided communication lies in its asynchronous nature: Unlike in point-to-point (or two-sided) communication where the sender and receiver explicitly call send and receive functions, in one-sided communication only the origin process calls the data-transfer function (put or get), and data transfer takes place without the target process explicitly calling any function to transfer the data. This model allows parallel programs to be less synchronizing and allows communication hardware to move data from one process to another with maximal efficiency. Nonetheless, some synchronization mechanism is needed in the programming model for the target process to indicate when its memory is ready for being read or written by a remote process and to specify when the data transfer is completed.

Because of the growing popularity of one-sided communication, the MPI Forum defined a specification for one-sided communication in MPI-2 [8]. MPI defines three data-transfer functions for one-sided communication: put (remote write), get (remote read), and accumulate (remote update). These data-transfer functions must be used together with one of three synchronization mechanisms—fence, post-start-complete-wait, and lock-unlock—as shown in Figure 1. Many

B. Di Martino et al. (Eds.): EuroPVM/MPI 2005, LNCS 3666, pp. 415–424, 2005.

```
Process 0                          Process 1
MPI_Win_fence(win)                 MPI_Win_fence(win)
MPI_Put(1)                         MPI_Put(0)
MPI_Get(1)                         MPI_Get(0)
MPI_Win_fence(win)                 MPI_Win_fence(win)
```

a. Fence synchronization

```
Process 0              Process 1                Process 2
                       MPI_Win_post(0,2)
MPI_Win_start(1)                                MPI_Win_start(1)
MPI_Put(1)                                      MPI_Put(1)
MPI_Get(1)                                      MPI_Get(1)
MPI_Win_complete(1)                             MPI_Win_complete(1)
                       MPI_Win_wait(0,2)
```

b. Post-start-complete-wait synchronization

```
Process 0                  Process 1                Process 2
MPI_Win_create(&win)       MPI_Win_create(&win)     MPI_Win_create(&win)
MPI_Win_lock(shared,1)                              MPI_Win_lock(shared,1)
MPI_Put(1)                                          MPI_Put(1)
MPI_Get(1)                                          MPI_Get(1)
MPI_Win_unlock(1)                                   MPI_Win_unlock(1)
MPI_Win_free(&win)         MPI_Win_free(&win)       MPI_Win_free(&win)
```

c. Lock-unlock synchronization

Fig. 1. The three synchronization mechanisms for one-sided communication in MPI. The numerical arguments indicate the target rank.

MPI implementations, including all vendor MPIs, support one-sided communication, with varying levels of optimization [1,2,4,6,7,9,10,12,14,15]. Nonetheless, Gabriel et al. [3] found that, because of the synchronization overhead in one-sided communication, regular point-to-point communication performs better than one-sided communication in five MPI implementations: NEC, Hitachi, IBM, Sun, and LAM. The only exceptions were NEC and Sun MPI when window memory allocated with the special function MPI_Alloc_mem is used. Clearly, it is necessary to study the costs associated with the synchronization mechanisms and optimize the implementations. In this paper, we analyze the semantics of the synchronization mechanisms, discuss options for implementing them, and analyze the overhead. This information is useful to MPI implementers in deciding which implementation method to use for a particular machine environment.

2   Fence Synchronization

Figure 1a illustrates the fence method of synchronization. In MPI, the memory region that a process exposes to one-sided communication is called a *window*, and a collection of processes create a *window object* that is used in subsequent one-sided communication functions. MPI_Win_fence is collective over the communicator associated with the window object. A process may issue one-sided operations after the first call to MPI_Win_fence returns. The next call to fence completes the one-sided operations issued by this process as well as the operations targeted at this process by other processes. An implementation of fence synchronization must support the following semantics: A one-sided operation cannot access a process's window until that process has called fence, and the second fence on a process cannot return until all processes needing to access that process's window have completed doing so.

2.1 Implementing Fence

In general, an implementation has two options for implementing fence: *immediate* and *deferred*.

Immediate Method. This method implements the synchronization and communication operations as they are issued. A simple implementation of this option is to perform a barrier in the first fence; perform the puts, gets, and accumulates as they are called; and perform another barrier at the end of the second fence after all the one-sided operations have completed. The first barrier ensures that all processes know that all other processes have reached the first fence and that it is now safe to access their windows. The second barrier ensures that a process does not return from the second fence until all other processes have finished accessing its window.

On a distributed-memory environment without hardware support for barriers, a barrier can be implemented by using the dissemination algorithm [5] with 0-byte messages. If p is the number of processes and α is the latency (or startup time) per message, this algorithm costs $(\lg p)\alpha$. The immediate method requires two barriers, which cost $2(\lg p)\alpha$. This method is expensive in environments where the latency is high, such as on clusters and networks running TCP. It is appropriate for environments where barriers can be fast, such as shared-memory systems or machines with hardware support for barriers, such as the Cray T3E and IBM BG/L.

Deferred Method. This method [12] takes advantage of the MPI feature that puts, gets, and accumulates are nonblocking and are guaranteed to be completed only when the following synchronization function returns. In the deferred method, the first fence does nothing and simply returns. The ensuing puts, gets, and accumulates are simply queued up locally. All the work is done in the second fence, where each process first goes through its list of queued one-sided operations and determines, for every other process i, whether any of the one-sided operations have i as the target. This information is stored in an array, such that a 1 in the ith location of the array means that one or more one-sided operations are targeted to process i, and a 0 means otherwise. All processes then do a reduce-scatter sum operation on this array (as in MPI_Reduce_scatter). As a result, each process knows how many processes will be performing one-sided operations on its window, and this number is stored in a counter in the window object. Each process then performs the data transfer for its one-sided operations and ensures that the counter at the target is decremented when all the one-sided operations from this process to that target have been completed. As a result, a process can return from the second fence when the one-sided operations issued by that process have completed locally and when the counter in its window object reaches 0, indicating that all other processes have finished accessing its window.

This method thus eliminates the need for a barrier in the first fence and replaces the barrier at the *end* of the second fence by a reduce-scatter at the *beginning* of the second fence before any data transfer. After that, all processes can do their communication independently and return when they are done (asynchronously). This method also enables optimizations such as message reordering, scheduling, and aggregation, which the immediate method does not.

On a distributed-memory system, a reduce-scatter operation on an array of p short integers (2 bytes) costs $(\lg p)\alpha + 2(p - 1)\beta$ [13], where β is the transfer time per byte between two processes. Because of the lower latency term, this method is preferred over the immediate method on systems where the latency is relatively high, such as on clusters.

2.2 Performance

To determine how MPI implementations perform for fence synchronization, we measured the cost of two `MPI_Barriers`, one `MPI_Reduce_scatter`, and two `MPI_Win_fences` on a Myrinet-connected Linux cluster at Argonne and a Sun SMP at the University of Aachen in Germany. On the Linux cluster, we used a beta version of MPICH2 1.0.2 with the GASNET channel on top of GM. On the Sun SMP, we used Sun MPI. We performed the operations several times in a loop, calculated the average time for one iteration, and then the maximum time taken by all processes. We used `MPI_Alloc_mem` to allocate window memory and passed assert `MPI_MODE_NOPRECEDE` to the first fence and (`MPI_MODE_NOSTORE` | `MPI_MODE_NOPUT` | `MPI_MODE_NOSUCCEED`) to the second fence.

Figure 2 shows the results. On the Linux cluster, the cost of two barriers is far more than the cost of a single reduce-scatter. Therefore, the deferred method is the preferred option, which is what MPICH2 uses in this case. We see that the cost of two fences is almost the same as that of a reduce-scatter. On the

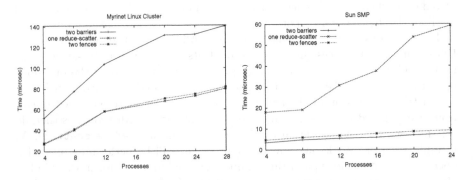

Fig. 2. Time taken for two barriers, one reduce-scatter, and two fences on a Myrinet-connected Linux cluster using MPICH2 (left) and on a Sun SMP using Sun MPI (right)

Sun SMP, Sun MPI has a very fast implementation of barrier, and therefore the immediate method is the preferred implementation strategy for fence. From the graph, it appears that Sun MPI does use the immediate method, because the time for two fences is only slightly higher than the time for two barriers.

3   Post-Start-Complete-Wait Synchronization

Fence synchronization, being collective over the entire communicator associated with the window object, results in unnecessary overhead when only small subsets of processes actually communicate with each other. To avoid this drawback, MPI defines a second synchronization mechanism in which only subsets of processes need to synchronize, as shown in Figure 1b. A process that wishes to expose its local window to remote accesses calls MPI_Win_post, which takes as argument an MPI_Group object that specifies the set of processes that will access the window. A process that wishes to perform one-sided communication calls MPI_Win_start, which also takes as argument an MPI_Group object that specifies the set of processes that will be the target of one-sided operations from this process. After issuing all the one-sided operations, the origin process calls MPI_Win_complete to complete the operations at the origin. The target calls MPI_Win_wait to complete the operations at the target.

An implementation of post-start-complete-wait synchronization must take into account the following semantics: A one-sided operation cannot access a process's window until that process has called MPI_Win_post, and a process cannot return from MPI_Win_wait until all processes that need to access that process's window have completed doing so and called MPI_Win_complete.

3.1   Implementing Post-Start-Complete-Wait

We again consider the immediate and deferred methods for implementing post-start-complete-wait. A few intermediate options also exist [4,8] but, for simplicity, we do not consider them here.

Immediate Method. In this method, MPI_Win_start blocks until it receives a message from all processes in the target group indicating that they have called MPI_Win_post. Puts, gets, and accumulates are performed as they are called. MPI_Win_complete waits until all one-sided operations initiated by that process have completed locally and then sends a done message to each target process. On the target, MPI_Win_wait blocks until it receives the done message from each origin process. Assuming that the size of the origin and target groups is g, the overhead of this method is $2g\alpha$. This method is appropriate in environments with low latency and native support for one-sided communication, such as shared memory, so that the data transfer can be initiated as soon as it is called.

Deferred Method. This method defers data transfer until the second synchronization call [12]. In MPI_Win_post, if the assert MPI_MODE_NOCHECK is not specified, the process sends a zero-byte message to each process in the origin group to indicate that MPI_Win_post has been called. It also sets the counter in its window object to the size of this group. On the origin side, MPI_Win_start does nothing and simply returns. All one-sided operations are simply queued up locally. In MPI_Win_complete, the origin process first waits to receive the zero-byte messages from the processes in the target group. It then performs all the one-sided operations and ensures that the window counter at the target gets decremented when all the one-sided operations from this process to that target have been completed. MPI_Win_complete returns when all its operations have locally completed. On the target, MPI_Win_wait simply blocks and invokes the progress engine until its window counter reaches zero, indicating that all origin processes have finished accessing its window.

Thus the only synchronization in this method is the wait at the beginning of MPI_Win_complete for a zero-byte message from the processes in the target group, and this too can be eliminated if the user specifies the assert MPI_MODE_NOCHECK to MPI_Win_post and MPI_Win_start (similar to MPI_Rsend). If the size of the origin and target groups is g, the overhead of this method is $g\alpha$. Therefore, the deferred method is faster in environments where latency is high.

4   Lock-Unlock Synchronization

In the lock-unlock synchronization method, the origin process calls MPI_Win_lock to obtain either shared or exclusive access to the window on the target, as shown in Figure 1c. After issuing the one-sided operations, it calls MPI_Win_unlock. The target does not make any synchronization call. When MPI_Win_unlock returns, the one-sided operations are guaranteed to be completed at the origin *and* the target. MPI_Win_lock is not required to block until the lock is acquired, except when the origin and target are one and the same process. Implementing lock-unlock synchronization when the window memory is not directly accessible by all origin processes requires the use of an asynchronous agent at the target to cause progress to occur because one cannot assume that the user program at the target will call any MPI functions that will cause progress periodically [8].

4.1   Implementing Lock-Unlock

We consider the immediate and deferred methods of implementing lock-unlock.

Immediate Method. In this method, MPI_Win_lock sends a lock-request packet to the target and waits for a lock-granted reply. Puts, gets, and accumulates are performed as they are called. MPI_Win_unlock waits until all one-sided operations initiated by that process have completed locally and then sends an unlock request to the target. It also waits to receive an acknowledgment from the target that all the one-sided operations issued from this process have completed at the target, as required by the semantics of MPI_Win_unlock. The cost for acquiring the lock is 2α, and the cost for releasing the lock is also 2α. Therefore, the total cost of this method is 4α, assuming no lock contention.

Deferred Method. In this method [12], MPI_Win_lock does nothing and simply returns. All one-sided operations are simply queued up locally. In MPI_Win_unlock, the origin sends a lock-request packet to the target and waits for a lock-granted reply. It then performs the one-sided operations. When these operations have completed locally, it sends an unlock request to the target and, in the general case, waits for a reply from the target indicating that the operations have completed at the target.

The deferred method also costs 4α in the general case, but it permits several optimizations that the immediate method does not. One optimization is that if any of the one-sided operations is a get, it can be reordered and performed last. Since the origin must wait to receive data in the get, when the get completes, it implies that the one-sided operations have also completed at the target (assuming ordered completion). In this case, an additional acknowledgment is not needed from the target, thereby reducing the cost to 3α. Another optimization in the case of a single put, get, or accumulate is that the origin can perform it as an atomic lock-(put/get/accumulate)-unlock request without having to wait for a lock-granted reply. If the operation is a get, even the additional completion acknowledgment from the target is not needed. These optimizations reduce the cost of lock-unlock to α and 0, respectively, because the lock request becomes part of the data transfer.

5 Analysis for Shared-Memory Environments

Shared-memory environments offer unique opportunities for optimizing MPI one-sided communication because of their support for atomic operations for fast synchronization and the ability to directly copy data to/from the user's buffer on the target if the window memory was allocated with MPI_Alloc_mem. We analyze in further detail the implementation of one-sided communication on shared-memory environments with lock-unlock synchronization. Consider this simple example that puts n longs into the memory window on the process specified by rank:

```
MPI_Win_lock(MPI_MODE_EXCLUSIVE, rank, 0, win);
MPI_Put(buf, n, MPI_LONG, rank, 0, n, MPI_LONG, win);
MPI_Win_unlock(rank, win);
```

We assume that the window memory was allocated with MPI_Alloc_mem and is directly accessible by a remote process. If we ignore error checking of function parameters, an MPI implementation need perform only the following steps for each of the above functions.

MPI_Win_lock
1. Make a routine call with four arguments
2. Convert win into an address (if not already an address)
3. Look up the address of the lock at the target (indexed access into win)
4. Check shared or exclusive access
5. Remote update for the lock

`MPI_Win_put`

1. Make a routine call with eight arguments
2. Convert `win` into an address (if not already an address)
3. Check that the remote memory is directly accessible
4. Get the base address of the remote memory (indexed access into `win`)
5. Get the displacement unit (indexed access into `win`)
6. Determine whether origin data is contiguous and get length (access datatype and multiply count by datatype size)
7. Determine whether target data is contiguous
8. Perform the copy of local to remote memory

`MPI_Win_unlock`

1. Make a routine call with two arguments
2. Convert `win` into an address (if not already an address)
3. Look up the address of the lock at the target (indexed access into `win`)
4. Remote update for the unlock

While the number of steps may seem large, they in fact involve relatively few instructions. However, the access to remote memory, either for the lock accesses or for the memory copy at the end of the `MPI_Put` step, may require hundreds of processor cycles. To determine the cost of performing the above steps, we wrote a shared-memory program using OpenMP [11] in which one thread performs the equivalent of lock, put, and unlock on another thread. We wrote four versions of this program:

1. A single routine OpenMP program where the lock, put, and unlock are all performed in the main routine by simply setting and clearing a flag for the lock and unlock steps and using `memcpy` to move the data.
2. The lock, put, and unlock operations are performed in separate routines, thus adding function-call overhead.
3. An `MPI_Win`-like structure is added so that the addresses of the lock and the memory at the target have to be obtained by indexing into the structure.
4. The routines use the same arguments as the corresponding MPI functions. This version adds more arguments to the routines and requires an extra lookup in the `win` structure for the displacement unit at the target.

We ran these programs on a Sun SMP (Sun Fire E6900, 1.2 GHz Ultra Sparc IV) at the University of Aachen with the Sun OpenMP compiler. We also ran the MPI version of the program with Sun MPI and a beta version of MPICH2 1.0.2 with the sshm (scalable shared memory) channel. For windows allocated with `MPI_Alloc_mem`, this version of MPICH2 uses the immediate method to implement all three synchronization methods in the sshm channel.

Table 1 shows the time taken to move 8, 256, 1024, and 64K `longs` (4 bytes) by these programs. The results show that the fastest MPI version is slower by a factor of 1.4 to 2 than the OpenMP version with all MPI features. We plan to investigate the cause of this difference in further detail, but preliminary studies indicate that the cost of `MPI_Win_lock` followed by `MPI_Win_unlock` is

Table 1. Time in seconds to perform a lock-put-unlock operation on a Sun SMP. n is the number of `longs` (4 bytes) moved. OpenMP 1 is a simple OpenMP implementation of this operation. The other three OpenMP programs add more features. OpenMP 4 mimics the steps an MPI implementation must implement. The last two columns show the times for two MPI implementations: Sun MPI and MPICH2.

n	OpenMP 1 (simple)	OpenMP 2 (routines)	OpenMP 3 (win struct)	OpenMP 4 (all MPI args)	Sun MPI	MPICH2
8	4.5e-7	4.8e-7	4.9e-7	5.4e-7	1.1e-6	1.3e-6
256	1.1e-6	1.1e-6	1.1e-6	1.2e-6	1.7e-6	1.9e-6
1024	4.7e-6	4.9e-6	5.0e-6	5.1e-6	5.2e-6	6.6e-6
64K	2.5e-4	2.6e-4	2.6e-4	2.7e-4	4.3e-4	5.4e-4

itself roughly twice as large as the equivalent steps in the OpenMP version. This may be due to the difference in the handling of thread and process locks in the operating system, and we plan to investigate this issue further.

6   Conclusions and Future Work

MPI one-sided communication has the potential to deliver high performance to applications. However, MPI implementations need to implement it efficiently, with particular emphasis on minimizing the overhead added by the synchronization functions. In this paper, we have analyzed the minimum requirements an implementation must meet to honor the semantics specified by the MPI Standard. We have discussed and analyzed several implementation options and recommended which option to use in which environments. Our analysis of the performance of lock-put-unlock on the Sun SMP demonstrates that MPI implementations are not too far off from delivering what can be delivered by using direct shared memory, although there is room for improvement. We plan to investigate in further detail where the additional gap lies and how much of it can be reduced with clever implementation strategies.

Acknowledgments

This work was supported by the Mathematical, Information, and Computational Sciences Division subprogram of the Office of Advanced Scientific Computing Research, Office of Science, U.S. Department of Energy, under Contract W-31-109-ENG-38. We thank Chris Bischof for giving us access to the Sun SMP machines at the University of Aachen.

References

1. Noboru Asai, Thomas Kentemich, and Pierre Lagier. MPI-2 implementation on Fujitsu generic message passing kernel. In *Proc. of SC99: High Performance Networking and Computing*, November 1999.

2. S. Booth and E. Mourão. Single sided MPI implementations for SUN MPI. In *Proc. of SC2000: High Performance Networking and Computing*, November 2000.
3. Edgar Gabriel, Graham E. Fagg, and Jack J. Dongarra. Evaluating dynamic communicators and one-sided operations for current MPI libraries. *Int'l Journal of High-Performance Computing Applications*, 19(1):67–80, Spring 2005.
4. Maciej Golebiewski and Jesper Larsson Träff. MPI-2 One-Sided Communications on a Giganet SMP Cluster. In *Proc. of the 8th European PVM/MPI Users' Group Meeting*, pages 16–23, September 2001.
5. Debra Hensgen, Raphael Finkel, and Udi Manbet. Two algorithms for barrier synchronization. *International Journal of Parallel Programming*, 17(1):1–17, 1988.
6. Weihang Jiang, Jiuxing Liu, Hyun-Wook Jin, Dhabaleswar K. Panda, Darius Buntinas, Rajeev Thakur, and William Gropp. Efficient implementation of MPI-2 passive one-sided communication over InfiniBand clusters. In *Proc. of the 11th European PVM/MPI Users' Group Meeting*, pages 68–76, September 2004.
7. Weihang Jiang, Jiuxing Liu, Hyun-Wook Jin, Dhabaleswar K. Panda, William Gropp, and Rajeev Thakur. High performance MPI-2 one-sided communication over InfiniBand. In *Proc. of the 4th Int'l Symp. on Cluster Computing and the Grid (CCGrid 2004)*, April 2004.
8. Message Passing Interface Forum. MPI-2: Extensions to the Message-Passing Interface, July 1997. http://www.mpi-forum.org/docs/docs.html.
9. Elson Mourão and Stephen Booth. Single sided communications in multi-protocol MPI. In *Proc. of the 7th European PVM/MPI Users' Group Meeting*, pages 176–183, September 2000.
10. Fernando Elson Mourão and João Gabriel Silva. Implementing MPI's one-sided communications for WMPI. In *Proc. of the 6th European PVM/MPI Users' Group Meeting*, pages 231–238, September 1999.
11. OpenMP. http://www.openmp.org.
12. Rajeev Thakur, William Gropp, and Brian Toonen. Optimizing the Synchronization Operations in MPI One-Sided Communication *Int'l Journal of High-Performance Computing Applications*, 19(2):119–128, Summer 2005.
13. Rajeev Thakur, Rolf Rabenseifner, and William Gropp. Optimization of collective communication operations in MPICH. *Int'l Journal of High-Performance Computing Applications*, 19(1):49–66, Spring 2005.
14. Jesper Larsson Träff, Hubert Ritzdorf, and Rolf Hempel. The implementation of MPI-2 one-sided communication for the NEC SX-5. In *Proc. of SC2000: High Performance Networking and Computing*, November 2000.
15. Joachim Worringen, Andreas Gäer, and Frank Reker. Exploiting transparent remote memory access for non-contiguous and one-sided-communication. In *Proc. of the 2002 Workshop on Communication Architecture for Clusters*, April 2002.

A Comparison of Three MPI Implementations
for Red Storm

Ron Brightwell

Scalable Computing Systems, Sandia National Laboratories*
P.O. Box 5800 Albuquerque, NM 87185-1110
rbbrigh@sandia.gov

Abstract. Cray Red Storm is a new distributed memory massively parallel computing platform designed to scale to tens of thousands of nodes. Red Storm has a custom network designed around the Cray SeaStar network interface and router. In this paper, we present an evaluation of three different MPI implementations for Red Storm: the vendor-supported MPICH2 implementation, and two other implementations based on MPICH 1.2.6. We discuss the differences in these implementations and show how various implementation strategies impact performance and scalability.

Keywords: MPI, Red Storm, XT3, Portals, Performance.

1   Introduction

Cray Red Storm is a new distributed memory massively parallel computing platform designed to scale to tens of thousands of processors. The Cray XT3 is the official product from Cray that differs slightly from the Red Storm platform that has been installed at Sandia National Laboratories in Albuquerque, New Mexico, USA. The XT3 has a three-dimensional torus network, while the Red Storm system is torus only in one direction. This limitation allows the Red Storm system to support more easily switching large sections of the machine between classified and unclassified computing. Other than this feature, the hardware and software environment of Red Storm is identical to the XT3. Henceforth, we will simply refer to Red Storm, since that is the platform on which all of our experiments were performed.

Like any other distributed memory parallel computing platform, the performance of the network and the performance of the MPI implementation are critical to the overall performance, scalability, and, ultimately, the success of the machine. For Red Storm, Cray has designed and implemented a custom network interface and router chip, called the Cray SeaStar [1], specifically to meet the demands of a large-scale distributed memory scientific computing machine. The network performance requirements for Red Storm were ambitious when they were first proposed in 2002. The network is required

---

* Sandia is a multiprogram laboratory operated by Sandia Corporation, a Lockheed Martin Company, for the United States Department of Energy's National Nuclear Security Administration under contract DE-AC04-94AL85000.

to deliver 1.5 GB/s of network bandiwidth into each compute node and 2.0 GB/s of link bandwidth. The one-way MPI latency requirement between nearest neighbors is 2 μsec and is 5 μsec between the two furthest nodes.

In this paper, we discuss the design, implementation, and performance of three different MPI libraries for Red Storm. Each of these libraries has different features that have different performance and scalability implications. We describe these differences and show performance results from several communication micro-benchmark tests. We also compare the performance of MPI to the performance of the lowest-level communication mechanism employed by the SeaStar. Our results show that the overhead of the MPI implementation for point-to-point message passing operations is less than 0.5 μsec.

The rest of this paper is organized as follows. The following section provides details of the hardware and software environment on Red Storm. Section 3 describes the three different implementations of MPI for Red Storm. A performance comparison of the three implementations is presented in Section 4, and relevant conclusions of this work are summarized in Section 5. Section 6 outlines plans for continued research and development activities surrounding MPI on Red Storm.

2   Red Storm

The following describes the hardware and software environment of the Red Storm system that was used for our experiments. A more detailed description of Red Storm can be found in [2].

2.1   Hardware

The Red Storm system installed at Sandia is composed of 10,368 compute nodes in 108 cabinets. The network is configured in a 27 x 16 x 24 mesh topology. Each compute node contains a 2.0 GHz AMD Opteron processor with 2 GB of main memory. Each node also contains a SeaStar network interface and router chip attached via a Hyper-Transport (HT) link. In addition to independent send and receive DMA engines, the SeaStar has an embedded 500 MHz PowerPC 440 processor for offloading network protocol processing activities and 384 KB of local scratch memory. The physical links in the network support up to 2.5 GB/s of data payload in each direction. The interface to the Opteron uses 800 MHz HT, which provides a theoretical peak of 3.2 GB/s per direction. After protocol overhead, the link is expected to deliver a peak payload rate of 2.8 GB/s.

2.2   Software

Compute nodes in Red Storm run a third-generation lightweight kernel called Catamount. Catamount is a follow-on to the Puma/Cougar [3] lightweight kernel that ran on Sandia's ASCI Red [4] machine. Catamount has been enhanced to provide running applications in full 64-bit mode on the Opteron and has also undergone some small changes to integrate it into Cray's system management infrastructure. Red Storm service nodes run SuSE Linux.

The software environment for the SeaStar is based on the Portals [5] network programming interface developed jointly by Sandia and the University of New Mexico. Portals provides one-sided data movement operations between disjoint processes. However, unlike most one-sided programming interfaces, the target of a remote operation is not a virtual address or memory key. Rather, the ultimate destination of an incoming message is determined solely by the receiver when the message arrives. The receiver is responsible for putting Portals objects together in a way that meets the needs of the upper-level protocol. Portals are very much like protocol building blocks that can be combined to meet a variety of needs. In the case of Red Storm, every service that uses the network does so via Portals, whether it be loading a job onto a compute node, network-based file systems, or MPI communication.

Portals is currently an active area of development for Red Storm. The current implementation of Portals for Red Storm handles much of the protocol processing on the Opteron and does not use the PowerPC to its fullest capabilities. This approach allows for a single instance of firmware for the PowerPC that supports both the physically contiguous memory model of Catamount and the non-contiguous physical memory model for Linux. When a new message arrives at the network interface, the SeaStar interrupts the host processor, which then processes the message header, traverses the Portals data structures, and programs the DMA engines on the SeaStar appropriately. The results that we present in Section 4 are using this interrupt-driven mode. The results are quite encouraging, given the cost of using interrupts. We expect an implementation that offloads all of the protocol processing to the PowerPC on the SeaStar to be available within the next few months, and it will be interesting to compare those results as well.

3   MPI Implementations

In this section we describe the three different MPI libraries that have been implemented for Portals on Red Storm.

3.1   MPICH2-0.97

The Cray supported version of MPI for Red Storm is based on MPICH2 [6]. They have created a Portals device for MPICH2 (version 0.97) that supports all of the MPI-2 functionality except for the dynamic process creation functions and the connect/accept functions. The organization of the Portals structures and protocols used for this implementation to implement the MPI point-to-point communication operations are essentially identical to those describe in [7]. However, Cray has made a few small changes to their implementation. Rather than having each non-blocking send and receive operation use a separate Portals event queue, this implementation uses only two event queues total: one for unexpected messages and one for all other types of communication events. Cray likely took this strategy to reduce the complexity of the implementation and to reduce the amount of memory needed for event queues on the SeaStar. A drawback to this approach is that the time needed to complete an operation is no longer specific to that operation. For example, when waiting on a message to arrive for a posted receive, this implementation must handle all other events that occur while waiting for the message to arrive. This strategy assumes that there is outstanding work to be done while waiting for communication operations to complete.

3.2 MPICH-1.2.6

This implementation was developed in support of the previous version of Portals that ran on the Cplant [8] Linux cluster at Sandia. It was ported forward to the current version of Portals that runs on Red Storm. Like the MPICH2 implementation from Cray, the Portals structures and protocols are essentially identical to the those described in [7], with one exception. This implementation has an optimization for very short messages that is similar to message "copy blocks" or "bounce buffers" used in other MPI implementations [9,10].

Previously, the short message protocol was implemented by creating a Portals memory descriptor over the region of memory to be sent and then invoking the Portals put operation to deliver this data to the destination. In order to avoid the overhead of creating a new memory descriptor each time a short message is sent, the MPI library creates a large memory descriptor during initialization. It divides this memory up into several short message buffers. When a short message is to be sent, the MPI library copies the message into one of these buffers and sends it. From the user point of view, the send is complete because the user buffer is free to modify the buffer. From a Portals point of view, the put operation may not have completed, since the events signifying completion may still be pending. The completion events for very short messages will be consumed whenever the library is blocked waiting for an operation to complete or whenever a short message is to be sent and there are no free short message buffers available. So, in addition to avoiding the overhead of creating and destroying a memory descriptor for each short message, this optimization attempts to hide the cost of consuming events associated with short messages. This approach also helps reduce the number of memory descriptors used for non-blocking short messages. The very short message optimization can be disabled via an environment variable, making it easy to measure the performance gained by using this strategy.

3.3 MPICH-1.2.6 Using SHMEM

Since Portals provides one-sided operations, it can easily support the Cray SHMEM [11] programming model (provided the operating system maps static variables at identical locations in separate processes, as Catamount does). Thus, it is possible to use the MPI implementation developed for SHMEM [12] on Red Storm. This is not a complete implementation of SHMEM (which Cray plans to provide for the XT3), but rather a small subset of SHMEM interface.

This implementation has a few advantages over the others in terms of Portals resource usage. Unlike the other two implementations, the number of Portals data structures that MPI uses is fixed. The SHMEM subset library creates eight memory descriptors and one event queue. These structures are created during MPI initialization and remain persistent throughout the life of the process. The other implementations create and destroy memory descriptors, match entries, and sometimes event queues as the process runs. This can be an advantage for applications that have large numbers of outstanding operations where the limited amount of memory on the SeaStar is a problem. In addition, the SHMEM implementation implements flow control at the user-level, so there is no way to exhaust the resources for handling unexpected messages. The other implementations allocate a fixed amount of space during initialization for unexpected message, and if this space is exceeded, the application process is terminated with a resource exhaustion error.

The main drawback of this implementation is that it performs all of the matching semantics of MPI at the user-level. The MPI library is responsible for maintaining the posted receive queue and traversing it each time a new message arrives. The other implementations take advantage of the matching semantics of Portals.

4   Micro-Benchmarks and Results

In order to measure the performance of the three MPI libraries, we use two micro-benchmarks. The first is a standard ping-pong latency and bandwidth benchmark developed at Sandia. This benchmark measures the ideal case where a receive is pre-posted. We also have a version of this benchmark that measures latency and bandwidth at the Portals level so that we can measure the overhead incurred by MPI.

The second benchmark used is the NetPIPE [13] benchmark. We used the standard MPI-1 module that comes with the distribution and measured latency and bandwidth using the standard ping-pong, ping-pong with pre-posting, bi-directional ping-pong, and streaming modes. We also developed a Portals module for NetPIPE so that we could again measure the overhead of MPI in these various modes.

Figure 1 shows the results of the Sandia benchmark for Portals, MPICH 1.2.6, MPICH 1.2.6 without the short message optimization, MPICH2, and MPICH 1.2.6 using SHMEM. The zero-length latency is 5.23 μsec, 5.64 μsec, 6.25 μsec, 8.12 μsec, and 11.74 μsec respectively. The MPI overhead for zero-length messages starts out relatively small, only 0.41 μsec, but eventually steadies at around 1.17 μsec for 64 bytes and beyond. Interestingly, the very short message protocol in MPICH 1.2.6 only ends up being a win for messages smaller than 320 bytes. The very short message switch point is currently set at 8 KB, so this number will need to be tuned as Portals development continues.

Figure 2 shows the NetPIPE latency performance for both the default mode of operation and for the mode that insures that a receive is pre-posted. For this test, the latency

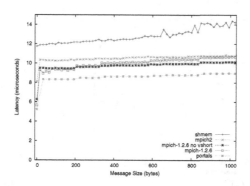

Fig. 1. MPI latency performance

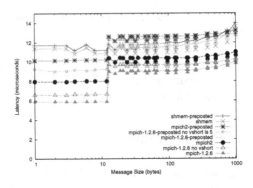

Fig. 2. NetPIPE latency performance

for a one-byte message for MPICH 1.2.6 is 5.9 μsec, 6.6 μsec for MPICH 1.2.6 without the very short message optimization, and 7.97 μsec for MPICH2. This graph clearly shows a jump at 12 bytes, which is the largest amount user data that can fit in a single Portals header packet and be serviced by a single interrupt on the SeaStar. Messages larger than 12 bytes require two interrupts to be serviced. For this test, pre-posting receives does not offer a performance gain.

Figure 3(a) shows the bandwidth results for NetPIPE. Both MPICH 1.2.6 and MPICH2 perform similarly up to the long message protocol crossover point of 128 KB. At that point, MPICH2 continues to perform well, but MPICH 1.2.6 falls off to only about 700 MB/s. The long message protocol for both implementations is eager, which means that a message will be sent across the wire twice if the message is unexpected. For this particular test, MPICH 1.2.6 gets the initial message to the receiver before the receive is posted, but since MPICH2 is slightly slower, it gets the message there after the receive has been posted. If we insure that a receive is always pre-posted, MPICH 1.2.6 performs as well as MPICH2. The asymptotic bandwidth is a little over 1.1 GB/s.

Figure 3(b) shows the bi-directional bandwidth results for NetPIPE. There is virtually no difference between MPICH 1.2.6 and MPICH2 for bi-directional bandwidth.

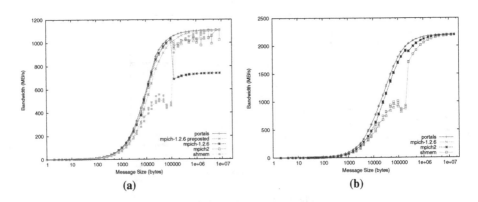

Fig. 3. NetPIPE uni-directional **(a)** and bi-directional **(b)** bandwidth

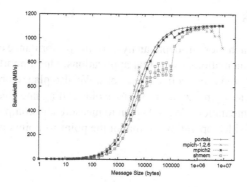

Fig. 4. NetPIPE Stream Bandwidth

Both are able to achieve an asymptotic bandwidth of a little more than 2.2 GB/s. Even the SHMEM implementation is able to sustain this level at very large message sizes.

Figure 4 shows the streaming bandwidth results for NetPIPE. The interesting result in this data is that the very short message optimization in MPICH 1.2.6 actually allows for greater performance than what the raw Portals performance can provide. The Portals module in NetPIPE waits for completion events after each message is sent, while the very short message optimization does not. This overhead is reflected in messages smaller than 8 KB. Beyond 8 KB, the performance of MPICH 1.2.6 and MPICH2 are nearly identical. However, there is an unexplained drop in performance for MPICH 1.2.6 for messages greater than 1 MB. This is most likely attributed to some of the streaming messages being sent across the wire twice due to a matching receive not being pre-posted.

5   Conclusions

In this paper, we have compared the performance using micro-benchmarks of three different implementations of MPI on the Cray Red Storm platform. Despite the current interrupt-driven implementation of Portals for Red Storm, MPI zero-byte half round trip latency is a little more than 5 μsec. While this number is more than twice the requirements for Red Storm, we expect this will improve significantly once a full offload implementation of Portals is completed. The bandwidth numbers indicate that the Cray SeaStar is able to deliver more than 1.1 GB/s of uni-directional bandwidth and is able to maintain that level of performance in both directions simultaneously. The MPICH 1.2.6 version slightly outperforms the MPICH2 implementation in terms of latency and bandwidth. We have measured the overhead of MPI on top of Portals to be a little more than a microsecond for short messages and have shown that a short message optimization can also provide an increase in streaming bandwidth performance.

6   Future Work

We expect to do a much more in-depth analysis of the performance of the MPICH 1.2.6 and MPICH2 implementations using real applications. One important area of performance yet to be analyzed is collective operations. We also plan to prototype some additional features, such as a rendezvous mode to avoid sending unexpected long messages twice, in both implementations. We also plan to measure and compare the performance of the MPICH2 one-sided operations to that of the point-to-point operations.

References

1. Alverson, R.: Red Storm. In: Invited Talk, Hot Chips 15. (2003)
2. Camp, W.J., Tomkins, J.L.: Thor's hammer: The first version of the Red Storm MPP architecture. In: In Proceedings of the SC 2002 Conference on High Performance Networking and Computing, Baltimore, MD (2002)
3. Shuler, L., Jong, C., Riesen, R., van Dresser, D., Maccabe, A.B., Fisk, L.A., Stallcup, T.M.: The Puma operating system for massively parallel computers. In: Proceeding of the 1995 Intel Supercomputer User's Group Conference, Intel Supercomputer User's Group (1995)
4. Timothy G. Mattson, David Scott, S.R.W.: A TeraFLOPS Supercomputer in 1996: The ASCI TFLOP System. In: Proceedings of the 1996 International Parallel Processing Symposium. (1996)
5. Brightwell, R., Hudson, T.B., Maccabe, A.B., Riesen, R.E.: The Portals 3.0 message passing interface. Technical Report SAND99-2959, Sandia National Laboratories (1999)
6. Gropp, W.: MPICH2: A new start for MPI implementations. In Kranzlmuller, D., Kacsuk, P., Dongarra, J., Volkert, J., eds.: Recent Advances in Parallel Virtual Machine and Message Passing Interface: 9th European PVM/MPI Users' Group Meeting, Linz, Austria. Volume 2474 of Lecture Notes in Computer Science., Springer-Verlag (2002)
7. Brightwell, R., Maccabe, A.B., Riesen, R.: Design, implementation, and performance of MPI on Portals 3.0. International Journal of High Performance Computing Applications **17** (2003) 7–20
8. Brightwell, R., Fisk, L.A., Greenberg, D.S., Hudson, T.B., Levenhagen, M.J., Maccabe, A.B., Riesen, R.E.: Massively Parallel Computing Using Commodity Components. Parallel Computing **26** (2000) 243–266
9. Chaussumier, F., Desprez, F., Prylli, L.: Asynchronous Communications in MPI – the BIP/Myrinet Approach. In Dongarra, J., Luque, E., Margalef., T., eds.: Proceedings of the EuroPVM/MPI'99 conference. Number 1697 in Lecture Notes in Computer Science, Barcelona, Spain, Springer Verlag (1999) 485–492
10. Dimitrov, R., Skjellum, A.: An efficient MPI implementation for Virtual Interface (VI) Architecture-enabled cluster computing. In: Proceedings of the Third MPI Developers' and Users' Conference. (1999) 15–24
11. Cray Research, Inc.: SHMEM Technical Note for C, SG-2516 2.3. (1994)
12. Brightwell, R.: A new MPI implementation for Cray SHMEM. In: Recent Advances in Parallel Virtual Machine and Message Passing Interface: 11th European PVM/MPI Users' Group Meeting. (2004)
13. Snell, Q.O., Mikler, A., Gustafson, J.L.: NetPIPE: A network protocol independent performance evaluator. In: Proceedings of the IASTED International Conference on Intelligent Information Management and Systems. (1996)

Probing the Applicability of Polarizable Force-Field Molecular Dynamics for Parallel Architectures: A Comparison of Digital MPI with LAM-MPI and MPICH2

Benjamin Almeida[1], Reema Mahajan[2], Dieter Kranzlmüller[3], Jens Volkert[3], and Siegfried Höfinger[1]

[1] Novartis Institutes for BioMedical Research,
IK@N Operations and In Silico Sciences,
Brunnerstraße 59, A-1235, Vienna, Austria
{benjamin.almeida, siegfried.hoefinger}@novartis.com
http://www.nibr.novartis.com
[2] Department of Chemical Engineering,
Indian Institute of Technology, Delhi,
Hauz Khas, New Delhi-16, India
reema.mahajan@gmail.com
http://www.iitd.ernet.in
[3] Joh Kepler University Linz,
Altenberger Straße 69, A-4040, Linz, Austria
{kranzlmueller, volkert}@gup.uni-linz.ac.at
http://www.gup.uni-linz.ac.at

Abstract. Polarizable Force Fields have a great potential in improving the quality of Molecular Dynamics simulations. Especially for the description of solvents such kind of simulations are greatly appreciated because of the important role solvation plays in all kind of biomedical research. The open source package TINKER is taken for parallelization of a routine made for the proper usage of Polarizable Force Fields. Profiling with the GNU tool *gprof* identifies the relevant target subroutines to parallelize. MPI is used for the parallelization. Several different parallel platforms and several different versions of MPI are tested and compared. Parallel scalability with reasonable amounts of parallel operating CPUs seems to be limited beyond a factor of 2.5. Appraisal of the present MPI-implementation of the TINKER Molecular Dynamics program using Polarizable Force Fields for large scale computation will heavily depend on the relative availability of computing resources within a given production environment.

Keywords: MPI, Polarizable Force Fields, Molecular Dynamics, MPICH2, LAM-MPI.

1 Introduction

Drug discovery is influenced by a great variety of physico-chemical factors that either favour or hinder the binding of drug-like molecules — the medicine — to

B. Di Martino et al. (Eds.): EuroPVM/MPI 2005, LNCS 3666, pp. 433–440, 2005.

important bioactive sites within the human body. Among many others, solvation is one such key-player in the set of determinants that will affect the process of drug ligation to its biological target. Therefore, a better understanding of solvation in general as well as the underlying molecular principles in particular are most desirable subject-matters of greatest interest to pharmaceutical research and medicinal chemistry of today [1].

Atomic scale detail of the behaviour of solvents may be obtained from Molecular Dynamics simulations (MD). Here, a whole assembly of individual solvent molecules — in combination with the target biomolecule and the putative drug — become subject to computer simulations aimed at a realistic and dynamic description of all the individual components and their interplay. MD in this respect may be seen as the theoretical machinery behind this molecular system that tries to model all the occurring physico-chemical interactions as accurate as possible. MD in particular will carry out in each of the simulated time steps a sum over all arising pairs of atoms and apply some inter particle potential, that usually incorporates empirically derived parameters — the so called Force Field (FF) [2]. In addition to such static interactions, the kinetic energy is also covered from individual velocities assigned to each atom, which in turn will constitute the thermic disorder. Recent advances in the development of FFs have been made with the introduction of atomic polarization effects — *Polarizable Force Field Molecular Dynamics* (PFFMD) [3] [4]. Initial results suggest a markedly improved quality of the description with PFFMD especially when it comes to solvation effects. Unfortunately PFFs are computationally intensive and conceptually more complex than traditional MD simulations that make use of static atomic partial charges instead. It therefore appears to be only a natural attempt to pursue optimization strategies in PFFMD via parallel computing. Especially with regard to demanding scientific applications more advanced GRID systems allow the user to allocate whole bundles of compute nodes, that may be employed in parallel to solve a common computational task. However, such parallel computing in the GRID will need to clearly prove its advantage over the conventional serial approach in order to justify the increased allocation of computer resources.

In the present article we describe the possibility of parallel PFFMD by using the *Message Passing Interface*, MPI [5]. We focus on the PFFMD module coming with the open source package TINKER [6]. The next section summarizes profiling data obtained from the runtime behaviour of a typical PFFMD run. In the subsequent section an outline of the parallel algorithm and its principal limitation is given. Finally, in the last section parallel performance data obtained with different versions of MPI on a realistic test case are presented, compared and analyzed.

2   Profiling of a Typical PFFMD Application with GNU's gprof

A water box consisting of 216 individual water molecules was set up and minimized, annealed and equilibrated at 298 K thereby using the AMOEBA PFF [3] coming with the TINKER package for molecular modeling [6]. The *dynamic* program was re-compiled on a single CPU of type Intel/PIII 1 GHz with the profiling flag -pg invoked, which will result in code that may be run-time analyzed with the profiling tool *gprof* from GNU. A test run of 100 steps of MD was profiled and analyzed. The corresponding call graph is represented in Fig.1. As becomes evident from the graph, subroutines *induce0a* and *image* are the functions that consume most of the total execution time of the program (87.6 %). Further branches in the call graph that also claim significant fractions of the CPU time are *empole1a* (6.5 %), *torque1* (3.4 %) and *ehal1a* (2.4 %). Taken together this subset of mentioned routines will make up already 99.9 % of the total execution time. A graphical representation of the relative consumption of the CPU time is given in Fig. 2. Since subroutine *image* is called from the inner part of subroutine

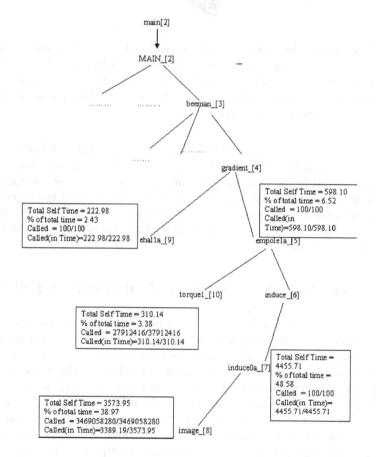

Fig. 1. Call graph and partial CPU time occupancy of various subroutines during execution of 100 steps of PFFMD with the TINKER program *dynamic* as observed with the GNU profiling tool *gprof*. 5th level subroutine *induce0a* and 6th level subroutine *image* are reported to be the major consumers of the total CPU time. Further optimization should therefore focus on these routines.

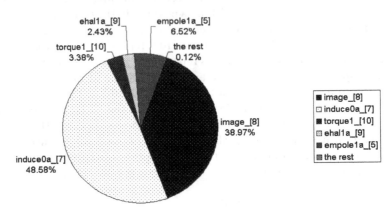

Fig. 2. Summary of the profiling results obtained with the GNU profiling tool *gprof*. The data is due to the run time behaviour of 100 steps of PFFMD with the TINKER program *dynamic* of a water box containing 216 water molecules. The identified main routines of CPU time consumption are *induce0a* and *image* both making up approximately 87.6 % of the total CPU time.

induce0a multiple times, the most obvious first trial of parallelization was to split the outermost loop in subroutine *induce0a* into fractions and compute each of these fractions in parallel. However care must be taken that the splitting of the loop will occur at an appropriate point in the code outside the region where the call to subroutine *image* will actually take place, so that each parallel task can still continue to call subroutine *image* independently from its own interior part a number of times.

3   Parallel Algorithm and Basic Constraints in PFFMD

Following the insights gained from the profiling data represented in the previous section, subroutine *induce0a.f* became the primary target of all MPI parallelization efforts. The central algorithm to be parallelized consists of a two-stage process. In the initial phase the calculation of the induced dipole moments due to all permanent multipoles is performed. This is done within a double loop over all inducible sites, i.e. all atomic centres. Once referring to a certain centre (outer loop) the action of all the other permanent multipoles residing on all the other centres is summed together (inner loop) and accumulated in a field-variable. In a subsequent single loop these field values become multiplied with the polarizability and thus establish a first set of directly induced dipoles at each atomic centre. Then the second stage of the algorithm is reached. Again a double loop over all inducible sites is performed. However, the active elements of this second inner loop now are formed by the induced dipole moments computed in the initial phase (transient multipoles). All this leads to another field which is considered within a subsequent single loop over all atomic sites, where at each atom position the dipole moment becomes modified from these second component of the field — "mutual" induced dipoles. The process is still continued once the initial set of induced dipoles has been modified a first time. Indeed after the first modification the calculation of the field values is repeated over and over again and in the course of these iterative refinements a final self-consistent solution to all the "mutual" induced dipoles is obtained [3]. The two-step logic of the algorithm posed a major challenge for the parallel variant of the program. This is because the individual threads working in parallel cannot do their job independently but need to communicate and share data in the beginning and end of each iteration. Therefore a centralized code design adopting a traditional master/slave model seemed to be a reasonable choice for a first parallel version. In so doing the outermost loop in subroutine *induce0a*, which runs over all atomic centres, was split and subsections of this loop were distributed over a number of parallel threads. Each of these parallel threads computed fragments of the field values and sent back their fractional results to the master, who had to properly assemble these partial results and re-construct a global array of field values. All data communication between master and slaves was done in blocking mode. Due to its iterative nature strong dependency on inter-node communication times was anticipated.

4   Parallel Performance with Different Versions of MPI

30 steps of PFFMD were carried out on a realistic test case of a huge water box containing 4000 water molecules. Our MPI implementation (as described in the previous section) of the TINKER *dynamic* program was employed on

- An Alpha cluster consisting of 4 processor nodes of type ES40 67, 667 MHz, OS TRU64, Digital MPI,
- An AMD/Opteron cluster consisting of 2 processor nodes of type AMD Opteron 1.8 GHz, OS SUSE LINUX 9.1, LAM-MPI 7.0.3,
- The same AMD/Opteron cluster but using MPICH2 v1.0 instead of LAM-MPI.

A summary of the obtained timings with increasing numbers of CPUs is presented in Table 1. A graphical representation of the measured speedup data is given in Fig. 3.

5   Discussion

In the present study a real world application of widespread use in the computational chemistry community has been analyzed for its parallel performance characteristics when using MPI to parallelize the most demanding parts of the program, i.e. subroutine *induce0a*. The observed scaling with increasing size of the parallel machine deserves further exemplification.

Table 1. Observed Speedup for MPI-Parallel PFFMD on a huge water box of 4000 water molecules using different versions of MPI. Selected parallel platforms were a cluster of Alphas ES40, 667 MHz EV67 as well as an AMD/Opteron cluster, 1.8 GHz. The MPI versions used were Digital MPI, LAM-MPI and MPICH2.

Nr. CPUs working in parallel	Exe Time Alpha 667 MHz Digital-MPI [s]	Exe Time Opteron 1.8 GHz LAM-MPI [s]	Exe Time Opteron 1.8 GHz MPICH2 [s]	Speedup Alpha 667 MHz Digital-MPI	Speedup Opteron 1.8 GHz LAM-MPI	Speedup Opteron 1.8 GHz MPICH2
1	9883	3427	3450	—	—	—
2	7765	2781	2792	1.273	1.232	1.236
3	6179	2315	2316	1.600	1.480	1.490
4	5203	2007	2027	1.900	1.708	1.702
5	4601	1842	1844	2.148	1.861	1.871
6	4248	1701	1721	2.327	2.015	2.005
7	3821	1588	1629	2.587	2.158	2.118
8	3915	1527	1545	2.524	2.244	2.233
9	3490	1480	1502	2.832	2.316	2.297
10	3305	1436	1454	2.990	2.387	2.373
11	3150	1408	1425	3.138	2.434	2.421
12	3073	1383	1393	3.216	2.478	2.477
13	3099	1348	1365	3.189	2.542	2.528
14	2992	1332	1363	3.303	2.573	2.531
15	3226	1311	1334	3.064	2.614	2.586
16	—	1305	1316	—	2.626	2.622

When using similar conditions for the AMD/Opteron cluster but changing just the version of MPI, i.e. LAM-MPI versus MPICH2, the parallel performance data looks rather similar. This holds true for absolute execution times (see Table 1 columns 3 and 4) as well as speedup factors (see Fig. 3 bottom two curves of hollow and filled triangles). Rare exceptions are most likely the effect of non-ideal load balance and/or competing system tasks that start up spontaneously and limit the actually available compute resources to a certain node. The quasi-identic behaviour of LAM-MPI with MPICH2 is remarkable in two respects. At first the LAM-MPI traffic was routed over a dedicated network switch, while MPICH2 was using a more general NFS switch and was therefore somewhat generally limited as far as communication was concerned. Next our present MPI implementation of the TINKER *dynamic* program did not make use of any typical MPICH2 features, such as parallel I/O, remote memory access, or dynamic process management, so the comparison is really focussing on the raw performance of either MPI version.

The inferior absolute performance on the Alpha ES40 cluster is somewhat compensated by a better scaling characteristics, although at certain points the increase in the number of parallel CPUs leads to a sudden off-leveling in per-

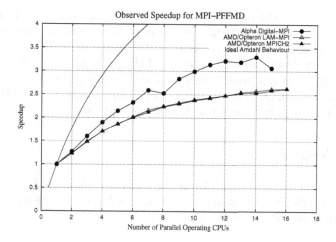

Fig. 3. Speedup data measured for a 30 step PFFMD run on different parallel platforms with different versions of MPI (test system: a huge water box containing 4000 water molecules). Ideal behaviour according to Amdahl's Law with a serial fraction of 0.124 (see section 2) is also included.

formance (e.g. compare 7 CPUs to 8 or 14 CPUs to 15 in Fig. 3). Again this is almost certainly an effect coming from the OS, that sometimes starts up unrelated system tasks which will consume a fraction of the allocated CPU.

Amdahl's Law reveals a different scaling behaviour to the one observed on either parallel platform (see Fig. 3 leftmost curve). In fact, it appears as if the present parallel scalability is growing below than the square of the number of CPUs employed. Thus within a cost-effective resource management it seems appropriate not to advise the present MPI implementation for large production runs. However, if hardware resources do play less a role within a certain computing facility, then a twofold acceleration factor may be easily reached from the present parallel approach using 6 to 8 CPUs.

A possible explanation to the non-Amdahl trend could certainly lie in the heavy demand on data communication (see section 3). In addition to that the second most costly routine *image* is also called from other subroutines in the *dynamic* program, for example *ebond*, *eangle*, *etors* and *elj*, each of which have not been taken into account for parallel execution within our present MPI implementation of the TINKER *dynamic* program.

6 Conclusion

The subroutine *induce0a* of the TINKER *Molecular Dynamics* (MD) program *dynamic* has been parallelized with MPI. This way a thorough investigation of the usage of MD based on *Polarizable Force Fields* (PFF) on parallel platforms could be undertaken. A variety of different MPI implementations, i.e. Digital

MPI, LAM-MPI, MPICH2 was tested and showed rather similar trend in scalability. The parallel *induce0a* module exhibits poor speedup characteristics beyond a factor of 2.5 most likely because of a strong dependence on inter-node communication times. Recommendation of the present MPI-implementation of the TINKER program *dynamic* using PFF is thus only justified when computing resources are abundant, as is the case in certain formations of the GRID. Twofold acceleration is roughly achieved with parallel employment of 6 to 8 CPUs.

Acknowledgment

The authors would like to thank Wolfgang Michael Terényi and Roland Felnhofer from Novartis IK@N Vienna and Pascal Afflard from Novartis IK@N Basel for support with the HPC-infrastructure.

References

1. Kubinyi, H.: Drug research: myths, hype and reality. Nat. Rev. Drug Discovery **2** (2003) 665–668
2. Kollman, P., Dixon, R., Cornell, W., Fox, T., Chipot, C., Pohorille, A.: The Development/Application of a 'Minimalist' Organic/Biochemical Molecular Mechanic Force Field using a Combination of ab Initio Calculations and Experimental Data. in Computer Simulation of Biomolecular Systems, Eds. van Gunsteren, W.F., Weiner, P.K., Wilkinson, A.J. **3** Escom, The Netherlands, (1997) 83–96
3. Ren, P., Ponder, J.W.: Polarizable Atomic Multipole Water Model for Molecular Mechanics Simulation. J. Phys. Chem. B **107** (2003) 5933–5947
4. Ponder, J.W., Case D.A: Force Fields for Protein Simulation. Adv. Prot. Chem. **66** (2003) 27–85
5. Dongarra, J. et.al.: MPI: A Message-Passing Interface Standard; (1995) http://www-unix.mcs.anl.gov/mpi/
6. Ponder, J.W.: http://dasher.wustl.edu/tinker/ (2005) version 4.2

Symmetrical Data Sieving for Noncontiguous I/O Accesses in Molecular Dynamics Simulations*

M.B. Ibáñez, F. García, and J. Carretero

Universidad Carlos III de Madrid,
Departamento de Informática,
Grupo de Arquitectura de Computadores,
Madrid, 28911, España
{mblanca, fgarcia, jcarrete}@arcos.inf.uc3m.es

Abstract. This article analyzes the impact of different solutions of I/O performance problem on Molecular Dynamics (MD) simulation. MD accesses data in a non-contiguous way, with unpredictable patterns that change as the simulation progresses. Such application drastically affect the performance of Parallel File Systems. We show that there is a strong relation between the distribution of the atoms on the simulation space and MD I/O performance. We propose a variant of data sieving approach that uses the symmetry of the MD's access pattern to minimize the number of I/O operations, and behaves considerably better than other tested solutions.

Keywords: Irregular applications, Data sieving, Noncontiguous I/O, Molecular Dynamics.

1 Introduction

Molecular Dynamics (MD) belongs to an important class of parallel scientific applications that are irregular and dynamic. They are dynamic because their access patterns change during the simulation, and irregular because the patterns of data access and computation are unknown until run time. Accesses to data often have poor spatial and temporal locality, which leads to ineffective use of the memory hierarchy [4]. Because both processor and memory technology is evolving rapidly, I/O has become the bottleneck in high-performance computing for these type of applications.

Parallel file systems such as PVFS provide a high-performance I/O infrastructure for storage of applications with contiguous access of data. Nevertheless, applications with noncontiguous I/O access patterns exhibit an unacceptable performance for large data sets. Current solutions to this performance problem involve calling the file system for every contiguous file region, using collective

---

* This work has been supported by the Spanish Ministry of Science under the TIN2004-02156 contract.

B. Di Martino et al. (Eds.): EuroPVM/MPI 2005, LNCS 3666, pp. 441–448, 2005.

I/O techniques or reducing I/O accesses by describing any noncontiguous I/O patterns [2], [6], [7]. In order to perform efficiently, these techniques require regularities in the data access patterns that are not present in MD simulations. The aim of this paper is to analyze under which conditions the former approaches can be applied to molecular dynamics and to propose new methods for improving I/O performance on these applications based on the prediction of spatioal data access patterns.

The rest of this paper is organized as follows. In Section 2, we overview different solutions given to the noncontiguous data access pattern problem that we consider for this work. In Section 3 we describe the Molecular Dynamics application and show compact and sparse access patterns that a simulation can exhibit. In Section 4 we analyze the performance improvement that solutions presented in Section 2 cause over a substance at its two extreme states: balanced and unbalanced. Based on these results and on the particular access pattern of MD simulations, we present a new solution to the noncontiguous data access pattern problem (Section 5). Section 6 concludes the paper.

2   Different Approaches for Noncontiguous I/O

There are a limited variety of interfaces that describe the I/O patterns of scientific applications. The I/O methods support do not match efficiently with the types of I/O operations that scientific applications perform [3]. Parallel file systems provide interfaces such as multiple I/O, data sieving I/O and list I/O for improving performance of applications. Following, we describe these interfaces and in which contexts they have proven to be useful.

In parallel file systems, access to each noncontiguous file data region requires a separate I/O request. Handling noncontiguous access is called multiple I/O [2]. The drawback of this approach is the large cost of transmitting and processing many individual data requests. As the number of processors increases, the I/O problem becomes worse.

Data sieving is a client-based collective I/O method that intends to reduce the number of I/O calls to the underlying file system by moving large regions of continuous data from file to a memory buffer and then extracting the desired regions according to the needs of the application. If the noncontiguous regions of different processors show spatial locality, this approach can avoid many I/O requests.

List I/O is a discontiguous access method useful for describing noncontiguous access patterns of data in memory and in file. Each access can be handled in a single I/O request.

It is well known that multiple I/O should not be considered for large-scale scientific applications with noncontiguous accesses patterns [2]. For applications where most of the noncontiguous regions of different processors show spatial locality, data sieving produces good results [2]. Finally, list I/O interface can perform a noncontiguous I/O data access with fewer I/O calls in an optimized implementation [3].

3 Molecular Dynamics

In this section we examine the relevant characteristics of MD simulations in terms of their I/O behavior. This will guide our work in the rest of the paper.

MD simulations consider the interaction of particles within a defined volume, usually a parallelepiped. Each particle interacts with the others within a specified cutoff radius. At each time step, it is necessary to compute forces and update positions and velocities of all particles. In the integration of the motion equations, the bodies can move independently, leaving one area of the space and entering to another one. In order to adapt to these changes, the application requires periodic recalculation of which particles can interact with which.

We carried out our experiences with a modified Plimpton's code [5]. The code uses an $n \times n$ matrix F (the F_{ij} element represent the force on atom i due to atom j) and X, a vector of length n that stores the position of each atom. Each iteration consists of the steps showed at Figure 1.

(1) Construct neighbor list every 20 steps
(2) Compute elements of F
(3) Update atom positions in X using F
(4) Compute temperature, energy and pressure of the system

Fig. 1. General program structure

Computations are inherently parallel, every process is responsible of computing the positions, velocities and forces of a subset of atoms (steps two, three and four) and need the current positions of all the atoms at step one.

Because the dynamic nature of the problem, the access pattern used to compute forces, temperature, energy and pressure of the system in MD program is irregular (see Figure 2). The indirectly referenced loop bounds of the inner j loop vary across iterations of the outer i loop. The construction of *nlist* and *nnlist* is done at step one of the algorithm (see Figure 1).

```
DO i = 1, nlocal
   DO j = nnlist(i), nnlist(i + 1) − 1
      G = G + function(x(nlist(j)))
   END DO
END DO
```

Fig. 2. Irregular pattern of MD program

Typically this generic access pattern causes a bad performance behavior when parallel I/O is used.

We are interested on analyzing the impact of the atom distribution on the I/O performance. To do this, we classify the substances depending on how sparse are the neighbors of their atoms. A balanced substance has the neighbors of its atoms nearby, its access pattern is compact as the one showed at Figure 3(left).

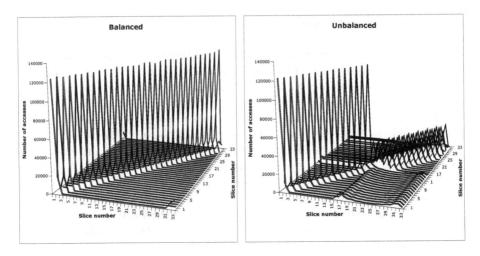

Fig. 3. Access pattern of a balanced and an unbalanced substances

An unbalance substance has the neighbors of its atoms on several regions of the simulation space (see Figure 3(right)); its access pattern is not only irregular but also sparse. Most MD simulations start with a balanced substance and evolve to a substance in an unbalanced state.

4   Preliminary Experiments on the Noncontiguous Data Access Pattern Problem

In this section we present and analyze a set of experiments we performed for two substances, one balanced and other unbalanced. We use the classical solutions to the noncontiguous access pattern problem previously described in Section 2, for these substances.

The experiments were carried out on a cluster of 8 dual nodes connected with a Gigabit Ethernet communication network and a fast Ethernet network. We used only the Gigabit Ethernet for our testing purposes. Each compute node includes 2 Pentium III running at 1.0 GHz, 1GB of RAM and two disks of 200GB and 40GB. The system runs Debian/GNU Linux with a 2.4 series kernel. The Parallel File System used was PVFS with the MPI-IO interface through ROMIO. We worked with various numbers of compute nodes and 8 PVFS I/O nodes.

4.1   Multiple I/O

The data access pattern of Molecular Dynamics simulations involves an important amount of small I/O requests; those are produced each time an atom must interact with its neighbors. Figure 4 shows how the write bandwidth scales linearly with the size of the problem while the read bandwidth exhibits very low values. This bad performance is a consequence of the irregularity of the substance's access pattern.

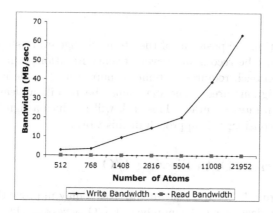

Fig. 4. Read and Write Bandwidth for different sizes of the problem. These results are obtained by using 8 processors and balanced substances.

4.2 Data Sieving I/O

Data Sieving method moves a continuous region of a file to a memory buffer. its success depends on the spatial locality of the access, i.e. successive data demands will access neughboring elements in the buffer.

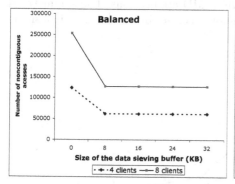 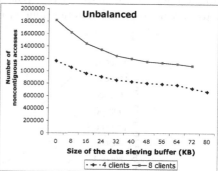

Fig. 5. Number of I/O accesses using different buffer sizes. Balanced and Unbalanced substances of 21952 atoms.

Figure 5 show the variation of the number of I/O accesses for two substances with the size of the buffers. For the balanced substance, the number of small I/O requests decreases when a memory buffer of 8KB is used. In spite of the increase of the size of the buffers, the number of I/O calls is not the minimum. On the other hand, the unbalanced substance improves slightly its behavior to the extent of the memory buffer but the number of I/O calls remains high.

4.3   List I/O

In a MD simulation the positions of the atoms change every step, the neighbors of each atom must be recomputed every twenty iterations. On the other hand, the List I/O approach requires to build a map of all the file accesses in order to identify contiguous areas. The map must be rebuilt at least every twenty iterations to guarantee accuracy. This task will produce a significant overhead, thus we do not consider this approach in this work.

5   Symmetrical Data Sieving I/O

According to figure 5, the data sieving approach may improve the application's performance by decreasing the number of I/O accesses. Over balanced substances, the method causes a moderate improvement up to a certain size of the memory buffer, after which the amount of I/O remains invariable. On the other hand, the performance improvement is gradual and proportional to the buffer size over unbalanced substances, but the number of I/O accesses stays high. The reason of this behavior is that for balanced substances, the neighbors of an atom are in its close environment, both to its right and to its left side in the file. While for unbalanced substances, the neighbors of a given atom are distributed on a broad file region. With the data sieving I/O approach, it is not possible to eliminate totally the small I/O requests on MD applications because the method does not consider the symmetric nature of the access pattern.

We propose a new approach that combines data sieving optimization with a prediction of the spatial access pattern of MD applications. The symmetrical data sieving approach handles noncontiguous accesses by moving data in a single

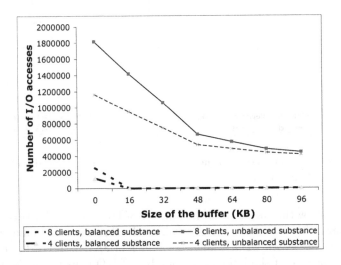

Fig. 6. Results of the symmetrical data sieving method. Number of I/O accesses using different buffer sizes. Balanced and unbalanced substances of 21952 atoms.

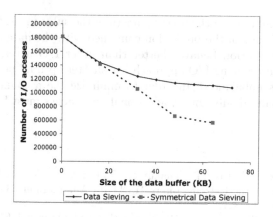

Fig. 7. Comparison of symmetrical data sieving and data sieving methods over unbalanced substances. Substance of 21952 atoms.

chunk into a temporary buffer. The buffer comprises two equal size file intervals: one starting at the first requested byte b_s up to the last requested byte b_e, and the interval preceding b_s. For balanced substances, the buffer will contain an atom and most of its neighbors. Thus, the data placed into the buffer will be extensively used by the client.

The results described in Figure 6 showed the tendency expected for symmetrical data sieving I/O. For the balanced substance and with a small symmetrical buffer, the number of I/O accesses is the minimum. In this case, as the client has all the data it needs to carry out its work, the performance is not disturbed by the I/O access problem. By contrast, for the unbalanced substance, the increase of the symmetrical buffer size, does not fully remove the I/O access problems. Nevertheless, Figure 7 shows that symmetrical data sieving method behaves better that data sieving method even for unbalanced substances.

6   Conclusions

As Molecular Dynamics simulations access its data in a noncontiguous way, I/O has become the bottleneck in high-performance computing for these applications. Our experiments show that current solutions to this issue improve I/O performance but the problem still persists.

In this work we present two MD's relevant characteristics that must be taken into account to solve the I/O performance problem. First, computing atom parameters implies accessing information that is for atoms both to the right as to the left of the atom. Second, as the simulation evolves, the distribution of the atoms changes and the accesses become more sparsed. It is worth noting that these characteristics are shared by N-body problems in general.

We propose a new solution to the noncontiguous access problem called Symmetrical Data Sieving I/O, based on a collective I/O implementation technique and a prediction of the spatial access pattern of MD applications. The prediction considers the symmetry in the access pattern of the substances. Our initial evaluation of the efficacy of the prediction combined with the data sieving method, shows that our solution behaves better than other I/O access methods. The Symmetrical Data Sieving I/O approach has shorter optimal performance when the substance is unbalanced. In order to minimize the amount of atomic I/O access for unbalanced substances, additional approaches must be explored.

References

1. Carns, P., Ligon, W., Ross, R., Thakur, R.: PVFS: A Parallel File System for Linux Clusters. Proc. of the Extreme Linux Track: 4th Annual Linux Showcase and Conference. (2000) 317–327.
2. Ching, A., Choudhary, A., Liao, W.: Noncontiguous I/O through PVFS. Proceedings of the IEEE International Conference on Cluster Computing. (2002) 405–414.
3. Ching, A., Choudhary, A., Coloma, K., Liao, W.: Noncontiguous I/O Accesses Through MPI-IO. Proceedings of the 3rd IEEE/ACM International Symposium on Cluster Computing and the Grid. (2003) 104–111.
4. Mellor-Crummey, J., Whalley, D., Kennedy, K.: Improving Memory Hierarchy Performance for Irregular Applications. Proceedings of the International Conference of Supercomputing. (1999) 425–433.
5. Plimpton, S.: Fast Parallel Algorithms for Short–Range Molecular Dynamics. Journal of Computational Physics. 117 (1995) 1–19.
6. Thakur, R., Choudhary, R. Borawekar, R., More, S. Kuditipudi, S.: Passion: Optimized I/O for Parallel Applications. IEEE Computer. 29(6) (1996) 70–78.
7. Thakur, R., Gropp, W., Lusk, E.: Data Sieving and Collective I/O in ROMIO. Proc. of the 7th Symposium on the Frontiers of Massively Parallel Computation. (1999) 182–189.

Simulation of Ecologic Systems Using MPI

D. Mostaccio, R. Suppi, and E. Luque

Computer Architecture and Operating Systems Department,
University Autonoma of Barcelona,
08193 Bellaterra, Barcelona, Spain
Diego.Mostaccio@aomail.uab.es,
{Remo.Suppi, Emilio.Luque}@uab.es

Abstract. The simulation of ecological models for individual oriented models presents multiple advantages for the biologist since it enables the more accurate reproduction of species behaviour. However, the drawback of this type of simulation is the large amount of computing needed to accomplish real simulations (thousands of individuals). Distributed simulation is an excellent tool for solving this type of problem. In the present paper, a model of these characteristics (Fish Schools) is analyzed and the corresponding distributed simulator based on MPI and using conservative algorithms is developed. In order to verify the goodness of this simulation, a set of experiments is accomplished and the speedup with respect to a sequential simulator is computed.

1 Introduction

Simulations of such complex systems as, for example, certain ecological systems, require a high processing capacity. **DES** (Distributed Event Simulation) enables users to obtain efficient solutions in acceptable time periods as long as the parallelism of the simulated model allows for a reasonable degree of concurrency.

Fish-Schools were the ecological model used for this work, which enabled us to simulate fish movements in an open environment. Two models can be used for the simulation of this type of system: aimed at population and aimed at individual. In the first, the modelling of the system is a complex task although the computing requirements are fewer. In the individual oriented, have a reduced cost in the modelling phase and gives better (more accurate) results, but the simulation time increases considerably when the quantity of individuals is increased.

In this sense, DES reduces the simulation times in individual oriented models through the distribution of the simulated individuals in different computing nodes. Simulation distribution implies communication between the different processes in order to exchange data and update the global state of the simulation. Experiences with ecological model simulation using PVM [1] as a communications library presented certain problems with simulation scalability and the size of the population to be simulated. This work presents the design, implementation and validation of a Fish-School DES simulator for individual oriented models using MPI [2] as a communication library.

Event synchronization in a distributed simulation (DES) can be accomplished in two possible ways: optimistic simulation and conservative simulation. In accordance

B. Di Martino et al. (Eds.): EuroPVM/MPI 2005, LNCS 3666, pp. 449–456, 2005.
© Springer-Verlag Berlin Heidelberg 2005

with the previous experiences of the group, the present work is aimed at distributed and conservative simulation, since the excessive optimism (generally in the frontier zone of the model) of the first algorithms prejudices the efficiency of the simulation. [3, 4, 10]

This paper is organized as follows: section two concentrates on Individual oriented Models (IoM). Section three focuses on Distributed Event Simulation as well as the design and implementation of the Fish-School DES simulator. Section four presents the experiments carried out to validate the results of the simulation and shows the performance obtained. Section five summarizes the conclusions obtained and outlines future work.

2   Individual Oriented Model (IoM)

Biologists and ecologists need to study and analyze the behaviour and system dynamics of ecosystem populations. The ecosystems studied are essentially dynamic systems with feedback behaviour. Ecological systems contain auto-regulation mechanisms but the many different studies of them have not obtained results comparable with those obtained for other disciplines (automatic control, traffic networks...).

Lotka and Volterra [5-7] represent the interaction between preys and predators and the most common ecologic models are developed using these definitions. These models are based on the interaction between species and describe their behaviour through differential equations, obtaining very good results for very limited, small models.

Individual Oriented Models (IoM) are an acceptable solution with considerable reference to the previous models. These models are based on the individual's behaviour and not on the community or group. IoM is based on the individual being considered through simple rules and the observation of the interaction between them in an ecosystem. There are two important advantages to these models: they are independent of the quantity of individuals and can simulate a very complex ecosystem from a simple element (individual) without describing the community analytically.

Moreover, in the Volterra's model when the individuals' number is increased, the equations to be resolved are highly complex. IoM scalability is only limited by the computing capacity of the system and we can o obtain very good results in acceptable times. Recent advances in cluster technology and communications provide us with the large computing capacity for IoM simulation.

2.1   Characteristics of Fish Schools

One of the most representative applications of IoM is used to describe the movement of given species. IoM use allows us to determine the movement of a group of species by using the movement of each member.

IoM uses very simple, biologically definite rules that are applied to each member to obtain the movement of the colony. Movements in Fish-Schools are governed by three basic postulates from the point of view of the individual: a) avoiding collisions, b) speed coupling and c) obtaining a position in the centre of the group.

These rules express both the individual's need for survival and its instinct for protection (the need to escape from predators). Each fish in the model is represented as a point in a three-dimensional space with an associated speed and each fish changes position and speed simultaneously after a certain period Δt. The actions that the model describes for each fish are:

- Each fish chooses up to X neighbour fish (X=4 seems sufficient for most schools), which will be those nearest and with direct vision. (Figure 1).
- Each fish reacts in accordance with the direction and distance of each neighbour. Three influence radios and three possible reactions are established. The final reaction will be the average of the reactions experienced by each neighbour.
  - If the neighbour is found within the smaller radio, the fish will carry out an "opposed to address" movement – repulsion action – (to avoid collisions).
  - If the neighbour is within a second influence radio, the fish will adopt the same direction as the neighbour.
  - If the neighbour is within a third radio, the fish will move towards it.
- Each fish calculates its news position according to the new direction.

This generates a very simple model that enables the description of complex behaviour. However, very high computing power is necessary, since the complexity algorithm is $O(N^2)$, where N is the number of fish (each fish attempts to find the neighbouring fish by inspecting all other fish in the school). [7, 8]

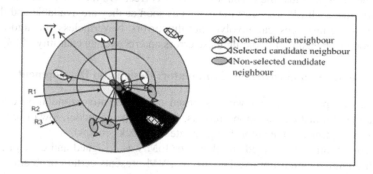

Fig. 1. Neighbour selection in the Fish Schools model

Each fish F_i is defined by its position p_i and velocity v_i, and chooses its potential neighbours by watching concentric zones of increasing radio until finding X Fish. The potential neighbours are chosen using the algorithm of front priority.

Each fish F_i, once they have selected the X neighbour, must then determine the reaction (rotation of the v_i) to each fish F_j. β_{ij} will be the F_i reaction with respect to F_j expressed in spherical coordinates. Each fish F_j can be found within one of three possible areas of influence with respect to F_i (R_1, R_2, R_3):

- If $Dist(F_j, F_i) \leq R_1$, F_i has a repulsion reaction with respect to F_j.
- If $R_1 < Dist(F_j, F_i) \leq R_2$, F_i adopts a parallel position with respect to F_j.
- If $R_2 < Dist(F_j, F_i) \leq R_3$, F_i is guided toward F_i.

Finally, reaction β (mean value for all β_{ij}) and v_i is rotated according to β value.

3 Distributed Event Simulation (DES)

The development of the simulation involved determination of a set of logical processes (LP) inside the architecture of the distributed system. Each LP generates and shares events that are interchanged in the computing systems as messages. The information contained in these messages is event type, specific data of this event and the time of occurrence.

Each LP simulates a section of the simulation world and this LP is assigned to a computing node in the distributed computing system. Fish movement in the same section does not generate events for other LP (there is no message exchange) but if a fish goes from one section to another, there is a migration message from the LP when the fish leaves for the LP where the fish enters.

This situation of event exchange (messages) with time-stamped can generate causality errors (the LP simulated time at which a message arrives - event - is superior to the time-stamped of the new event). In order to solve this problem, there are two mechanisms in the DES: Conservative and optimistic algorithms. The conservative approach uses synchronization to avoid causality errors. In this algorithm, the events are processed when it is sure that the execution order is correct. On the other hand, in optimistic algorithms, each LP processes the events as soon as they are available and this execution, in some instances, can involve causality errors. Nevertheless, the algorithm has a detection mechanism for avoiding these errors and recovering causality. [9, 3]

3.1 Conservative and Distributed Simulator: Design and Development

As mentioned previously, our work is based on conservative simulation due to the fact that the optimistic simulation presents, for this type of model, low efficiency due to excessive optimism (the algorithm generates rollbacks chains).

The conservative distributed simulator for IoM was designed and developed using the MPI communication library. The use of PVM presents certain problems with the scalability (size of colonies) and simulator stability for a huge number of individuals (thousands of individuals) due to the size of PVM daemon buffers and the PVM daemon synchronization overhead. [1, 2]

A very important aspect of these models is individual distribution in the distributed computing architecture. Loret et al. [7] propose a model distribution that assigns static group partitions of fishes to each processor and the selection of candidate neighbours is implemented using centralized data structures.

Our solution is based on the dividing the problem division into a set of logical processes (LP), which will be executed in the different processors. For each LP, an initial partition of the problem (number of fish) and this number will change dynamically during the simulation. The LPs have a physical zone of the problem to simulate (spatially explicit simulation) and the fish movement will imply migrations between the LPs.

Two possible interactions are used in the simulation run that consider fish position: information exchange and migration. The actual design implements three messages types (EvRequest, EvAnswer, EvMigration) and two internal events (*NextStep, EvResume*) in order to control the simulation state machine (figure 2).

The event sequence is: simulation start consuming the *EvNextStep* event (initial event is added to the event list during simulation start up and initial fish distribution). In the simulation loop, the new fish positions are calculated but it is necessary to know the position and speed of all potential neighbours (the possible neighbours in the same LP or in the LP of the contiguous block).

The request for information from the neighbour block (if necessary) is performed by sending an EvRequest message and in the conservative simulation this LP is blocked waiting for an answer (Wait for Answer). The answers are generated for the neighbour's LPs as EvAnswer and the simulation is restarted in the waiting LP by consuming an EvRessumeStep event and then EvNextStep.

When the fish position update generates a situation in which the placement of this fish is in another block (LP), a migration event is produced (EvMigration) and a sequence of operations is called in order to send a message (fish) to the contiguous block (fish migration) and to update the data structures of the current LP. The *EvLastStep* event is generated in order to end the simulation in the DES.

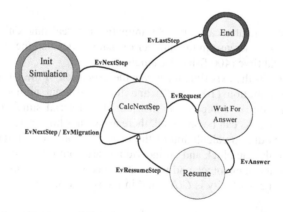

Fig. 2. Machine state of the DES Fish Schools

4   Experimental Results

The verification and validation of the model and simulator were made using a Beowulf cluster with Linux and Fast Ethernet. In order to analyze the performance of the distributed simulator versus a serial simulator, a sequential simulator has been developed. This simulator has only one event list and runs on a single computer (one processor). As a performance measure, frame time generation was selected. This measurement is the time necessary to compute the next position and speed for all individuals in the simulated world.

The experimentation framework was made using a different configuration of the simulator parameters: fish number (100 to 25,000 individuals), computing nodes and LP processes (1 to 32), and ecosystem size (300x300x300) and fish density are constants for all experiment sets. Figure 3 shows the frame time using the sequential simulator with the fish population in the range of 100-25000 individuals.

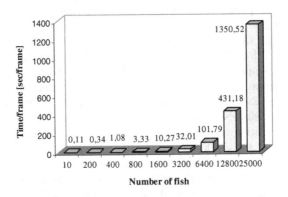

Fig. 3. Time per frame in the sequential simulator

The sequential simulator can carry out animations in real time only for 100 fishes (≈10 frames/sec). An increase in the number of fish implies that the simulator can not generate data in real time (400 fish ≈ 1 frame/sec).

The second step of the experiment set was the execution of the DES for a different fish number configuration (100-25000) and for a different number of computing nodes (2-32). In order to compare the distributed and serial simulator, the speedup was obtained (fig. 4). As can be observed, there are values higher than the linearity.

This situation is due to calculation of the neighbour position. In the sequential algorithm there is only one block and all individuals are computed, but in the DES only the individuals of the same block and the contiguous blocks are computed. In the first case, the algorithm complexity is $O(N^2)$ and in the second it is $O(N^2/n)$ where **n** is the number of LPs.

Figure 4 shows the performance using the processing time only (without communications) and figure 5 shows the time per frame obtained considering the communication time and the LP blocking time (waiting for the answer from the neighbour). The time per frame in the sequential simulator for a colony size of 6400 individuals (or smaller) is always smaller than in the DES and independent of the number of processors. The advantage of DES is the huge number of individuals and computing nodes (i.e. 8 processors and 25000 individuals). The simulation time in the DES was reduced by a factor of 3.45 with respect to the sequential simulator (considering communication and waiting time) for the same (simulated) frame number.

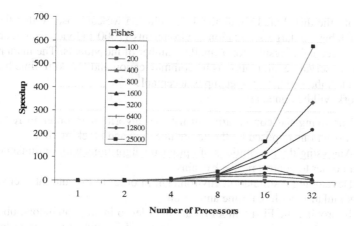

Fig. 4. Performance using the processing time only (without communications)

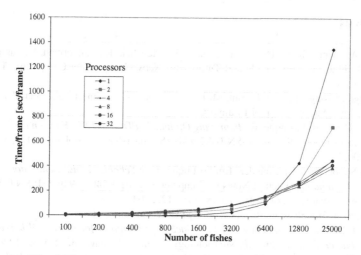

Fig. 5. Time per frame including communication and waiting time

5   Conclusions

The ecological systems simulation is a field that requires large computing capacity powers when individual oriented models are used. The Distributed Event Simulation (DES) is an excellent tool for solving this type of problem.

The present work analyzes the DES in Individual oriented models (specifically fish movement) using a distributed simulation based on conservative algorithms. The results obtained demonstrate that it is a viable option but that conservative algorithms impose a strong limitation on the method possibilities.

In order to measure the benefits of DES a set of experiments has been performed. These experiments were performed using a sequential simulator and the results were compared with the distributed simulator. In conclusion we can observe that the DES is recommendable for large simulation environments (7000 individual or more) but there are also acceptable results for a smaller number of individuals. The main restriction of DES conservative simulation is its communication and blocking time but even with these times, the simulation speedup is acceptable.

Future work will be aimed at:

- Improving the conservative distributed algorithm in order to reduce the restrictions imposed by the communication and blocking time.
- Analysing the possibilities of optimistic algorithms with optimism control in order to control rollback chains.
- Developing a simulation environment in order to execute interactive DES simulation with real time animation.
- Improving the Fish School model in order to include predators, obstacles, more than one species, energy control, and individual ages in order to be able to serve the different analysis needs of biologists and ecologists.

References

1. Geist, A., Beguelin, A., Dongarra, J, Jiang, W., Manchek, R., Sunderman, V. Parallel Virtual Machine. A Users' Guide and Tutorial for Networked Parallel Computing. The MIT Press, 1994.
2. Message Passing Interface Forum. MPI: A Message-Passing Interface standard. Technical report, 1995. http://www.mpi-forum.org.
3. Suppi, R., Cores, F, Luque, E., *Improving Optimistic PDES in PVM Environments*, Lecture Notes in Computer Science ISSN 0302-9743, Springer-Verlag, Vol. 2329(1), pages 107-116, 2002.
4. Remo Suppi, Daniel Fernández, Emilio Luque. *Fish Schools: PDES Simulation and Real Time 3D Animation*. Lecture Notes in Computer Science 3-540-21946-3, ISSN 0302-9743 Springer-Verlag EC. Vol.: 3019, pages: 505-512, 2004.
5. Smith, M. J. *Models in Ecology*, Cambridge University Press, 1994.
6. Kreft, J. Booth, G, Wimpenny, W. T. J. *BacSim, a simulator for individual-based modelling of bacterial colony growth*, Microbiology Journal, 144, pages 3275-3287, 1998.
7. Lorel, H, Sonnenschein, M., *Using Parallel Computers to simulate individual oriented models: a case study*, Proceedings of the 1995 European Simulation Multiconference (ESM), pages 526-531, June 1995.
8. Husth, A., Wissel, C. *The simulation of movement of fish schools*, Journal of Theoretical Biology, 156, 365:385, 1992.
9. Serrano, M., Suppi, R., Luque, E. *Parallel Discrete Event Simulation, State of art.* Computer architecture and Operating Systems Department. University Autonoma of Barcelona. 2000. http://pirdi.uab.es
10. Langlais, M., Latu G., Roman J. and Silan P.. *Parallel numerical simulation of a marine host-parasite system*. Europar'99 Parallel Processing. pp. 677-685. LNCS 1685 - Springer Verlag. 1999.

Load Balancing and Computing Strategies in Pipeline Optimization for Parallel Visualization of 3D Irregular Meshes

Andrea Clematis[1], Daniele D'Agostino[1], and Vittoria Gianuzzi[2]

[1] IMATI-CNR, Via de Marini 6, 16149 Genova, Italy
{clematis, dago}@ge.imati.cnr.it
[2] DISI University of Genova, Via Dodecaneso 35, 16146 Genova, Italy
gianuzzi@disi.unige.it

Abstract. Parallel visualization is assuming an increasing role in the deployment of Web and Grid based systems for scientific applications. The visualization process consists of a set of filters or components that are executed in a pipelined assembly that should be adaptively configured on the basis of user requirements, processing architecture and network characteristics. In this paper we focus our attention on the visualization of 3D irregular meshes produced by the interrogation of volumetric data using an isosurface extraction algorithm. We consider a simplified pipeline consisting of two components: isosurface extraction, and mesh simplification. We show that also in this simple case an in-deep analysis is necessary in order to optimize the whole pipeline. In fact different implementation and load balancing strategies are possible for each single component, but the whole pipeline optimization could be achieved combining non-optimal implementation of individual stages. Moreover the quality of the produced mesh should be considered in the selection of an adequate component implementation. The proposed analysis permits to point out trade-offs and algorithmic requirements that should be considered in the design of a complete visualization system for advanced Grid applications.

1 Introduction

In the scientific community, the interrogation and visualization of 3D data, produced by complex simulations or by acquisition instruments, is of paramount importance. In many cases scientists from different disciplines would like to access remote data repositories, to perform their interrogation and analysis, and to get the results visualized on their local computer [1,2,3,4].

In this paper we are interested in addressing problems and possible solutions in the design of an adaptive pipeline of operations for parallel visualization of 3D irregular meshes. The pipeline is the core of a visualization system [5] that allows the interrogation of 3D data collections, and the proper visualization of results on local clients.

B. Di Martino et al. (Eds.): EuroPVM/MPI 2005, LNCS 3666, pp. 457–466, 2005.
© Springer-Verlag Berlin Heidelberg 2005

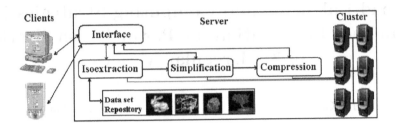

Fig. 1. A data processing pipeline for remote visualization

Figure 1 provides an example of such pipeline. The isosurface extraction module performs 3D data interrogation and produces a Triangulated Irregular Network that may be processed by the simplification and/or by the compression module in order to permit progressive visualization and to perform latency tolerant interaction with remote clients. In fact meshes are often composed by so many triangles that rendering is very difficult, and the size of models heavily impacts also on transmission time, because the available communication bandwidth can be a considerable bottleneck.

Depending on the size of the produced isosurface, the kind of the client and other architectural parameters, the pipeline will consist of different subsets of components.

A parallel implementation of the isosurface extraction, simplification and compression algorithms is often necessary in order to reduce the execution time. Considering each component by itself it is possible to find the most efficient algorithm. However, it is not possible to affirm that the best result for the whole pipeline is obtained combining such algorithms, in fact different, not optimal algorithms could lead to a more efficient pipeline execution. Also the quality of the produced mesh must be considered when selecting component implementation. In this paper we discuss load balancing and computing strategies for a simplified pipeline composed by the isoextraction and the simplification modules.

The paper is organized in the following way: in Section 2 and 3 we shortly describe the two components of the pipeline, and their parallelization; in Section 4 we discuss the issue of a global optimization of the pipeline, presenting experimental results, followed by conclusions and future works in Section 5.

2   Domain Partitioning and Load Balancing for Isosurface Extraction

Pseudocode 1.1 presents an high level description of the Marching Cubes algorithm [6], the classical approach to extract isosurfaces from volumetric data sets. A volumetric data set consists of values corresponding to points disposed over a regular grid, and it is usually structured in a set of planes, called slices.

Listing 1.1. Marching cubes algorithm for isosurface extraction

```
for each cell in the volume
  if the cell is traversed by the isosurface (active cell)
    compute local triangulation
```

For our purposes it is important to underline that the test operation, necessary to determine active cells that are traversed by the isosurface, has a unitary cost that is considerably lower than computation of the local triangulation. However since the test is executed on all the cells of the data set its total cost is normally the highest one. If we define $numC$ the number of cells of a data set and $numAC(v) \leq numC$ the number of active cells for the isovalue v, a simplified analytic formulation of the computational cost of the algorithm is obtained as the sum of the time necessary to check all the cells (T_{Scan}) and the time necessary to construct the triangulation (T_{Constr}):

$$T_{IsoExtr}(v) = T_{Scan} + T_{Constr}$$
$$T_{Scan} = C_1 * numC \tag{1}$$
$$T_{Constr} = C_2 * numAC(v), \qquad C_2 > C_1$$

We have experimentally determined the value of the constants $C_1 = 8.3 * 10^{-8}$ sec. and $C_2 = 3.5 * 10^{-7}$ sec. on a PC whose characteristics are described in Section 4.

A first straightforward parallelization strategy for this algorithm is to adopt the "processor farm" paradigm and to partition the original data set into a set of independent tasks, each one made by contiguous slices, that are assigned to different workers. In order to obtain an efficient implementation we should divide the domain in a number of *tasks* considerably greater than the number of workers. In fact volume partitions intersected by the isosurface produce tasks that are more costly to evaluate with respect to empty ones, and load balancing is obtained using fine grained domain partitioning.

The use of fine grained tasks, that is necessary to implement an efficient processor farm, has some drawbacks, the most important is the production of over-partitioned isosurface like in Figure 2(a). This point may cause some important problems especially when a further processing step, such as simplification, is necessary. However the processor farm strategy is auto-balancing, and the execution time needed with fine grained tasks, to extract the isosurface on a cluster of n processors is:

$$T_{Farm} \simeq \frac{1}{n} * T_{IsoExtr}(v) + T_{Overhead\_Farm} \tag{2}$$

with a normally low value of $T_{Overhead\_Farm}$.

Other parallelization strategies are possible, in fact the Marching cubes algorithm can be re-arranged in order to obtain a data parallel version that permits to produce an output evenly distributed on the processors and with better topological characteristics. In fact, the fine grained partitioning of the resulting mesh

(a) Mesh partitioning for the farm algorithm

(b) Mesh partitioning for the balanced data parallel algorithm

Fig. 2. The different partitions of a mesh using four processes. The left one is produced by the farm version with a number of tasks equal to 4 times the number of workers, the right one is produced by the balanced data parallel version.

can be avoided at the cost of an additional load balancing step. In [7] we proposed this strategy, described in the following.

Looking at code of Listing 1.1, the general loop can be re-arranged so that the whole data set is scanned in order to find out active cells, then the isosurface is extracted considering only slices containing active cells. Thus, the isosurface extraction can be decomposed in 2 steps, where the first step is the *scanning* of the active cells, and the second step is the *construction* of the triangles composing the isosurface. Considering an execution distributed on n processors, the data set is initially equally partitioned among the processors. Then each process analyses the cells and records the number of intersection points for each slice. Such information is sent to a coordinator process that re-partition the active slices among the processes in order to balance the number of intersection points. Finally, each processor re-analyses its active cells and computes the isosurface. In Figure 2(b) we present an example of volume partitions among four processes, after the load balancing stage. Each worker receives the part of the volume between two lines. The number of active cells (and consequently triangles and edges) in each partition is about the same. We called this version of the algorithm *Balanced Data Parallel (Bal_DP)*.

Considering formula 1, the time T_{Bal_DP} needed to execute this algorithm is:

$$T_{Bal\_DP} \simeq \max_i(\frac{1}{n}T_{Scan} + \frac{1}{n}T_{Constr} + T_{Scan} * \alpha_i) + T_{Overhead\_LB} \qquad (3)$$

where i ranges on the set of all processes, α_i is the percentage of the active cells re-scanned by the i-th processor during the second step of the algorithm, and $T_{Overhead\_LB}$ is due to load balancing.

With a suitable choice of the fragmentation size, the execution time T_{Farm} results to be lower than T_{Bal_DP}, due to the overhead imposed by the re-scanning of the data set cells and load balancing (See Section 4 for experimental results). However, the mesh obtained using the Data Parallel version is not over partitioned.

3 Parallel Simplification

The goal of the simplification step is to reduce the number of triangles that compose the isosurface preserving its main characteristics, and hence its quality. This operation is very important for different reasons, for example in order to speed up the transmission of very large meshes.

Several proposals were made to simplify triangular meshes. For a survey see [8]. We chose the Garland-Heckbert algorithm proposed in [9] because it produces high quality simplified meshes. Pseudocode 1.2 presents an high level description of the algorithm

Listing 1.2. Garland-Heckbert algorithm for mesh simplification

```
for each vertex of the mesh
    compute the quadric

for each manifold edge of the mesh
    evaluate error and insert in the heap

while numTriangles > threshold do
    select the edge with the minimum error
    edge collapse
    update the topology and the error of neighbour edges
```

The algorithm evaluates all the manifold edges of the mesh on the basis of the *quadric error metrics*. The cost assigned to each edge represents the amount of error introduced in the mesh after its deletion with respect to the original mesh. This is the reason why this algorithm is able to produce high quality results. The *edge collapse* operation consists in the contraction of the selected edge in a vertex and the deletion of the incident triangles.

In our pipeline we preserve mesh characteristics as holes, so only manifold non-boundary edges can be collapsed. If we define $numT$ the number of triangles of the mesh, $perc$ the percentage of triangles to leave in the simplified mesh, then the number of edges to collapse is $numER = \frac{(1-perc)*numT}{2}$. Denoting T_{ValE} the time to build quadrics for the vertices, to evaluate the edges and to insert in the heap, and T_{ESel} the time to collapse an edge, modify the topology and the error of adjacent edges, a simplified analytic formulation of the computational cost of the algorithm is:

$$T_{Simpl} = T_{ValE} + T_{ESel}$$
$$T_{ValE} = numE * (log_2 numE * K1 + K2) \tag{4}$$
$$T_{ESel} = numER * (log_2 numE * K3 + K4)$$

where numE is the number of edges of the mesh. In the above formulas the $log_2 numE$ factor represents the cost related to the heap management. For a detailed discussion of the cost of the algorithm the reader may refer to [10]. In that report it is also shown that it exists a good accordance between cost formulas and measured times.

When investigating for a parallelization strategy for this algorithm, a first point is that the "processor farm" paradigm is not suitable if the purpose is to keep the quality of the simplified mesh, because data cannot be processed independently. In fact using processor farm a mesh is subdivided in independent tasks, and the same amount of simplification will be required for each task, but the quality of the resulting mesh would be generally very poor. As an example, let us consider a mesh representing a mountain and a plain partitioned between two processes: one process receives the mountain and the other the plain. If both processes simplify the same percentage of triangles, the speed up of the algorithm would be nearly linear, but the resulting mesh is too much detailed in the plain and too few detailed in the mountain. Furthermore there is the problem of edges that lie on the border between two processes. If only a process collapses a shared edge, the resulting mesh is not topologically coherent and it needs a costly post-processing step to rearrange data.

In [11] we proposed a parallel algorithm based on the data parallel approach with a coordinator process. The input mesh is subdivided among N worker processes (different from the coordinator). Each process evaluates independently its edges and extracts an ordered list of candidate edges. These candidates are the edges the process would collapse if it works independently. The coordinator receives all these lists and globally selects the $numER$ edges with the minor error. Each process then removes only the edges indicated by the coordinator.

In this manner it is possible to achieve a good trade off between speed up and quality. In the current version of the algorithm we don't allow the simplification of shared edges, but we plan to introduce it in an additional evaluation step. It is to notice that the number of edges to collapse will not be equal for each process. For this reason the number of selected edges $numES$ will be greater than the number of collapsed edges $numER$. In this manner, in the case of a load balanced situation, the coordinator is able to select more than $\frac{numER}{N}$ edges for some processes and less for others.

4   Pipeline Optimization

In the previous Sections the possible parallelization strategies of each pipeline component are described. Now we address the assembly of the different component versions, and we consider the whole pipeline performance and the resulting mesh quality as well as the figures of merits to evaluate the different implementations.

In Table 1 the characteristics of the data sets used for the experiments are listed. In the same table the times for the sequential isosurface extraction and simplification operations are shown. The percentage of simplification used is

Table 1. This table summarizes the size of input data and the time for both the sequential isosurface extraction and simplification components. Isosurfaces were simplified to the 10% of the original triangles. Data in brackets are estimates obtained using Equation 4 after experimental evaluation of costants K1,...,K4.

Data set Id - Isovalue	Input file size (Mbyte)	% of Active cells	Output size (Triangles)	Isoextraction (sec.)	Simplification (sec.)
Bonsai 2	16 MB	11.7	3,896,985	3.4	(142.5)
Frog 1	30 MB	1.7	1,073,360	2.7	36.1
XmasTree 180	499.5 MB	0.9	4,514,539	30.2	(166.3)
VisFemale 1210	867 MB	1.5	13,642,014	54.5	(529.6)

10%, that is only 10% of triangles are kept in the final mesh. For large data sets the simplification time is not available, since the sequential code cannot handle large meshes, but an estimate determined using a cost model based on Equation 4 is proposed.

Sequential computing times have been collected using a Linux PC equipped with a 2.66 GHz Pentium processor, 512 MB of Ram and two EIDE disks interfaced in RAID 0. Parallel computing times have been collected using a cluster of 16 PCs with the previous characteristics, interconnected through an Ethernet - Gigabit switch and having input data sets stored in a PVFS 1 parallel filesystem.

The more efficient parallel solution for the isosurface extraction component is the "processor farm" paradigm, as shown in Table 2, since the load balancing step of the Balanced Data Parallel version represents an overhead.

Table 2. Measured speed up for the Farm and the Balanced Data Parallel versions of the isosurface extraction algorithm

Processors	4		8		16	
Data Set	F	Bal_DP	F	Bal_DP	F	Bal_DP
Bonsai 2	3.2	2.7	5	3.2	4.9	3.4
Frog 1	3.8	3.7	5.2	4.7	5.8	5.1
XmasTree 180	3.2	3.1	5.7	4.9	8.6	6.4
VisFemale 1210	3.1	2.6	5.8	4.2	5.3	5.9

The parallel simplification algorithm is costly with respect to the isosurface extraction algorithm, as we can see in Table 1. The computational cost of the simplification depends on both the number of edges to evaluate ($numE$) and the number of edges to collapse ($numER$).

We may obtain three different pipeline configurations from the assembly of the previously considered implementation of components.

Pipeline 1: Balanced Data Parallel Isoextraction → Simplification
The result of the balanced data parallel version of the isosurface extraction is suitable for the simplification algorithm using the same number of processes. In

fact, balancing the isosurface extraction on the active cells has the consequence of producing an almost balanced output with nearly the same number of edges for each process. Moreover the mesh partitions are coarse grained, thus reducing the new borders. Finally if the two operations are executed by the same process it is possible to keep the produced isosurface into main memory, and no data movement is necessary between the two stages.

Pipeline 2: Farm Isoextraction → Simplification
The output of the farm algorithm for isosurface extraction, instead, is not suitable for the simplification step. First of all, it is not guaranteed that each worker produces the same number of edges. A worker, in fact, may examine a number of tasks without active cells. At the same time a region containing contiguous part of the isosurface can be splitted in a large number of tasks to obtain load balancing, thus creating a huge number of new borders.

Pipeline 3: Farm Isoextraction → Clusterization → Simplification
Both the problems of pipeline 2 can be solved with an intermediate step between the farm algorithm of the isosurface extraction and the simplification. This step has the purpose to reassemble the triangulation produced for each "active task" into a set of contiguous parts of the mesh balanced with respect to the number of triangles. In this case the resulting mesh has a quality that is similar to that of pipeline 1. Heavy data movements may be necessary during the clusterization stage.

While the solution proposed for pipeline 2 is not suitable because of the quality of the resulting surface, the merits of the other two strategies should be evaluated and compared also considering their efficiency.

Then we have:

$$T_{Pipe1} = T_{Bal\_DP} + T_{Simpl}$$
$$T_{Pipe3} = T_{Farm} + T_{Clust} + T_{Simpl}$$

(5)

The choice between the two pipelines, thus depends on the comparison between the overhead due to the Clusterization stage with respect to the additional rescanning required by the *Bal_DP* algorithm.

Table 3 reports speed up for the two pipelines. Missing values are due to the high number of page faults of the simplification algorithm. Values for "Frog 1" are computed considering the actual results of the sequential simplification component. In the other cases speed-up is evaluated considering an estimate of

Table 3. Measured speed up for the Pipeline 1 and 3

Data Set \ Workers	Bal_DP → Simpl			F → Simpl		
	4	8	16	4	8	16
Bonsai 2	-	5.4	7.2	-	4.7	6.8
Frog 1	3.2	6.2	7.2	2.9	5.6	6.8
XmasTree 180	-	5.1	8.3	-	4.7	7.4
VisFemale 1210	-	-	4.3	-	-	3.1

the sequential computing time for the simplification (Table 1). The lower speed up of the "VisFemale 1210" is due to memory trashing of the simplification algorithm for this large mesh. It is important to point out that parallel computing is in these cases an enabling technology that permits to obtain the desired results. Pipeline 1 provides better speed-up than pipeline 3. This despite the farm isoextraction is more efficient than load balanced data parallel.

5 Conclusions and Future Works

In this paper we have considered the trade-offs that arise in the optimization of a parallel visualization pipeline. The main point is that the choice of best performing algorithms depends on the optimization of a simple pipeline stage rather than on the optimization of the whole chain. In our evaluation we also consider the quality of the resulting mesh as one of the parameters that drive the proper algorithm selection.

In actual systems it is not possible to find out a single pipeline configuration that performs at the best in all the cases. For this reason the present study is just a starting point in order to properly define suitable criteria that drive the selection in actual system implementations. It is worthwhile to remember that the presented algorithmic analysis is just one aspect in order to implement a dynamic and adaptive pipeline. A general solution to this and similar problems is still far to be identified.

To this aim it is important the availability of convenient programming tools to support the deployment, for each pipeline stage, of several components, each one implementing a different algorithm, but all presenting the same interface. In such a way, it is possible to dynamically compose the most efficient application according to the needs of the client.

Acknowledgments

This work has been supported by MIUR programme L.449/97-00 High Performance Distributed Computing Platform, and by FIRB strategic project on Enabling Technologies for Information Society, Grid.it. The Christmas Tree data set was generated by the Department of Radiology, University of Vienna and the Institute of Computer Graphics and Algorithms. The VisFemale data set is by the Departments of Cellular and Structural Biology, and Radiology, University of Colorado School of Medicine, under the Human Visible Project grant.

References

1. E.W. Bethel and J. Shalf, Grid-distributed visualizations using connectionless protocols. In IEEE Computer Graphics and Applications, 23:2, 2003, pp. 51-59.
2. B. Hamann, W. Bethel, H.D. Simon and J.C. Meza, NERSC "Visualization Greenbook" : Future Visualization Needs of the DOE Computational Science Community Hosted at NERSC. In Int. Journal of High Performance Computing Applications, 17:2, SAGE, 2003, pp. 97-124.

3. J. Shalf and E.W. Bethel, The Grid and Future Visualization System Architectures. In IEEE Computer Graphics and Applications, 23:2, 2003, pp. 6-9.
4. Visapult, http://www-vis.lbl.gov/projects/visapult/
5. A. Clematis, D. D'Agostino, W. De Marco and V. Gianuzzi: A Web-Based Iso-surface Extraction System for Heterogeneous Clients. In Proceedings of the 29th Euromicro Conference, IEEE Computer Society Press, 2003, pp. 148-156.
6. W. Lorensen and H. Cline: Marching cubes: A high resolution 3-D surface construction algorithm. In Computer Graphics, Vol. 21, 1987, pp.163-169.
7. A. Clematis, D. D'Agostino and V. Gianuzzi: An Online Parallel Algorithm for Remote Visualization of Isosurfaces. In Proceedings of the 10th EuroPVM/MPI Conference, LNCS, No. 2840, 2003, pp. 160-169.
8. C. Gotsman, S. Gumhold and L. Kobbelt: Simplification and Compression of 3-D Meshes. In Tutorials on multiresolution in geometric modelling, A. Iske, E. Quak, M. Floater (eds.), Springer, 2002.
9. M. Garland, and Paul S. Heckbert: Surface Simplification Using Quadric Error Metrics. In Computer Graphics, Vol. 31, 1997, pp. 209-216.
10. D. D'Agostino: A Parallel Simplification Algorithm based on the Quadric Error Metrics. Technical Report IMATI-CNR-Ge, No. 04/2005.
11. A. Clematis, D. D'Agostino, V. Gianuzzi and M. Mancini: Parallel Decimation of 3D Meshes for Efficent Web based Isosurface Extraction. Parallel Computing: Software Technology, Algorithms, Architectures & Applications, in Advances in Parallel Computing series, 13, G.R. Joubert, W.E. Nagel, F.J. Peters, W.V. Walter, Eds., Elsevier, 2004, pp. 159-166.

An Improved Mechanism for Controlling Portable Computers in Limited Coverage Areas*

David Sánchez, Elsa M. Macías, and Álvaro Suárez

Grupo de Arquitectura y Concurrencia (GAC),
Department of Ingeniería Telemática, University of Las Palmas de Gran Canaria,
Campus Universitario de Tafira, 35017 Las Palmas de Gran Canaria, Spain
{dsanchez, emacias, asuarez}@dit.ulpgc.es

Abstract. A network formed by desktop and portable computers is a useful environment for doing parallel computing. In this infrastructure we implement Master/Slave parallel distributed programs which exhibit strict data dependencies among iterations and parallel calculations inside an iteration. In a previous work, we developed a load balancing strategy that uses performance information of slave computers supplied by a framework based on Simple Network Management Protocol (SNMP). This strategy considers the received beacon strength in the portable computers for executing this kind of applications efficiently. However, when a portable computer is located in a limited coverage area, our framework considers that this resource is unavailable, and therefore it can't be used for parallel computing. In this paper we present a mechanism based on the extension of our SNMP framework that allows us to use the computers while there is a wireless network connection.

1 Introduction

Nowadays, the proliferation of high performance portable computers and the recent advances in wireless technologies allow combining wireless local area networks (WLAN) with traditional local area networks (LAN) for doing parallel and distributed computing [1], being a hot topic for next years [2].

We have demonstrated that a computing environment formed by fixed and portable computers can be efficiently used to implement Master/Slave applications with strict data dependencies among iterations. We consider the execution of this kind of applications using our LAMGAC [3] middleware over MPI-2. It is evident that the intrinsic heterogeneity of this natural but currently used computing environment (different processing power and communication bandwidths) makes that the efficient execution of these applications be a very difficult task. If we do not consider load balancing strategies, long idle times will be obtained.

We have developed a load balancing strategy that uses performance information of the slave computers for calculating the adequate data distribution, such that all slave processes finishes at the same time per iteration. This performance

* Research partially supported by Canary Government under Contract: PI:164/2004.

B. Di Martino et al. (Eds.): EuroPVM/MPI 2005, LNCS 3666, pp. 467–474, 2005.

information must be collected in background such that it does not degrade the performance of parallel application. SNMP [4] is used efficiently in [5] for collecting information of fixed resources that constitute a computing environment. In [6] we developed a non-intrusive framework based on SNMP to get performance parameters of the computers. These parameters are used for carrying out an efficient load balancing in presence of heterogeneous computing power and communication. Besides, we implemented an extension of LAMGAC to ease the programming of the above parallel applications where a load balancing strategy is necessary and the framework SNMP is used, which was presented in [7].

The dynamic behavior of portable computers implies to apply control techniques to take into account the wireless beacon strength and the battery energy level. Otherwise the master process can be infinitely locked if it is waiting for results from some slave running on a portable computer that moves out of coverage or its battery is empty. We modified LAMGAC and SNMP framework to consider these issues [8]. In that work, when some portable computer was located in a limited coverage area (weak wireless beacon strength) our load balancing strategy did not estimate the load to distribute to the slave process because that computer can go out of coverage with high probability. As a result, the overall processing power was reduced despite the fact that the slave process could continue calculating and sending results while there was connection.

The new contribution is an improvement of our previous software framework [8] to know in a precise way the portable computers with wireless network link when they are in a limited coverage area. In this way, the processing power of the environment is maximized, because each computer is used while it has a wireless connection. Also, we propose a controlled data reception scheme to avoid that the master process can lock when it waits for results that will not arrive.

The rest of the paper is organized as follows. In section 2 we briefly describe the architecture used for executing Master/Slave applications in a LAN-WLAN environment, as well as the main functions of LAMGAC. In section 3, we explain the improved mechanism. Next, in section 4 we show some experimental results. Finally, we sum up the conclusions.

2   Previous Work

Figure 1.a shows a graph of the combination of the LAN-WLAN. On one hand, there is a master computer where runs the master process that distributes the load to slave processes. Master computer can communicate with the LAN computers and WLAN computers through an access point. On the other hand, the slave computers of LAN are fix and they always have connection with the master computer. However, the computers of WLAN can change their physical location, even entering and going out of coverage.

We consider parallel applications in which the master process distributes (in each iteration) a particular amount of data to each slave process (figure 1.b) that is directly correlated with the calculation performance of the slave processors; it is estimated with our load balancing strategy. Data distribution has to be made

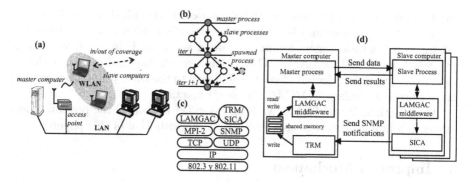

Fig. 1. a) Hardware architecture, b) Master/Slave application dependencies structure, c) Software architecture, d) SNMP framework and parallel program

from the master process because of the spawned processes at run time in the portable computers only communicate with it. When each slave process finishes its calculations it sends results to the master process. The master process must receive the results of all slave processes before sending them the new data.

We implement this kind of applications using our LAMGAC middleware, based on MPI-2, to manage the dynamic expansion of slave processes on the portable computers that enter and go out of coverage at run time and to implement a load balancing strategy (figure 1.c). We also use MPI for doing data communications. We control the variation of the number of portable computers in each iteration. Table 1 shows the main functions of LAMGAC.

Table 1. Main functions of LAMGAC

Functions	Description
LAMGAC_Update	Updates the number of processes that runs in desktop and portable computers
LAMGAC_Balance	Estimates the amount of data to be sent to the slave processes for balancing the execution time
LAMGAC_ItestBattery_beacon	Checks if there are portable computers in a limited coverage area or its battery will empty soon
LAMGAC_Store_info	Stores the effective computed load and the execution time spent by a slave process in each iteration

To carry out an effective load balancing, we have designed a framework based on SNMP that collects performance information about computers. We briefly resume the interaction between this framework and the parallel program (figure 1.d). In each slave computer there is running a *Slave Information Collection Agent* (SICA) that monitors some parameters of the computer performance (system load, network latency, wireless link, battery level, etc) and notifies them (with a *PDU InformRequest*) to *Traps Reception Manager* (TRM) when a significant event has happened (a new slave process has started, the battery will empty soon or the wireless beacon strength is weak). The TRM process is located in the master computer and is in charged of decoding the received performance information from each agent about computers and store it in the shared memory, which is queried by LAMGAC functions. For example, this information is used to estimate the adequate load to distribute to the slave processes when the master process invokes *LAMGAC_Balance()* function.

The functionality of TRM have been extended to control the portable computers in a limited coverage area which is the contribution of this paper.

3   Improved Mechanism

In this section we present our improved mechanism, which is based on the control of portable computers in limited coverage area and in a controlled data reception.

3.1   Control of Portable Computers

The aim of the new contribution is to use the portable computers located in a limited coverage area until there is not network connection. This way, we take advantage of processing power of computers for all the time that it be possible.

In order to implement the improvement mechanism we extended our SNMP framework. In particular, we enlarged the TRM process. The enlargement consists on creating a thread, by TRM process, when a portable computer is within a reduced coverage area (thread SnmpPing in figure 2.a). That is, a thread is created when TRM receives a notification with a value of *lbLinkLevel* below a threshold. It creates a thread by each portable computer in this situation.

Every thread verifies the wireless network link state between the master computer and the portable computer. For that, each thread implements a SNMP *GetRequest* operation, which is periodically invoked to query *lbLinkLevel* object in the management information base, LBGAC-MIB, of the portable computers.

Every thread is running while the portable computer is located in the limited coverage area. That is, the thread finishes when the wireless beacon strength is above a threshold (good coverage area) or when there is not network link between master and slave computer (timeout of several consecutive *GetRequest* operations expires). When the thread finishes, it writes in shared memory the situation of portable computer: online or offline. This value is queried by the beacon and battery function (*LAMGAC_ItestBattery_beacon()*) to inform to the master process about slave processes running in unavailable portable computers.

The implementation of this improvement mechanism modifies the description of *LAMGAC_ItestBattery_beacon()* function as follows (we emphasize in italic format the modification with regard to the description shows in the table 1): Checks if there are portable computers *with a wireless network connection with the master process* or its battery will empty soon.

On the other hand, due to dynamic behaviour of the wireless channel, processor load and location of portable computers, the time elapsed from the SNMP query is realized to the response is received can vary in a considerable way at any moment. It represents a tradeoff for choosing a value for *GetRequest* operation

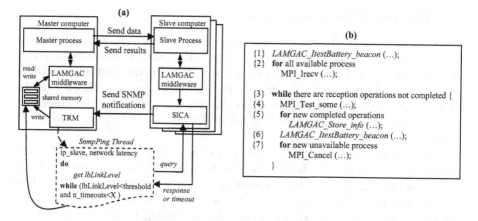

Fig. 2. Improved mechanism: a) SnmpPing thread, b) Controlled reception scheme

timeout. A high timeout value implies a long wait time when the portable computer is not available (out of coverage, channel failures, etc). In this situation, beacon and battery function can indicate that there is a network connection when it is not possible (it is invoked when SNMP operation timeout does not have expired, and therefore, the information stored in shared memory represents a false value). A low timeout value can imply that the SNMP query does not have its SNMP response (for example, due to the congestion on the wireless channel, timeout expires before the response arrives to destination). In this situation, the function can return a value that indicates not network connection when there is. Keeping in mind this tradeoff, we conclude the appropriate timeout value has to be lightly above the time elapsed for implementing a successful query. We calculate the timeout value for doing a query to the computer i (c_i) as follows:

$$timeout(c_i) = 2 \times t\_lat(c_i) + \frac{(sizeof(send) + sizeof(recv))}{B} + t\_get(c_i) \quad (1)$$

where:

- $t\_lat(c_i)$ is the network latency between the master and the slave portable computer. For simplicity, we assume that this value is always equal for both communication directions. This value and B parameter are estimated by SICA when *lamd* daemon starts (LAM-MPI distribution), and sent to TRM via notifications.
- $sizeof(send)$ is the package size of *GetRequest* operation (92 bytes).
- $sizeof(recv)$ is the package size of *Response* operation (93 bytes).
- B is network throughput between the master and the slave computer.
- $t\_get(c_i)$ is the time taken by SICA for decoding the query, calculate and return the results to the TRM. This value depends several factors as: processor speed, memory size, system load, etc. Therefore, in order to calculate the above metric, we estimate this value in a empiric way. We realized many measures about of time elapsed for a success SNMP *Get-Request* operation, and we conclude that this value is approximately 0.7 ms. (in the computing environment specified in the table 2 in section 4).

3.2 Controlled Reception Scheme

Up to now, when the master process initiates the data reception, we suppose that all receptions will be completed. However, if a portable computer is out of coverage, the master process can be infinitely locked if the portable computer does not come back to WLAN, because the master process is waiting for results that will not be sent. In order to avoid this problem, we propose a controlled reception scheme, which is based on LAMGAC and MPI functions. Next, we explain the scheme.

Before master process initiates the reception operations, *LAMGAC_ItestBattery_beacon()* is called to check if there is some unavailable portable computer (step 1 in figure 2.b). If so the reception operation is not implemented for slave process running in that computer. Then, the master process initiates not blocking reception (step 2). Next, the master loops waiting results (step 3). Continuously, the master process tests if some reception operation has completed (step 4). When a reception operation is completed, *LAMGAC_Store_info()* function is invoked to store performance information about process at the current iteration in the shared memory (step 5). This is necessary for load balancing tasks. Next, the master process tests again if there are some new portable computers that can not communicate with the master computer (step 6). If it occurs, the reception operations corresponding to the processes running in those computers are cancelled (step 7). Therefore, it has to be kept in mind by the programmer for calculating again the results not received. This loop is repeated until there are not results to receive.

4   Experimental Results

We have realized several experiments in a combination of IEEE 802.3 and IEEE 802.11 networks of computers under Linux operating system with the specifications of table 2, without and with the new improvement presented in this paper. Every simulation was repeated 10 times obtaining a low standard deviation. Sequential simulation was developed on the faster processor.

Table 2. Computing resources characteristics

Processor / Memory Size	Network Card (Mbps)	Latency (ms)
PIII 450Mhz/128 MB (master)	100	
2 desktop computers PIV 2.4 Ghz/512 MB	100	0.05
2 portable computers PIV 2.4 Ghz/512 MB	11	0.95

In order to carry out the experiments we used as parallel application a Hw/Sw Codesign tool [6] using the programming model specified in [7]. This tool estimates the best Hw/Sw resources combination for a given VHDL input specification of a voice recognition system. Before each data distribution master process calculates, with a recursive procedure, all the possible combinations of Hw/Sw

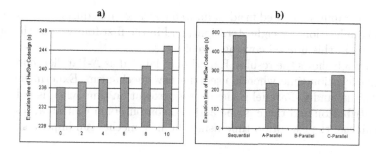

Fig. 3. Experimental results: a) Thread overhead, b) Execution time improvement

resources for implementing a VHDL process of the system. As this procedure is not parallelized the speedup is considerably reduced (figure 3.b).

4.1 Thread Overhead

A SnmpPing thread introduces overhead in the master computer, portable computer and network. Therefore it is necessary to study its overhead in the execution time of parallel application. Overhead is due to:

- Master computer: create thread, build *GetRequest* operation, send query and store data in shared memory.
- Portable computer: decode SNMP query, collect and send information.
- Network: UDP packets for get and response operations.

First and second item depends on performance and current load of computers. The latter depends on network characteristics, as network latency and throughput, and besides, it is affected by wireless network performance (congestion, coverage, shared medium, etc). Keeping in mind this considerations, results presented in this section can vary from a computing environment to another.

To evaluate the thread overhead, we realized several experiments in the worst possible scenario. That is, we forced the execution of threads from the beginning of parallel application until it finishes. We only use two portable computers, therefore we replicated the threads in these computers for doing measurements.

Figure 3.a shows the average execution time of Hw/Sw Codesign tool. As it can see, execution time of parallel application follows approximately a linear relationship with the number of executed threads, and it increases about nine seconds for ten threads executed during all the execution of application. In any case this time will have a minimum influence on the overall execution time of the application, and therefore, its overhead is negligible. Besides, in a real scenario the thread overhead is below to the results shown in this section, because the thread is only created when the computer is inside limited coverage area.

4.2 Execution Time Improvement

In order to demonstrate the improvement of execution time when it is used the new mechanism presented in this paper, we realized several simulations with the portable computers located inside a limited coverage area during some iterations. Parallel application has eight iterations, and in the first four we put the portable computers in this limited coverage area.

Figure 3.b shows the results. Experiment labeled as A-Parallel represents the average execution time when is used the new mechanism. Experiments labeled as B-Parallel and C-Parallel is the average execution time when is not used the new contribution and there is one and two computers in that area, respectively. As it can see, the execution time using the new mechanism improves.

5   Conclusions

In a previous work, our SNMP framework considered the portable computers as unavailable when were located in a limited coverage area, because those computers could go out of coverage with high probability. As a result, the overall processing power was reduced despite the fact that the computers could continue calculating and sending results while there was connection. In this paper, we presented a lightweight improvement of our SNMP framework to check the wireless network link between the master and slave computers located in that area. The collected information is used by a new data reception scheme to avoid that the master process can be infinitely locked when it is waiting for results that will not be sent from those computers. This new contribution allows us to use the portable computers while there is a wireless network connection, and therefore the processing power of the computing environment is maximized.

References

1. Cheng, L. Wanchoo, A., Marsic, I.: Hybrid Cluster Computing with Mobile Objects. 4th IEEE Conference on HPC. Beijin, China (2000) 909–914
2. Zomaya, A.: Mobile Computing: Opportunities for Parallel Algorithms Research. 15th IEEE IPDPS, USA (2002) 144–147
3. Macías, E., Suárez, A.: Solving Engineering Applications with LAMGAC over MPI-2. 9th EuroPVM/MPI. Linz, Austria. LNCS 2474. Springer Verlag (2002) 130–137
4. Subramanian M.: Network Management: Principles and Practice. Addison-Wesley (2000)
5. Busby, R., Nielsen, M., Andresen, D.: Enhancing NWS for Use in an SNMP Managed Internetwork. 14th IEEE IPDPS, Cancún, Mexico (2000) 506–511
6. Sánchez D., Macías E., Suárez A.: Anticipating Performance Information of Newly Portable Computers on the WLAN for Load Balancing". 5th PPAM. Czestochowa, Poland. LNCS 3019. Springer-Verlag (2003) 946–953
7. Sánchez D., Macías E., Suárez A.: A Library for Load Balancing in Master/Slave Applications on a LAN-WLAN Environment. 12th IEEE Euromicro PDP. A Coruña, Spain (2004) 168–175
8. Sánchez D., Macías E., Suárez A.: Load Balancing Detecting Battery Energy Level and Wireless Beacon Strength. 12th IASTED PDCS, M del Rey, USA (2003)268–273

An MPI Implementation for Distributed Signal Processing

J.A. Rico Gallego, J.C. Díaz Martín, and J.M. Álvarez Llorente

Departament of Computer Science, University of Extremadura,
Avda. de la Universidad s/n, 10071, Cáceres, Spain
{jarico, juancarl, llorente}@unex.es
http://gsd.unex.es

Abstract. Video, image and signal processing applications show a high computational complexity and real-time restrictions. Hence they demand task distribution on DSP multi-computers. Unfortunately, DSP vendors offer these platforms nowadays with proprietary kernels and hardware specific communication libraries. As a result, current applications are hardware targeted and not portable at all. We understand that the use of MPI in the signal processing world should help to improve this scenery. This article describes a preliminary implementation of MPI for Sundance machines. Size and performance figures are given.

Keywords: Digital signal processing, digital signal processors, DSP multicomputers, MPI, communication middleware.

1 Introduction and Goals

MPI [1] is an industry de-facto parallel programming standard based on the message-passing paradigm. MPI is widely used to distribute complex applications, particularly in the scientific arena. On other hand, DSP processors deploy architectures that are specialised on digital signal processing. DSP applications, particularly those of real-time video processing demand a computational complexity that if is far from being satisfied by current processors either DSP or general purpose ones. The partition of the application in independent tasks and its further distribution in DSP multicomputers seems to be the way to go, as industry developments confirm. The state of the art in DSP multicomputers is well represented by the developments of Motorola, Sundance [2] or Hunt Engineering. Our development hardware environment are Sundance SMT310Q PCI multicomputer carrier boards with four Texas Instruments [3] TMS320C6000 digital signal processors. Unfortunately, vendors offer these kinds of platforms nowadays with proprietary kernels and hardware specific communication libraries. DSP/BIOS [4], Virtuoso, VxWorks, OSE or 3L Diamond [5] are well known examples to name but a few. As a result, the common pattern in current applications is hardware targeting, and they are not portable at all. Given this state of the art, our aim is developing interfaces that allow building distributed applications achieving the highest degree of portability without sacrifice performance. We understand that

B. Di Martino et al. (Eds.): EuroPVM/MPI 2005, LNCS 3666, pp. 475–482, 2005.

the Message Passing Interface standard (MPI) in the signal-processing world should help to improve this scenery. A light and fast implementation of MPI is a must to meet this goal. This paper describes the design principles and architecture of eMPI (standing for embedded MPI), a preliminary implementation of MPI for Sundance multicomputers with these features.

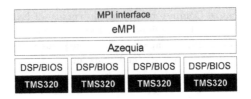

Fig. 1. Multi-computer PCI SMT310Q by Sundance

Fig. 1 shows that our MPI implementation consists of three layers of software, a commercial off the shelf RTOS, the Azequia distributed middleware and eMPI. The rest of the paper is structured as follows. Section 2 presents Azequia, a communication framework for DSP-based multi-computers. Section 3 introduces eMPI, our MPI implementation. Section 4 gives size and performance figures.

2   Azequia: A MPI-Centric Middleware

A fast implementation of MPI should run over a fastdistributed communications framework. That is why we developed Azequia [6]. Azequia runs currently on the TMS320C6000 family of digital signal processors. Texas Instruments supplies these DSPs with a small footprint stand-alone real-time operating system known as DSP/BIOS. Azequia has been developed upon DSP/BIOS for easier portability between the TMS320 processor family. The Diamond distributed RTOS case study will put Azequia in perspective. Under Diamond, a complete application is a collection of one or more concurrently executing tasks. A task is a multithreaded C program. Every task has a vector of input ports and a vector of output ports that are used to connect tasks together and that are passed to main. For instance, this sentence sends the upper character to "the output port 0": chan_out_word (toupper (c), out_ports [0]); A program called the configurer running in the PC host combines task image files to form the executable file. A usersupplied textual configuration file drives the configurer. It specifies the hardware -available processors and physical links connecting them, the software -tasks and connections between them, and how tasks are assigned to processors. The key issue here is that no dynamic addressing is involved, what makes tasks communication transparent to specific locations. We understand that static configuration solves most of current practical problems, but it fails to face technical challenges such as runtime reconfiguration, task migration or fault tolerance in the DSP world. Azequia is our contribution in that address. It proposes and researches an original mechanism of process management that enable the creation

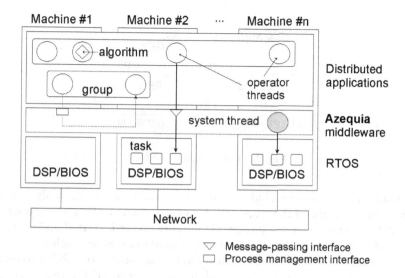

Fig. 2. The Azequia middleware

of remote tasks at run-time, as well as a stack of message passing interfaces that make possible true location-transparent MPI-like communication in DSP multi-computers.

Typically, signal processing leads to an algorithm that is applied to the current window of one or more data streams in an infinite loop. In our model, a loop iteration receives the input windows, does the computing task, and sends the produced window. This simple activity pattern of an algorithm makes it suitable for a single thread of execution. A thread that runs a DSP algorithm is known as an operator. Hence, we assume a DSP distributed application as a group of operators running in one or more machines that communicate through the Azequia middleware as Fig. 2 shows. The addressing scheme is one of the key features of a distributed system. Each thread in the system has assigned an address that distinguishes it from the rest in a global scope. Threads are the end points of a communication. The Azequia address is transparent to the thread location. Its type is Addt_t, a structure with the pair [group, rank]. There should not be two groups with the same identifier. The rank identifies an operator inside a group, ranging from 0 up to the maximum number of operators in the group.

Being a stand-alone RTOS, DSP/BIOS is not aware of other CPUs in a multi-computer environment. Notwithstanding, Azequia uses only the tasking (TSK primitives) and semaphores (SEM primitives) services of DSP/BIOS, for basic concurrent support. Migration of Azequia to Pthreads or other RTOS is straightforward. Azequia just builds a run-time local process management facility upon which building a distributed one. Azequia is composed by six public interfaces, shown by Fig. 3. KER, the kernel, provides the local management and OPR the distributed one. It allows a thread to create an operator in a given machine, as well as to destroy, start and kill it. Each operator type has a wellknown name. The so named register is an internal kernel module that keeps

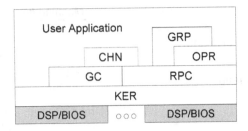

Fig. 3. The Azequia interfaces

the features of the operators linked in memory, i.e., the name, the body function, the parameters size and the stack size. It is used when creating a new operator. The GRP interface operates groups. Groups are created, started and destroyed as a whole. Upon these kernel primitives, Azequia builds two higher level, user oriented communication libraries, called group communication (GC) and remote procedure call (RPC). GC facility, much more similar to MPI, provides location transparency. KER communication primitives come inspired by the basic MPI ones, though exhibiting bigger and lower level functionally. They conform a simple, but yet powerful and flexible interface:

```
int     send    (int sync, char *buffer, int cnt, int mchn,
                 Addr_t src, Addr_t dst, int tag, Rqst_t *rqst,
                 uns timeout);
int     recv    (int sync, char *buffer, int cnt, int mchn,
                 Addr_t src, int tag, Rqst_t *rqst, Status *status,
                 uns timeout);
int     waitany (int count, Rqst_t *rqst, int *index, Status
                 *status);
int     waitall (int count, Rqst_t *rqst, Status **status);
void    test    (Rqst_t rqst, int *flag, Status *status);
```

Note that, in contrast with its MPI counterparts, it is possible to send a message to any thread, not only to those in the same group. This allows invoking RPCs with other services or machines in order to create remote operators, for instance. Besides, communication deadlines are provided as relative time-outs. The sync parameter determines if send and recv operate either in synchronous (blocking) or asynchronous (non-blocking) mode. Send primitive sends count bytes of buffer buffer to dst operator, labeled with the tag tag. Recv primitive is similar. The rqst object is returned only in asynchronous mode. Further waitany and waitall suspend the invoking operator until the communication requests get satisfied. Azequia provides the semantics of Diamond channels by means of the CHN channel interface. Channels are implemented as an independent library upon the GC interface. There are two kinds of channels, input and output ones. Inside an operator, channels of the same sense are known by order number 0, 1,

2, The programmer just sends data to output channel and data arrives to the connected operators.

3   EMPI

EMPI (embedded Message Passing Interface) is an in progress implementation of MPI-2 standard. Most of current MPI implementations, including the most widely used as MPICH [7] or LAM execute upon hardware platforms with lots of processors and extensive use of dynamic memory. P4 ([8], [9]) is the base of MPICH for clusters of UNIX workstations. We ported it to the C6000 family, but its assumption of an underlying UNIX system makes it too heavy in order to achieve the high efficiency demanded by a realtime environment. EMPI, in contrast, is based on Azequia. Its more outstanding features are the assumption of a static application model, and a single-process multi-threaded environment. Though still a partial implementation, eMPI includes enough functionality to implement virtually any MPI application. Besides of initialisation and termination primitives, it includes the point-to-point and collective communications as broadcasting and barrier synchronization. Also predetermined and user defined MPI collective operations, as well as management of groups and communicators. Low power consumption concerns make memory a much more scarce resource in DSP platforms. Besides, real-time performance precludes run-time dynamic allocation of memory. Azequia and eMPI run-time systems do not use malloc. All data memory is reserved at compile time. Notwithstanding, applications are launched dynamically. Azequia groups are closed and represent a signal processing application. An eMPI application is composed by MPI groups, and an Azequia group acts as the container of these MPI groups. Therefore, an eMPI application runs in the context of an Azequia group. The eMPI library keeps a structure that statically stores a fixed number of applications as Fig. 4 shows.

An eMPI application is formed by a set of tasks and a set of groups. Each task stores his own communicators. The creation of a communicator implies a collective operation that sets-up its identifier. The pair composed by the application -the Azequia group number- and the communicator identifies a "communication context". Communicators with the same identifier in different tasks share a group. In most MPI implementations the process remains as the fundamental parallel entity. Messages are sent and received in the process context. Multithreading these processes demand an MPI re-entrant implementation. As current DSP architectures do not provide protection between address spaces the parallel entity must be the thread. Every Azequia thread has an independent global address with its own message queue. Azequia threads, in summary, behave as full right processes. Some other thread-based MPI implementations solve the problem of shared global variables by pre-processing, putting this variables in thread-specific data. This issue is left to the user in eMPI. Some implementations send each message in two separate transmissions, one for the envelope and other for the data. This is an unacceptable approach for our target DSP real-time applications. In eMPI tag, data type and communicator identifier are packed in

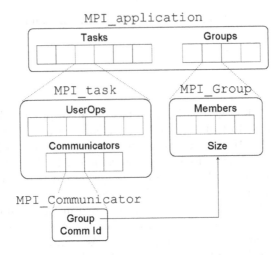

Fig. 4. Structure of an eMPI application

the Azequia tag parameter. Once done this, the whole message is sent. Azequia takes the message from the input queue on tag basis. A single communication is enough. For performance reasons, eMPI currently does not support the reception of messages lesser than expected. This is not a serious disadvantage in DSP applications.

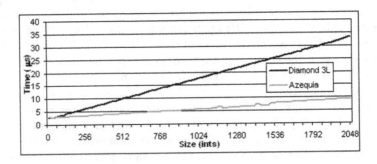

Fig. 5. Azequia vs. Diamond in shared memory

4   Size and Performance

Current eMPI text size is 18 Kb, while Azequia text size is 37 Kb. An eMPI applica-tion can be stored in less than 6 Kb of data memory. EMPI introduces around 1 ms overload upon Azequia. Fig. 5 shows the time it takes doing send/receive for different short and long messages under 3L Diamond and Azequia in a single TMS320C6416 processor. Note that Azequia and eMPI improve Diamond results.

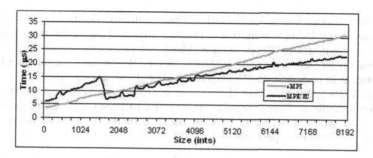

Fig. 6. eMPI vs. MPICH2 in shared memory.

Fig. 6 compares eMPI and MPICH2. A 1.73 GHz Pentium Centrino processor runs MPICH2, while a 720 MHz TMS320C6416 runs eMPI. As it can be seen, the performance of eMPI is quite similar to MPICH2, despite of the processors speed difference.

5   Conclusion

We have presented an MPI implementation for DSP multicomputers built upon a distributed middleware whose interfaces have been modeled after the MPI standard. Under this assumption, eMPI has resulted quite fast and very small. Besides the obvious portability advantages that it provides, we think that eMPI has revealed MPI as a suitable standard for parallel signal processing in current hardware. We are working on completing eMPI, improving its performance and porting it to PC environments.

Acknowledgements

SICUBO S.L., the "II Plan Regional de Investigación, Desarrollo Tecnológico e Innovación de Extremadura" under Project 2PR03A042 and the Spanish Plan Nacional I+D+I under Project TIC2003-08756 founded this work.

References

1. The MPI Forum. The MPI message-passing interface standard. June 1995.
2. http://www.sundance.com
3. http://www.ti.com
4. Texas Instruments: TMS320C6000 DSP/BIOS User's Guide. Literature Number SPRU303B, Texas Instruments (2002).
5. Diamond User Guide for Sundance Multiprocessor Technology Ltd. Version 2.2.1-F. July, 2003. 3L Limited.
6. Juan C. Díaz Martín, Juan A. Rico Gallego, Jesús M. Álvarez Llorente, Carmen Calvo Jurado, "On Interface Design for Distributed Signal Processing". 12th European Signal Processing Conference (EUSIPCO 2004). Vienna (Austria), September 6-10, 2004. ISBN 3-200-00165-8, pgs: 1365-1368.

7. William Gropp, Ewing Lusk, Nathan Doss, Anthony Skjellum, "A High Performance, Portable Implementation of the MPI Message Passing Interface Standard". Parallel Computing, 22, pp. 789-828 (1996).
8. Ralph Butler, Ewng Lusk: Users Guide to the P4 Parallel Programming System. Technical Report ANL-92/17, Argonne National Laboratory (1992).
9. Juan A. Rico, Juan C. Díaz Martín, José M. Rodríguez García, Jesús M. Álvarez Llorente, Juan L. García Zapata, "Porting P4 to Digital Signal Processing Platforms". EuroPVM/MPI 2003 (Venice, Italy). Lecture Notes in Computer Science, ISSN 2-540-20149-1, pgs: 362-368

A Parallel Exponential Integrator
for Large-Scale Discretizations
of Advection-Diffusion Models[*]

L. Bergamaschi[1], M. Caliari[2], A. Martínez[1], and M. Vianello[3]

[1] Department of Mathematical Methods and Models for Scientific Applications,
University of Padua, Italy
[2] Department of Computer Science, University of Verona, Italy
[3] Department of Pure and Applied Mathematics, University of Padua, Italy

Abstract. We propose a parallel implementation of the ReLPM (Real
Leja Points Method) for the exponential integration of large sparse
systems of ODEs, generated by Finite Element discretizations of 3D
advection-diffusion models. The performance of our parallel exponential
integrator is compared with that of a parallelized Crank-Nicolson (CN)
integrator, where the local linear solver is a parallel BiCGstab accelerated with the approximate inverse preconditioner FSAI. We developed
message passing codes written in Fortran 90 and using the MPI standard.
Results on SP5 and CLX machines show that the parallel efficiency raised
by the two algorithms is comparable. ReLPM turns out to be from 3 to
5 times faster than CN in solving realistic advection-diffusion problems,
depending on the number of processors employed.

1 Finite Element Discretization of the Advection-Diffusion Model

We consider the classical evolutionary advection-diffusion problem

$$\begin{cases} \dfrac{\partial c}{\partial t} = \text{div}(D\nabla c) - \text{div}(cv) + \phi & x \in \Omega,\ t > 0 \\ c(x,0) = c_0(x),\ x \in \Omega; \\ c(x,t) = g_{\mathrm{D}}(x,t),\ x \in \Gamma_{\mathrm{D}};\ \langle D\nabla c(x,t), \nu \rangle = g_{\mathrm{N}}(x,t),\ x \in \Gamma_{\mathrm{N}};\quad t > 0 \end{cases} \quad (1)$$

with mixed Dirichlet and Neumann boundary conditions on $\Gamma_{\mathrm{D}} \cup \Gamma_{\mathrm{N}} = \partial\Omega$,
$\Omega \subset \mathbb{R}^3$. Equation (1) represents, e.g., a simplified model for solute transport
in groundwater flow (advection-dispersion), where c is the solute concentration,
D the hydrodynamic dispersion tensor, $D_{ij} = \alpha_{\mathrm{T}}|v|\delta_{ij} + (\alpha_{\mathrm{L}} - \alpha_{\mathrm{T}})v_iv_j/|v|$,
$1 \leq i,j \leq d$, v the average linear velocity of groundwater flow and ϕ the source.

[*] Work supported by the research project CPDA028291 "Efficient approximation
methods for nonlocal discrete transforms" of the University of Padova, and by the
MIUR PRIN 2003 project "Dynamical systems on matrix manifolds: numerical methods and applications" (co-ordinator L. Lopez, Bari).

B. Di Martino et al. (Eds.): EuroPVM/MPI 2005, LNCS 3666, pp. 483–492, 2005.

The standard Galerkin Finite Element (FE) discretization of (1) with nodes $\{x_i\}_{i=1}^N$ and linear basis functions gives a large scale linear system of ODEs like

$$\begin{cases} P\dot{\mathbf{c}} = H\mathbf{c} + \mathbf{b}, & t > 0 \\ \mathbf{c}(0) = \mathbf{c}_0 \end{cases} \tag{2}$$

where $\mathbf{c} = [c_1(t), \ldots, c_N(t)]^{\mathrm{T}}$, $\mathbf{c}_0 = [c_0(x_1), \ldots, c_0(x_N)]^{\mathrm{T}}$, P is the symmetric positive-definite mass matrix and H the (nonsymmetric) stiffness matrix. Boundary conditions are incorporated in the matrix formulation (2) in the standard ways.

2   Exponential Integration via Polynomial Approximation

In the sequel we consider stationary velocity, source and boundary conditions in (1), which give constant H and $\mathbf{b}$ in system (2), which is the discrete approximation of the PDE (1). As known, the solution can be written explicitly in the exponential form

$$\mathbf{c}(t) = \mathbf{c}_0 + t\varphi(tP^{-1}H)\left[P^{-1}H\mathbf{c}_0 + P^{-1}\mathbf{b}\right] , \tag{3}$$

where $\varphi(z)$ is the entire function $\varphi(z) = (e^z - 1)/z$ if $z \neq 0$, $\varphi(0) = 1$.

Applying the well known mass-lumping technique (sum on the diagonal of all the row elements) to P, we obtain a diagonal mass matrix P_{L}. Now system (3) (with P_{L} replacing P) can be solved by the exact and explicit exponential time-marching scheme.

$$\mathbf{c}_{k+1} = \mathbf{c}_k + \Delta t_k \varphi(\Delta t_k H_{\mathrm{L}})\mathbf{v}_k, \quad k = 0, 1, \ldots,$$
$$H_{\mathrm{L}} = P_{\mathrm{L}}^{-1}H, \qquad \mathbf{v}_k = H_{\mathrm{L}}\mathbf{c}_k + P_{\mathrm{L}}^{-1}\mathbf{b}. \tag{4}$$

Exactness of the exponential integrator (4) entails that the time-steps Δt_k can be chosen, at least in principle, arbitrarily large with no loss of accuracy, making it an appealing alternative to classical time-differencing integrators (cf. [6,5]).

However, the practical application of (4) rests on the possibility of approximating efficiently the exponential propagator $\varphi(\Delta t H_{\mathrm{L}})\mathbf{v}$, where $\mathbf{v} \in \mathbb{R}^N$. To this aim, we adopt the Real Leja Points Method (shortly ReLPM), recently proposed in the framework of FD spatial discretization of advection-diffusion equations [5], and extended to FE in [2]. Given a matrix A and a vector $\mathbf{v}$, the ReLPM approximates the exponential propagator as $\varphi(A)\mathbf{v} \approx p_m(A)\mathbf{v}$, with $p_m(z)$ Newton interpolating polynomial of $\varphi(z)$

$$p_m(A) = \prod_{k=0}^{m-1} (A - \xi_k I), \qquad m = 1, 2, \ldots \tag{5}$$

at a sequence of Leja points $\{\xi_k\}$ in a compact subset of the complex plane containing the spectrum (or the field of values) of the matrix A. Following [5,2], an algorithm for the approximation of the advection-diffusion FE propagator $\varphi(\Delta t H_L)\mathbf{v}$ can be easily developed, by means of Newton interpolation at "spectral" Leja points. In the sequel, the compact subset used for estimating the spectrum of H_L in (4) will be an ellipse.

Algorithm ReLPM (Real Leja Points Method)

1. INPUT: H_L, $\mathbf{v}$, Δt, tol
2. Estimate the spectral focal interval $[\alpha, \beta]$ for $\Delta t H_L$, by Gershgorin's theorem
3. Compute a sequence of Fast Leja Points $\{\xi_j\}$ in $[\alpha, \beta]$ as in [1]
4. $d_0 := \varphi(\xi_0)$, $\mathbf{w}_0 := \mathbf{v}$, $\mathbf{p}_0 := d_0\mathbf{w}_0$, $m := 0$
5. WHILE $e_m^{\text{Leja}} := |d_m| \cdot \|\mathbf{w}_m\|_2 > \text{tol}$
   (a) $\mathbf{z} := H_L\mathbf{w}_m$
   (b) $\mathbf{w}_{m+1} := \Delta t\,\mathbf{z} - \xi_m\mathbf{w}_m$
   (c) $m := m + 1$
   (d) compute the next divided difference d_m
   (e) $\mathbf{p}_m := \mathbf{p}_{m-1} + d_m\mathbf{w}_m$
6. OUTPUT: the vector $\mathbf{p}_m$: $\|\mathbf{p}_m - \varphi(\Delta t H_L)\mathbf{v}\|_2 \approx e_m^{\text{Leja}} \leq \text{tol}$

The ReLPM algorithm turns out to be quite simple and efficient. Indeed, being based on two-term vector recurrences in real arithmetic, its storage occupancy and computational cost are very small, already with one processor. For implementation details not reported here, we refer to [5].

2.1 Parallel ReLPM (Real Leja Points Method)

A standard data-parallel implementation of ReLPM has been performed. The cost of computing the Leja points is negligible with respect to the rest of the algorithm ([1]) and hence we decided that every processor performs step 3 separately, without exchanging data. To perform an efficient parallel implementation of the ReLPM we choose to partition the matrix H_L by rows and the vectors involved in algorithm ReLPM consequently. In this way the daxpy operations in 4, 5b and 5e are performed without any communication among processors. Moreover, estimation of the focal interval (step 2) and computation of the 2-norm of a vector (to check the exit test) needs that the processors exchange only a scalar (the result of their local computation). The matrix vector product of step 5a requires the processors to communicate a number of elements of vector $\mathbf{w}_m$. We employed the parallel sparse matrix vector routine, successfully experimented in [4], which will be described in 3.3.

3   Parallel Implementation of Crank-Nicolson

3.1   Crank-Nicolson (CN) Method

Crank-Nicolson (CN) is a robust method, widely used in engineering applications, and a sound baseline benchmark for any advection-diffusion solver. In the case of the relevant ODEs system (2) (with stationary $\mathbf{b}$), its variable step-size version writes as

$$\left(P - \frac{\Delta t_k}{2} H\right) \mathbf{u}_{k+1} = \left(P + \frac{\Delta t_k}{2} H\right) \mathbf{u}_k + \Delta t_k \, \mathbf{b}, \quad k = 0, 1, \ldots, \quad \mathbf{u}_0 = \mathbf{c}_0 \ , \quad (6)$$

where, for estimation of the local truncation error and step-size control, we have used standard finite-difference approximation of the third derivatives in $\|\ddot{\mathbf{c}}(t_k)\|_2 \Delta t_k^3 < 12 \, \text{tol}$.

The large and sparse linear system in (6) is solved the BiCGstab iterative method [9], preconditioned at each step, since the system matrix depends on Δt_k and hence varies from step to step. To accelerate the iterative solver, we consider the "approximate inverse preconditioners". They explicitly compute an approximation to A^{-1} and their application needs only matrix vector products, which are more effectively parallelized than solving two triangular systems, as in the ILU preconditioner. We selected the FSAI (Factorized Sparse Approximate Inverse) preconditioner proposed in [7], whose construction is more suited to parallelization than other approaches [3].

3.2   FSAI Preconditioning

Let A be a symmetric positive definite matrix (SPD) and $A = L_A L_A^T$ be its Cholesky factorization. The FSAI method gives an approximate inverse of A in the factorized form $H = G_L^T G_L$, where G_L is a sparse nonsingular lower triangular matrix that approximates L_A^{-1}. To construct G_L one must first prescribe a selected sparsity pattern $S_L \subseteq \{(i,j) : 1 \leq i \neq j \leq n\}$, such that $\{(i,j) : i < j\} \subseteq S_L$, then a lower triangular matrix $\hat{G}_L$ is computed by solving the equations $(\hat{G}_L A)_{ij} = \delta_{ij}, (i,j) \notin S_L$. The diagonal entries of $\hat{G}_L$ are all positive. Defining $D = [\text{diag}(\hat{G}_L)]^{-1/2}$ and setting $G_L = D\hat{G}_L$, the preconditioned matrix $G_L A G_L^T$ is SPD and has diagonal entries all equal to 1.

The extension to the nonsymmetric case is straightforward; however the solvability of the local linear systems, and the nonsingularity of the approximate inverse, are only guaranteed if all the principal submatrix of A are non singular (which is the case, for instance, if $A + A^T$ is SPD). In the nonsymmetric case two preconditioner factors, G_L and G_U, must be computed. We limit ourselves to nonsymmetric matrices with a symmetric nonzero pattern (which is the common situation in matrices arising from FE discretization of PDEs), and set the sparsity patterns for G_U factor as $S_L = S_U^T$. The preconditioned matrix reads $D = G_L A G_U D^{-1}$, with $D = diag(G_L) = diag(G_U)$.

We set the sparsity patterns of the lower and upper triangular factors to allow nonzeros corresponding to nonzeros in the lower and upper triangular part of A^2, respectively. Next we perform a postfiltration step of the already constructed factors by using a small drop–tolerance parameter ϵ. The aim is to reduce the number of nonzero elements of the preconditioner, in order to decrease the arithmetic complexity of the iteration phase together with the communication complexity of multiplying the preconditioner by a vector.

For deeper implementation and performance details of parallel FSAI the author is referred to [3].

3.3   Efficient Matrix-Vector Product

Following [4], we now briefly describe our implementation of the matrix-vector product, which is tailored for application to sparse matrices and minimizes data communication between processors. Within the ReLPM or CN algorithms, the vector $\mathbf{y} = B\mathbf{v}$ has to be calculated for $B = A, G_L, G_U$. Assume that the $N \times N$ matrix B is uniformly partitioned by rows among the p processors, so that $n \approx N/p$ rows are assigned to each processor. The same is done for the vector $\mathbf{v}$. The subset P^r containing the nonzero elements belonging to processor r can be subdivided into two disjoint subsets $P_1^r = \{b_{ij} \in P^r, \ (i-1)n+1 \leq j \leq in\}$ and $P_2^r = P^r \backslash P_1^r$. Define the sets C_k^r, R_k^r of indices as: $C_k^r = \{j : b_{ij} \in P_2^r, k = ((j-1) \text{ div } n) + 1\}$; $R_k^r = \{i : b_{ij} \in P_2^r, k = ((j-1) \text{ div } n) + 1\}$. Processor r has in its local memory the elements of the vector $\mathbf{v}$ whose indices lie in the interval $[(r-1)n+1, rn]$. Before computing the matrix-vector product processor r: for every k such that $R_k^r \neq \emptyset$ sends to processor k the components of vector $\mathbf{v}$ whose indices belongs to R_k^r; gets from every processor k such that $C_k^r \neq \emptyset$, the elements of $\mathbf{v}$ whose indices are in C_k^r. At this point every processor is able to complete locally its part of the matrix-vector product.

4   Parallel Experiments and Results

4.1   Description of the Test Cases

We now discuss in detail two examples (cf. [2]), concerning FE discretizations of 3D advection-dispersion models like (1).

Example 1. The domain is $\Omega = [0,1] \times [0,0.5] \times [0,0.1]$, with a regular grid of $N = 161 \times 81 \times 41 = 534\,681$ nodes and $3\,072\,000$ tetrahedral elements. Here, $\phi \equiv 0$ and $c_0 \equiv 1$. Dirichlet boundary conditions are $c = 0$ on $\Gamma_{\mathrm{D}} = \{0\} \times [0.2, 0.3] \times [0,1]$, while the Neumann condition $\partial c / \partial \nu = 0$ is prescribed on $\Gamma_{\mathrm{N}} = \partial \Omega \backslash \Gamma_{\mathrm{D}}$. The velocity is $v = (v_1, v_2, v_3) = (1, 0, 0)$, the transmissivity coefficients are piecewise constant and vary by an order of magnitude depending on the elevation of the domain, $\alpha_{\mathrm{L}}(z) = \alpha_{\mathrm{T}}(z) \in \{0.0025, 0.025\}$.

Example 2. Same problem as of Example 1. However, the domain is discretized with a regular grid of $N = 161 \times 81 \times 161 \approx 2.1 \times 10^6$ nodes and about 12 millions of tetrahedral elements. Matrix of discretization H_L has roughly 2.1×10^6 rows and 3.1×10^7 nonzero elements.

In these examples the boundary conditions and vanishing sources lead to a zero steady state. The two integrators are employed on a time interval which produces a decrease of two orders of magnitude of the initial solution norm. While for CN the local time-step is selected adaptively, in order to guarantee a local error below the given tolerance, for the exponential integrator there is no restriction on the choice of Δt_k, since it is exact for autonomous linear systems of ODEs. To follow with some accuracy the evolution of the solution, we propose as in [5] to select the local time-step in (4) in such a way that the relative variation of the solution be smaller than a given percentage η, that is

$$\|\mathbf{c}_{k+1} - \mathbf{c}_k\|_2 \leq \eta \cdot \|\mathbf{c}_k\|_2, \quad 0 < \eta < 1 . \tag{7}$$

If condition (7) is not satisfied, the time step Δt_k is halved and $\mathbf{c}_{k+1}$ recomputed; if it is satisfied with $\eta/2$ instead of η, the next time-step Δt_{k+1} is doubled. Clearly, smaller values of η allow better tracking of the solution.

4.2   Parallel Programs and System's Architecture

The parallel programs are fortran 90 message passing codes, written using the MPI standard [8]. The message passing programming model is a distributed memory model with explicit control parallelism. Message passing codes written in MPI are obviously portable and should transfer easily to clustered SMP systems, which are gradually becoming more prominent in the HPC market. We run the codes on two supercomputers located at the CINECA Supercomputer center of Bologna, Italy (http://www.cineca.it).

IBM SP5 supercomputer, an IBM SP cluster 1600, made of 64 nodes p5-575 interconnected with a pairs of connections to the Federation HPS (High Peformance Switch). Globally the machine has 512 IBM Power5 processors, capable of 4 double precision floating point operations per clock cycle, and 1.2 TBs of memory. Each microprocessor is supported by 36 MB of Level 3 cache. The peak performance of SP5 is 3.89 Tflops. Each p5-575 node contains 8 SMP processors POWER5 at 1.9 GHz, with 16GB of memory each. The HPS switch is capable of a bandwidth of up to 2GB/s unidirectional.

IBM Linux Cluster (CLX), made of 512 2way IBM X335 nodes. Each computing node contains 2 Xeon Pentium IV processors. All the compute nodes have 2GB of memory (1GB per processor). Most processors of CLX are Xeon Pentium IV at 3.06 GHz with 512MB of L2 cache and the remaining ones, bought at the beginning of 2005, are Xeon Pentium IV EM64T at 3.00GHz with 1024MB of L2 cache. All the CLX processors are capable of 2 double precision floating point operations per cycle, using the INTEL SSE2 extensions. All the nodes are interconnected to each other through a Myrinet network (http://www.myricom.com), capable of a maximum bandwidth of 256MB/s between each pair of nodes. The global peak performance of CLX is of 6.1 TFlops. Parallel programming on the CLX is mainly based on the MPICH-GM version of MPI (myrinet enabled MPI).

4.3   Results Concerning Example 1

We show in this section the timings of the two MPI codes when solving the problem described in Example 1. In the SP5 machine the running times were obtained by using the nodes in dedicated mode, hence reserving to our own use the entire node (8 processors) even to measure CPU times with 1,2, and 4

Table 1. Timings and speedups for Example 1 solved with CN with BiCGstab accelerated by diagonal and `mixed` preconditionings on the IBM SP5

			Diagonal					`mixed` $\epsilon = 0.05$		
p	iter	T_p	T_{sol}	CPU	S_p	iter	T_p	T_{sol}	CPU	S_p
1	36224	19.8	5765.7	5872.2		14694	180.6	3870.7	4137.0	
2	36140	10.1	2911.36	2968.5	2.0	14653	94.3	1907.0	2048.4	2.0
4	36382	4.4	1040.18	1078.5	5.4	14696	46.9	813.4	878.0	4.7
8	36196	1.7	471.4	480.8	12.2	14689	23.7	409.6	440.4	9.4
16	36276	0.9	254.5	260.5	22.5	14665	12.6	202.6	226.5	18.3

Table 2. Timings and speedups for Example 1 solved by parallel ReLPM on the IBM SP5

		$\eta = 0.02$				$\eta = 0.05$		
p	Steps	iter	CPU	S_p	Steps	iter	CPU	S_p
1	239	17332	1282.2		98	18201	1343.3	
2	239	17332	634.9	2.0	98	18201	663.2	2.0
4	239	17332	228.9	5.6	98	18201	237.1	5.7
8	239	17332	108.9	11.8	98	18201	113.4	11.8
16	239	17332	58.2	22.1	98	18201	63.1	21.3

processors. We did not take any advantage of shared memory inside the node. In the CLX cluster only one of the two processors in each node were used, to optimize memory accesses performance.

CN has been run with variable stepsize, leading to 479 time steps to complete the simulation. To avoid the cost of constructing the FSAI preconditioner at each time-step, we chose to compute it selectively, depending on the variation of Δt_k. Besides, an improved preconditioning strategy (`mixed`) is proposed which consists in using Jacobi for the (well-conditioned) first steps, and FSAI for the remaining. The switch between the two accelerators takes place the next timestep after the solver first employs a number of iterations larger than a fixed value (40). BiCGstab iterations are stopped when the residual r_k satisfies $\|r_k\| \leq 10^{-4}\|b\|$. We report in Tables 1 and 2 the results of the codes running on the SP5 with a number o processors $p = 1, \cdots, 16$. As for CN the number of BiCGstab iterations (`iter`), and the CPU times for computing the preconditioner (T_p), for the iterative solver (T_{sol}) and the overall CPU time are given, whereas for ReLPM we provide the number of steps, the number of total inner iterations (`iter`) and the overall CPU time. The `mixed` preconditioning strategy results in a reduction of number of linear iterations and CPU time with respect to the diagonal preconditioner.

The ReLPM has been run using $\eta = 0.02$ and 0.05, with similar performances. Obviously, the value $\eta = 0.02$ allows a better tracking of the solution.

In Table 3 we report the summary of the results of the same runs on the CLX machine. Here, the speedup values are between 11 and 16 hence yielding a parallel efficiency of at least 70% on 16 processors. The timings results demonstrate

Table 3. Summary of results on the CLX machine

	Crank Nicolson				ReLPM			
	Diagonal		mixed		$\eta = 0.02$		$\eta = 0.05$	
p	CPU	S_p	CPU	S_p	CPU	S_p	CPU	S_p
1	8771.9		6484.1		1627.1		1425.7	
16	810.2	10.8	552.8	11.7	99.7	16.3	119.2	11.9

that both codes scale well with increasing number of processors. Moreover, they show a superspeedup when using more than 2 processors due to cache effects, since only for $p \geq 4$ the local matrix resides entirely in the Level 3 cache when performing the matrix vector product.

4.4 Results Concerning Example 2

As in the previous example, CN has been run with variable stepsize, leading to 479 time steps to complete the simulation. We used the mixed preconditioning strategy and set the limit number for Jacobi preconditioning to 60 iterations. The same exit test as in Example 1 was used.

We report in Table 4 the timings concerning Example 2 on the SP5 with $p = 1, \cdots, 64$. As for CN, the code could not run for $p = 1, 2$ and 4. This is so because the limit of 1.667 GB of available memory of the SP5 nodes is not sufficient to hold the local system and preconditioner matrices when less than 8 processors are used. As in the previous example, both codes scale well with increasing number of processors again achieving a superspeedup for $p \geq 16$ due to cache effects. We recall that both examples rely on the same differential problem; however the discretization of Example 2 is made on a finer grid, which yields an algebraic problem roughly four times larger than that of Example 1. We note that CN is on the average four times slower than our ReLPM.

The summary of the performance results are reported in Figure 1. For ReLPM on the SP5 we obtain perfect speedups up to 8 processors. For $p > 8$ the curves of both CN and ReLPM are above that of the ideal speedup. Figure 1 demonstrates that the scaling of the codes in the CLX machine decreases when using more

Table 4. Timings and speedups for Example 2 solved with CN accelerated with the mixed preconditioner and ReLPM with $\eta = 0.02$ on the SP5. Symbol †stands for "out of memory". The speedups for CN have been computed as $S_p^* = 8 * T_8 / T_p$.

	Crank Nicolson					ReLPM			
	mixed $\epsilon = 0.1$					$\eta = 0.02$			
p	iter	T_p	T_{sol}	CPU	S_p^*	steps	iter	CPU	S_p
1	†	†	†	†		238	17710	5308.2	
2	†	†	†	†	†	238	17710	2672.6	2.0
4	†	†	†	†	†	238	17710	1342.2	4.0
8	18050	62.9	2337.0	2453.7	8.0	238	17710	661.4	8.0
16	18006	31.8	1020.0	1073.8	18.3	238	17710	237.2	22.4
32	18105	16.6	433.5	457.8	42.9	238	17710	132.1	40.2
64	18086	10.4	250.2	267.9	73.3	238	17710	66.4	79.9

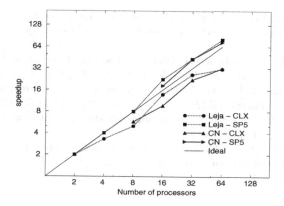

Fig. 1. Speedups vs p for ReLPM and CN on the SP5 and CLX supercomputers

than 32 processors. We think that a potential source of this poor scaling in the CLX machine is the smaller bandwidth and higher latency of the interconnection network with respect to the HPS of the SP5 machine.

5   Conclusions

A parallel implementation of two algorithms for the solution of advection-diffusion equations on 3D domains has been proposed. Parallelization of ReLPM revealed almost straightforward, being based on matrix-vector products and not requiring linear system solutions. The CN solver has been carefully parallelized, with special emphasis on the selection of an efficient parallel preconditioner. Results on two supercomputers in the solution of a problem of more than 2 million unknowns show the very good scalability of the two codes, enhancing at the same time the efficiency of ReLPM both in terms of CPU time and computer storage.

References

1. J. Baglama, D. Calvetti, and L. Reichel. Fast Leja points. *Electron. Trans. Numer. Anal.*, 7:124–140, 1998.
2. L. Bergamaschi, M. Caliari, and M. Vianello. The ReLPM exponential integrator for FE discretizations of advection-diffusion equations. In M. Bubak, et al., editors, *ICCS 2004, Proceedings, Part IV, LNCS 3036*, pages 434–442. Springer, 2004.
3. L. Bergamaschi and A. Martínez. Parallel acceleration of Krylov solvers by factorized approximate inverse preconditioners. In M. Daydè et al., editor, *VECPAR 2004, LNCS 3402*, pages 623–636, Springer, 2005.
4. L. Bergamaschi and M. Putti. Efficient parallelization of preconditioned conjugate gradient schemes for matrices arising from discretizations of diffusion equations. In *Proceedings of the Ninth SIAM Conference on Parallel Processing for Scientific Computing*, March, 1999. (CD–ROM).

5. M. Caliari, M. Vianello, and L. Bergamaschi. Interpolating discrete advection-diffusion propagators at spectral Leja sequences. *J. Comput. Appl. Math.*, 172(1):79–99, 2004.
6. M. Hochbruck, C. Lubich, and H. Selhofer. Exponential integrators for large systems of differential equations. *SIAM J. Sci. Comput.*, 19(5):1552–1574, 1998.
7. L. Yu. Kolotilina and A. Yu. Yeremin. Factorized sparse approximate inverse pre-conditionings I. Theory. *SIAM J. Matrix Anal. Appl.*, 14:45–58, 1993.
8. MPI Forum. *MPI: A message passing interface standard*, 1995. also available online at `http://www.mpi-forum.org/`.
9. H. A. van der Vorst. Bi-CGSTAB: a fast and smoothly converging variant of Bi-CG for the solution of nonsymmetric linear systems. *SIAM J. Sci. Statist. Comput.*, 13(2):631–644, 1992.

Parallel Grid Adaptation and Dynamic Load Balancing for a CFD Solver

Christoph Troyer[1], Daniele Baraldi[2], Dieter Kranzlmüller[1],
Heinz Wilkening[2], and Jens Volkert[1]

[1] GUP, Joh. Kepler University Linz,
Altenbergerstr. 69, A-4040 Linz, Austria
ctroyer@gup.uni-linz.ac.at
http://www.gup.uni-linz.ac.at
[2] Institute for Energy, Joint Research Centre,
PO Box 2, 1755 ZG Petten, The Netherlands
http://www.jrc.nl

Abstract. In this paper we present the parallel version of a CFD solver which works on unstructured 3-dimensional grids. The parallelization was achieved using the MPI library. The program includes a method for dynamically adapting a grid on a parallel computer. Grid adaptation is one of the reasons that lead to a poor workload distribution. Therefore a dynamic load balancing method was implemented in order to make sure that the available hardware is utilized in an efficient way. The paper concludes with the presentation of a performance test.

1 Introduction

The CFD solver under examination is called REACFLOW and was developed at European Commission DG Joint Research Centres in Ispra and Petten [1,7]. REACFLOW is a finite volume code that runs on 2-D and 3-D unstructured grids and is mainly used for simulating gas explosions on large industrial scales. This is often the only way to study explosions that would be either too expensive or too dangerous to be studied by physical experiments.

The size and complexity of the examined geometries, the unsteadiness of flow phenomena and the high spatial resolution needed to model effects on a range of geometrical scales put enourmous requirements on the numerical methods and on the hardware. Depending on the problem size a complete simulation can take several weeks if not months on a powerful single processor machine. Sometimes the problem would not even fit into the memory of such a computer.

The parallelization of the code can solve these problems by allowing the program to be executed on parallel hardware. The following section gives an overview of how the most basic part of REACFLOW was parallelized. In Section 3 and 4 we present a method for adapting the grid in parallel and how the workload can be distributed evenly among the processors during a simulation. An example showing the performance of the parallel version of REACFLOW is presented in the last section.

B. Di Martino et al. (Eds.): EuroPVM/MPI 2005, LNCS 3666, pp. 493–501, 2005.

2   Basic Parallelization

The parallel version of REACFLOW was created by using domain decomposition. This means that the problem data is split into pieces which are then assigned to the processors of a parallel computer. We use either Metis [3] or Jostle [6] to find the initial decomposition of the mesh. Neighbouring processor subdomains overlap each other.

This overlap allows a processor to carry out all calculations of a computational phase without the need of communicating with other processors. Only after each phase the intermediate results have to be exchanged for all grid nodes in the overlap regions. The computational phases are:

- Calculation of the gradients;
- Convective solver;
- Diffusive solver;
- Turbulent solver and calculation of chemical source terms.

The inter-processor communication is done via message passing. We use the MPI library [5], mainly for two reasons: It is a widely spread standard and it makes it possible to develop code that can be executed on a range of different hardware architectures.

3   Parallel Grid Adaptation

In many simulations the interesting parts of the flow domain are small compared to the whole geometry. These are the regions with the steepest variable gradients, for example a shock wave. In order to get a sufficient computational accuracy to resolve the steep gradients the grid has to be very fine in such regions whereas it can be coarse everywhere else. The problem is that phenomena like shock waves and flame fronts are moving through the whole domain, which means that a static grid has to be fine everywhere. Increasing the grid resolution leads to higher memory requirements and longer computation times. Therefore it is advantageous to have a dynamic adaptive grid capability, which allows to refine or coarsen the grid dynamically.

In the scalar version of REACFLOW dynamic grid adaptation works as follows. The adaptation variables of two neighbouring control volumes are compared and if the difference is above a given limit the two nodes are flagged for refinement. If the difference is below a certain threshold then the nodes are flagged for coarsening. After all control volume pairs have been checked the elements are examined. If at least one of the nodes of an element is marked as refinable then the element is marked as refinable as well. In case the first node of an element is planned to be removed and none of the other nodes of this element are flagged for refinement then the element is marked as removeable.

Once all elements have been examined the actual refinement and coarsening can be initiated. Elements are refined by inserting an additional node between the two endpoints of its longest edge. The longest edge is chosen in order to avoid getting very stretched elements which would lead to problems with the flow solvers. Furthermore the additional node can only be inserted if the edge, which is going to be divided, is the longest one for all elements that have this edge in common. If one of these surrounding elements has an even longer edge, then the algorithm tries to insert the new grid node at this newly found edge which again has to be the longest one for all elements that are grouped around it. This recursive procedure can lead to the insertion of grid nodes several elements away from the position where it was intended to take place. It has turned out that this can be seen as a useful feature since the grid gets already refined ahead of the shock wave.

The scalar version of the grid adaptation code is hard to parallelize because of its non-local nature. The refinement of one element can lead to the refinement of a remote element. Whenever the algorithm moves away from an original refinable element in search of a terminal edge it would cause communication between processors in case this search involves crossing inter-processor domain boundaries. Therefore a different way of refining the grid was chosen for the parallel version of REACFLOW.

The main difference between the parallel and the scalar grid adaptation is the way how terminal edges are found. The parallel version maintains for each edge in the grid a flag whether it is the longest one for all of its surrounding elements. Hence it is no longer necessary to examine elements in a recursive manner. If an element is flagged as refinable then we can immediately determine whether this element can actually be divided simply by checking the above mentioned edge flag. The subsequent bisection of the element works again like in the scalar version. This solution requires of course to update the edge flags of elements which were affected by refinement or coarsening respectively.

The bisection of an element is illustrated in Figure 1. The element ABCD was marked as refinable and AB is its longest edge. By looking at the edge flag of AB we can tell that it is a terminal edge and immediately insert a new node E between the nodes A and B and then divide all elements which are grouped around AB. Two counters are associated with each edge. One is the number of elements that have this edge in common (Nc), and the other one stores the number of elements for which the edge in question is the longest one (Nl). Apparently if the two counters are equal then the edge is a terminal edge.

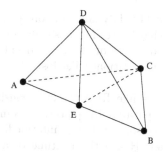

Fig. 1. Bisection of an element

After refining or coarsening of elements all that has to be done is to update the two counters for all affected edges. For each child element the longest edge is determined and Nl of this edge is increased by one. The edge AB is replaced by AE and a new edge EB is added. Both edges have Nl=0 and Nc is inherited from the former edge AB. The new edges ED and EC have either Nc=2 or Nc=4 depending on whether they are located on the boundary of the problem geometry or not. For the edge CD the counter Nc is increased by one and for the rest of the edges Nc doesn't change. At this point all necessary information is updated.

The only information that needs to be exchanged between the processors are the two counters after each grid adaptation step. The grid coarsening works exactly as in the scalar version and is completely local. The parallel version simply prohibits the removal of grid nodes in the boundary region between two processors and therefore saves some communication overhead. Elements in the boundary regions are still coarsened in the end, because the boundaries are not static due to load balancing.

4   Dynamic Load Balancing

In a parallel environment it is important that the workload is evenly distributed among the processors in order to minimize the total runtime of a calculation. There are several factors that have a negative impact on the workload distribution. Due to grid adaptation the number of grid points within a processor subdomain can increase significantly which puts additional burden on the affected processors. The chemical solver of REACFLOW is another cause for a poor workload distribution, because it treats different control volumes in a different way. The CPU time required for a single volume depends for example on the temperature or the ratio of fuel to oxygen at this certain volume. The most obvious reason for an imbalanced workload is the fact that a parallel computer is shared among several users who can all run their applications at the same time. All reasons for a poor workload distribution can be summarized as follows.

- Varying user load on different processors;
- Inhomogeneous hardware environment;
- Varying sizes of subdomains;
- Varying computational requirements of a subdomain.

By redistributing the workload dynamically one can assure to make the most efficient use of the available hardware. The basic idea of a load balancing strategy is to measure the runtime of parts of the code, decide whether rebalancing is necessary, invoking a graph repartitioner like Jostle [6] or Parmetis [4] and finally migrate the data.

Unfortunately CFD codes like REACFLOW consist of multiple computational phases, which are separated by phases of communication. During the communication phases processors exchange intermediate results and therefore have to wait for each other. Hence the runtime of a computational phase is always determined by the processor with the biggest workload as illustrated in

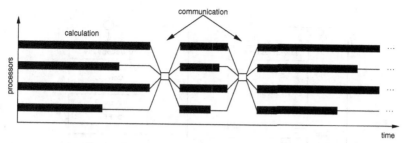

Fig. 2. Calculation phases in a SPMD program

Figure 2. The problem is that different phases can have different needs regarding the workload distribution. The time needed for the convective solver for example roughly depends on the number of control volumes whereas the runtime of the chemical solver also depends on the state of the volumes. This means that a distribution which is optimal regarding one phase can be unsuitable regarding another one.

There are different solutions for this multi-phase problem. Each phase can be balanced independently from others, but this requires to change partitions between the phases which results in a considerable communication overhead. Another possibility is to take the requirements of all phases into account by putting more weight onto the more time consuming phases, thus finding a compromise. If there is a single phase that consumes considerably more time than all of the other phases then one can concentrate on this phase when redistributing the data.

It turned out that it was possible to split some of the phases of REACFLOW into parts and reorder and recombine them in such a way that the result was a single large phase and a set of rather small ones. The load balancing was then optimized for this main phase, accepting that the resulting distribution would be suboptimal for the small phases. Figure 3 shows the flow chart of the timestep loop before and after the reordering of phases. The spacings between the blocks are the points where the processors have to exchange messages.

The decision whether to rebalance or not is based on the so called load imbalance factor. This factor is calculated as follows.

$$LIF = \frac{L_{max}}{L_{avg}} = N \frac{L_{max}}{L_{total}} \tag{1}$$

N is the number of processors whereas L_{max}, L_{avg} and L_{total} are the maximum, average and total workloads for N processors. L_{max} and L_{total} are obtained by simply measuring the real time spent inside the main computational phase using the function *gettimeofday*.

Once it has been decided to rebalance, the weights of the gridpoints are needed as input for the mesh repartitioner. The weight of a node represents the costs for doing all calculations for this node. Therefore the weight can be equated with the node-related calculation time. In this case we use the C-function *clock*, which returns the CPU-time. This is important, because the influence of varying

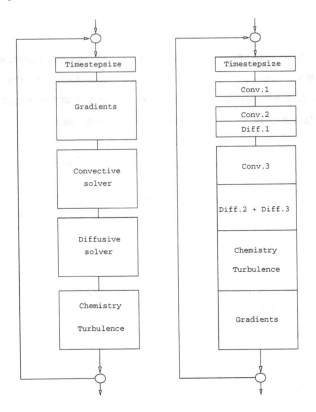

Fig. 3. Flow charts of the time loop before and after reordering

user load on different processors has to be eliminated in order to guarantee that the node weight does not depend on the subdomain it belongs to.

Each node weight consists of a fixed part, which is related to the non-chemistry calculations and a variable part which is related to the combustion model. Within this model 3 classes of nodes can be identified: cold, burning and burnt. Cold nodes have a temperature below a given threshold and are not taken into account by the chemistry solver and their variable weight can be set to zero. Burning and burnt nodes differ in their ratios of concentrations of fuel and oxygen and their weights are determined in a statistical manner. Each timestep the numbers of burning and burnt nodes are counted and the total time needed for the combustion model is measured. This triplet of values can be interpreted as a point in 3-dimensional space. After a couple of timesteps one can see that all these points roughly form a plane and by using linear regression the parameters of this plane, which equal to the variable weights of burning and burnt nodes, can be calculated.

Once the node weights are available either Jostle or Parmetis is used to find a new partitioning of the mesh. The actual data migration has to be done by REACFLOW itself.

5   Results

The test case presented here is a simulation of a detonation experiment in the Russian RUT Facility [2], an underground tunnel filled with a combustible gas mixture. The simulations ran on a 128 processor SGI Origin 3800.

The parallel calculations were done once with and once without dynamic load balancing. In this test case the main reason for a poor workload distribution is the dynamic refinement of the grid. Initially the grid contains only 3.600 nodes and this number is increased up to 185.000 in the scalar version and up to 161.000

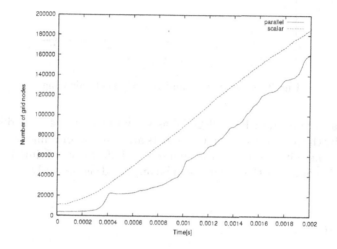

Fig. 4. Increasing number of grid points due to grid adaptation

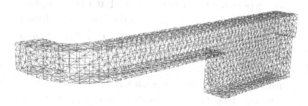

Fig. 5. Initial coarse grid

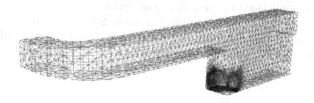

Fig. 6. Refined grid after several timesteps

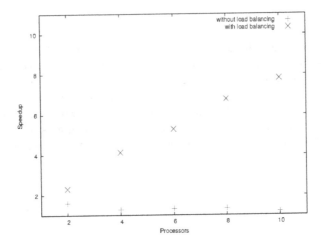

Fig. 7. Speedup with and without load balancing

in the parallel version of REACFLOW as shown in Figure 4. The difference is explained by the different ways how elements are selected for refinement. Figure 5 and 6 show the initial grid and the adapted grid after several timesteps.

Figure 7 shows that the speedup can be improved significantly by rebalancing the workload.

6   Conclusion

We have presented a communication-efficient method of adapting a grid in parallel calculations. The experimental results show that grid adaptation requires the workload to be redistributed during the simulation. Therefore a dynamic load balancer was implemented which monitors the load distribution and repartitions the problem data if necessary.

The parallel adaptation algorithm generally inserts less grid points compared to the scalar version. In some cases this can have an impact on the results of the calculation. It is possible to get a better approximation of the scalar grid adaptation by adjusting a set of adaptation parameters, but it would be better to have a method which is not based on trial and error. A possible solution for this problem could be to force the refinement of elements which are in the immediate neighbourhood of the refinement region.

The load balancer can be improved as well. Sometimes two or even more redistribution steps are necessary until a balanced state is reached. This is a problem especially in cases where the workload changes rapidly.

References

1. M. Arienti, T. Huld and H. Wilkening: An Adaptive 3-D CFD Solver for Modeling Transient Turbulent Deflagrations on Large Scale. ECCOMAS 98: Fourth Computational Fluid Dynamics Conference, Athens, Greece (1998)
2. W. Breitung, S.W. Dorofeev, A.A. Efimenko, A.S. Kochurko, R. Redlinger and V.P. Sidorov: Large Scale Experiments on Hydrogen-Air Detonation Loads and their Numerical Simulation. Proc. Int. Topical Meeting on Advanced Reactor Safety (ARS '94), Pittsburgh, USA, Vol II, (1994) 733–745
3. G. Karypis and V. Kumar: METIS: Unstructured Graph Partitioning and Sparse Matrix Ordering System. Technical report, Department of Computer Science, University of Minnesota, (1995)
4. G. Karypis, K. Schloegel and V. Kumar: ParMetis - Parallel Graph Partitioning and Sparse Matrix Ordering Library. University of Minnesota, Department of Computer Science and Engineering, Army HPC Research Center, Minneapolis, MN 55455 (2003)
5. Message Passing Interface Forum. MPI: A Message Passing Interface Standard. Int. J. of Supercomputer Applications 8 (1994)
6. C. Walshaw and M. Cross: Dynamic Mesh Partitioning and Load-Balancing for Parallel Computational Mechanics Codes. In B. H. V. Topping, editor, Computational Mechanics Using High Performance Computing. Saxe-Coburg Publications, Stirling, (2002) 79–94 (Invited Chapter, Proc. Parallel and Distributed Computing for Computational Mechanics, Weimar, Germany, 1999).
7. H. Wilkening and T. Huld: An Adaptive 3-D CFD Solver for Modeling Explosions on Large Industrial Environment Scales. Combustions Science and Technology 149 (1999) 361–387

4th International Special Session on: Current Trends in Numerical Simulation for Parallel Engineering Environments ParSim 2005[*]

Carsten Trinitis[1] and Martin Schulz[2]

[1] Lehrstuhl für Rechnertechnik und Rechnerorganisation (LRR),
Institut für Informatik,
Technische Universität München, Germany
Carsten.Trinitis@in.tum.de
[2] Center for Applied Scientific Computing,
Lawrence Livermore National Laboratory,
Livermore, CA
schulzm@llnl.gov

The use of parallel programming and architectures has become essential for simulating practical problems in engineering disciplines. The remarkable progress in CPU power, system scalability, and interconnect technology, as well as the introduction of new paradigms like computational Grids or E-Services, continues to provide new opportunities, as well as new challenges. These trends are paralleled by progress in numerical simulation techniques and the integration of software used to support a large variety of engineering applications.

Since its introduction at EuroPVM/MPI 2002, ParSim is dedicated to providing a forum for interdisciplinary cooperations in this important field. It brings together researchers with different backgrounds to discuss current trends in parallel simulation. In contrast to traditional conferences, emphasis is put on the presentation of up-to-date results with a short turn-around time. It is our hope that this offers a unique opportunity to present new aspects in this dynamic field and discuss them with a wide, interdisciplinary audience. The EuroPVM/MPI conference series, as one of Europe's prime events in parallel computation, serves as an ideal surrounding for ParSim. This combination enables the participants to present and discuss their work within the scope of both the session and the host conference.

This year, 12 papers were submitted to ParSim and we selected five of them. They cover both computer science aspects, including object oriented programming and cooperative and interactive frameworks, as well as experiences with special applications from various fields, including electrical engineering, civil engineering and computational fluid dynamics. We are confident that this resulted in an attractive program and we hope that this session will be an informal setting for lively discussions and for fostering new collaborations.

---

[*] Part of this work was performed under the auspices of the U.S. Department of Energy by University of California Lawrence Livermore National Laboratory under contract No. W-7405-Eng-48. UCRL-ABS-213386.

Several people contributed to this event. Thanks go to Jack Dongarra, the EuroPVM/MPI general chair, and to Beniamino Di Martino and Dieter Kranzlmüller, the PC chairs, for their encouragement and support to continue the ParSim series at EuroPVM/MPI 2005. We would also like to thank the numerous reviewers, who provided us with their reviews in such a short amount of time (in most cases in just a few days) and thereby helped us to maintain the tight schedule. Last, but certainly not least, we would like to thank all those who took the time to submit papers and hence made this event possible in the first place.

We hope this session will fulfill its purpose to provide new insights from both the engineering and the computer science side and encourages interdisciplinary exchange of ideas and cooperations. We hope that this will continue ParSim's tradition at EuroPVM/MPI.

Applying Grid Techniques to an Octree-Based CSCW Framework

R.-P. Mundani[1], I.L. Muntean[1], H.-J. Bungartz[1], A. Niggl[2], and E. Rank[2]

[1] Institut für Informatik, Technische Universität München, Germany
[2] Lehrstuhl für Bauinformatik, Technische Universität München, Germany

Abstract. Many simulation tasks are intended as a mere stand-alone application. Hence, integrating them into some embedding framework often fails due to missing interfaces for data and information interchange. Furthermore, keeping track of consistency among all embedded tasks and participating experts can be very difficult and, thus, a lot of effort has to be invested.

Within our approach, an octree-based CSCW framework for processes arising in civil engineering both provides appropriate interfaces for process integration and assures global consistency in a distributed cooperative working environment. To some extent, by completely embedding simulation tasks into this framework, a so-called problem solving environment has been established.

To foster parallel processing and to further exploit the inherent hierarchy of this approach, grid techniques seem perfectly suited to adopt the full potential of distributed and parallel computing. This not only allows us to tackle large scenarios such as ensembles of buildings, it also gives us the advantage of sophisticated level-of-detail studies without busting capacities of the underlying hardware ressources.

1 Motivation

There exist a lot of specialised solutions or systems for simulation tasks related to the fields of mechanical or civil engineering. Such systems have to fulfil different requirements from all involved applications. Thus, beside the necessity of efficiently bridging the gap between CAD and simulation also consistency among all participants has to be assured. The drawback of these customised systems lies in their lack of flexibility to integrate further applications and their often difficult porting to be run as high-performance application on a variety of clusters and/or supercomputers using distributed computing services as provided by grid techniques.

In [1,2], we presented an octree-based framework for CSCW and process integration of different applications from the field of civil engineering. Furthermore, in [3] we have shown a hierarchical approach for organising computations based on a finite element discretisation of p-version type. Due to this approach, even huge problems can be efficiently processed as well as the amount of necessary recomputations can be dramatically reduced in case the geometric model changes.

B. Di Martino et al. (Eds.): EuroPVM/MPI 2005, LNCS 3666, pp. 504–511, 2005.

A further exploitation of this hierarchy combined with the idea of distributed processing of the underlying simulation tasks inevitably points in the direction of grid computing.

In this paper, we present the necessary changes of our approach to be run as grid application, taking advantage of both distributed storage and distributed computations on several high-performance computers. Based on the Globus Toolkit software we address the advantages of distributed storage (OGSA-DAI) related to the underlying geometric model as well as to the results of a hierarchical nested-dissection approach for solving statics' simulations based on the p-version of finite elements.

2 Octree-Based CSCW Framework

The core architecture of our framework is designed as a client-server system. All relevant geometric data including attributes (material parameters or boundary values for simulation, for instance) are stored in a relational database management system (RDBMS), only accessible by an octree-structure called control tree (see [4] for details). These two components are referred to as server. Clients can check out parts of the geometry from the server both as surface-oriented and volume-oriented model, thus, both worlds can be served by this framework. The latter one, a linearised octree coded as binary stream [1], is generated in real time or *on-the-fly* due to a highly efficient algorithm using half-spaces (see [5] for details).

Whenever a client wants to check in altered data the server performs an octree-based collision detection, rejecting everything that interferes with other parts of the model. Hence, global consistency can be achieved. For the moment, several tasks have been integrated or embedded, resp., into this framwork, ranging from CAD, CFD (computational fluid dynamics) [6], CSD (computational structural dynamics) to visualisation, shortest-path algorithms [7], and evacuation simulation [8]. Figure 1 depicts a schematic overview.

Here, integration means that a process can access shared data from the server via services provided by the framework. In contrast to that, embedding means a process has been made a part of the server, thus, it can be accessed by other processes (from outside) as service, too. This is the first step towards a system that provides all the computational facilities necessary to solve a target class of problems, a so-called problem solving environment [9]. As shown in Fig. 1, CSD has already been embedded into the framework, allowing a fast and efficient processing of model variants and detailed studies with sophisticated p-version type finite element codes due to a hierarchical approach.

By organising the elements of a finite element discretisation via an octree, efficient hierarchical solvers can be applied [3]. This not only reduces the computational effort whenever changes of the underlying geometric models occur as only those parts directly influenced have to be recomputed, also the efficient processing of large data sets comes into reach by exploiting the octree's hierarchy. Furthermore, this approach is predestinated for a parallel processing on

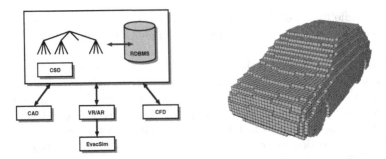

Fig. 1. Schematic overview (left-hand side) of our octree-based CSCW framework with all participating tasks so far – CAD, visualisation combined with shortest-path algorithms (VR/AR), evacuation simulation (EvacSim), computational fluid dynamics, and computational structural dynamics. On the right-hand side a sample octree model of a car as used for CFD is shown.

clusters and supercomputers as issues related to domain decomposition and load balancing can be easily addressed by the octree, too.

3   Computing and Storage Services for Grids

One of the most common toolkit software for building grid solutions and applications is the Globus Toolkit [10]. It allows sharing computing power, databases, and other resources between different applications run at various sites under the control of different institutions. It favors uniform access to these resources and cooperative work done among several experts [11,12]. The toolkit includes components and libraries for security, information infrastructure, resource management, data management, communication, fault detection, and portability, required for the development of distributed systems [13]. Most of the software components of the current version of the toolkit implement web services[1] as their basic infrastructure for distributed systems.

In order to satisfy the needs of an application for working with large sets of data, GT4 (Globus Toolkit 4) provides the Open Grid Services Architecture – Data Access and Integration (OGSA-DAI) component. For accessing the computing power of high performance machines connected to grid, web services can be developed on top of GT4 using corresponding MPI implementations (such as MPICH-G2[2]).

OGSA-DAI[3] is middleware that supports the access and manipulation of data from both XML and relational databases in grids. The services and functionality it provides are designed for high level data integration. It supplies interfaces to operate with web services (WS-I), the open grid services infrastructure (OGSI),

---

[1] http://www.globus.org/ogsa/

[2] http://www3.niu.edu/mpi/

[3] http://www.ogsadai.org.uk

and with the web services resource framework (WS-RF). Data access and manipulation requirements are expressed in XML documents. Requests and their responses are enclosed in perform documents or response documents, resp.

MPICH-G2 is a grid-enabled implementation of the MPI standard. It couples different high performance computers using Globus services. Thus, it enables the user to embed parallel paradigms in grid applications. Therefore, computationally intensive tasks, which have high potential for parallelisation, can be executed as grid applications.

4   Migrating the Octree-Based Framework to the Grid

Two main issues we address in this paper are related to the needs of storing and handling very large CAD models coming from civil engineering applications, on the one hand, and to the usage of available computation resources (i. e. high performance clusters) for carrying out various simulation tasks based on these models, on the other hand. Both needs can be typically satisfied by grid architectures.

4.1   The Service-Based Architecture of Our Framework

In Fig. 2, we present the new architecture of our octree-based CSCW framework. An OGSA-DAI grid data service (DMS) is designed to be responsible for data operations within our framework. The DMS service can access now different databases, physically distributed on different database servers, whereas client applications (such as CAD or CSD) are not aware of the physical location of data. Queries for retrieving components/parts of a geometric model are formulated by client applications enclosed in perform documents and sent to the DMS. The latter one processes the XML documents, executes the queries, and delivers response documents with XML formating back to clients.

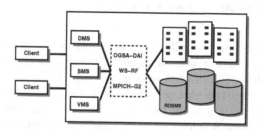

Fig. 2. The new service-based architecture of our CSCW framework from Fig. 1. The provided services are: data management service (DMS), simulation management service (SMS), and visualisation management service (VMS).

For accessing the computational resources in the grid, we have designed a service called simulation management service (SMS). It implements the computationally intensive parts of the CSD component. It also allows the parallel execution of the CSD solver on different high performance systems (clusters or constellations) and the storage of the results in databases. The visualisation management service (VMS) performs various post-processing operations on the simulation data. Hence, computationally intensive simulation tasks are now accessible by other tasks and experts as grid services for a parallel and distributed processing. Similar services can be developed for CFD or VR components as well.

4.2   Technical Changes/Transformations

Due to the service-oriented architecture, some major technical transformations/changes are applied to the architecture from Fig. 1. One of the changes refers to the communication between the CSCW framework components. We shifted it from a socket-based to a http-based one. For the implementation of the grid services, we use WS-RF of GT4. The DMS, SMS, and VMS expose their interfaces to client programs. These invoke the web services using grid clients API (such as OGSA-DAI client API).

If previously the queries were expresed in SQL format and executed using database-specific client API, they now conform to the OGSA-DAI client API. As mentioned before, the SQL queries are packed into perform documents. An example of a perform document used for asking our grid data service to retrieve an object labelled *GEOM_00000007* of a geometric model named *uniqua_3s_p4* is provided below. Here, the new manner of writing the queries can be seen.

```
<?xml version="1.0" encoding="UTF-8"?>
<perform xmlns="http://ogsadai.org.uk/namespaces/2005/03/types">
   <flow name="getCompModel"> <sqlQueryStatement name="compFaces">
      <expression>
         select ID_Face, ID_Edge, Direction from face_idx
         where ID_Object = "GEOM_00000007" AND Model = "uniqua_3s_p4"
      </expression>
      <webRowSetStream name="compFacesOutput"/>
   </sqlQueryStatement> </flow> </sequence>
</perform>
```

Another modification that affects clients refers to the internal mechanism of cooperation between framework components. Previously, the components interacted directly, without intermediate stages. With the new architecture, when data interactions with the grid come into play, clients ask from a DAI service group registry information about our data service. An instance of the service is created by a Grid Data Service Factory, a handle is returned to the clients, and finally the cooperation client–service can proceed.

The types of technical transformations mentioned above are the most relevant ones in the current evolution stage of our CSCW framework.

5   A First Illustrative Example: CSD

Based on a finite element discretisation using hexahedral elements, in our approach all those elements are ordered in a hierarchical way by efficiently storing the relevant data (stiffness matrices, loading vectors, and degrees of freedom (DOF)) to an octree. Thus, hierarchical solvers such as the nested dissection algorithm [14] can be applied. The main advantage of this approach is to exploit the locality of geometric changes and, thus, to re-compute only those parts of the tree where changes occur while still using the already computed results for the rest. Furthermore, due to the tree's hierarchy distributed and parallel processing can easily be achieved as intended for the grid version of this services.

The entire computational process can be divided into two main tasks. In a first bottom-up step – called assembly – local unknowns, i. e. unknowns fully described at a certain tree level, can be eliminated, thus, forming new systems of linear equations (SLE) handed to the next higher level. Once the root of the tree is reached, the resulting SLE can now be solved. In a second top-down step the (partial) solution is passed to all of a node's sons, which themselves now can compute the solution for the formerly eliminated unknowns (see [3] for details).

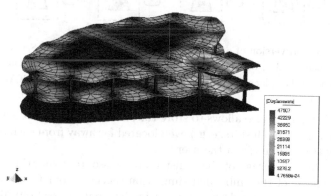

Fig. 3. Displacement field (shown exaggerated) as computed by our simulation service

One example processed with the current (serial) simulation service comprises two floors of an office tower located in Vienna (see Fig. 3). The model consists of 4171 elements and was computed for polynomial degrees 1 to 4. After an initial assembly/solution step some elements (stiffness matrices and loading vectors) were replaced by altered ones and the entire model was re-computed. As Fig. 4 shows, the necessary time for the re-assembly is significant smaller than for the initial one and, thus, a lot of computational effort can be saved. The relative costs compared to a complete assembly step are shown in the last column (costs).

As most of the time (approx. 99 %) during the solution step is spent at the root level, this part has been outsourced to a multiprocessor server (Quad Itanium). Thus, using OpenMP the SLE can be processed by running four threads of a cg solver in parallel. Further enhancement can be achieved when running the assembly step in parallel, too. By cutting the tree under the root node it falls apart into eight disjoint parts called subtrees. Each subtree can be further decomposed into smaller parts if necessary. As communication only takes place from a subtree's root node to the father's root node once during assembly and once during the solution step, each part can be processed nearly independent

p	DOFs	Assembly	Solution	Re-Assmb.	Costs [%]
1	23,856	10.795 s	2.661 s	0.660 s	6.114
2	84,660	192.193 s	19.355 s	9.375 s	4.878
3	145,464	795.036 s	53.439 s	36.853 s	4.635
4	255,606	2725.943 s	143.631 s	123.836 s	4.543

Fig. 4. Results for initial assembly, solution, and re-assembly of our simulation service for the sample model (two floors of an office tower) with 4171 elements shown in Fig. 3

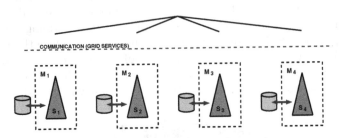

Fig. 5. Future grid version of our simulation service: each subtree (S_i) is processed on a different machine (M_i) while storing all relevant data to an RDBMS. Data access and communication takes place via grid services and is managed by the CSCW framework.

from all others. That also allows to store the subtrees on different systems (multiprocessor servers or clusters, e. g.) even located far away from each other, keept completely transparent from the user.

By handling the usage of these distributed ressources via grid technologies, the management of the embedded simulation service of our CSCW framework can be efficiently processed as described in the previous section. To minimize memory usage all data stored to a subtree's nodes can be moved to a database, thus, the "naked" tree structure is all remaining in memory. Again, the database is accessed via grid services (OGSA-DAI) and, hence, has not to be close or local to the machine processing a subtree. From the user's point of view, communication still only takes place with the framework by accessing one or more of the provided services. Thus, consistency among all user can still be assured by the framework at any time.

As being no longer restricted to one system (with small resources) even larger models with higher polynomial degrees of the underlying FEM discretisation can be processed by this approach. Furthermore, being able to zoom in on parts of the model and re-computing them with finer resolutions without the need of re-computing the rest of the model, this will help to understand statics' problems in a much more vivid way.

6   Conclusion

In this paper, we presented an octree-based CSCW framework for processes from the field of civil engineering. Beside classical approaches for the parallelisation of the underlying simulation tasks, grid techniques seem perfectly suited for a distributed processing related to both computation and storage. As we are still at the beginning of our work, further research is necessary to evaluate the benefit of this approach and to show its relevance for large engineering applications. Nevertheless, current results sound very promising. Hence, this is the first step in the direction of embedded simulation processes in a distributed environment for cooperative working.

References

1. Mundani, R.P., Bungartz, H.J.: An octree-based framework for process integration in structural engineering. In: Proc. of the 8th World Multi-Conf. on Systemics, Cybernetics and Informatics. Volume II., Int. Institute of Informatics and Systemics (2004) 197–202
2. Niggl, A., Romberg, R., Rank, E., Mundani, R.P., Bungartz, H.J.: A framework for concurrent structure analysis in building industry. In: Proc. of the 5th European Conf. on Product and Process Modelling in the Building and Construction Industry, A.A. Balkema Publishers (2004)
3. Mundani, R.P., Bungartz, H.J., Rank, E., Niggl, A., Romberg, R.: Extending the p-version of finite elements by an octree-based hierarchy. To appear in Proc. of the 16th Int. Conf. on Domain Decomposition Methods (2005)
4. Mundani, R.P., Bungartz, H.J.: Octrees for cooperative work in a network-based environment. In: Proc. of the 10th Int. Conf. on Computing in Civil and Building Engineering, VDG Weimar (2004)
5. Mundani, R.P., Bungartz, H.J., Rank, E., Romberg, R., Niggl, A.: Efficient algorithms for octree-based geometric modelling. In: Proc. of the Ninth Int. Conf. on Civil and Structural Engineering Computing, Civil-Comp Press (2003)
6. Brenk, M., Bungartz, H.J., Mehl, M., Mundani, R.P., Düster, A., Scholz, D.: Efficient interface treatment for fluid-strcuture interaction on cartesian grids. To appear in Proc. of Coupled Problems 2005 (2005)
7. Drexl, T.: Entwicklung intelligenter Pfadsuchsysteme für Architekturmodelle am Beispiel eines Kiosksystems (Info-Point) für die FMI Garching. Diplomarbeit, Institut für Informatik, Technische Universität München (2003)
8. Mundani, R.P., Bungartz, H.J., Giesecke, S.: Integrating evacuation planning into an octree-based cscw framework for structural engineering. To appear in Proc. of the 22nd CIB-W78 Conf. on Information Technology in Construction (2005)
9. Gallopoulos, E., Houstis, E., Rice, J.R.: Computer as thinker/doer: Problem-solving environments for computational science. IEEE Computational Science & Engineering 1 (1994) 11–23
10. Foster, I.: A globus primer toolkit (2005)
11. Foster, I., Keselman, C., Tuecke, S.: The anatomy of the grid: Enabling scalable virtual organizations. Int. J. Supercomputer Applications 15 (2001)
12. Foster, I., Keselman, C., Nick, J., Tuecke, S.: The physiology of the grid: An open grid services architecture for distributed systems integration (2002)
13. Foster, I., Keselman, C.: The Grid: Blueprint for a Future Computing Infrastructure. Morgan Kaufmann Publishers (1999)
14. George, J.A.: Nested dissection of a regular finite element mesh. SIAM Journal on Numerical Analysis 10 (1973) 345–363
15. Kra, D.: Six strategies for grid application enablement, part 1: Overview (2004)

Parallel Modeling of Transient States Analysis in Electrical Circuits

Jaroslaw Forenc[1], Andrzej Jordan[1,2], and Marek Tudruj[2,3]

[1] Bialystok Technical University, Faculty of Electrical Engineering,
45D Wiejska St., 15-351 Bialystok, Poland
{jarekf, jordana}@pb.bialystok.pl
[2] Polish-Japanese Institute of Information Technology,
86 Koszykowa St., 02-008 Warsaw, Poland
[3] Institute of Computer Science, Polish Academy of Sciences,
21 Ordona St., 01-237 Warsaw, Poland
tudruj@pjwstk.edu.pl

Abstract. In this paper a speculative method for parallel modeling of transient states analysis of electrical circuits in distributed systems is presented. Solving systems of linear or nonlinear ordinary differential equations that describe the transient states is a purely sequential process. The proposed speculative computation method converts involved sequential computations into intensively parallel ones. The general idea of this method is based on decomposition of the analysed time domain into sub-intervals in which parallel solving is done based on speculatively assumed initial conditions. Parallel computations in subsequent time sub-intervals are conducted with the use of sequential numerical Runge-Kutta method. Application of the method for simulation of functioning of a DC motor with a controlled integration step size is shown.

1 Introduction

Numerical analysis of transient states in large and complicated electrical circuits is usually very time consuming and, thereby, costly. In such a case, achieving high accuracy of computation results in a short time requires applying high performance computer systems [1]. Transient states in electrical circuits are often described by systems of ordinary differential equations (ODEs), linear or nonlinear. Most of numerical methods for solving these kinds of systems of equations are typically sequential [2]. Several approaches towards the parallel solution of ODEs have been developed, e.g. extrapolation methods [3], multiple shooting [4], relaxation techniques [5] and Runge-Kutta methods, which are iterated Runge-Kutta methods [6,7]. They are based on implicit Runge-Kutta methods and classical embedded Runge-Kutta methods. A good overview of parallel methods of ODE solving can be found in papers [8,9,10] and monographs [11,12].

The speculative method that we propose in this paper originates in the approach to the parallel analysis of transient states in electrical circuits first published in [13,14]. The main aim of this method is to reduce the time of the

B. Di Martino et al. (Eds.): EuroPVM/MPI 2005, LNCS 3666, pp. 512–519, 2005.

analysis that is the time of the ODEs system solving. The general idea of the speculative method is based on decomposition of the time domain. Computations in the particular sub-intervals of time are conducted in parallel with the use of one of the well-known sequential numerical methods for ODEs systems solving. Parallel computations require knowledge of initial conditions at the beginning of each sub-interval. In the speculative method, instead of one condition, a set of speculative initial conditions is determined for each state variable. Therefore, the computations in particular sub-intervals are conducted repeatedly but with different initial conditions. The final solution in the entire analysis interval consists of selected solutions from particular sub-intervals.

In this paper we use a modified speculative method where several initial conditions are determined, but the computations in each sub-interval are performed only once with the use of one processor [15,16]. The novelty of this paper is the application of the fourth-order Runge-Kutta method with a step size control, as sequential numerical method of ODEs system solving in the modified speculative method. As an example of the application of this method, the analysis of the transient state in a nonlinear model of a DC-motor powered by a solar voltage generator after a global linearization of three nonlinear equations which describe this state, will be shown.

2   The Modified Speculative Method Algorithm

The algorithm of the modified speculative method is composed of two main stages. In the first stage, the total time interval of the transient analysis is divided into sub-intervals and the initial conditions, necessary for parallel computations, are determined. In the second stage, main computations are conducted and the final solution is determined.

The first stage of the modified speculative method is executed sequentially by the master processor. In this stage the total time interval of the transient analysis, i.e. the time interval (t_0, t_N), is divided into a given number of N sub-intervals $(t_i, t_{i+1}), i = 0, 1, \ldots, N - 1$. These sub-intervals will be also denoted by $P_i, i = 0, 1, \ldots, N - 1$ (Fig. 1).

In the case of the application of the numerical method with a controlled integration step size, the total time interval of the transient analysis should be divided in the way which ensures similar number of integration steps and thereby, similar time of computations in each sub-interval. In the fourth order Runge-Kutta method with an automatic control of a step size, the size of the integration step h, thus the number of steps in particular sub-intervals, is considerably dependent on the nature of the plot. In this case, applying the simplest, equal, division of the total time interval of the transient analysis is not optimal. In order to get similar time of computations in each sub-interval, computations are executed in the whole interval (t_0, t_N) with assumed low accuracy ε_L (e.g. $\varepsilon_L = 10^{-1} \div 10^{-3}$) and approximate solution $x_0^p, x_1^p, \ldots, x_m^p$ at points $t_0^p, t_1^p, \ldots, t_m^p$ (where $t_0^p = t_0$ and $t_m^p = t_N$) is obtained. Then, the obtained number of points m of approximate solution is divided by the number of

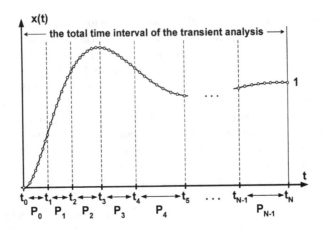

Fig. 1. Division of the total time interval of the transient analysis into sub-intervals; 1 - approximate solution determined with a low accuracy ε_L

sub-intervals N, obtaining, according to equation (1), points $t_1, t_2, \ldots, t_{N-1}$ which are the division of the total time interval of the transient analysis (Fig. 1).

$$t_i = t^p_{(i \cdot \frac{m}{N})}, \qquad i = 1, 2, \ldots, N-1 \tag{1}$$

In this way, the total time interval of the transient analysis (t_0, t_N) has been divided into non-equal sub-intervals. The sub-intervals in which state variables oscillate, have a shorter length comparing sub-intervals in which the function is even.

In order to execute parallel computations in particular sub-intervals, it is necessary to know the initial conditions at the beginning of each sub-interval. At the point t_0, these conditions are known from the assumption. For the rest of the points $t_1, t_2, \ldots, t_{N-1}$, instead of one condition, a set of initial conditions is determined for each state variable. For this purpose, previous computations with a low accuracy ε_L are used. In the Runge-Kutta method with an automatic control of the step size, the determination of the solution at each point requires one execution of computations with step h and two with step $h/2$ [2]. In this way, two solutions at each point are obtained. The absolute value of the difference of computations with step h ($x_{i,j,h}, i = 1, 2, \ldots, N-1, j = 1, 2, \ldots, n$, where n is the number of state variables) and $h/2$ ($x_{i,j,h/2}$), at points $t_1, t_2, \ldots, t_{N-1}$, multiplied by safety coefficient k (assumed "a priori"), determines the length of the section ($\Delta x_{i,j}$) in which the initial conditions for the particular sub-intervals will be presented:

$$\Delta x_{i,j} = |x_{i,j,h/2} - x_{i,j,h}| \cdot k, \qquad i = 1, 2, \ldots, N-1, \quad j = 1, 2, \ldots, n \tag{2}$$

As the beginning of each section ($x^s_{i,j}$), the values computed with step $h/2$ are assumed:

$$x^s_{i,j} = x_{i,j,h/2}, \qquad i = 1, 2, \ldots, N-1, \quad j = 1, 2, \ldots, n \tag{3}$$

while the end of the section $(x_{i,j}^e)$ is the shift for his length in the convergence direction to the exact solution:

$$x_{i,j}^e = x_{i,j,h/2} \pm \Delta x_{i,j}, \qquad i = 1, 2, \ldots, N - 1, \quad j = 1, 2, \ldots, n \qquad (4)$$

The values of the state variables computed with both steps, h and $h/2$, determine the convergence direction to the exact solution. If $x_{i,j,h} > x_{i,j,h/2}$, then in equation (4) sign "$-$" occurs, otherwise sign "$+$" occurs. The obtained section divided (for each state variable) into a particular number of sub-intervals determines the sets of initial conditions (Fig. 2) from which one will be finally the initial condition used in computations. In order to determine this number, the master processor executes computations in the first sub-interval (t_0, t_1) with a high accuracy ε_H (the same as for main computations in the second stage of the algorithm). Next, it compares the obtained state variables values at time point t_1 with the initial conditions at the same time point and registers the numbers of the initial conditions (separately for each state variable), which are closest to the values computed with a high accuracy ε_H. At the rest of the time points $t_2, t_3, \ldots, t_{N-1}$, such points are chosen that have the same numbers as determined above.

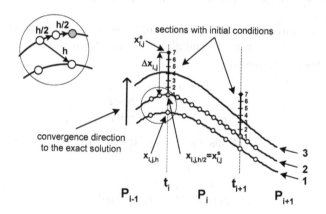

Fig. 2. Evaluation of the initial conditions in the modified speculative method; 1 - approximate solution determined with step h, 2 - approximate solution determined with step $h/2$, 3 - unknown exact solution

The second stage of the modified speculative algorithm is executed in parallel. The number of processors used for computations is smaller by one than the number of sub-intervals N because computations in the sub-interval (t_0, t_1) were already executed in the first stage of the algorithm. Before the computations start, the master processor sends the initial data (i.e. initial conditions determined in the first stage of the algorithm) to slave processors. All processors execute parallel computations in their sub-intervals with a high accuracy ε_H. The master processor executes computations in the sub-interval (t_1, t_2), the first slave processor - in the sub-interval (t_2, t_3), the second slave processor - in the sub-interval (t_3, t_4), etc. When the computations are finished, the slave processors send the obtained results to the master processor, which saves the final solution on its local hard disk.

3   Numerical Example

As an example of the application of the presented method, the analysis of transient state in a nonlinear model of a DC-motor powered by a solar generator [17] will be presented. The transient state in this motor is described by the system of three nonlinear ODEs:

$$\dot{x}_1 = -a_1 e^{ax_1} - a_2 x_2 + u$$
$$\dot{x}_2 = a_3 x_1 - a_4 x_2 - a_5 x_3$$
$$\dot{x}_3 = a_6 x_2 - a_7 x_3 \tag{5}$$
$$x_1(0) = x_{1,0}, \quad x_2(0) = 0, \quad x_3(0) = 0$$

where $x_1(t)$ is the generator voltage, $x_2(t)$ is the current of the DC-motor and $x_3(t)$ is the rotation speed of the motor. The rest of coefficients on the right hand side of equations (5) result from the parameters of the DC-motor and the solar generator.

We applied global linearization [18,19] of these equations in order to examine the dynamics of the system. In the global linearization method, the change of state variables is introduced:

$$z_1 = x_1$$
$$z_2 = -a_1 e^{ax_1} - a_2 x_2 \tag{6}$$
$$z_3 = a_3 x_1 - a_4 x_2 - a_5 x_3$$

and after basic transformations a system of linear equations is obtained:

$$\dot{z}_1 = z_2 + u$$
$$\dot{z}_2 = z_3 + \bar{g}_2(x, u) \tag{7}$$
$$\dot{z}_3 = b_1 z_1 + b_2 z_2 + b_3 z_3 + \bar{g}_3(x, u)$$
$$z_1(0) = x_{1,0}, \quad z_2(0) = -a_1 e^{ax_{1,0}}, \quad z_3(0) = a_3 x_{1,0}$$

where

$$\bar{g}_2(x, u) = aa_1 e^{ax_1}(a_1 e^{ax_1} + a_2 x_2 - u) - (1 - a_2) \cdot (a_3 x_1 - a_4 x_2 - a_5 x_3) \tag{8}$$
$$\bar{g}_3(x, u) = a_6 x_1 - a_1 a_4 e^{ax_1} - (a_2 a_4 + a_5 a_6) x_2 + a_5 a_7 x_3 + a_3 u$$

Coefficients $b_i, i = 1, 2, 3$ were chosen in such a way as to ensure stability of obtained equations: $b_1 = -a_6, b_2 = a_3 - a_4, b_3 = -a_4$.

The required state variables are determined by means of inverse transformation:

$$\tilde{x}_1 = z_1$$
$$\tilde{x}_2 = -\frac{1}{a_2}(z_2 + a_1 e^{az_1}) \tag{9}$$
$$\tilde{x}_3 = \frac{1}{a_5}(-z_3 + a_3 z_1 + \frac{a_4}{a_2}(z_2 + a_1 e^{az_1}))$$

Accurate solution of the linear equations (7) requires a very small integration step [15]. In such a case, the application of the modified speculative method can be helpful and allows to reduce the computations time.

The computations were carried out with the use of the cluster of five workstations. Each node in the cluster contains Intel Xeon 2.66 GHz processor, 1 GB of RAM and 80 GB of disk space. Individual nodes are connected by Gigabit Ethernet. The software environment is Fedora Core 1 Linux and LAM/MPI 7.03.

For the modified speculative algorithm, the following parameters of computations were set up: on the basis of computations with a low accuracy $\varepsilon_L = 10^{-2}$, the total time interval of the transient analysis (t_0, t_N) - $0.0 \div 0.09s$ was divided into 6 non-equal sub-intervals with the length: (t_0, t_1) - $9.22 \cdot 10^{-4}s$, (t_1, t_2) - $1.72 \cdot 10^{-3}s$, (t_2, t_3) - $7.06 \cdot 10^{-3}s$, (t_3, t_4) - $7.25 \cdot 10^{-3}s$, (t_4, t_5) - $9.50 \cdot 10^{-3}s$, (t_5, t_6) - $6.35 \cdot 10^{-2}s$. For each sub-interval 10 initial conditions were assigned. The main computations were carried out with the assumed high accuracy $\varepsilon_H = 10^{-7}$.

Fig. 3 and Fig. 4 present the obtained results for state variables $\tilde{x}_1(t)$ - the generator voltage and $\tilde{x}_2(t)$ - the current of a DC-motor, respectively. These results were obtained in the case of division of the total time interval of the transient analysis into 6 sub-intervals. In order to compare the results obtained by the modified speculative method, the system of nonlinear equations (5) was solved with application of the sequential algorithm of the fourth-order Runge-Kutta method with fixed integration step size $h = 10^{-9}$ s. The solution of the modified speculative method is convergent with the solution obtained by the sequential algorithm.

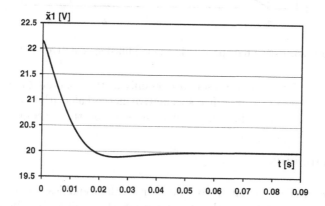

Fig. 3. Solution of the state variable $\tilde{x}_1(t)$

Fig. 5 presents the value of the estimated speedup. In order to determine the speedup, the total time interval of the transient analysis was divided into 3, 4, 5 and 6 sub-intervals (in the case of 2 sub-intervals only one processor was used). During this estimation only the number of the Runge-Kutta method calls, which means the evaluation of state variables values according to the fourth-order Runge-Kutta method formulas, was taken into consideration. The time of communication between processors was omitted. It must be stressed that the

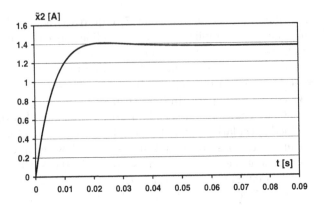

Fig. 4. Solution of the state variable $\tilde{x}_2(t)$

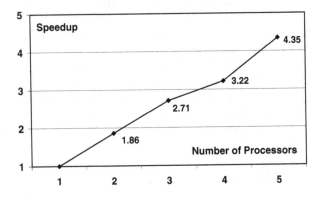

Fig. 5. Estimated speedup in the modified speculative method

communication between processors occurs only at the beginning and at the end of the second stage of the modified speculative method algorithm. For this reason the communication time has only a little influence on the speedup.

4   Conclusions

Application of the speculative computation method enables a reduction of the computation time in the analysis of transient states described by systems of ordinary differential equations. The application of the method with an automatic control of integration step size considerably reduces the time of computations comparing the methods with a fixed integration step. In the presented example, a good speedup and good accuracy of the solution were obtained.

    This work has been supported by the KBN Grant No.: 3T10A 066 27 and internal PJIIT grants.

References

1. Machowski, J., Bialek, J.W., Bumby, J.R.: Power System Dynamics and Stability, John Wiley & Sons, New York (1997)
2. Press, W.H., Flannery, B.P., Teukolsky S.A., Vetterling, W.T.: Numerical Recipes in C: The Art Of Scientific Computing, Camridge University Press (1993)
3. Lustman, L., Neta, B., Gragg W.: Solution of Ordinary Differential Initial Value Problems on an Intel Hypercube, Computer and Mathematics with Applications 23 (1992) 65–72
4. Khalaf, M. S., Hutchinson D.: Parallel Algorithms for Initial Value Problems: Parallel Shooting, Parallel Computing 18 (1992) 661–673
5. Prince, P.J., Dormand, J.R.: High Order Embedded Runge-Kutta Formulae, Journal of Computational and Applied Mathematics 7 (1981) 67–75
6. Van der Houwen, P. J., Sommeijer B. P.: Iterated Runge-Kutta Methods on Parallel Computers, SIAM Journal on Scientific and Statistical Computing with Stepsize Control, Journal of Computational and Applied Mathematics 29 (1990) 111–127
7. Rauber, T., Runger, G.: Iterated Runge-Kutta Methods on Distributed Memory Multiprocessors, Euromicro Workshop on Parallel and Distributed Processing (1995) 12–19
8. Burrage, K.: Parallel Methods for Initial Problems, Applied Numerical Mathematics 11 (1993), 5–25
9. Gear, C.W.: Massive Parallelism Across Space in ODEs, Applied Numerical Mathematics 11 (1993) 27–43
10. Gear, C.W., Xuhai, X.: Parallelism Across Time in ODEs, Applied Numerical Mathematics 11 (1993) 45–68
11. Petcu D.: Parallelism in Solving Ordinary Differential Equations, Mathematical Monographs 64, Tipografia Universitatii (1998)
12. Burrage, K.: Parallel and Sequential Methods for Ordinary Differential Equations, Oxford Science Publications (1995)
13. Forenc J., Jordan A., Tudruj M.: A Survey of Speculative Methods for Transient State Analysis, PARELEC'2002, IEEE Computer Society (2002) 353–358
14. Jordan A., Forenc J., Tudruj M.: Speculative Parallel Processing Applied to Modelling of Initial Problems. COMPEL, Vol. 24, No. 1 (2005) 127–144
15. Forenc J., Jordan A.: The Modified Speculative Method for the Transient States Analysis. PARELEC'2004, IEEE Computer Society (2004) 189–193
16. Forenc J.: A New Approach to the Speculative Method for the Transient States Analysis, SPIE 5775 (2005) 406–413
17. Jordan, A., Benmouna, M., Bensenane, A., Borucki, A.: Optimal Linearization of Non-Linear State Equations, Automatique, Systems Analysis and Control (1987) 263–271
18. Kaczorek, T., Jordan, A., Forenc, J.: Global Linearization of a Non-Linear Model of a DC Drive System, The Second Grant Conference: Numerical Methods in Modelling of Electric Devices, Warsaw (2002) 7–16
19. Kaczorek, T.: Lie algebra, Seminar of the Department of Theoretical Electrotechnics and Metrology, Bialystok (2003), (unpublished paper)

The COOLFluiD Parallel Architecture

Dries Kimpe[1,3], Andrea Lani[2], Tiago Quintino[2,3], Stefaan Poedts[1],
and Stefan Vandewalle[3]

[1] Centre for Plasma-Astroyphysics, K.U.Leuven, Celestijnenlaan 200B,
B-3001 Leuven, Belgium
Dries.Kimpe@wis.kuleuven.be
[2] Von Karman Institute, Aerospace Dept., Waterloose steenweg 72,
B-1640 Sint-Genesius-Rode, Belgium
[3] Scientific Computing Group, K.U.Leuven, Celestijnenlaan 200A,
B-3001 Leuven, Belgium

Abstract. This paper discusses the parallel design of COOLFluiD (Computational Object Oriented Library for **Flui**d Dynamics), a state-of-the-art C++ framework for multi-physics simulations using multiple numerical methods on unstructured grids. By using advanced techniques and specific design patterns, flexibility and modularity are assured. COOLFluiD was recently adapted to support parallel computations on distributed memory machines. For this, a parallel layer was added, designed to minimize impact on both users and software developers, while maintaining high performance. From the user's point of view, parallelisation is fully transparent. The techniques making this possible will be discussed. Also presented is a technique for reconciling generic programming with libraries requiring explicit type information.

1 Introduction

Since long, scientific simulation codes have supported parallel architectures. Usually, such a code only supports one parallel model, although sometimes compilation without parallelisation is still possible. However, in the latter case parallelisation is generally very intrusive, polluting the code with `#ifdef MPI` (or equivalent) blocks[8,9]. Others[3] require the user to link with a special emulation library which creates the illusion of a single-cpu parallel machine. Combined with the dynamical nature of scientific codes, these practices often reduce the lifetime of such projects to a few years, before they eventually break down under their own complexity.

In 2001, the COOLFluiD project was started. Its goals were to be clean, modular and maintainable and to be able to serve as a base code for a long period. It is written in C++ and uses object orientation to facilitate code reuse and modularity. Generic programming and advanced template techniques are used to regain some of the speed lost due to the object orientation.

The target audience of COOLFluiD consists mainly of two groups. The first group are the end users, running the software to simulate physical phenomena. They want COOLFluiD to deliver high performance and require it to be easy to use. The second group consists of the developers extending the functionality of COOLFluiD. They want the code to be clean, modular and stable.

B. Di Martino et al. (Eds.): EuroPVM/MPI 2005, LNCS 3666, pp. 520–527, 2005.

Upon introducing the parallelisation, the expectations of both groups had to be respected. In this paper, we will focus on the design changes that made this possible. More information concerning COOLFluiD's general design can be found in [4] and [5].

2 Design of the COOLFluiD Parallel Layer

The task of the parallel layer consists mainly of hiding specific implementation details. It should conceal the parallelisation from the rest of the framework, promote code reuse between serial and parallel versions and decouple the framework from a specific parallel paradigm. This is realised by exporting a number of high-level *concepts* to the layers above.

It is common practice to add an abstraction layer to make a code independent of the underlaying layers. However, every indirection layer introduces overhead. The COOLFluiD parallel layer is an exception to this, as it is only present at source code level. By using advanced C++ techniques, the compiler is able to remove almost all traces of this indirection, and only inlined model specific code remains after compilation.

2.1 Parallel Models

Within the parallel layer, the *parallel model* represents a certain parallel implementation (e.g. MPI[1], PVM[2]). The used parallel model is only known to the parallel layer. In the code, each concept is provided by a template class, of which the template parameter identifies the parallel model. This class is subsequently specialised for every model. There is almost absolute freedom in how this is done, as there is no inheritance relationship between the general concept and its specialisation. As is common in generic programming, each specialisation only has to "look like" the original (also referred to as *duck typing*[12]). As this does not rely on virtual functions, the exact specialisation used is completely known at compile time, giving enormous freedom to the optimiser to generate efficient code.

Every model has a *tag class* associated with it. This class contains no variables or member functions, and is never instantiated. The tag only is used for overloading the different concepts. Most concepts also have a suitable default for their template parameter. At compile time, this default is aliased to one of the tag classes by taking into account the available models and the preference of the user. For example, the reduction concept defaults to the tag GLOBAL. Outside of the parallel layer, only these aliases are used. When compiling with MPI support, the configuration system will alias GLOBAL to MPI, the tag class of the MPI model implementation. When compiling without parallel support, GLOBAL will alias NONE, the tag class of the serial implementation. This way, client code using the concepts is independent of the selected parallel model.

This scheme has a number of advantages compared to the more common alternative of selectively including header files (containing the specific code written for one model). Although our scheme does not allow run-time selection of the model (as this would always create some runtime overhead), it does allow easy sharing of code between different models. It even allows combining different models in the same program. Certain modules and plug-ins could be compiled both with and without MPI support, allowing COOLFluiD – when appropriate – to select optimal serial algorithms even when it was configured and compiled with support for parallel execution!

2.2   The Parallel Environment

The Parallel Environment (PE) is a first example of such a concept. This singleton class represents the execution environment. It initialises the parallel computer and – when necessary – takes care of transparent registration of data types and operations. The PE class is the only source of explicit parallelisation details (e.g. the number of available CPU's).

2.3   The Datastorage

Some completely new operations, previously unexisting in the serial version were necessary. The synchronisation of the unknowns between the different CPU's is an example of this. These operations are handled, whenever possible, by extending existing high level structures and implementing them as parallel concepts. Because of the modular nature of the framework, a facility had to be created that allowed sharing of data between unrelated modules. For this, the datastorage was created. It allows storing and registering data under a name, by which other modules can retrieve it. In COOLFluiD, mesh information (including unknowns) is stored in such a fashion.

In a parallel simulation, these unknowns need to be synchronised between iterations. To enable this, the datastorage was promoted to a *concept*. This means the every parallel model should provide a datastorage implementation which, although is not exactly alike at the type level, looks and feels the same to the programmer. For performance reasons, no default template parameter is is provided for the datastorage. As the datastorage is heavily used, the overhead of allowing distributed access to every instance would be too large. The programmer explicitly has to specify if a distributed, global datastorage or a purely local one is needed. This is done by providing the LOCAL or GLOBAL tag as one of the template arguments.

Global data can be accessed by other CPU's in an easy model independent way without requiring explicit parallelisation details as rank or communicator. Each model provides an optimized datastorage implementation, making full use of any advanced capabilities the model supports. The serial model has the old datastorage as its implementation, extended by empty inlined synchronisation methods.

LOCAL always aliases the the old non-parallel datastorage. When the user preferred MPI support, the global datastorage is provided by the MPI model. However, when configured without parallelisation, both GLOBAL and LOCAL alias the serial model, allowing all code to be compiled without changes and without any performance overhead.

2.4   Implementing a New Parallel Model

Every parallel model has to provide a specialised version of each concept. By careful code design, their number could be kept to an absolute minimum. For explicit simulations, next to model specific workarounds – such as the MPI type mapping – only three concepts need to be coded when adding a new parallel model.

COOLFluiD was designed for distributed memory machines. It relies on domain decomposition to divide the workload between processors. Still, implementing a shared memory model is possible, although more research is necessary to see if a fully shared datastorage implementation is feasible.

2.5 Example

When modifying COOLFluiD to support parallel architectures, the code had to be changed in surprisingly few places. One of these was in the computation of the residual. This is a typical mesh computation, in which a numerical value is computed locally at each mesh point, after which all values are to be summed into one numerical result. In COOLFluiD, this is done by a *functor* (a class implementing operator()). In a parallel simulation, the residual has to be gathered from all CPU's. Hence, when compiling with MPI support, the code should transform into a call to MPI_Allreduce. However, on systems lacking parallel support, the same code has to compile without overhead. Listing 1 shows the neccesairy modifications that make this possible.

Listing 1 The old and new residual code (simplified)

```
// Old code
class ComputeResidual { // CalcResidual is provided elsewhere
  double operator () { return CalcResidual (); }
};
// Modified code
class ComputeResidual : public Parallel::GlobalReduce {
  typedef double RESULTTYPE;
  double operator () { return GR_GetGlobalValue(); }
  RESULTTYPE GR_GetLocalValue() { return CalcResidual (); }
  void GR_Combine(RESULTTYPE & T1, RESULTTYPE & T2, RESULTTYPE & Out)
    { Out = T1 + T2; }
};
```

The GlobalReduce concept is the key to this transformation. To use it, the Compute-Residual class only needs to be extended with two small functions and a *typedef*. The typedef is neccesary to expose the resulting data type of the residual (a double or float in this case) to the GlobalReduce class. The first function, GR_GetLocalValue() is just the old operator() returning the local residual. The second function, GR_Combine() should return the combination of its two input parameters. Operator() is adapted to return the value of GR_GetGlobalValue(), a function inherited from the GlobalReduce class.

These small changes make it possible for the ComputeResidual class to do the right thing without even knowing the concrete parallel model: the hidden extra template parameter, defaulting to GLOBAL, will make sure the right model specific code is used. When compiled with MPI support, the code fragment returns the combined residual. In the serial implementation of the concept, GR_GetGlobalValue() merely returns GR_GetLocalValue(). As such, when compiling without parallel support, the call to GR_GetGlobalValue() results in the inlining of GR_GetLocalValue() inside `operator()`, resulting in exactly the same code as existed before the parallelisation of COOLFluiD.

3   Hiding MPI

COOLFluiD strongly relies on generic programming techniques to allow both code modularity and efficiency to coexist. Most concepts are generic in nature: for instance, the reduce concept demonstrated above should handle any C++ type, including user defined ones. However, some libraries[1][2][13], such as the MPI library, force the user to specify the type of the data involved. Special care is required when calling these libraries from generic code.

3.1   Goals

Because of their explicit type parameter the MPI C (or C++) language bindings do not interact well with generic programming. Therefore, the COOLFluiD MPI Parallel Model needs a type mapping facility, which transparently translates C++ datatypes to their corresponding MPI type definitions. This facility has two main tasks: it has to map C++ *types* to their corresponding library type *value*, and also has to automatically register all other C++ types, including user defined types. To ensure the highest performance, this mapping should be done at compile time whenever possible. In all other cases, overhead should be as low as possible. In order to respect the modularity and layer goals, no code modification outside of the parallel layer is allowed.

3.2   The COOLFluiD Automatic Type Mapping Facility

All of the above goals were obtained by using a combination of *type traits*[6], template classes and self-registering singletons[7]. The type system is accessed through a template function with one template parameter and no "ordinary" function parameters. It returns the MPI type corresponding to its template parameter, which has to be specified by the programmer. A helper function taking a normal function parameter is also provided. It enables the compiler to deduce the corresponding C++ type, relieving the programmer of this task, and hence diminishing the opportunity for mistakes.

The mapping function is really a family of functions of which the exact function contents differ greatly depending on the structure of the type involved. By using type traits, the layout of a given type is detected at compile time. If the type has a predefined library equivalent (e.g. `MPI_DOUBLE`), the mapping is performed at compile time by an inline function which only returns the corresponding library type. Most optimisers are able to "see through" small inline functions, enabling them to replace the function call by the returned value.

New types are, on first usage, constructed appropriately by using the correct MPI type constructors. For example, in case of an array, its size and base type are deduced, and code is inserted to register the type at runtime by a call to `MPI_Type_vector`. The base type is handled in a recursive way, allowing full mixing of arbitrary types. However, because of current language limitiations (C++ is lacking reflection support), structures and classes cannot be deduced automatically. In these cases the option is given to have them automatically registered as an array of bytes, which is not a problem in homogenous clusters. Alternatively, the user can declare its own registration routine, which will take precedence over the automatic routines. This is also useful for registering types that supersede the C++ type system (e.g. non-contiguous types). The newly registered type is stored, and subsequent requests to map this type will immediately return the stored type. All created types are also remembered, and are released when the type system is shut down. As all type properties are deduced from the type itself, no external code needs to be modified.

3.3 Overhead

By making use of inline functions, the automatic type registration system is able to achieve zero overhead for any native MPI type. The overhead for user defined types is limited to exactly one if-test. Even this last test can be avoided if the user guarantees the type will have been preregisterered. Listings 2 and 3 show the generated assembler code for native MPI code and the code generated by the COOLFluiD type system for both simple types and user types. All tested compilers (versions 3.3 and 3.4 of the GNU C++ compiler and versions 7 and 8 of the Intel compiler) were able to perform these optimisations. As listing 2 shows, the automatic type mapping generates optimal code, indistinguishable from hand-coded calls. The example lists the generated assembler code when calling a function expecting an MPI data type.

Listing 2 "Native" MPI code and automatic type mapping for built-in types

```
# Direct specification of the type: Accept(MPI_DOUBLE);
movl    MPI_DOUBLE, %edx
movl    %edx, (%esp)
call    Accept


# Automatic type mapping: Accept(MPIDataTypeHelper<double>::GetType());
movl    MPI_DOUBLE, %edx
movl    %edx, (%esp)
call    Accept
```

Listing 3 Using the type mapping system with a user-defined type (simplified)

```
# Accept(MPIDatatypeHelper<UserType>::GetType());
cmpb    $0, guard variable for GetType<UserType>()::Type
jne .L3
movb    $1, guard variable for GetType<UserType>()::Type
# code for constructing the MPI representation of UserType omitted
.L3:    movl    GetType<UserType>()::Type, %edx
movl    %edx, (%esp)
call    Accept
```

3.4   Related Work

MPI C++ Bindings. The MPI C++ bindings are only a thin layer on top of MPI. For generic programming, they provide no added value but grouping the many MPI functions inside an appropriate namespace and class. Transfer functions still require an explicit parameter specifying the type of the data involved.

C++2MPI. Although C++2MPI[11] can be used to generate the C++ type to MPI type mappings, the tool is not useable in a framework like COOLFluiD. It requires special #pragma markers around every user type that will be transferred through MPI. As such, it requires changes in the core framework and in modules not related to the parallelisation. This is not acceptable, as this clearly violates the design goals stated at the start of the project. Moreover, C++2MPI requires preprocessing of the source code by a seperate tool. This tool would again need to be run on almost every file of the whole project (as each module could declare a type which could subsequently be transferred to another cpu).

OOMPI. Object Oriented MPI (OOMPI) is a class library specification that encapsulates the functionality of MPI into a functional class hierarchy. Unlike the MPI C++ bindings, it does not closely adhere to the MPI C bindings. It is a layer on top of MPI to provide an object oriented interface. As such, it is comparable to the MPI type abstraction system present in COOLFluiD. In OOMPI, all data transfer functions accept an OOMPI_Message class that describes both the data location and the datatype. The library provides constructors for the built-in types, so a native type can be passed anywhere a OOMPI_Message is expected. The compiler automatically promotes this type to the class type required by the transfer functions. Although this promotion is very lightweight[10], it happens on every invocation. The OOMPI documentation states that:

> Message objects could be eliminated entirely by declaring each communication function in terms of every base data type. However, this would result in an enormous number of almost identical member functions.

However, this is exactly what is done by the COOLFluiD type system: for every type passed to MPI, the compiler generates code to register and free it – if not a built-in type – and creates a small inlined function to return the actual MPI data type.

4   Conclusions and Future Work

COOLFluiD proves that a flexible, maintainable parallel simulation code can be created by using modern software techniques and a careful design. Moreover, parallelisation can be (almost) completely shielded from end users and developers. It includes a type mapping system, designed to ease integrating MPI into generic libraries.

Work has started on loosely coupled simulations, in which two or more completely seperated COOLFluiD instances (running different physical models) exchange simulation data. This will also be handled by employing the techniques presented in this paper. For this, a third parallel model alias (REMOTE) will be introduced. It will enable COOLFluiD to use shared memory, MPI or other paradigms as inter-simulation communication channels, independently of the selected intra-simulation communication model.

References

1. Message Passing Interface Forum, *MPI: A message-passing interface standard*, http://www.mpi-forum.org.
2. A. Geist, A. Beguelin, J. Dongarra, W. Jiang, R. Manchek and V. Sunderam, *PVM: Parallel virtual machine: a users' guide and tutorial for networked parallel computing*, MIT Press. 1995.
3. Argonne National Laboratory, *PETSc. Portable, Extensible Toolkit for Scientific Computation*, http://www-unix.mcs.anl.gov/petsc, 2004.
4. A. Lani, T. Quintino, D. Kimpe, H. Deconinck and S. Vandewalle, *The COOLFluiD Framework: Design Solutions fo High Performant Object Oriented Numerical Simulations of Partial Differential Equations*, International Conference on Computational Science 2005, 279-286.
5. P. De Ceuninck, T. Quintino, S. Vandewalle, H. Deconink. *Object-Oriented Framework for Multi-Method Parallel PDE software*, ECOOP'03 - European Conference for Object-Oriented Programming in the workshop Parallel/High-Performance Object-Oriented Scientific Computing.
6. Nathan Myers, *A New and Useful Template Technique: Traits*, C++ Report, June 1995.
7. J. Beveridge, *Self-Registering Objects in C++*, Dr. Dobb's Journal, August 1998b.
8. G. Toth, *A General Code for Modeling MHD flows on Parallel Computers: Versatile Advection Code*, IAU Colloquium No. 153 on Magnetodynamic Phenomena in the Solar Atmosphere, Makuhari, Japan Proceedings p. 471-472, 1995.
9. G. Gray, T. Kolda, *APPSPACK 4.0: Asynchronous Parallel Pattern Search for Derivative-Free Optimization*, Sandia National Laboratories, http://software.sandia.gov/appspack.
10. P. Rijks, J. Squyres, A. Lumsdaine, *Performance Benchmarking of Object Oriented MPI*, Technical report, University of Notre Dame Department of Computer Science and Engineering, 1999.
11. R. Hillson, M. Iglewski, *C++2MPI: A Software Tool for Automatically Generating MPI Datatypes from C++ Classes*, 2000 International Conference on Parallel Computing in Electrical Engineering (PARELEC 2000), 13-17, 2000.
12. A. Koenig, B. E. Moo, *Templates and Duck Typing*, C/C++ Users Journal June, 2005.
13. The National Center for Supercomputing Applications, *HDF5 Homepage*, http://hdf.ncsa.uiuc.edu/HDF5/.

Calculation of Single-File Diffusion Using Grid-Enabled Parallel Generic Cellular Automata Simulation

Marcus Komann, Christian Kauhaus, and Dietmar Fey

Friedrich-Schiller-University Jena, Germany
marcus.komann@web.de, kauhaus@inf.uni-jena.de,
dietmar.fey@uni-jena.de

Abstract. Parallel execution of simulation runs has become indispensable in different research areas recently. One of the most promising and powerful models in science are cellular automata (CA). This paper describes the approach to migrate parallel execution of a cellular automata simulation program called ParCASim via grid-capable message passing from cluster to grid environments. The applicability of the program is demonstrated for the calculation of sophisticated single-file diffusion problems and exploits the improved compute power of the underlying grid in order to deliver results faster and more precise than on classic cluster architecture. In this paper, we discuss ParCASim as a tool, and the utilised cluster-to-cluster interconnection is outlined. Furthermore, the paper describes single-file diffusion and how it is calculated via the CA model. Finally, time and efficiency measurements are taken which show the applicability of the approach.

1 Introduction

The importance of parallel and distributed computing has increased much in the last years due to the inability of single computers to solve some sophisticated computational problems efficiently. Some problems are even impossible to solve on a single computer because their size exceeds the computer's physical memory. Scientists have come across these problems in many different areas, for example meteorology (e. g. weather forecast [MLB01]), medicine (myocardium phenomena research [CzG04]), microsystems design (photo detector engineering [FKK04]), fluid dynamics (channel flow [DFP03]) etc.

To solve these problems, the use of cluster computers has raised much interest because it provides plenty of compute power for a much lower price than dedicated supercomputers. The production of cheap high performance network hardware has contributed to this development because processor power already was available for acceptable prices before. This fast network hardware can now be used along with powerful processors to form an efficient execution platform for scientific research and simulation.

A lot of successful research effort has been made recently in homogeneous computing environments like clusters along with the development of middleware and programming languages. But with increasingly demanding applications and much idle

B. Di Martino et al. (Eds.): EuroPVM/MPI 2005, LNCS 3666, pp. 528–535, 2005.

computing power among especially desktop computers the need for the connection and concurrent usage of systems with different hard- and software is constantly rising.

A cluster computer is able to speedup a parallel program by using more compute power on the parts of the program which could be calculated concurrently. Nonetheless, a cluster only has limited resources even if these limits can be high. They exist due to the cluster's design and architecture. If someone wants to execute a program which does not fit in the cluster's memory or which would use too much time on an existing cluster these limitations have to be overcome. This can be done by extending the cluster with new computational nodes, where each node might exhibit different capabilities regarding processor, memory performance, interconnects, external networking facility, operating system and so on. Especially if these nodes are idle nodes like unused desktop computers in a company or a university it would be nice to utilise their additional compute power to gain scientific results faster or with higher precision.

Many physical, biological, chemical, social or economical processes problems can be described using cellular automata and therefore the demand for high speed simulation programs is large. There are already some parallel CA simulators like CARPET/CAMEL. But they are only able to run on single clusters. One special problem which needs particular high-speed research is the single-file diffusion problem which is one of our current research projects. Because of this demand and because CA are very well scalable and can be parallelised relatively easy we built a generic parallel CA simulation program which is capable not only of running on a cluster but also in a grid environment. This program can simulate a vast variety of cellular automata and – due to its parallel implementation – is able to calculate huge setups very fast.

This paper is organised as follows. After the introduction we shortly outline cellular automata and the simulation program ParCASim along with the single-file diffusion problem and how it is solved by using CA in chapter 2. Chapter 3 describes the setup of the utilised cluster and grid and the middleware used. Finally we present taken measurements and results and end with conclusions.

2 Cellular Automata

2.1 Introduction

One might say that the basic laws of physics relevant to everyday phenomena are all found and understood completely (not considering quantum mechanics). But there are some common natural systems whose complex structure and behaviour make it very difficult or even impossible to describe them analytically. For example, the laws that specify the freezing of water and the conduction of heat have long been known, but analysing their consequences for the intricate patterns of snowflake growth has not yet been possible. Many large and complex systems might be broken down into identical parts of which each follows simple rules. But when these components are put together they also might show a much more complicated and sophisticated behaviour because of interaction of the several components.

One could try to simulate the whole system by simulating each of the components separately. That approach can work for some of the systems but in most cases it will fail. In order to find out about the system, one has to try to filter the mathematical essence by simulation and observation of the complete system instead. The desire in such an approach would be to identify fundamental analytical mechanisms that are common to various natural systems and could therefore be used more generally. In order to find and analyse the mathematical basis for our generation of high complexity, one must identify simple mathematical systems that capture the essence of the process. Cellular automata (CA) are a candidate class of such systems.

2.2 The Model

CA are discrete dynamical universes whose behaviour is completely specified in terms of local transition rules. Local extension is represented by a uniform grid consisting of local objects called cells which contain only few data each. Time advances in discrete steps and (depending on the local rule) at each step each cell computes its new state from that of its adjacent neighbours. Therefore, the system's laws are local and uniform but it is possible to gain knowledge about long-term behaviour of the whole complex system.

It should be noted that it is currently unknown whether there are better algorithms than straightforward simulation for getting to know the state of a designated cell at one given point in time although this is the desired aim. This problem is complete for **P** [GHR95]. Besides, it might be important to know the state of the cells after every transition and this is why there is no way of dodging step-by-step simulation.

There is lots of literature around on the subject of cellular automata. A very detailed look on CA along with some elaborately information can be found in [Wol02a] and [Ila02]. A closer look to questions of complexity is given in [Wol02b].

A classic, standard cellular automaton $A=(L, Q,, N, \delta, Q_0)$ is defined by a regular lattice geometry L in which the single cells are embedded, a finite set of states Q (sometimes also called alphabet), a local, uniform and finite neighbourhood N, a transition function $\delta: Q^N x\, Q \rightarrow Q$ which applies for each cell and an initial setting Q_0 of the lattice cells.

The states of all cells change according to the transition function to new states of Q after discrete time steps. The state of the whole automaton changes spatially parallel and synchronously to the global clock and is heavily dependant on the starting configuration.

2.3 ParCASim

ParCASim means parallel cellular automata simulator. It uses MPI as communication facility on a cluster and IMPI [GHD00] in grid environments. It was designed to simulate a vast variety of different CA. ParCASim uses an input file where the different attributes of the CA are specified (like dimension, size, number of states, radius, neighbourhood, edge type and the amount of iterations). In order to give the user the possibility to write his own generic transition function ParCASim is able to use either a transition table from the input file or a generic rule file which can be written by the user. It was designed to simulate huge lattices in possibly high dimensions. There are already existing CA simulators which can be used for small setups. ParCASim is intended for complex and highly-demanding scientific applications where it is important to have the possibility to utilise generic and/or non-deterministic transition functions.

2.4 Single-File Diffusion

Single-file diffusion refers to the one-dimensional motion of particles in materials which are that narrow that mutual passage of particles is excluded. Since the sequence of particles remains the same, this leads to strong deviations from normal diffusion, e.g. an increase of the particle mean-square-displacement as $t^{1/2}$. Besides theoretical interest, the single-file diffusion problem is encountered in various fields (one-dimensional hopping conductivity, ion transport in biological membranes, channelling in zeolithes).

This type of diffusion can be modelled very well using CA. To do so, one has to create a one-dimensional automaton with the desired amount of particles referring to cells in the CA. The ordering of the particles doesn't change but their state might modify according to certain influences especially by their neighbours. This can be simulated by applying the local transition rule to the cells. The unforeseeable behaviour of the single particles can then be observed by simulating the lattice as we do with ParCASim.

3 Cluster and Grid Measurement

In order to compute the single-file diffusion problem efficiently we decided to try to execute simulations both on cluster and on grid. This chapter describes the approach, the hardware setups and the middleware used. The results are shown at the end.

3.1 The Testing Environment

The testing environment is situated at the department of computer architecture at the Friedrich-Schiller-University in Jena/Germany. In order to proof the feasibility of ParCASim in a grid environment we have built a heterogeneous setup which consists of two different computer systems.

Sun. The first one is a SUN Sparc Enterprise 2500. It has 5 equal Sparc processors and 3 GB of RAM and is running Solaris 8. This machine is connected to the internet via a 100MBit/s SUN network adapter.

Cluster. The second part of our heterogeneous system is a Beowulf-style [Ste99] cluster. This cluster consists of eight worker nodes and one master node. The eight nodes have Intel Pentium IV processors with two gigahertz cpu frequency and 512 MB RAM. The master node consists of a 2.4 gigahertz hyperthreading Pentium IV with 2 GB RAM. The whole system is interconnected via Gigabit Ethernet and has a 100 megabits-per-second connection to the internet on the head node. Only this PC has a public IP address. All other nodes have internal IP addresses and are not directly accessible from the outside. Because the head node uses two different network interface controllers for inside and outside communication an adequate middleware has to deliver some routing infrastructure to enable nodes from outside the cluster to communicate with the internal nodes.

Grid Middleware. We tried several message-passing middleware systems like PACX [PAC03], MPICH/Madeleine [Aum02] etc, of which most did not work correctly or did not work at all. Most of these systems use gateway nodes to connect different clusters to one another. The usual way to provide heterogeneity is by enabling the programmer to use a new set of libraries for linking his program and new program start-up mechanisms. These libraries are typically built on so-called devices which are stated below the MPI communication library. This approach requires to solve many sophisticated problems like message routing, effective implementation of collective operations, reliable ordered message submission within different communication methods, packet bundling strategies, protocol conversion or addressing issues in possibly incompatible networks, dealing with firewalls and maybe even effective load balancing.

In the end we found the LAM implementation of IMPI [GHD00, SLG00] to be most suitable for our purposes. The architecture has four parts: a server where implementations synchronise at start-up, a client which acts as a representative for one MPI implementation, one or more hosts which belong to a specific client (the cluster nodes) and processes which run on a specific host. The clients take care of inter-cluster communication via providing inter-cluster nodes TCP/IP connections.

While testing IMPI on our heterogeneous environment we encountered the problem that IMPI routing did not operate as intended. Processes from within the cluster (only internal and no public IP address) were not able to send messages to the outside. The forwarding through the head node (with public IP address) did not work out as proposed in the specification. The IMPI router process was unable to do a network address translation when sending messages from one of the worker nodes to the SUN nodes.

To solve this problem we enhanced our heterogeneous environment by adding openMosix [Bar03] to the cluster. This is a Linux kernel extension which allows users to look at a cluster as one system (also called *single system image*). It is able to migrate processes from a busy node to another one running the same kernel. This could be used for load balancing and for preventing memory swapping (resource sharing) and results in improved overall performance. Our approach is the following: We start all cluster processes on the head node and use openMosix to migrate them to other nodes in the cluster. Hence, openMosix takes care of communication of intra-cluster nodes.

3.2 Measurement Results

We only provide a sample measurement result here to express proof of concept because achievable speedup is heavily dependant on the specific calculated CA. Especially in the grid environment, where message latency might be high and if the time nodes compute is relatively slow in comparison to message time, then speedup might be very low. Hence, parallel execution only makes sense if the time consumed by computation is high in comparison to message passing time. But for large, complex lattices this is a given fact and this is what ParCASim is intended for. The reason for this lies in the structure of a partitioned CA simulation. Every node gets one part of the complete lattice and has to communicate a certain amount of data with it's neighbours before each iteration step to receive the neighbour cells itself doesn't compute. If these steps themselves only take very short time the overhead in communication kills the speedup gained with higher amount of nodes and compute power.

We started with calculating a problem on one of the SUN processors and added the other processors step-by-step until five afterwards. When we reached this maximum amount of Sparcs, and thus fastest possible execution on this architecture, we went to grid execution by attaching cluster nodes one by one up to nine.

Fig. 1 and Figure 2 show the results achieved when calculating a lattice with 1 million and with 10 million cells in 100 iterations and a simple rule file. These simulations took 312 and 3171 seconds on a single processor of the SUN computer. First, it can be seen that parallel execution is useful both on homogeneous and on heterogeneous systems but especially on grid. The SUN alone needs 69 and 682 seconds with maximum amount of processors. Second, moving to grid lets us gain even more. Using all available nodes the calculation only took 56 and 203 seconds in the end. More complex rule files or larger lattices would add to this fact because the ratio between communication time and computation time would be smaller what would lead to less overhead and even faster execution times.

The small rise in Fig. 1 between five and six (moving to grid) processors is due to increased communication overhead in grid which appears because of slower underlying network hardware in the heterogeneous system. This overhead was 23 seconds in our examples and is larger than computation power gained by adding one of the cluster nodes. Using further cluster nodes pays off with faster execution but we reach saturation very fast in this example as we do not decrease execution time significantly later on by adding more processors. Fig. 2 does not show this behaviour because overhead is smaller than gained computation power and adding nodes further on speeds up computation.

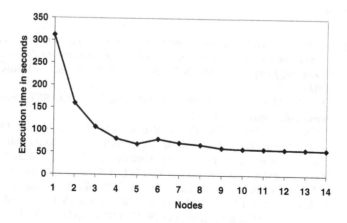

Fig. 1. Results with 10^6 cells and 100 iterations

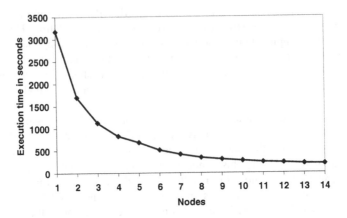

Fig. 2. Results with 10^7 cells and 100 iterations

4 Conclusions

We presented a working fast generic cellular automata simulator for complex and possibly non-deterministic lattices and showed how single-file diffusion problems can be solved by using it. Calculating these diffusion phenomena is a current subject of research in Jena and will be expanded in future. We also outlined the setup of a heterogeneous computing environment using IMPI and openMosix to connect a Linux cluster and a SUN Sparc machine and how this grid can be used to speed up the parallel CA simulation. Measurements were taken on cluster and on grid which showed the feasibility of the approach. Complex cellular automata can be simulated in much shorter amount of time by parallelising them and by furthermore executing the parallel program not only on cluster but on grid.

References

[Aum02] O. Aumage: *Heterogeneous multi-cluster networking with the Madeleine III communication library*, In Proceedings of 16th International Parallel and Distributed Processing Symposium (IPDPS 2002), page 85, Washington – Brussels – Tokyo, 2002

[Bar03] M. Bar, Maya, Anu, Asmita, Snehal, Krushna (MAASK): *Introduction to openMosix*, Linux Congress, 2003

[CzG04] P. Czarnul, K. Grzeda: *Parallel Simulations of Electrophysiological Phenomena in Myocardium on Large 32 and 64-bit Linux Clusters*, In D. Kranzmüller, P. Kaczuk and J. Dongarra (editors): Recent Advances in Parallel Virtual Machine and Message Passing Interface, number 3241 in LNCS, pages 234 - 241, Springer, 2004

[DFP03] J. Denev, T. Frank, K. Pachler: *Large Eddy Simulalow using a PC-cluster Architecture*, In I. Lirkov, S. Margenov, J. Wasniewski, Y Plamen (editors): Large-Scale Scientific Computing, number 2907 in LNCS, Springer, 2004

[FKK04]   D. Fey, M. Komann, C. Kauhaus: *A framework for optimising parameter studies on a cluster computer by the example of micro-system design*, In D. Kranzmüller, P. Kaczuk and J. Dongarra (editors): Recent Advances in Parallel Virtual Machine and Message Passing Interface, number 3241 in LNCS, pages 436 – 441, Springer, 2004

[GHD00]   W. L. George, J. G. Hagedorn, J. E. Devaney: *IMPI: Making MPI Interoperable* and complete IMPI specification, number 105 in Journal of Research of the National Institute of Standards and Technology, 2000

[GHR95]   R. Greenlaw, H. J. Hoover, W. L. Ruzzo: *Limits to Parallel Computation*, Oxford: Oxford University Press, 1995

[Ila02]   A. Ilachinski: *Cellular Automata: a discrete universe*, Singapore (et. al.): World Scientific, 2002

[MLB01]   J. Michalakes, R. Loft, A. Bourgeois: *Performance-Portability and the Weather Research and Forecast Model*, HPC Asia 2001

[PAC03]   R. Keller: *PACX documentation*, HLRS Stuttgart, Germany, 2003

[SLG00]   J. M. Squyres, A. Lumsdaine, W. L. George, J. G. Hagedorn, J. E. Devaney: *The Interoperable Message Passing Interface (IMPI) Extensions to LAM/MPI*, Proceedings of MPI Developer's Conference (MPIDC) 2000, Cornell University, NY, 2000

[Ste99]   T. L. Sterling: *How to build a Beowulf: a guide to the implementation and application of PC cluster*, Cambridge, Mass (et. al.): MIT Press, 1999

[Wol02a]   S. Wolfram: *A new kind of science*, Champaign, Ill: Wolfram Media, 2002

[Wol02b]   S. Wolfram: *Cellular automata and complexity: collected papers*, Boulder, Co: Westview-Press, 2002

Harnessing High-Performance Computers for Computational Steering

Petra Wenisch[1], Oliver Wenisch[2], and Ernst Rank[1]

[1] Technical University Munich, Arcisstrasse 21, 80290 Munich, Germany
{wenisch, rank}@bv.tum.de
http://www.inf.bauwesen.tu-muenchen.de
[2] Leibniz Computing Center, Barerstrasse 21, 80333 Munich, Germany
wenisch@lrz.de
http://www.lrz.de

Abstract. This study presents a computational steering project together with optimization approaches to counteract problems of interactive use when coupling a supercomputer with a visualization interface, allowing for user interaction during a running computational fluid dynamics (CFD) simulation. The underlying Lattice Boltzmann-based CFD computation and grid generation is performed on a supercomputer while the steering and data visualization of the simulation is done on a graphical front-end application on an external system. The interaction during a simulation run comprises not only the variation of parameters, but also the modification of the geometry and hence the computational grid. The simulation kernel shows good parallel efficiency on the Hitachi SR8000 supercomputer at the Leibniz Computing Center (LRZ) and therefore enables an online simulation which instantly reacts to user manipulations.

1 Introduction

Numerical CFD simulations are gaining increasing importance as a valuable supplement to classical wind-tunnel experiments. However, an accurate computation of a fluid flow scenario is still a time-consuming process and requires resources of powerful clusters or supercomputers to keep the computation time reasonably short. The pre-design phase of buildings or even huge ships usually lasts only a few weeks and later changes to the design incur a dramatic increase in costs [1]. Therefore, an interactive tool is desired to perform several case studies in a very short time to achieve preliminary investigations, possibly followed by just a few carefully selected simulations with more details. The present interactive CFD application was developed with this kind of issues in mind. It concentrates on indoor air flow simulations and allows the user to interact with flow parameters, boundary conditions, and geometry within the simulated scene during a simulation run.

The first part of this paper gives a brief introduction to the Lattice Boltzmann method. Subsequently, the main aspects of computational steering, viz. the simulation kernel, grid generation and, most importantly, the communication concept will be discussed.

B. Di Martino et al. (Eds.): EuroPVM/MPI 2005, LNCS 3666, pp. 536–543, 2005.

2   The Lattice Boltzmann Method

The Lattice-Boltzmann method (LBM) has emerged as a complementary technique for the computation of fluid flow phenomena [2,3]. Common numerical methods for the simulation of fluid dynamics are based on a discretization of the Navier-Stokes nonlinear partial differential equations. The LBM represents an alternative approach consisting of a discrete microscopic model combined with statistical physics. By computing the dynamics of particle densities for a discrete number of velocities and directions at each grid point appropriately, quantities such as mass and momentum are conserved to fulfill the hydrodynamic laws. The Lattice-Boltzmann algorithm computes the 'collision' of microscopic, 'virtual particles' and updates the velocity distribution functions within each simulation time-step. Collision is represented by the evaluation of the new velocity distribution functions and, luckily, does not require data exchange with adjacent grid nodes. 'Propagation' refers to the migration of these distribution functions to neighboring cells. Typically, the LBM is implemented on uniform Cartesian grids making this approach to solving the Navier-Stokes equation particularly well-suited for taking advantage of the parallelization and vectorization capabilities of high-performance supercomputers like the 2 TFlop/s Hitachi SR8000 at the LRZ in Munich [4].

3   Computational Steering

Usually, large and compute-intensive CFD simulations are run non-interactively as batch processes on queuing systems of high-performance computers. After the pre-processing step, i.e. after mapping the CAD data onto a computational grid and defining the boundary conditions, a file or a database describing this setup information is generated and submitted to a batch queue. As soon as the required resources (a sufficient number of free processors and amount of memory) are available, computation starts and saves its output to disk. The user evaluates the simulation results during the subsequent post-processing step. These steps are often carried out on different hardware architectures, e.g. desktop PC, supercomputer and graphics workstation, respectively.

This kind of workflow is practical for problems when one is interested in obtaining detailed information in a fluid flow investigation. For the wide variety and the explorative environment needed in case studies, however, the user would like to interact with a running simulation and to visualize the corresponding physical reaction immediately. This is the basic idea behind computational steering [5]. It requires the integration of the above-mentioned steps, viz. pre-processing, computation and post-processing, into a single environment.

To fulfill the requirement of immediate or at least low latency response to user interaction during a running simulation, a fast CFD solver must be run on a supercomputer or cluster and has to be coupled to a steering and visualization workstation by an efficient communication concept. In addition, the computation kernel has to allow for a fast grid modification during the simulation, and both steering and visualization should be available to the user within a single environment. These aspects will be described in the following.

3.1 Kernel Optimization and Grid Generation

To establish the basis for interactive numerical computation, special care has been taken to take full advantage of the vectorization and parallelization capabilities of the specialized Hitachi SR8000 hardware.

The SR8000 is a pseudo-vector machine with 8 directly available CPUs on each SMP node [11]. Therefore, the fluid domain computation has been parallelized in slices on several nodes via MPI. In addition, the computing loops were parallelized using the 8 CPUs within each node (in COMPAS mode) by OpenMP directives.

Figure 1 illustrates the performance and scaling of the simulation kernel after the optimizations on the SR8000. The benchmarks were performed with a constant grid size of 100x280x180 per slave (weak scaling) and with a total grid size of 300x280x180 (strong scaling).

The improvements were accomplished by specifically rewriting the main computation loops to enable the optimizing Hitachi compiler to use its vectorizing and pre-fetching capabilities much more efficiently. Details can be found in [12].

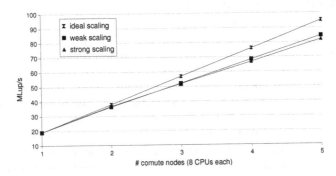

Fig. 1. Offline simulations without steering and visualization: At a high level of absolute performance the application shows high efficiency for strong and weak scaling with 86 and 89 %, respectively

Finally, fast grid generation is also an essential requirement for an interactively steerable CFD application to allow the modification of the geometry, e.g. by inserting or deleting obstacles. In our application, the user can load arbitrary triangulated, CAD-generated geometries which the grid generator (described in detail in [13]) transforms onto a uniform Cartesian grid representation required by the Lattice-Boltzmann method (see section 2). The corresponding voxelization algorithm for an optimized grid generation is based on the hierarchical space-partitioning concept of an octree.

3.2   Communication Concept

In the following, we concentrate on an appropriate design of communication to meet the special technical challenges of computational steering. The visualization workstation provides the functionality to display the current data and interact with the simulated scene at the same time. Following user interaction, the modifications are forwarded to the supercomputer. There the new configuration is incorporated immediately and new results are sent to the visualization client, where the user can observe the adaptation of the fluid flow. Fig 2 shows the different components of the application and the corresponding data flows.

The slave processes (SIM-S) on the supercomputer mainly perform the LB computation. To achieve good performance, it is essential to use vendor-optimized intra-machine MPI for the communication with other processes. As long as no interaction has occurred, they send current results at user-defined, regular intervals and check for updated computational grids. In the event of interaction, a new grid is received and new results can be sent after just a few time-steps to give the user fast initial feedback depending on his manipulations. Consequently, the transmission intervals are not necessarily regular throughout the run.

On the visualization side (VIS), an additional communication thread is started to check for incoming results and to send user modifications without interrupting steering and post-processing. To keep the data transfer as short and infrequent as possible, only modifications to the setup are forwarded. Therefore, the

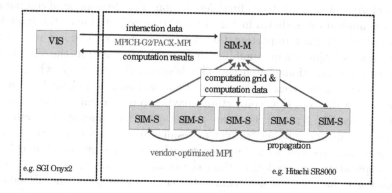

Fig. 2. Application scheme showing the visualization (VIS) and the simulation consisting of a master node (SIM-M) and computation slaves (SIM-S). At the LRZ, the VIS process is usually run on a separate graphics workstation, whereas the simulation uses several nodes on the Hitachi SR8000. Within the SR8000, all processes communicate via a vendor-optimized version of MPI. Inter-machine communication between VIS and SIM-M is implemented using either Globus MPICH-G2 or PACX MPI. The simulation side transfers its most recent pressure and velocity fields to the visualization at short intervals during which the user can analyze these results. The user is able to change the scene geometry or simulation parameters throughout the ongoing computation.

transmission process is not triggered until after the user has completed any modifications. Since the results are not necessarily sent at regular intervals either, the receipt of data is, in essence, an event-driven process in both directions.

Usually, the visualization is done on an external graphics workstation with a different hardware architecture from that of the supercomputer. To enable the slaves to use vendor-optimized MPI, an appropriate message-forwarding library for heterogeneous communication has to be used. In this respect, we have tested the Globus MPICH-G2 [6] and PACX-MPI [7] libraries. With its multifunctional Globus framework, MPICH-G2 opens the application to the world of grid computing and is accordingly becoming more popular on supercomputer sites. Unfortunately, our efforts to port MPICH-G2 to the SR8000 with vendor-MPI enabled have not been successful so far. Therefore, we are unable to present performance data for this library. In contrast, PACX-MPI can easily be used in conjunction with vendor-optimized MPI for internal communication on the Hitachi. One drawback of PACX-MPI is the fact that it starts two extra MPI processes on each system for internal issues. Since the Hitachi has just 8 interactive nodes, it was only possible to use 6 of them for the application. In addition, we observed that performance decreased by approximately 5 percent due to the overhead caused by PACX-MPI, which is still acceptable [8].

To uncouple communication due to interactive steering and computation, an additional process (SIM-M) has been introduced on the supercomputer. We will refer to this as the master node which communicates with the steering terminal. It checks in both directions whether new data is available, and pre-processes the data before forwarding it to its final destination. The decoupling of computation and communication is shown in Fig. 3 for a trace collected with the Intel Trace Analyzer [9]. Another important task for the master is to collect results from the slaves and send them, combined in a single message, to the visualization client to avoid additional latencies. This is especially important when the network connection between supercomputer and visualization client is limited by routers, firewalls, and slow connections maybe even in 'competition' with other users. Finally, Fig. 3 reveals the main advantage of introducing a collector node: During the time-consuming transfer of results (from process 1 to 0), the slave processes (2-4) are able to overlap computation with communication as long as the computation time is longer than the communication time. It is necessary to point out that using non-blocking MPI communication on the SR8000 (and several other MPI implementation, cf [10]) does not allow this overlap. Fig. 4 shows that the performance of the application decreases dramatically without a collector node.

In particular, we found that the network connection of the SR8000 to any other computer is fairly unsatisfactory (25 MB/s maximum throughput on its single Gigabit Ethernet line). To demonstrate the quality of the communication strategy in a hypothetic high-performance network, the application was run entirely on the SR8000 to measure the scalability at the communication thread of the visualization (see left graph in Fig. 4). The Hitachi interconnect provides a maximum throughput of 780 MB/s between two nodes, which is satisfactory for

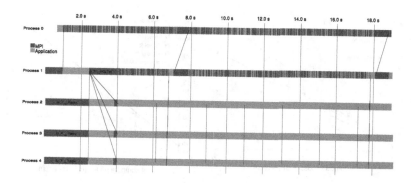

Fig. 3. This trace depicts the distribution of computation and communication. In this case, we recorded 5 processes, namely visualization (process 0), simulation master (process 1), and 3 slaves (processes 2-4). The timeline is given on the abscissa and covers 11 time-steps starting from the application's initialization. Red marks represent MPI function calls while green shows periods of computation or other application-specific processing. Frequent checks for user interaction and time-consuming communication to an external machine can be taken off the computation processes by introducing a master node.

this kind of application. More recent systems such as the SGI Altix 3700 Bx2 (installed in 2005 at the LRZ) offer a similar bandwidth for external up- and downlink (10 GE interfaces). The graph shows the efficiencies for strong scaling for various transmission intervals in forwarding results for a constant total problem size of 300x280x180 grid nodes. Even in the case of this rather highly resolved grid we achieve a frame update rate of 0.5 Hz. Since the communication volume increases with larger grids, weak scaling measurements are not relevant for this type of application.

To demonstrate the performance gain using a master node, the measurements have been carried out with the computation running on the Hitachi SR8000 at LRZ and the visualization on an external Opteron PC at the Chair for Bauinformatik (TUM). True scaling of the application could only be achieved using a master node in combination with long update intervals of the visualized scene (see right graph in Fig. 4).

To achieve high update rates even for high-resolution grids and external visualization we propose using compression for external communication. If possible, reducing the size of the data sets could also improve performance. However, we estimate that together with the corresponding (but unavoidable) overhead for describing and reconstructing the reduced data set the performance gain may again be equalized. Finally, moving the postprocessing and graphics computation onto the supercomputer (maybe even in a parallel version) seems the most promising solution for this dilemma, but may not always be applicable. This is especially true for the Hitachi SR8000, which is not well-suited to support postprocessing or visualization due to its bad performance in non-vectorizable code. Thus, we consider porting the application to another HPC system with

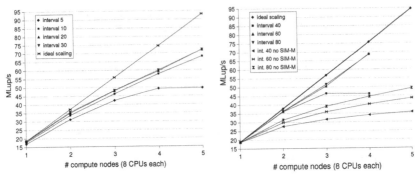

Fig. 4. These two graphs show the scaling behavior for various update intervals in dependence on the number of computation nodes used. Performance is measured in MLup/s (million lattice site updates per second) in the visualization process. The left-hand panel refers to runs with both visualization and simulation processes on the SR8000 to emulate a 'high-performance network situation'. Runs with update intervals of 10 or more time steps show good scaling already, whereas shorter intervals have a negative influence. On the right-hand panel, visualization is done externally. Runs using more computation nodes show a similar scaling behavior as before, although in this case, the update interval has to be significantly longer (60 instead of 10 time steps). In contrast, the application shows poor scaling efficiency (<50%) in all cases when the master node (SIM-M) is omitted.

improved networking and postprocessing capabilities thereby disregarding a (minor) performance degradation for the simulation kernel. This could be, e.g., the combination of the new SGI Tornado and Prism systems, which will be installed at the LRZ in 2006.

4   Conclusion and Outlook

We have presented approaches of optimization for a computational steering CFD environment. While the CFD kernel is processing continuously in the background, it updates simulation results for the visualization client at regular intervals. At the same time, the simulated scene can be modified interactively and the kernel will instantly incorporate the changes. Due to the efficient grid generation, the modification of complex geometries is well supported. Performing the simulation on the Hitachi SR800 and the visualization on an external graphics workstation the achieved frame update rate using moderately fine grids is quite satisfactory. To overcome the bottleneck arising from the downloading of large data from the supercomputer, future developments will focus on ways to transfer part of the postprocessing onto the HPC system or using current technology like high-bandwidth interconnects in SGI HPC and visualization systems.

Further improvements will cover local grid refinement at the boundaries as well as the integration of turbulence and thermal phenomena into the kernel making the performance requirements of the code even more demanding.

5   Acknowledgements

The authors would like to express their gratitude to Dr. Matthias Brehm (LRZ) for assisting during the optimization process, Irene Geiseler (also LRZ) for help with the complex scheduling of the interactive application. For assistance with optimizing and using PACX and MPICH-G2 on different architectures, we would like to thank Thomas Zeiser (RRZE), Rainer Keller (HLRS) and Helmut Heller (LRZ). This project has received financial support from KONWIHR (funded by the state of Bavaria), which is gratefully acknowledged.

References

1. Henkel S.: Ventilation and Smoke Simulation in Cargo Holds and Engine Rooms on board of RORO-Ferries, Proceedings RoomVent2004, Coimbra, Portugal (2004)
2. Krafczyk, M.: Gitter-Boltzmann Methoden: Von der Theorie zur Anwendung, Professoral dissertation, LS Bauinformatik, TU München (2001)
3. Treeck, C. v.: Gebäudemodell-basierte Simulation von Raumluftströmungen, PhD Thesis, LS Bauinformatik, TU München (2004)
4. Pohl, T., Deserno, F., Thürey, N., Rüde, U., Lammers, P., Wellein, G., Zeiser, T.: Performance Evaluation of Parallel Large-Scale Lattice Boltzmann Applications on Three Supercomputer Architectures, Proceedings SC04, Pittsburgh, USA (2004)
5. Mulder, J. D., Wijk, J. van, Liere, R. van: A Survey of Computational Steering Environments, Future generation computer systems, 15(2), (1999)
6. http://www3.niu.edu/mpi
7. http://www.hlrs.de/organization/pds/projects/pacx-mpi
8. Keller, R., High Performance Computing Center Stuttgart: Personal Comm. (2005)
9. http://www.intel.com/software/products/cluster/tcollector/index.htm
10. White III, J. B. and Bova, S. W.: Where's the Overlap? An Analysis of Popular MPI Implementations, MPIDC'99
11. http://www.lrz-muenchen.de/services/compute/hlrb/hardware-en
12. Wenisch, P., Wenisch, O., and Rank, E.: Optimizing an Interactive CFD Simulation on a Supercomputer for Computational Steering in a Virtual Reality Environment, in print, Springer, Germany (2005)
13. Wenisch, P. and Wenisch, O.: Fast octree-based Voxelisation of 3D Boundary Representation-objects, Technical Report, LS Bauinformatik, TU München (2004)

Author Index

Lecture Notes in Computer Science

For information about Vols. 1–3599

please contact your bookseller or Springer

Vol. 3655: A. Aldini, R. Gorrieri, F. Martinelli (Eds.), Foundations of Security Analysis and Design III. VII, 273 pages. 2005.

Vol. 3654: S. Jajodia, D. Wijesekera (Eds.), Data and Applications Security XIX. X, 353 pages. 2005.

Vol. 3653: M. Abadi, L. de Alfaro (Eds.), CONCUR 2005 – Concurrency Theory. XIV, 578 pages. 2005.

Vol. 3652: A. Rauber, S. Christodoulakis, A M. Tjoa (Eds.), Research and Advanced Technology for Digital Libraries. XVIII, 545 pages. 2005.

Vol. 3649: W.M.P. van der Aalst, B. Benatallah, F. Casati, F. Curbera (Eds.), Business Process Management. XII, 472 pages. 2005.

Vol. 3648: J.C. Cunha, P.D. Medeiros (Eds.), Euro-Par 2005 Parallel Processing. XXXVI, 1299 pages. 2005.

Vol. 3646: A. F. Famili, J.N. Kok, J.M. Peña, A. Siebes, A. Feelders (Eds.), Advances in Intelligent Data Analysis VI. XIV, 522 pages. 2005.

Vol. 3645: D.-S. Huang, X.-P. Zhang, G.-B. Huang (Eds.), Advances in Intelligent Computing, Part II. XIII, 1010 pages. 2005.

Vol. 3644: D.-S. Huang, X.-P. Zhang, G.-B. Huang (Eds.), Advances in Intelligent Computing, Part I. XXVII, 1101 pages. 2005.

Vol. 3642: D. Ślezak, J. Yao, J.F. Peters, W. Ziarko, X. Hu (Eds.), Rough Sets, Fuzzy Sets, Data Mining, and Granular Computing, Part II. XXIII, 738 pages. 2005. (Subseries LNAI).

Vol. 3641: D. Ślezak, G. Wang, M. Szczuka, I. Düntsch, Y. Yao (Eds.), Rough Sets, Fuzzy Sets, Data Mining, and Granular Computing, Part I. XXIV, 742 pages. 2005. (Subseries LNAI).

Vol. 3639: P. Godefroid (Ed.), Model Checking Software. XI, 289 pages. 2005.

Vol. 3638: A. Butz, B. Fisher, A. Krüger, P. Olivier (Eds.), Smart Graphics. XI, 269 pages. 2005.

Vol. 3637: J. M. Moreno, J. Madrenas, J. Cosp (Eds.), Evolvable Systems: From Biology to Hardware. XI, 227 pages. 2005.

Vol. 3636: M.J. Blesa, C. Blum, A. Roli, M. Sampels (Eds.), Hybrid Metaheuristics. XII, 155 pages. 2005.

Vol. 3634: L. Ong (Ed.), Computer Science Logic. XI, 567 pages. 2005.

Vol. 3633: C. Bauzer Medeiros, M. Egenhofer, E. Bertino (Eds.), Advances in Spatial and Temporal Databases. XIII, 433 pages. 2005.

Vol. 3632: R. Nieuwenhuis (Ed.), Automated Deduction – CADE-20. XIII, 459 pages. 2005. (Subseries LNAI).

Vol. 3631: J. Eder, H.-M. Haav, A. Kalja, J. Penjam (Eds.), Advances in Databases and Information Systems. XIII, 393 pages. 2005.

Vol. 3630: M.S. Capcarrere, A.A. Freitas, P.J. Bentley, C.G. Johnson, J. Timmis (Eds.), Advances in Artificial Life. XIX, 949 pages. 2005. (Subseries LNAI).

Vol. 3629: J.L. Fiadeiro, N. Harman, M. Roggenbach, J. Rutten (Eds.), Algebra and Coalgebra in Computer Science. XI, 457 pages. 2005.

Vol. 3628: T. Gschwind, U. Aßmann, O. Nierstrasz (Eds.), Software Composition. X, 199 pages. 2005.

Vol. 3627: C. Jacob, M.L. Pilat, P.J. Bentley, J. Timmis (Eds.), Artificial Immune Systems. XII, 500 pages. 2005.

Vol. 3626: B. Ganter, G. Stumme, R. Wille (Eds.), Formal Concept Analysis. X, 349 pages. 2005. (Subseries LNAI).

Vol. 3625: S. Kramer, B. Pfahringer (Eds.), Inductive Logic Programming. XIII, 427 pages. 2005. (Subseries LNAI).

Vol. 3624: C. Chekuri, K. Jansen, J.D.P. Rolim, L. Trevisan (Eds.), Approximation, Randomization and Combinatorial Optimization. XI, 495 pages. 2005.

Vol. 3623: M. Liśkiewicz, R. Reischuk (Eds.), Fundamentals of Computation Theory. XV, 576 pages. 2005.

Vol. 3622: V. Vene, T. Uustalu (Eds.), Advanced Functional Programming. IX, 359 pages. 2005.

Vol. 3621: V. Shoup (Ed.), Advances in Cryptology – CRYPTO 2005. XI, 568 pages. 2005.

Vol. 3620: H. Muñoz-Avila, F. Ricci (Eds.), Case-Based Reasoning Research and Development. XV, 654 pages. 2005. (Subseries LNAI).

Vol. 3619: X. Lu, W. Zhao (Eds.), Networking and Mobile Computing. XXIV, 1299 pages. 2005.

Vol. 3618: J. Jedrzejowicz, A. Szepietowski (Eds.), Mathematical Foundations of Computer Science 2005. XVI, 814 pages. 2005.

Vol. 3617: F. Roli, S. Vitulano (Eds.), Image Analysis and Processing – ICIAP 2005. XXIV, 1219 pages. 2005.

Vol. 3615: B. Ludäscher, L. Raschid (Eds.), Data Integration in the Life Sciences. XII, 344 pages. 2005. (Subseries LNBI).

Vol. 3614: L. Wang, Y. Jin (Eds.), Fuzzy Systems and Knowledge Discovery, Part II. XLI, 1314 pages. 2005. (Subseries LNAI).

Vol. 3613: L. Wang, Y. Jin (Eds.), Fuzzy Systems and Knowledge Discovery, Part I. XLI, 1334 pages. 2005. (Subseries LNAI).

Vol. 3612: L. Wang, K. Chen, Y. S. Ong (Eds.), Advances in Natural Computation, Part III. LXI, 1326 pages. 2005.

Vol. 3611: L. Wang, K. Chen, Y. S. Ong (Eds.), Advances in Natural Computation, Part II. LXI, 1292 pages. 2005.

Vol. 3610: L. Wang, K. Chen, Y. S. Ong (Eds.), Advances in Natural Computation, Part I. LXI, 1302 pages. 2005.

Vol. 3608: F. Dehne, A. López-Ortiz, J.-R. Sack (Eds.), Algorithms and Data Structures. XIV, 446 pages. 2005.

Vol. 3607: J.-D. Zucker, L. Saitta (Eds.), Abstraction, Reformulation and Approximation. XII, 376 pages. 2005. (Subseries LNAI).

Vol. 3606: V. Malyshkin (Ed.), Parallel Computing Technologies. XII, 470 pages. 2005.

Vol. 3605: Z. Wu, M. Guo, C. Chen, J. Bu (Eds.), Embedded Software and Systems. XIX, 610 pages. 2005.

Vol. 3604: R. Martin, H. Bez, M. Sabin (Eds.), Mathematics of Surfaces XI. IX, 473 pages. 2005.

Vol. 3603: J. Hurd, T. Melham (Eds.), Theorem Proving in Higher Order Logics. IX, 409 pages. 2005.

Vol. 3602: R. Eigenmann, Z. Li, S.P. Midkiff (Eds.), Languages and Compilers for High Performance Computing. IX, 486 pages. 2005.